Six Sigma Software Quality Improvement

Six Sigma Software Quality Improvement

Success Stories from Leaders
in the High Tech Industry

Vic Nanda

Jeffrey A. Robinson, Ph.D.

New York   Chicago   San Francisco
Lisbon   London   Madrid   Mexico City
Milan   New Delhi   San Juan
Seoul   Singapore   Sydney   Toronto

The **McGraw·Hill** Companies

Cataloging-in-Publication Data is on file with the Library of Congress

McGraw-Hill books are available at special quantity discounts to use as premiums and sales promotions, or for use in corporate training programs. To contact a representative, please e-mail us at bulksales@mcgraw-hill.com.

Six Sigma Software Quality Improvement: Success Stories from Leaders in the High Tech Industry

1 2 3 4 5 6 7 8 9 0   DOC/DOC   1 7 6 5 4 3 2 1

ISBN   978-0-07-170062-7
MHID   0-07-170062-5

The pages within this book were printed on acid-free paper.

Sponsoring Editor
Judy Bass

Copy Editor
Lisa McCoy

Editorial Supervisor
Stephen M. Smith

Proofreader
Claire Splan

Production Supervisor
Richard C. Ruzycka

Art Director, Cover
Jeff Weeks

Acquisitions Coordinator
Michael Mulcahy

Composition
TypeWriting

Project Manager
Patricia Wallenburg, TypeWriting

To:

My father, Mr. Hari Om Nanda, who passed away during the writing of this manuscript.
Dad, we miss you.
and
My family—especially my two tornadoes, Hersh and Ellora—
for their constant support, patience, and understanding
in allowing me to work on the manuscript.
Kids, I'm back and I love you!

—Vic Nanda

Dr. Lewis Edwin Robinson, who shared with me his passion for learning
and
Barbara Jean Robinson, who shared her passion for people.

—Jeff Robinson, Ph.D.

CONTENTS

Foreword xi
Preface xiii
Acknowledgments xvii
About the Editors xix

CHAPTER 1   Executive Overview of Six Sigma 3

PART ONE
DMAIC Projects

CHAPTER 2   DMAIC Primer 27
Jeffrey A. Robinson, Ph.D.

CHAPTER 3   How Motorola Minimized Business Risk before
Changing Business-Critical Applications 47
Vic Nanda, Motorola

CHAPTER 4   TCS Reduces Turn-Around Time for Software Change Requests 69
Siddharth Kawoor, Tata Consultancy Services

CHAPTER 5   Defect Reduction 97

CHAPTER 5A   How TCS Helped a Major Global Bank Reduce
Customer Complaints 99
Deepak Ramadas, Tata Consultancy Services

CHAPTER 5B   Improving Test Effectiveness the Six Sigma Way 119
José Antonio Mechaileh and Alecsandri Dias, Motorola

CHAPTER 6   Help Desk Improvement 143

CHAPTER 6A   EMC: Improving Development Productivity 145
Rich Boucher, EMC[2]

CHAPTER 6B   **Infosys Helps a Global Bank Reduce Business Risk
                    from Errors in Online Banking Applications** 159
                    Prakash Viswanathan and Anshuman Tiwari, Infosys

CHAPTER 6C   **Infosys Significantly Reduces Telecom
                    Client's Operational Expenses** 173
                    Prakash Viswanathan and Anshuman Tiwari, Infosys

CHAPTER 7   **Productivity Improvement** 189

CHAPTER 7A   **TCS Improves Fraud Detection for a Global Bank** 191
                    Asheesh Chopra, Tata Consultancy Services

CHAPTER 7B   **Infosys Improves Software Development Productivity
                    for a Large Multinational Bank** 213
                    Prakash Viswanathan and Anshuman Tiwari, Infosys

CHAPTER 8   **DMAIC Conclusions and Lessons Learned** 231

PART TWO
Lean Six Sigma Projects

CHAPTER 9   **Lean Primer** 235
                    Jeffrey A. Robinson, Ph.D.

CHAPTER 10   **Leaning Six Sigma Projects: How to Run a
                    DMAIC Project in Five Days** 253
                    Jeffrey A. Robinson, Ph.D.

CHAPTER 11   **How IBM Reduced Help Desk Escalations and Overhead Activities** ... 273
                    Timothy Clancy, IBM

CHAPTER 12   **Motorola Realizes Significant Cost Avoidance
                    by Streamlining Project Documentation** 291
                    Vic Nanda, Motorola

CHAPTER 13   **Boiling the Ocean with Value Streams, Kaizens, and Kanbans** 311
                    Timothy Clancy, IBM

CHAPTER 14   **How a Global Retailer Improved the Reliability of
Software Development and Test Environments** **339**
Sanjay Dua and Ambuli Nambi Kothandaraman,
Tata Consultancy Services

CHAPTER 15   **Lean Conclusions and Lessons Learned** 361

PART THREE
Design for Six Sigma Projects

CHAPTER 16   **DFSS Primer** 365
Eric C. Maass, Ph.D., Medtronic, and Patricia D. McNair, Raytheon

CHAPTER 17   **How to Radically Streamline Your Business Processes** 379
Bill Cooper, Motorola

CHAPTER 18   **How Motorola Reduced the Effort Required for
Software Code Reviews.** 409
Fernanda de Carli Azevedo Oshiro and
Luiz Antonio Bernardes, Motorola

CHAPTER 19   **Predictive Engineering to Improve Software Testing.** 429
Patricia D. McNair and Eric C. Maass, Ph.D., Motorola

CHAPTER 20   **Improving Product Performance Using Software DFSS** 443
Patricia D. McNair, Edilson Albertini da Silva, and
Alex Garcia Gonçalves, Motorola

CHAPTER 21   **High-Speed Product Development at Xerox** 463
Robert Hildebrand, Xerox

CHAPTER 22   **How Seagate Technology Reduced Downtime and
Improved Availability to 99.99 Percent** 491
Peter J. Clarke, Seagate

CHAPTER 23   **DFSS Conclusions and Lessons Learned** 509

PART FOUR
Six Sigma Programs

CHAPTER 24 **Cisco Successfully Reinvents Its Six Sigma Program** 513
Jason Morwick, Cisco, and Evan H. Offstein, Frostburg State University

CHAPTER 25 **Six Sigma Practice at Thomson Reuters** 523
Jian Chieh Chew, Thomson Reuters

CHAPTER 26 **How Convergys Injected Six Sigma into the Company DNA** 537
Manisha Kapur, Convergys

CHAPTER 27 **Bumps in the Road** 553

APPENDIX A **Chapter Tools Matrix** 561

APPENDIX B **Computing Return on Investment** 569

Glossary 581

Contributor Biographies 595

Company Profiles 601

Index 605

FOREWORD

Motorola (renamed Motorola Solutions, Inc. in January 2011) has the distinction of being both the "inventor" of Six Sigma and a major sponsor of its world-wide adoption. Over the last three decades, scores of companies have improved the quality of their products, reduced their costs, and grown their customer satisfaction using the core principles of Six Sigma. A library of business books attests to the success of the methodology. But Six Sigma was introduced 25 years ago. The world in which we operate has changed as the need for speed and agility is accelerating the pace of modern business, and as consumer-driven experience is replacing traditional product performance as the litmus test for customer satisfaction. So we are right to ask "Is Six Sigma still relevant in today's world?"

That is the real question that Vic Nanda and Jeff Robinson take on in this book, which isn't just stories about "goodness," but rather a guide to understanding how Six Sigma has evolved from its manufacturing roots into a comprehensive mindset and management system. The case studies presented here don't help us make better widgets. Instead, they provide a framework for creating an environment that fosters effective problem solving and successful innovation in the software industry. As more and more products have software as the differentiating component, they demonstrate that data mining and complexity reduction are at the heart of today's Six Sigma, and they show how to apply those lessons in actual business environments. All of this is "real world," none of the businesses profiled are perfect, and not everything goes as planned. The power of Six Sigma has always been in helping business leaders cope with these undesirable and unexpected outcomes ("defect reduction" in traditional quality terms), but the high-tech industry adds an additional layer of challenge. Software in particular is caught between competing directives: be innovative, be intuitive, be first-to-market, *and* be defect-free. It's a conundrum that software developers and software quality professionals continue to struggle with every day.

I wish that I could say that there is a magic bullet that definitively solves the riddle. There is no "one size fits all" methodology that guarantees flawless results. One of the great strengths of Vic Nanda and Jeff Robinson's work is that they don't claim that any one approach—Six Sigma or otherwise—can be successfully applied across all areas and address all problems. What they do, however, is demonstrate that many leaders in the high-tech industry have used Six Sigma to improve their operational results and profitability when they applied it judiciously.

Having had the privilege of working with both of the editors, I can attest that one of their greatest strengths—which they bring to this volume—is knowing where and when to apply the Six Sigma methodology. They show us how to avoid falling victim to the fallacy that if a little of something is good, then a lot of it must be better! The editors use the case studies in this book to make it crystal clear that the maximum value is extracted by a limited, but very targeted, application of Six Sigma principles. The winning formula isn't to make everything in your portfolio a Six Sigma project; it is identifying the key issues you need to address and then truly committing to those programs in terms of talent, time, and focus. And in my experience, there is also an additional benefit from this approach. If you understand and apply the principles of Six Sigma in major programs, a very interesting secondary effect occurs. Across the board, people become data-driven. The right questions start appearing about every undertaking, not just those that are designated as Six Sigma projects. There is a perceptible shift in the mindsets of managers who evaluate projects and developers who create them. The end result is that performance is increased across all programs, not just those that are managed under the official rubric of Six Sigma. Paradoxically, as Six Sigma is applied more sparingly, it becomes more powerful as a paradigm. And that may well be the greatest contribution of Six Sigma in today's fast-paced world of software development.

Leslie Jones
Senior Vice-President and Chief Information Officer
Motorola Solutions, Inc.

PREFACE

Why Did We Write This Book?

This book is unique, a first of its kind. It is not another Six Sigma textbook. There are plenty of books in the market that describe Six Sigma in theory but they are silent on *how to apply* Six Sigma to solve common product and process improvement challenges in the software and IT industry.

Theory is one thing; practice another. People learn best by a combination of both. A good cooking show is one that has a reputable chef, a selection of appetizing foods that an audience would be interested in preparing, and excellent demonstrations of the recipes. In the Six Sigma world, it means illustration of the Six Sigma methodology to solve real-world problems—not some unique problem specific to a company, but everyday problems in high-tech companies, problems that most of us can relate to.

This book is about reputable chefs (Six Sigma experts from today's leading high-tech companies), popular dishes (common process and quality improvement challenges in the high-tech industry), and their recipes (how to solve them using Six Sigma). In other words, this book is about Six Sigma in action, and it is Six Sigma in action in the software and IT industry.

The compelling collection of Six Sigma success stories in this book debunks the myth that Six Sigma is less relevant for software process and quality improvement. With the help of 25 Six Sigma experts from Motorola, IBM, Cisco, Seagate, Xerox, Thomson Reuters, TCS, EMC, Infosys, and Convergys, we get behind corporate walls and get a firsthand account of corporate Six Sigma programs and learn how these companies are successfully leveraging Six Sigma for software process and quality improvement.

Who Is the Audience for This Book?

We wrote this book primarily for two types of audiences: business executives and software quality professionals in the high-tech industry. If you fall into one of these categories, here is why you should read this book:

▲ If you are a business executive and you:
  ▼ Have heard about Six Sigma but are unsure if it can help with software process and quality improvement, then this collection of Six Sigma success stories should convince you.
  ▼ Have a Six Sigma program but you continue to struggle with intractable software process and quality challenges such as long cycle times, too much process complexity,

too many software defects, too much documentation, and more, then learn how other companies have successfully overcome these challenges using Six Sigma.

▼ Are interested in trying out Six Sigma or have a Six Sigma program that needs to be revitalized, then learn about the essential elements of a robust Six Sigma program. Learn from the firsthand accounts of companies that have successfully implemented Six Sigma for demonstrable business and financial impact.

▲ If you are a software quality professional and you are leading a software process or quality improvement project, then learn how another company may have already solved the same or a similar problem using Six Sigma.

How This Book Is Organized

This book begins with an executive overview of Six Sigma and the rest of the book is divided into four parts.

▲ **Part One** begins with a primer on Define, Measure, Analyze, Improve, and Control (DMAIC) methodology and contains eight DMAIC success stories from Motorola, TCS, EMC, and Infosys. Topics include business risk reduction, cycle time reduction, defect reduction, help desk improvement, productivity improvement, and test efficiency improvement.

▲ **Part Two** starts with a primer on Lean Six Sigma and contains five Lean Six Sigma success stories from IBM, Motorola, and TCS. Topics include cycle time reduction, documentation streamlining, defect reduction, and help desk improvement.

▲ **Part Three** offers a primer on Design for Six Sigma (DFSS) and Define, Measure, Analyze, Design, and Verify (DMADV) methodology and includes six DFSS/DMADV success stories from Xerox, Seagate, and Motorola. Topics include process redesign, process improvement, productivity improvement, new product design, test effectiveness improvement, and system availability improvement.

▲ **Part Four** contains a description of Six Sigma programs, including lessons learned and implementation advice from Six Sigma leaders at Cisco, Thomson Reuters, and Convergys.

How to Read This Book

The success stories in Parts One, Two, and Three have two reading tracks for two types of audiences.

▲ **Business Executives:** If you are an executive who wants to get the gist of each success story without getting buried in the details, then read only the executive summaries to understand the problem statement, get an overview of how Six Sigma was used to solve the problem, and see what improvement and financial benefit were realized.

▲ **Software Quality Professionals:** If you are a practitioner and want all the details on how to overcome problems that confront high-tech companies, then read the complete success stories. Each success story is presented in sufficient detail to enable software quality professionals to relate to challenges within their own organizations and learn how other companies have dealt with similar challenges using Six Sigma. Recognizing that software quality professionals are interested in knowing exactly how a problem was solved, the success stories provide step-by-step instructions and demonstrate use of Six Sigma statistical tools in sufficient detail so as to allow quick replication in companies that wish to solve similar problems.

To allow for easy readability, all chapters were written using a common template and edited for consistent voice, as if written by a single author. We recognize that perhaps a vast majority of the readers are not Six Sigma experts; therefore, we do not assume advanced knowledge of Six Sigma. Six Sigma concepts are explained in sufficient detail to the extent necessary and feasible in this book. You may want to keep a good Six Sigma textbook handy if you want to develop a greater understanding of Six Sigma concepts not detailed in this book.

If you are a business executive, read the first chapter, primers, and executive summaries in Parts One through Three, and Part Four of the book. Learn how to set up a new Six Sigma program or improve your existing program, and learn how other companies have overcome software process and quality challenges that you may face as well. For such challenges, direct your software quality and process improvement leader to read the detailed success story to understand exactly how another company solved the same or similar problem.

If you are a software quality professional, read the entire book, including the detailed success stories.

In closing, while Six Sigma is slowly gaining traction in the high-tech industry, it is our belief that this book will help expedite that process. We hope that the success stories showcased in this book will help convince the nonbelievers and skeptics of the undisputable benefits of implementing Six Sigma for software process and quality improvement, and help companies that have already implemented Six Sigma learn from each other.

Vic Nanda and Jeff Robinson, Ph.D.

ACKNOWLEDGMENTS

This book is the result of tireless efforts of several individuals to whom we are truly indebted.

First and foremost, we would like to express our sincere thanks and deep gratitude to all the contributors to this book. It has been a pleasure to work with each one of them, and without their contributions this book would not have been possible.

We would also like to thank each of the companies for their willingness to share their Six Sigma success stories in detail, providing us fascinating insight into their Six Sigma projects and precious advice on how to deal with similar challenges in our companies. Thanks to Motorola, IBM, Infosys, TCS, Xerox, EMC, Seagate, Convergys, Thomson Reuters, and Cisco—their generosity truly helps move the state of the art of Six Sigma forward.

We would like to recognize several reviewers whose feedback and suggestions were immensely helpful in improving the chapters and the book overall. Thanks to Anshuman Tiwari, Bruce Hayes, Bruce Spiro, Chris Samuelson, Dion Dunn, Jeff Smith, Kandy Senthilmaran, and Marge Olmstead.

We would also like to sincerely thank Judy Bass at McGraw-Hill, who believed in this book from the initial proposal stage and worked diligently with us to bring it to fruition. Thanks also to the others at McGraw-Hill involved with this book for their tireless efforts behind the scenes. Special thanks to Patricia Wallenburg from TypeWriting for her valuable work and excellent typesetting and a job well done.

Portions of the input and output contained in this book are printed with permission of Minitab Inc. All material remains the exclusive property and copyright of Minitab Inc. (all rights reserved). Likewise, certain definitions are reproduced in the glossary from ASQ publications with permission. We would like to thank Eston Martz and Cate Twohill at Minitab Inc. and Valerie Ellifson at ASQ for providing these permissions.

Last but not least, we are grateful to our family and friends for their sacrifices, understanding, and patience as we toiled long hours on this book. We certainly would not have been able to accomplish this monumental task without their constant support.

ABOUT THE EDITORS

Vic Nanda is a Senior Manager of Process & Quality Improvement in IT at Motorola Solutions, Inc. He has more than 15 years of experience in the software and IT industry, including various roles in software quality assurance, system testing, and software development. He has extensive experience with Six Sigma, SEI Capability Maturity Model (CMMi), IT Infrastructure Library (ITIL), TL 9000, and ISO 9000.

He is the author of *Quality Management System Handbook for Product Development Companies* (CRC Press, 2005) and *ISO 9001—Achieving Compliance and Continuous Improvement in Software Development Companies* (Quality Press, 2003; Spanish translation by AENOR, 2005).

Mr. Nanda has authored several peer-reviewed journal papers, magazine articles, and book chapters on process and quality improvement, and is a frequent speaker at process and quality conferences. He is a member of the editorial boards of the *Software Quality Professional Journal* and the *International Journal of Performability Engineering*, a member of the reviewer board for *IEEE Software* magazine, a member of the Standing Review Board of ASQ Quality Press, and a member of the ASQ National Awards Committee.

He is a certified Six Sigma Black Belt (by Motorola), a Certified Manager of Quality/Organization Excellence (CMQ/OE), a Certified Quality Auditor (CQA), a Certified Software Quality Engineer (CSQE), ITIL Foundations Certified, and a Certified ISO 9000 Lead Auditor.

Mr. Nanda was awarded the prestigious Feigenbaum Medal by the ASQ in 2006 for "displaying outstanding characteristics of leadership, professionalism, and potential in the field of quality and also whose work has been or will become of distinct benefit to humankind." He had previously been awarded ASQ's Golden Quill Award in 2003. He was profiled in the Face of Quality in *Quality Progress* magazine in November 2003.

Mr. Nanda is a member of the steering committee of the Philadelphia Software Process Improvement Network (SPIN) and a senior member of ASQ. He has an MS in computer science from McGill University, and a bachelor's degree in computer engineering from the University of Pune, India.

Jeffrey A. Robinson, Ph.D., is an IT technologist and project and program manager who has worked in software development, computer integrated manufacturing, and process and quality for more than 25 years. He has been a CMM/CMMi assessor and Malcolm Baldrige Quality Assessor, and is a certified IT Infrastructure Library (ITIL) practitioner as well.

A former USMCR jet fighter pilot, air traffic controller, and semiconductor device physicist before he ventured into IT programming and information systems, he enjoys solving problems of all kinds.

Dr. Robinson has been teaching graduate and undergraduate courses for more than 21 years and has developed and taught numerous technology courses in computer science, programming, operating systems, quantitative statistics, database design, decision theory, project management, risk management, organizational design, networking, database administration, business intelligence, data mining, and multimedia graphics. He is a frequent lecturer and an author of more than 40 technical papers, and holds four software patents in manufacturing control theory as well.

As a certified Master Black Belt, he has been applying and teaching Six Sigma techniques for more than 15 years in a broad range of environments, from semiconductor manufacturing and medical device manufacturing to IT, automotive, and financial management systems. As a consultant, he has worked with numerous companies, developing and delivering Six Sigma courses to improve process and quality programs.

He is currently a principal consultant and is serving on the American Society of Mechanical Engineers (ASME) Subcommittee on Software V&V for NQA (Nuclear Quality Assurance).

Dr. Robinson has a BA in physics from Monmouth College, a BS in electrical engineering from the University of Illinois, an MBA from Central Michigan University, and a Ph.D. in information systems from Nova Southeastern University.

An avid reader, he has also written and published science fiction and has co-edited an online SF e-zine. He is married and has two grown sons, two teenage daughters, and too many pets.

Six Sigma Software Quality Improvement

"In God we trust; all others must bring data."
—W. EDWARDS DEMING

CHAPTER 1

Executive Overview of Six Sigma

"Software Errors Cost U.S. Economy $59.5 Billion Annually!"[1]

"In the Software Industry:

▲ Only 35% of all projects succeed

▲ 19% fail outright

▲ 46% were challenged [by:]

  ▼ Cost overruns

  ▼ Late

  ▼ Fewer than desired features"[2]

Software is pervasive in today's world. A vast majority of today's appliances, systems, operations, and processes are powered by software. Yet, high-tech companies continue to be challenged in overcoming intractable software quality problems and process inefficiencies, while at the same time trying to deliver (and maintain) software as per agreed scope, on schedule, within budget, and per defined quality criteria. The numbers cited at the beginning of this chapter are startling and undisputable. This, despite the fact that there are so many process and quality frameworks and maturity models that have been around for quite some time and have been widely adopted in the IT industry, such as CMMI, ISO 9001, TL 9000, ITIL, and others.

Let's face it. Just like the fact that an organization's products or services are likely to have defects (even the achievement of Six Sigma quality assures the near-elimination of defects but not the absence of defects), likewise, any organization, no matter how mature and what quality framework it uses, will face operational problems. These include but are not limited to product quality problems, process complexity problems, process effectiveness problems, process efficiency problems, cycle time problems, estimation problems, execution problems, and so on. This is because while the aforementioned maturity models and quality frameworks embody best practices that certainly enhance an organization's ability to meet its commitments, they do not provide, nor do they intend to provide, complete methodologies and comprehensive toolkits of statistical and quality tools to solve different types of problems. That is precisely what Six Sigma provides. Significant financial returns from Six Sigma are not limited to manufacturing environments. The following success stories illustrate how these principles can provide significant payback in the software and IT industries.

What Is Six Sigma?

Six Sigma is a proven, data-driven suite of improvement methodologies based on a common philosophy and supported by measurements and tools for process and product improvement. Simply put, it is management by facts, not opinion.

Sigma is a statistical term that measures how far a given process deviates from its goal. The main idea behind Six Sigma is that if you can measure how many "defects" you have in a process or product, you can systematically determine how to eliminate them and get as near as possible to "zero defects."

The Six Sigma Philosophy

The Six Sigma philosophy states that by reducing variation, that is, by getting a process or product to perform within customer specifications, one can eliminate defects, which results in improved customer satisfaction, reduced operating costs, and increased profitability. This is because defects are directly correlated with operating costs (higher rework cost or cost of poor quality results in higher operating costs and thus, lower profits) and inversely correlated with customer satisfaction (the greater the number of defects, the lower the customer satisfaction). Indeed, Six Sigma pioneers such as Mikel Harry argue that Six Sigma is fundamentally about improving profitability, although improved quality and efficiency are immediate by-products of Six Sigma.[3]

The Six Sigma Approach

In order to eliminate defects, Six Sigma focuses on minimizing variation, because variation results in inconsistency in meeting customer specifications (defects), which in turn leads to dissatisfied customers.

The *sigma level* corresponds to where a process or product performance falls when compared to customer specifications. In other words, the difference between the upper and lower bounds of the customer specification (denoted by the Lower Specification Limit, or LSL, and Upper Specification Limit, or USL, respectively, in Figure 1.1) represents the range within which the product or service must fall in order to meet customer specifications, with optimum design or target (T) at the center. This range is called the *design width*. The actual range in which the developed product or service falls, or the process spread, is called the *process width*.

A process that is centered has a normal distribution (or can be represented by a bell curve) with mean (μ) aligned with target (T), and the specifications placed six standard deviations to either side of the mean, as shown in Figure 1.1. Due to the natural drift that occurs in process execution, it is observed that over time the process mean drifts from the target by as much as 1.5 standard deviations. In the manufacturing world, such drift is typically due to tool wear and tear, while in the software world, it is due to reduced process adherence by humans over time,

Figure 1.1

The concept of
Six Sigma.[4]

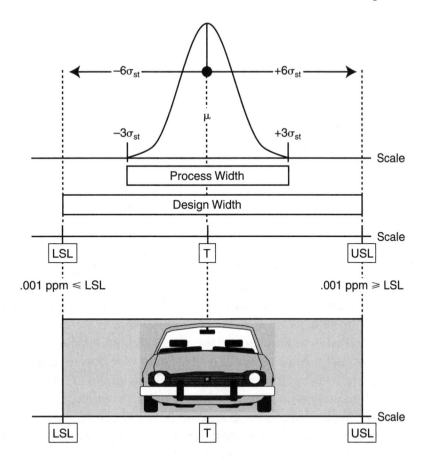

or changes in processes and tools. The consequence of this drift is that a small portion of the distribution extends beyond the specification limits—3.4 parts in 1 million to be precise.

Let's use a simple example from everyday lives to understand these concepts. As shown in Figure 1.1, think of a garage in which you park your car every day. Let's say you purchase an expensive new car and you are parking it in the garage for the first time. The first few days, you will obviously be very careful to make sure you center it nicely, away from the walls, so you don't accidentally scratch it (especially the side mirrors!) while getting it in and out of the garage. In this case, the width of the garage is your specifications limit, or design width, while the actual width of your car is the process width. Having gotten used to parking your new car in the garage for a few days, you will grow increasingly confident and most likely a little less cautious, and as a result you will no longer be centering the car on target, but perhaps drifting a bit to the left or right (although you can't afford to drift too much, else you again risk scratching the car). This is the concept of process drift that was illustrated earlier.

Another concept that is relevant to understanding variation is that of accuracy and precision. The process width, or the standard deviation of a distribution, is also referred to as *precision*. The difference between the mean and target is referred to as *accuracy*. Therefore, as a process spreads, its precision decreases. Likewise, the further away the mean is from the target, the less accurate the process is. We will pictorially depict the difference between accuracy and precision momentarily.

In Six Sigma, the quest to reduce variation consists of two primary goals to be achieved in the following sequence:

1. **Reduce spread:** Reduce process width so that less and less of the product or process falls outside of the specification limit. In other words, increase process precision (or reduce variation).
2. **Center the process:** Center the process mean on the center so that more and more of the process or product average (mean) falls on target (T). In other words, increase process accuracy.

To illustrate this, let us consider a man who is doing some shooting practice. As shown in Figure 1.2, the target area is represented by the circle, and the goal for the shooter is to shoot at the center of the target area—the bull's eye, represented by "T." In the current situation, the shooter is accurate because the bullet holes are close to the bull's eye (he can, of course, *improve* his accuracy by shooting closer to the bull's eye). However, notice that the shooter has poor repeatability; that is, as the shooter takes *repeated* shots, his shots are not clustered tightly together. The cluster would be precise if all the shots were close to each other, even if they were not close to the bull's eye. The shooter's first goal is to improve his precision, because in order to have high accuracy, it would require that the shooter's repeated shots cluster on or very close to the target. In other words, the shooter would not be able to improve his accuracy without first improving his precision. Once he has improved his precision, his second goal is to aim to hit the bull's eye by improving his accuracy.

In the analogy just described, think of each of the bullet holes as data points. Another way to monitor whether or not a process or product is within specification limits is by plotting these data points using a control chart, as shown in Figure 1.3. In addition to the customer specifications (USL and LSL), the control chart uses additional safeguards (represented by

Figure 1.2

Accuracy versus precision.

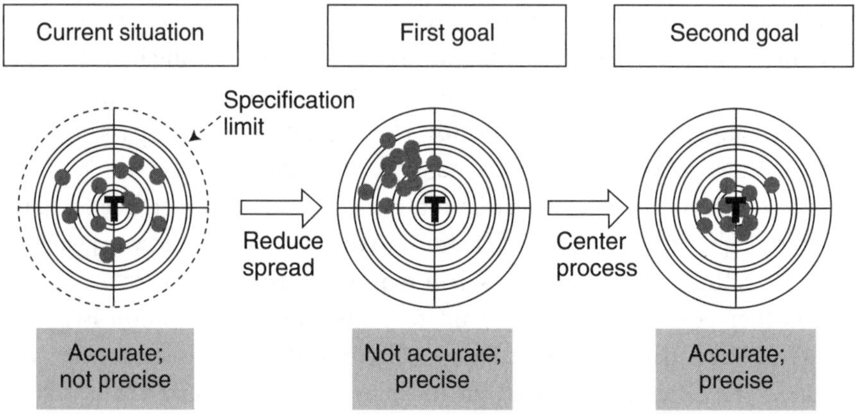

| Current situation | First goal | Second goal |

Specification limit

Reduce spread

Center process

Accurate; not precise

Not accurate; precise

Accurate; precise

Figure 1.3

Control chart.

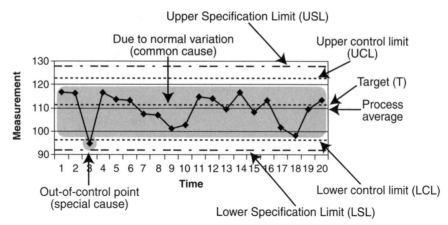

upper control limit, or UCL, and lower control limit, or LCL), which represent thresholds that, if breached, indicate "special cause" variation that, if left unchecked, will result in the customer specifications being breached (defect). This is the reason why the UCL and LCL fall within the USL and LSL. Such special-cause variation thus indicates the need for corrective action to bring the process back under control so that the customer specification limits are not breached and no defects result.

Indeed, from a Six Sigma perspective, when the control chart for a process or product is plotted, the first order of business is to look for any such "out of control" data points representing special-cause variation and to bring them back within the UCL and LCL control limits. In other words, reduce process spread and improve process precision. The next step is to look at the process average (or mean) to see how far from target it is in order to determine ways to improve the process average so that it falls on target. In other words, center the process and improve process accuracy.

Returning to our analogy of parking the car in the garage, our first objective, even before we purchase the new car, should be to make sure that it is not wider than the garage or is so wide that one would not be able to easily enter and exit the garage. This is because the width of the car is our "process width," and we know that once we have purchased the car, there is nothing we can do to reduce its process width. Once we have purchased a car that we know will easily fit in the garage, our second objective is to make sure that each time we park the car in the garage, we center it adequately so there is no risk of damage. And there you have it, the basic concept of Six Sigma!

The Concept of Sigma Levels

Higher levels of sigma correspond to fewer and fewer defects, and thus higher and higher levels of customer satisfaction. In fact, each additional sigma level results in an *exponential* reduction in defects. For example, a move from four sigma to five sigma requires 27 times improvement in performance, and a move from five sigma to six sigma requires more than 60 times improvement in performance. Six Sigma (6σ) represents the near elimination of defects at 99.9997 percent goodness, or only 3.4 defects per million opportunities (DPMO)—a quality level that is synonymous with world-class quality (see Table 1.1).

Table 1.1 Practical Impact of Process Capability[5]

Sigma Level	DPMO	Cost of Poor Quality (% of sales)	Benchmark
6	3.4	<10%	World-class
5	233	10–15%	
4	6,210	15–20%	Industry average
3	66,807	20–30%	
2	308,537	30–40%	Noncompetitive
1	690,000		

To put it in perspective, here is why the industry average of four sigma is usually not good enough:

▲ If your electricity provider operated at four sigma, you would be without power for more than an hour every week.

▲ If your cell phone company operated at four sigma, you would have no cell phone service for more than four hours every month.

▲ If book publishers operated at four sigma, you would see one misspelled word in every 30 pages of text. By contrast, two sigma is 25 misspelled words in every page, and six sigma is one misspelled word in all the books of a small library.[6]

It ultimately boils down to an organizational decision on what sigma level is the most appropriate from a cost-benefit perspective. In other words, every process or product does not have to reflect Six Sigma quality. Generally, products and processes involving human safety, health, and money are most likely to strive for Six Sigma quality.[7] Indeed, the airline industry operates with less than one fatality per million travelers, which is better than the Six Sigma level of 3.4 fatalities per million travelers.

Six Sigma Improvement Methodologies and Tools

The suite of Six Sigma improvement methodologies comprises three key methodologies, each appropriate for use for a specific purpose:

▲ **DMAIC (Define, Measure, Analyze, Improve, Control)** for process improvement by reducing process variation and defects.

▲ **DFSS (Design for Six Sigma)** for designing new processes and products. This methodology has different variants, which we will read about in Chapter 16.

▲ **Lean Six Sigma (or simply, Lean)** for improving process efficiency and speed.

As a rule of thumb, a DMAIC project may result in anywhere from a modest to a significant improvement, typically up to 50 percent improvement. On the other hand, a DFSS or Lean Six Sigma project is more likely to result in a radical or breakthrough improvement, and it may deliver as much as 100 percent or more improvement to a process or product.

Key measurements used in Six Sigma include critical to quality (CTQ), mean (μ), standard deviation (σ), DPMO, and process capability (Cp, Cpk). For a definition of these terms, refer to Appendix C.

Tools used in Six Sigma include qualitative and quantitative (statistical) tools for data analysis, root cause analysis, root cause validation, and identification and selection of improvements.

▲ Qualitative tools include but are not limited to process mapping, fishbone diagram, cause and effect matrix, five whys, failure mode effects analysis (FMEA), and so on.

▲ Quantitative tools include but are not limited to Kruskal-Wallis, one- and two-sample T-test, analysis of variance, confidence intervals, F-tests, one- and two-proportion tests, Monte Carlo simulation, regression, Design of Experiments (DOE), and so on.

DMAIC Overview

Explained in simple terms, in DMAIC methodology, the purpose of the:

▲ Define phase is to define the problem or improvement opportunity (Big Y) and translate it into critical customer requirements (CCRs)
▲ Measure phase is to gather data on current process performance
▲ Analyze phase is to analyze the data to gather information on the causes that are resulting in the problem(s), or the factors that can be adjusted to improve performance (Small Xs)
▲ Improve phase is to identify the specific improvements to be implemented
▲ Control phase is to ensure that there is a plan in place to secure the gains and make them permanent, including mechanisms for detection of deviation in process or product execution from desired levels

The DMAIC phases are executed sequentially, although there may be some overlap and iterations between phases, as well as feedback from subsequent phases to previous ones.

Key activities in the Define phase are:

▲ Define the problem statement or improvement opportunity in a way that effectively articulates why the Six Sigma project is necessary and why it is necessary *right now*.
▲ Develop the project charter.
▲ Map the current process.
▲ Gather the voice of the customer (VOC).
▲ Form the project team.

Key activities in the Measure phase are:

▲ Identify the measurements to collect.
▲ Develop and execute the measurements collection plan.
▲ Develop and validate the measurement system.
▲ Identify baseline performance measurements as Cp/Cpk or DPMO (and sigma level).

Key activities in the Analyze phase are:

▲ Stratify data to identify the underlying problem(s).
▲ Identify root causes.
▲ Validate root causes.

Key activities in the Improve phase include the following:

▲ Identify potential solutions.
▲ Evaluate and select potential solutions.

Key activities in Control phase include the following:

▲ Pilot potential solutions (if needed).
▲ Evaluate pilot results (if applicable).
▲ Develop the control plan.
▲ Develop the change implementation plan.
▲ Develop procedures, standards, and training material.
▲ Deliver training.
▲ Communicate improvements.
▲ Implement improvements.

DFSS Overview

The DFSS methodology is appropriate for designing new products or processes, or redesigning existing ones if incremental improvements are not sufficient and breakthrough improvements become necessary.

The most widely used framework for DFSS is DMADV (Define, Measure, Analyze, Design, and Verify). In this methodology, the purpose of the:

▲ Define phase is to define the design goal or improvement opportunity (Big Y) and translate it into CCRs
▲ Measure phase is to identify and gather data for key metrics that best quantify the VOC
▲ Analyze phase is to identify key design factors that influence the CCRs, identify design alternatives, and select a design approach
▲ Design phase is to identify design parameters, flow down the CCRs to design parameters, assess the impact of variability in design parameters on CCRs, assess design gaps, and identify corrective actions
▲ Verify phase is to identify potential design failure modes, exercise the new design via pilots and prototypes, and prepare for deployment of the new design

Besides the DMADV framework, other popular DFSS frameworks include:

- ▲ Identify, Design, Optimize, Verify (IDOV)
- ▲ Concept, Design, Optimize, Verify (CDOV)
- ▲ Define, Measure, Analyze, Design, Optimize, Verify (DMADOV)

Lean Six Sigma Overview

Lean Six Sigma is focused on eliminating waste and reducing capital investment in an organization by focusing only on activities that create value. Principles of Lean include:

- ▲ Zero waiting time
- ▲ Zero inventory
- ▲ Scheduling (internal customer pull instead of a push system)
- ▲ Batch to flow (reduced batch sizes)
- ▲ Line balancing
- ▲ Reducing actual process cycle times

Given that Lean Six Sigma is focused on improving process efficiency and speed, one may wonder what Lean Six Sigma has to do with solving problems, improving performance, reducing defects, and improving sigma levels. Let us consider a couple of different perspectives.

One view is to consider Lean and its tools such as value stream mapping, cycle time analysis, and pull systems (kanban) as yet another toolkit that is part of the overall Six Sigma toolkit. This means that within the scope of a traditional DMAIC project, where the Big Y is to reduce cycle time, one can leverage Lean tools such as value stream mapping to map out the current process and perform a cycle time analysis to identify improvement opportunities that help achieve the Big Y of the project, and thus help solve a problem or improve performance.

Another view is that by eliminating nonvalue-added steps in the process, Lean Six Sigma helps reduce the number of opportunities, which in turn results in fewer defects and thus increases the "percentage good" for the same sigma level. As an example, if a 20-step process operating at four sigma is made lean by eliminating 10 steps, the percentage good improves from 88.29 percent to 93.96 percent. Now, if one were to apply the traditional DMAIC methodology of Six Sigma to improve the process to five sigma, the percentage good improves dramatically—to 99.768 percent![7]

Brief History of Six Sigma

As with any improvement methodology and quality framework, Six Sigma has evolved over the years. Six Sigma was developed by Motorola in the early 1980s in response to then-CEO Bob Galvin's challenge to the company to improve the quality of Motorola products tenfold within five years. As Motorola's executives explored ways to meet the challenging objective set by the CEO, an engineer by the name of Bill Smith worked behind the scenes to study the correlation between a product's field life and how often it had been repaired during the manufacturing process. In 1985, Smith presented a paper that concluded that if a product was found defective and corrected during the production process, other defects were bound to be missed and found later by the customer during early use of the product. On the other hand, if the product was assembled error free, it rarely failed during early use.[3]

Smith's research was the genesis of Six Sigma. In its early days, Six Sigma was simply a metric, a measure of goodness. The pursuit of Six Sigma quality at Motorola meant the pursuit of no more than 3.4 DPMO from a baseline of about four sigma, which was costing the company 5 to 10 percent in annual revenues, and in some cases, as much as 20 percent of revenues to correct.[3,8] By 1993, through systematically detecting and fixing defects in the manufacturing process and proactively improving the manufacturing controls and product design to prevent the introduction of defects, Motorola had achieved Six Sigma quality level at several of its manufacturing facilities.[3] In part due to its pioneering Six Sigma work, Motorola won the prestigious Malcolm Baldrige National Quality Award in 1988. Other companies began to sit up and take notice. The Six Sigma revolution was now well underway.

In the early 1990s, Motorola set up the Six Sigma Research Institute in collaboration with Honeywell (then Allied Signal), Texas Instruments, Eastman Kodak, and other early adopters. This is when Six Sigma started to evolve into an improvement methodology denoted by MAIC (Measure, Analyze, Improve, Control). In 1995, General Electric (GE) embraced Six Sigma, and its CEO Jack Welch became its leading advocate. That year, GE estimated the opportunity loss at 3.5 DPMO to be 10 to 15 percent of annual revenues, or about US$7 billion.[9] By 2001, as a direct result of pursuing Six Sigma, GE had realized net cumulative cost savings of approximately $4.5 billion.[10] During these years, as it adopted and used Six Sigma, GE also refined the MAIC methodology to add "Define" as the first phase, and the MAIC methodology thus evolved into today's DMAIC. In the early 2000s, DFSS and Lean were added to the suite of Six Sigma methodologies.

Since its initial development, Six Sigma is said to have had three waves of adoption:

▲ In the 1980s, due to Motorola's pursuit of Six Sigma quality levels at its manufacturing facilities, Six Sigma came to be viewed as an improvement methodology relevant only to the manufacturing industry.

▲ Beginning in the late 1990s, service industries such as financial, healthcare, insurance, and others started to adopt Six Sigma.

▲ In the early 2000s, high-tech companies started embracing Six Sigma for process and quality improvement in software, systems engineering, and IT.

Six Sigma Program Success Factors

Six Sigma has now been around for close to 25 years. Six Sigma as an improvement methodology, however, is more recent—only about 15 years old. Over these years, Six Sigma has been effectively deployed by companies such as Motorola, GE, Xerox, IBM, and other companies showcased in this book. There is a lot to gain from the collective experience and wisdom of these companies that have successfully implemented Six Sigma. By learning from them, we can emulate their successes and learn how to steer clear of pitfalls and roadblocks.

This section lists key Six Sigma program success factors and roadblocks culled from literature.[11,12,13] They are summarized here to help you plan for success—whether you are planning to launch a Six Sigma program or need to revitalize a languishing Six Sigma program. The firsthand accounts from Cisco, Thomson Reuters, and Convergys of their Six Sigma programs provide fascinating insight into what worked and what didn't work for these companies, what lessons they learned along the way, and how we can enhance our own Six Sigma programs.

Key Six Sigma program success factors are:

1. **View Six Sigma as a management system**

   Companies that have successfully deployed Six Sigma have taken a holistic view, wherein Six Sigma is seen as a management system as opposed to just being considered a metric or improvement methodology to solve problems. This requires a change in mindset and a cultural shift in the organization. It requires the willingness to gather data, to perform qualitative and quantitative analysis to arrive at the best solution, and to embrace Six Sigma as the "way we work." Quite often, people are inclined to jump straight to the solution with confident assertions of "I know how to fix *this* problem!" when, in fact, the solution they may be recommending is suboptimal, completely ineffective, or perhaps they are solving the *wrong* problem. This is where the importance of data comes in. Management by facts and data is always better than management by opinion. Data mining will provide you with the information to correctly formulate the problem statement (or validate a problem statement that is already formulated), and help drive corrective actions in the right direction.

2. **Management commitment and visible support**

   If an organization is going to make this fundamental shift in how it approaches problem solving and improvement opportunities, it needs the strong and visible support of senior management to make that transition. Companies that first embraced Six Sigma and implemented robust Six Sigma programs such as Motorola and GE were able to do so because of the active support of their CEOs—Bob Galvin at Motorola and Jack Welch at GE.

   In addition, for Six Sigma to take hold in an organization, middle managers are the key to success, because often they are project sponsors or project champions. Many companies have trouble getting middle management (who will be directly affected by Six Sigma measurements) to embrace Six Sigma. In the worst case, this can lead to active resistance to change, thus resulting in a failed deployment or a suboptimal deployment, with pockets in the organization that use Six Sigma and others that don't.[12]

   Senior and middle managers can show active support for the Six Sigma program by:
   ▲ Taking the training themselves
   ▲ Sponsoring Six Sigma projects
   ▲ Attending project reviews and asking probing questions

▲ Asking the Six Sigma class participants to communicate their expectations and emphasize the importance of the program to the success of the business and employee growth

▲ Publicly recognizing successful Six Sigma projects and teams

▲ Making Six Sigma a criterion for incentive compensation and promotions

3. Develop the infrastructure

An effective Six Sigma program requires a robust supporting infrastructure. Such an infrastructure includes but is not limited to:

▲ Definition of a process for the identification of Six Sigma projects so that the organization invests Six Sigma resources in solving the *right* problems

▲ Definition of a process for how Six Sigma projects are governed (initiated, reviewed, tracked, cancelled, assessed for financial benefits, and concluded)

▲ Investment in skilled resources

▲ Investment in tools for recording, tracking, and reporting on Six Sigma projects

▲ Specification of the organizational process and criteria* for achieving Green Belt (GB), Black Belt (BB), and Master Black Belt (MBB) certification

▲ Implementation of a Six Sigma training program

▲ Definition of how Six Sigma projects are to be executed (project roles, templates, extent of rigor, reporting, and project reviews)

4. Invest in skilled resources

Companies that have had good results with their Six Sigma programs are those that have invested in *skilled* resources to act as MBBs to train, coach, mentor, and support their BBs and GBs and to help groom them as future MBBs and BBs, respectively.

A good MBB is not just someone who is good with the use of statistical tools, but one who is an effective change agent and also has good business acumen. In fact, Volvo refers to them as future leaders, and GE views MBBs as likely candidates for future senior management positions.

Quite often, Six Sigma projects fail and Six Sigma either fails to take hold in an organization or loses momentum because project teams run into problems with using statistical tools or other barriers that they do not know how to circumvent. This is where a good MBB can be invaluable in showing the way past an organizational roadblock or out of a statistical quagmire.

An investment in MBBs, BBs, and GBs is one good investment with enormous payback—anywhere from US$100,000 to US$200,000 per improvement project, with a return on investment (ROI) often between five and ten times the original investment.[12] BBs, with 100 percent of their time allocated to BB projects, can execute five or six projects during a 12-month period, adding approximately $1 million in annual profits. In fact, with a cost

---

* Black Belt certification typically requires four weeks of training, generally spread over a few weeks, plus two completed projects, each with a benefit of at least US$175,000.[20] GB training typically consists of one week of training and one completed project of at least US$50,000 benefit.

of poor quality as a percentage of sales between 15 and 30 percent (for a company operating at four sigma or s Sigma level) (see Table 1.1), and given the fact that consequences of many quality failures such as customer dissatisfaction and loss of market share cannot be quantified, the opportunity lost for a company that does not invest in Six Sigma is staggering.

5. **Invest in training**

   Companies that have effectively implemented Six Sigma have made significant investments in training and made a concerted effort to do it right. This means not cramming scores of employees in Six Sigma training classes to hit some target goal of percentage of employees trained in Six Sigma. Instead, begin with carefully identifying who needs to be trained and provide them with the right training, with adequate rigor and ongoing support of MBBs so they are well equipped to successfully execute Six Sigma projects.

   Best practices with respect to Six Sigma training include but are not limited to:[12,13, 14,15]

   ▲ The training program should be a good combination of theoretical instruction and practical application using real-life examples relevant to the students' daily jobs.

   ▲ Do not focus just on statistical tools, but ensure the training content adequately covers customer orientation, project management, organizational change management, and behavioral techniques that are so vital to effect change in an organization.

   ▲ Ensure the instructors are Six Sigma experts with excellent communication skills.

   ▲ Customize the training material to your company and its needs, especially the examples. This may require changing and enhancing the standard training material.

   ▲ Spread the Six Sigma training program and make it *hands-on* to allow the students to take a real problem from their daily jobs and apply Six Sigma techniques to solve the problem.

   ▲ Instead of overwhelming all BBs and GBs with training on advanced statistical tools, ensure that the MBBs can help them with the use of those tools when necessary.

6. **Value results**

   It is important that the Six Sigma program be monitored with the right metrics. Right metrics are those that value results, that is, that show meaningful financial and operational benefit (refer to the success stories later in this book for several examples). Metrics are meaningless if they tout the number of employees trained on Six Sigma or the number of completed Six Sigma projects without an indication of quantified results. Metrics need to show which measures and objectives on the senior leadership's dashboard have been favorably affected by Six Sigma projects (see item 10: *Select projects that matter*).

7. **Recognize and quantify all financial benefits**

   Closely related to valuing results is quantifying *all* financial benefits as best as you can. If you are going to gain management attention and retain the management commitment that helped launch your Six Sigma program, you need to demonstrate that Six Sigma is helping to move the needle on financial measures. For this, you need to quantify not just the operational benefit (such as reduction in percentage defectives or improvement in cycle time), but also translate the improvement in currency terms and categorize it as cost savings, cost avoidance, or revenue gain.

▲ How many dollars of your current spending has the Six Sigma project been able to reduce (cost savings)?

▲ How many dollars in future cost has your Six Sigma project helped avoid (cost avoidance)?

▲ By designing a process or product (using DFSS) or improving it (using DMAIC), what is the estimated revenue opportunity for your company, and what is the best estimate of how much of that revenue can be realized (revenue gain)?

8. Focus on the customer

Businesses exist for one purpose—to make a profit. Simply cutting costs will not take a company far enough. Ultimately, it comes down to increasing revenue and gaining and retaining market share. Therefore, the traditional use of Six Sigma as a methodology to reduce cost of poor quality and improve the bottom line is limited to realizing only half of the benefits that Six Sigma can deliver. A Six Sigma program is realizing its true potential if it is making a meaningful difference to the customer. Favorable impact to the customer will help retain the customer, or win new business from the customer, or help gain new customers—the latter two directly result in top-line growth.

Back in the late 1990s when GE launched Six Sigma, it was used primarily for reducing variation and taking cost out of production. Customers felt left out. In the GE annual report of 1998, Jack Welch quoted the customers as asking "When do I get to see the benefits of Six Sigma?" This led to the introduction of the *outside-in* thinking concept at GE, with the goal of focusing on CTQs that mattered to the customer. In subsequent years, GE reported not only internal savings from Six Sigma, but customer benefit as well.[13]

In order to be customer focused, the improvement goal (CTQ) for a Six Sigma project needs to be customer focused and completed projects should be evaluated from a customer perspective.

9. Adapt Six Sigma to your organization's needs

When it comes to Six Sigma deployment, the age-old adage "one size does not fit all" is certainly true. Depending on your business, you will need to adapt Six Sigma to your needs.

▲ As stated earlier, you will need to customize the training to your organization.

▲ You will also need to review the Six Sigma tools to see which ones are more relevant to your business. Statistical tools that are more relevant to manufacturing processes or product design processes may be less relevant in transactional processes.

▲ You will need to adapt the Six Sigma program to your company's size. Many of the early adopters of Six Sigma in the software and IT industry are the larger and more established companies. Small and medium-sized companies do not need to implement a Six Sigma program that is as elaborate or sophisticated as at a larger company. For example, smaller companies have a smaller portfolio of Six Sigma projects and less complex organization structure; therefore, they may not need sophisticated tools to track, roll up, and report on projects. Likewise, they may not need the breadth of training curriculum and advanced statistical training classes that a larger company might offer its employees. In cases where the cost of licenses for certain statistical tools from larger vendors is prohibitive, smaller companies can explore cheaper and less sophisticated offerings by other vendors or use home-grown tools.

▲ You will need to consider which Six Sigma methodologies (including specialized DFSS methodologies described in Chapter 16) would be beneficial for your company. In most companies, a majority of the Six Sigma projects are DMAIC projects, along with a smaller number of Lean and DFSS projects.

10. **Select projects that matter**

Ultimately, Six Sigma is about results. It is about solving problems—solving the *right* problems. In any company, the number of project opportunities is usually far more than the resources available to work on them. The same is true of Six Sigma. Therefore, the Six Sigma program needs to be selective in what improvement projects to launch. Improvement opportunities need to be prioritized in order to select projects that matter.

Before we review the process for project selection, let's review the characteristics of a Six Sigma project. Remember, any improvement opportunity where Six Sigma tools may be applicable is *not* necessarily a Six Sigma project. There is overhead associated with planning, managing, and reporting on any project, not just Six Sigma projects, so think carefully if the improvement opportunity can be handled as a quick win with the targeted use of one or more (usually one) Six Sigma tools, or if a formal Six Sigma project needs to be initiated.

Characteristics of Six Sigma projects are:

▲ **Alignment:** The project is either directly connected to or has a clear linkage to senior management's documented goals and objectives.

▲ **Importance:** The project is of major importance to the company in terms of expected operational and financial impact.

▲ **Goal:** The project has a well-defined goal statement.

▲ **Scope:** The project has a reasonable scope that can be completed in three to six months.

▲ **Complexity:** Problems are one of two types: solution known and solution unknown. Six Sigma projects are aimed at problems with solution unknown. These include complex problems where many alternatives exist.

▲ **Impact:** Six Sigma projects should have a minimum financial impact as specified by the company. As an example, Black Belt projects at Motorola must have a minimum financial impact of US $250,000, and Green Belt projects must have a minimum financial impact of US $50,000.

▲ **Management Support:** The project has management support.

In order to ensure that the organization initiates projects that matter the most, it will need to follow a systematic approach to identifying and selecting projects:

▲ **Form a steering committee:** Begin by forming a project selection steering committee comprising senior management. You will need the involvement of these decision makers, because they are the best people to determine the relative priority of the management objectives for the year and to sift through the voice of the customer to identify improvement opportunities.

▲ **Determine improvement opportunities:** For each management objective, discuss major challenges and improvement opportunities. Review external and internal voice-of-the-customer information to identify customer issues. If there are a large number of objectives and customer issues and the team wants to focus on the higher-priority ones for generating project ideas, techniques such as multivoting may be used to eliminate lower-priority items from consideration.

▲ **Generate project ideas:** For the management objectives and customer issues under consideration, brainstorm project proposals and list them as shown in Table 1.2.

Establish a ranking scale that will be used to assign weights to each of the management objectives and customer issues. The committee may assign the ranking by working together as a group or individually, and then take an average. This is how Objective #2 and Issue #2 receive a ranking of 7, even though the scale only has scores of 0, 3, 5, 8, and 10.

Next, establish a proposal ranking scale to define the relationship between project proposal and management objective or customer issue. Again, the committee may establish the ranking scale collectively or individually.

Each committee member then uses the ranking system to score how well the proposal addresses the management objective or customer issue. The average for each cell is then multiplied by the corresponding importance weight, and all such values for the proposal are added to arrive at the overall project score.

▲ **Prioritize proposals:** The proposal list can then be sorted in descending order of project scores to produce a prioritized list of proposals (see Table 1.3).

The proposal priority list should be maintained on an ongoing basis, and it should be dynamic. As management objectives and customer issues are effectively addressed, they may be removed and new ones added. This may result in reprioritization of proposals, and in some cases, may result in lower-priority projects being put on hold or cancelled in order to allow for reassignment of resources to higher-priority ones.

11. Implement project reviews

One major pitfall companies need to guard against is letting Six Sigma projects languish, because that can quickly cause the overall Six Sigma program to sputter, lose momentum, and come to a grinding halt. The most direct benefit of a project review (with management participation) is that it keeps a project team on its toes—the steady and healthy pressure of regular project reviews assures that the train keeps moving in its tracks and does not stall or derail altogether. But there are several reasons why project reviews are critically important. They:

▲ Provide regular insight to the project sponsor with what is actually happening in the project

Table 1.2 Project Selection Matrix

Management Objective or Customer Issue	Objective #1	Objective #2	Objective #3	Issue #1	Issue #2	Issue #3	
Importance	8	7	3	10	7	5	
Project Proposal							**Project Score**
Name of Proposal #1	3	5	0	7	0	5	154
Name of Proposal #2	0	8	4	5	4	0	146
Name of Proposal #3	5	3	5	5	0	7	161
Name of Proposal #4	0	0	0	7	8	0	126
Name of Proposal #5	8	0	5	5	3	5	175
Name of Proposal #6	3	4	0	0	5	3	102
Name of Proposal #7	0	9	0	5	0	8	153
Name of Proposal #8	7	0	5	5	6	0	163
Name of Proposal #9	0	0	9	7	0	7	132
Name of Proposal #10	10	5	0	0	0	3	130

Importance (of Mgmt. Objective or Customer Issue)	Rating
0	Not important
3	Slightly important
5	Important
8	Very important
10	Critical

Proposal Rank	Rating
0	No correlation
3	Very little correlation
5	Some correlation
8	High correlation
10	Complete correlation

Table 1.3 Proposal Priority List

Management Objective or Customer Issue	Objective #1	Objective #2	Objective #3	Issue #1	Issue #2	Issue #3	
Importance	8	7	3	10	7	5	
Project Proposal							**Project Score**
Name of Proposal #5	8	0	5	5	3	5	175
Name of Proposal #8	7	0	5	5	6	0	163
Name of Proposal #3	5	3	5	5	0	7	161
Name of Proposal #1	3	5	0	7	0	5	154
Name of Proposal #7	0	9	0	5	0	8	153
Name of Proposal #2	0	8	4	5	4	0	146
Name of Proposal #9	0	0	9	7	0	7	132
Name of Proposal #10	10	5	0	0	0	3	130
Name of Proposal #4	0	0	0	7	8	0	126
Name of Proposal #6	3	4	0	0	5	3	102

▲ Provide an excellent opportunity to the project sponsor and senior management to visibly demonstrate their commitment to the project and the Six Sigma program

▲ Allow the project sponsor and other stakeholders to assess whether the project is following the Six Sigma methodology and using the correct tools to systematically arrive at the final solution

▲ Provide an excellent opportunity to the project leader to seek the sponsor's support to remove barriers and address risks that the project team may be encountering

Project reviews are meant to be short and crisp, with probing questions from the sponsor and other stakeholders that challenge the project team and enable them to remain focused on the end goal. Recommended questions for project reviews, depending on the current status of the project, are[17]:

▲ What is the problem statement and goal statement?

▲ Has your project charter been approved?

▲ Is your project on track per the schedule? Are there any risks or roadblocks you need help with?

▲ Based on your data analysis, what are the top four or five causes that account for 80 percent of the problem?

▲ Based on the root causes, what corrective actions do you recommend?

▲ What is the status of the implementation of corrective actions?

▲ What is the actual or anticipated improvement in performance as a result of these actions?

▲ What is the financial benefit from your project? Show how you arrived at this benefit.

▲ How will you ensure that the improvements resulting from this project will be permanent, and how will we be able to detect any deviation from expected performance levels?

12. **Develop a robust deployment plan**

There is no sure-fire and quicker way to a failed Six Sigma implementation than a poor deployment plan. Committed senior management is a prerequisite to institutionalize Six Sigma in a company, but it is not a guarantee for success. Therefore, it is important that the deployment team carefully strategize about how it will deploy Six Sigma. Fascinating stories from Cisco, Reuters, and Convergys appear later in the book, but here is some quick advice to get you started:

▲ Start with senior management and educate them on Six Sigma—what it is and how they can benefit from it. Be sure to include success stories and benefits realized by other companies in the same industry or, even better, from direct competitors.

▲ Train a small team of improvement specialists who will participate in piloting or proving out the relevance and benefits of Six Sigma to your company. Pilot projects should be those whose importance is readily apparent to the workforce, can be accomplished in a relatively short amount of time, and whose resulting operational and financial benefits can be easily quantified. Successfully concluded pilot project(s) will provide you with the credibility, management support, and impetus to move forward with the program (refer to Chapter 11 for an example of a pilot project).

▲ Expand the Six Sigma rollout either to the entire company or to functions that have been identified as early adopters (your senior management will need to decide this). Additional successful Six Sigma projects executed by these early adopters will enable Six Sigma to spread and take root in the rest of the company.

▲ Ensure a robust process, support infrastructure, resources, and training are in place to support the continued use of Six Sigma as it becomes a part of your organizational DNA.

13. **Communicate**

This is all about developing an effective communications campaign, alleviating anxiety about the change, minimizing resistance to change, enabling the organization to transition from its current way of working to the new way of working, and celebrating successes. Key elements of organization change management are:

▲ Communications begin with explaining what Six Sigma is; why the company is adopting Six Sigma; anticipated benefits of implementing Six Sigma; showing a vision of the future state; and allaying fears by explaining what the deployment plan is, how employees are affected, and what training and support would be available.

▲ During the pilot phase, communications shift to explaining what projects are being piloted and what benefits have been realized.

▲ Closer to deployment, communications pertain to the training plan, training events, and training status reporting.

▲ After deployment, communications are for the purpose of maintaining and enhancing the commitment to Six Sigma. This is accomplished by regular communications to socialize the successes of Six Sigma projects, publicly recognize the project team members, and celebrate major improvements.

Financial Benefits of Six Sigma

Fundamentally, Six Sigma is about improving a company's competitive advantage and boosting its *profitability*. Therefore, a key characteristic of good Six Sigma projects is well-documented, rigorous, and audited financial benefit approved by a person in the finance organization.

The success stories showcased in this book not only provide an excellent illustration of using Six Sigma for problem solving, but they also describe compelling financial benefits ranging from an average benefit of approximately US$600,000 in some cases (Chapters 3, 4, 5B, 6A, 7B, 14, and 18) to an average savings of US$5 million in other cases (Chapters 6B, 6C, 7A, 12, 19, 20, and 21).

In addition to the financial benefits of the Six Sigma projects showcased in this book, here are some other notable examples that provide additional proof of Six Sigma's relevance to the high-tech industry:

▲ The IT organization at Raytheon Aircraft saved US $500,000 from a single project in 2002.[18]

▲ The nine CIOs at Textron saved a total of US $5 million in six months.[18]

▲ Raytheon Aircraft's IT department used Six Sigma to improve claims processing and save the company $13 million.[19]

▲ Seagate's IT department booked direct savings from Six Sigma analyses of US $3.7 million in one year. Since instituting Six Sigma, the IT department has saved US $4.5 million overall in two years.[19]

Six Sigma and Software

The high-tech industry is fundamentally different from the manufacturing industry in many ways, which has a direct bearing on how Six Sigma is used in this industry. For example:

▲ Software product is nonphysical while a manufactured product is tangible, with physical attributes such as length, breadth, height, volume, color, and so on.

▲ Software development focuses on building a unique product. Once that product is developed, it can be replicated easily with no part-to-part variation. That is why software quality professionals focus their attention on the software development process rather than on the manufacturing process of reproducing the duplicate copies.[21]

▲ The software development process relies heavily on people and their skills, and the most important factor is human intelligence. Manufacturing activities are machine-intensive and targeted to minimize cognition.[22]

▲ The behavior of a software process is difficult to predict, while the behavior of a manufacturing process is predictable.[23]

▲ The software development process is invisible, and it has to be made visible with flowcharts, use cases, dataflow diagrams, and so on. The manufacturing process, on the other hand, is inherently visible.[21]

▲ Process variation in software development is affected by large differences in skills and experience from one software developer to another. Key sources of variation in manufacturing are differences between components and runs of assembly processes.[21]

▲ The overall process cycle time may be much longer for creating a software product than a manufactured item. Therefore, Six Sigma projects in software may take longer or may need to be conducted with greater risk, due to smaller amounts of data.[7]

Companies that are successfully using Six Sigma in the software and IT domains are those that have been innovative and creative in applying Six Sigma when the opportunity to apply the methodology and tools may not seem as intuitive to everyone. This is because they have been open-minded, flexible, and eager to adapt and apply. Along the way, they have overcome intractable problems and recognized stellar performance and financial gains. Let's see how they did it…

References

1. Study commissioned by the U.S. Department of Commerce's National Institute of Standard Technology, 2002.
2. The Trends in IT Value, Standish Chaos Project, Standish Group International, 2008.
3. Mikel Harry and Richard Schroeder, *Six Sigma: The Breakthrough Management Strategy Revolutionizing the World's Top Corporations*, Currency, 2000.
4. Mike Harry, *The Vision of Six Sigma: A Roadmap for Breakthrough*, Sigma Publishing Company, 1994.
5. Mike J. Harry, *Six Sigma: A Breakthrough Strategy for Profitability, Quality Progress*, May 1998.
6. Brian Cusimano, "How Dealers Can Put TQM to Work," *Water Technology Magazine*, July 2006.
7. Jeannine M. Siviy, M. Lynn Penn, and Robert W. Stoddard, *CMMI and Six Sigma*, Addison Wesley, 2008.
8. Ravi S. Behara, Gwen F. Fontenot, and Alicia Gresham, "Customer Satisfaction Measurement and Analysis Using Six Sigma," *International Journal of Quality & Reliability Management*, Vol. 12, No. 3, 1995, pp. 9–18.
9. Michelle Conlin, "Revealed at Last: The Secret of Jack Welch's Success," *Forbes*, Jan. 26, 1998, pp. 44.
10. George Eckes, *The Six Sigma Revolution: How General Electric and Others Turned Process into Profits*, New York, Wiley, 2001.
11. Mark Goldstein, "Six Sigma Program Success Factors," *Quality Progress*, Nov. 2001.
12. Lennart Sandholm and Lars Sorqvist, "12 Requirements for Six Sigma Success," *Six Sigma Forum Magazine*, Nov. 2002.
13. Gerald J. Hahn, "20 Key Lessons Learned," *Quality Progress*, May 2002.
14. Roger Hoerl, "Six Sigma Black Belts: What Do They Need to Know," *Journal of Quality Technology*, Vol. 33, No. 4, pp. 391–435.
15. Gerald J. Hahn, Necip Doganaksoy, and Christopher Stanard, "Statistical Tools for Six Sigma," *Quality Progress*, Sept. 2001, pp. 78–82.
16. Michael M. Kelly, "Three Steps to Project Selection," *Six Sigma Forum Magazine*, Nov. 2002.
17. Arun Hariharan, "CEO's Guide to Six Sigma Success," *Quality Progress*, May 2006.
18. Tracy Mayor, "Six Sigma Comes to IT Targeting Perfection," www.cio.com.au, Feb. 2004.
19. Edward Prewitt, "Quality Methodology: Six Sigma Comes to IT," www.cio.com, Aug. 2003.
20. James M. Lucas, "The Essential Six Sigma," *Quality Progress*, Jan. 2002.
21. Rupa Mahanti, "Six Sigma for Software," *Software Quality Professional*, Vol. 8, No. 1, 2005.
22. M. A. Lantzy, "Application of Statistical Process Control to the Software Process," In Proceedings of the 9th Washington Ada Symposium on Ada: Empowering Software Users and Developers (July):113–23, 1992.
23. R. V. Binder, "Can a Manufacturing Quality Model Work for Software?" *IEEE Software*, Vol. 14, No. 5: 1997, pp. 101–105.

PART ONE

DMAIC Projects

PART ONE OF THIS BOOK INTRODUCES projects that demonstrate the DMAIC Six Sigma methodology. As explained in the introductory primer, these types of projects focus on eliminating or mitigating problems with existing processes. Though the nature of the problems addressed varies from one success story to the next, these examples typify the types of techniques and tools that can be consistently and effectively employed to implement process improvement solutions for universal problems in the high-tech industry.

DMAIC projects can be used to remedy problems or to address more general improvement opportunities. When implemented properly, DMAIC projects can be fast, effective, and provide significant financial returns.

CHAPTER 2

DMAIC Primer

Jeffrey A. Robinson, Ph.D.

Introduction

The DMAIC process is the heart of Six Sigma. The name DMAIC (pronounced *day-MAY-ihk*) is an acronym standing for the five phases of this problem-solving lifecycle. The phases are:

- ▲ Define
- ▲ Measure
- ▲ Analyze
- ▲ Improve
- ▲ Control

Together, these phases lead Six Sigma practitioners through a series of activities that systematically address problems and provide viable solutions. We will discuss each of these phases in more depth in a moment, but first, we need to establish the foundation for this process and the context within which it is best applied.

The Problem with Problems

The DMAIC process is first and foremost an approach for attacking problems, but not just any kind of problems. Specifically, the DMAIC method addresses "process" problems, recurring problems, the kind that occur over and over again.

Random problems, one-off anomalies, or one-of-a-kind types of failures typically do not warrant the use of the DMAIC methodology. Odd or unusual occurrences that are not likely to ever happen again are best resolved using different approaches. However, it is the more aggravating and annoying problems, the ones that occur repeatedly, that are often the most serious and costly.

Problems that recur are special. The fact that they happen more than once is often a strong indicator that they are related to some underlying process or processes that create them or allow them to occur systematically. Resolving such types of problems permanently typically requires changes to these underlying processes or procedures. Unless such process changes are made, the problems will persist. DMAIC is thus a process improvement process focused on finding and fixing problems.

One of the precepts of Six Sigma is that in order to affect permanent corrective actions, one must understand the factors that contribute to the undesirable conditions or outcomes. Moreover, one must discern the root causes for the symptoms being observed before viable solutions can be implemented. Finally, in order to effectively and permanently fix the problems, one must repair, alter, improve, or change the underlying conditions or processes that create the problems.

DMAIC is a relatively straightforward methodology to identify, quantify, analyze, and correct process-related problems of almost any type.

The processes involved may be simple or complex. They may be short or long. They may be financial processes, manufacturing processes, or transactional processes. They may be processes for managing projects, processing loan applications, designing products, creating sales plans, interacting with customers, developing software, ordering parts, or managing inventories.

A process is any defined set of tasks that are performed repeatedly. They may be manual or automated or a hybrid of both. Processes may be formally documented sets of procedures, or they may be ad hoc activities that are performed with some degree of regularity.

The DMAIC methodology is best suited for dealing with complex problems and complex processes, where root causes may be difficult to identify or where solution sets are complex or costly. They are particularly valued in situations where corrective actions are critical and where solutions are expensive. When there is great complexity or uncertainty, the techniques embedded within DMAIC excel at separating real issues from extraneous data and noise. Almost as important, the use of DMAIC can greatly improve the confidence of those using it that they have identified the correct problem and selected the best solution.

Conversely, DMAIC projects should not be applied to quick wins, problems where the root cause is known or the solutions are obvious. Applying DMAIC or Six Sigma techniques to such situations is overkill. DMAIC projects should be limited to problems that are complex or where there are many possible solutions and great cost or risk.

However, DMAIC is not a panacea. It is not fool-proof. As is the case with any problem-solving method, practitioners can make bad choices and draw wrong conclusions. Preconceptions, invalid assumptions, preferences, and unconscious biases can divert the focus of a project toward inappropriate activities and solutions. Bad data can lead you down incorrect paths of reasoning and result in solutions that fail. People can infer the wrong meaning, or they may misinterpret data and draw the wrong conclusions. However, the quantitative nature of the techniques used within DMAIC and the many checkpoints within this Six Sigma methodology provide excellent checks to biases, personal preferences, prejudgments, and fallacious decision making that might be introduced to this problem-solving approach.

Quantitative Problem Solving

Perhaps the thing that most clearly distinguishes DMAIC from other problem-solving approaches is it focuses heavily on quantitative data. The importance of quantitative analysis was probably best stated by William Thompson, Lord Kelvin, in lectures he gave in the 1890s. He said:

> "When you can measure what you are speaking about, and express it in numbers, you know something about it; but when you cannot measure it, when you cannot express it in numbers, your knowledge is of a meager and unsatisfactory kind: it may be the beginning of knowledge, but you have scarcely, in your thoughts, advanced to the stage of science."
>
> —WILLIAM THOMPSON, LORD KELVIN

One might paraphrase this eloquent principle more simply, by saying, "If it isn't a number, it's an opinion."

Ultimately, what we mean by this is that expressing things quantitatively reflects a deeper and more valid understanding of the thing being described. If you cannot express a problem in numbers, then you really don't know what the problem is. You need to ask questions that will lead you to these numeric answers.

When you cannot describe things in numbers, you tend to use adjectives and adverbs instead. The problem might be "very bad" or "not too bad," or the solution might be "pretty good" or "okay." These are subjective assessments and reflect opinions rather than facts. How much more meaningful are statements to describe problems like "Thirty-three percent of all loan applications by customers are cancelled after 40 days, resulting in an annual loss of $2.6 million dollars in wasted labor for the bank"?

However, in order to find facts to adequately describe problems, one must ask questions, such as:

▲ How severe is the problem?
▲ How often does it occur?
▲ How much does it cost?
▲ How long has this been happening?
▲ Where and when does it happen?
▲ Where and when isn't it happening?
▲ When did it start?
▲ Is it increasing? If so, how fast?

Thus, the DMAIC process is often presented to those learning this methodology as a Socratic process, a process of asking questions and finding answers … and sometimes questioning the answers one finds.

Learning enough about a problem to describe it in numbers is one of the first steps to systematically addressing it. If you fail to apply this rigor, even to the very definition of the problem that you seek to fix, you risk addressing problems that are not real but may only be suspected or assumed. You risk addressing problems that are not worth fixing, and you may spend more to fix a problem than it originally cost. You also risk chasing the wrong problems altogether. Worse, however, is the risk of engineering and designing solutions that won't work (thus wasting a lot of time and effort), and cost even more than the original problem itself. Without the rigor and discipline of formal quantitative approaches, one can end up treating symptoms while leaving the underlying causes unaddressed.

In each phase of the DMAIC process, different quantitative tools are used to define, assess, and analyze data to develop a sharp focus on the real problem and to separate it from inconsequential or distracting data. Only by identifying and fixing the root cause can one hope to affect permanent corrective actions to problems, and only through the use of quantitative methods can one verify that solutions actually work.

DMAIC and Six Sigma Tools

An examination of Six Sigma will quickly reveal that it is not a single prescriptive model. It is not a straightforward checklist that you can use to solve all problems. That is one reason that this book is outlined the way it is. Just as there are different types of problems, so are there different approaches to problem solving (DMAIC, DMADV, or DFSS, Lean).

Depending on how you count them, there are about 80 or 90 different tools and techniques, best practices, tools, and procedures that are used with some regularity in these different Six Sigma methodologies (Crevling, 2003). It is, however, unlikely that any project would ever use them all. Most will only use a handful of quantitative methods during the course of these DMAIC projects, and the tools used may differ significantly from one project to the next.

If you think of the Six Sigma methodology as a hardware store with many tools, then DMAIC and DMADV are basically just two different toolkits that utilize some of these techniques. However, just as there are many types of problems, there are many different toolkits. Recognize that the tools used by a plumber will be different from the tools used by a carpenter or those used by an electrician. Some tools may be common to all three toolkits; some may be unique to the different specializations and the different kinds of problems these practitioners face.

In the same way, each DMAIC project may apply different tools to achieve its goals. The selection of the best tools and techniques requires considerable training and experience. You probably have to know about more tools than you will actually use in order to understand which ones are the most appropriate. However, you not only have to know how to use these tools, but when to use them as well.

The DMAIC Phases

The Define phase is the first phase in the DMAIC lifecycle (Figure 2.1). This is where the project is formally begun, the problem is defined, and the team is formed. If the problem is a process-

Figure 2.1

The phases of the DMAIC lifecycle.

related problem (one that recurs with annoying regularity), then the process should be clearly identified.

Once the ground rules for the project have been defined and the goals and the team selected, the real work begins with the Measure phase. This is where data is collected in preparation for more detailed analysis later. This is where the problem is defined in quantitative terms and where the nature of the problem is dissected into the factors that contribute to or produce the undesirable outcomes that are to be remedied. The purpose of this phase is to identify, confirm, or discover the factors that might lie at the root of the problem and to collect data that will accurately confirm and quantify the problem and its magnitude.

Once data has been gathered from various sources, generated, or collected specially for the project, the Analysis phase commences. It is during this phase that the widest range of quantitative tools and techniques are used. Statistics are rigorously applied to discern real issues from suspected ones and to answer questions about the importance of the many possible factors identified in the earlier phases. Hypothesis testing is almost always applied during the analysis to evaluate which factors are the most important contributors to the problem in question. Various statistical tools can reduce the total possible factors to the most significant one(s) that should be addressed.

After analysis is mostly complete, additional statistical tools and techniques are applied in the Improve phase. At this point, the key factors that underlie the problem have been identified and different solutions can be suggested, evaluated, and compared. It is not uncommon for robust process models to be constructed to measure the effectiveness of different alternative solutions and to predict the outcomes of different corrective actions. Where such modeling is not feasible, practitioners can implement formal design of experiments to test the efficacy of solutions before a specific course of action is committed to.

Finally, in the Control phase, permanent corrective actions are developed and implemented. Process controls are designed and established so that problems will be reduced or eliminated. Control plans will be established to prevent the problems in the processes from reappearing or to detect the problems expeditiously should they occur again.

In all phases of the DMAIC process, the focus of the project is to quantitatively understand the nature of the problem and to apply statistics to prioritize, select, confirm, and measure appropriate solutions.

Once again, the DMAIC lifecycle is not an appropriate problem-solving approach for all problems. If the problems are relatively simple and the solutions are clear or obvious, then the DMAIC methodology is overkill. However, when problems are complex or confusing, or solutions are expensive or risky, the application of the DMAIC problem-solving methodology can be effective and can reveal solutions that would not have otherwise been considered.

Let's review these phases in a little more detail.

Define

The DMAIC process begins with the Define phase. Again, this is the starting point, where the project is formally begun, the problem is defined, and the team is formed.

Several best practices are critical to any DMAIC project, but not all of these are quantitative in nature. For instance, one of the first project artifacts that is produced is the project charter. This is a document that clearly establishes several things before real investigatory work can begin.

A project charter contains several sections, and all of them are important. Creating a charter is an exercise in itself. Figure 2.2 shows an example of a charter for a DMAIC project.

1. **Business Case**

   The business case is the reason the project is important. The business case should tie to the problem or opportunity statement that follows and the goal statement. Very often, business cases link to key business goals, key performance indicators (KPIs), initiatives, or value propositions. Typically, the business case explains the urgency and need for the project and explains why the problem is important. The business case thus sometimes illustrates the "burning platform" that requires action of some sort.

2. **Problem Statement (or Opportunity Statement)**

   Defining the problem is one of the most critical steps in executing an effective DMAIC project. If you define the problem incorrectly or start with a bad problem statement, you can direct the project in the wrong direction and preclude solutions you might have otherwise considered.

Figure 2.2

A sample charter.

Business Case	Opportunity Statement
◆ This project supports the corporate goal of becoming the number 2 global financial services company by increasing customer retention and satisfaction.	◆ An opportunity to reduce customer defection (27% of applicants) and reduce cost may be achieved by improving our loan and lease processes. The loan and lease processes currently have an average cycle time of 9.2 days which is worse than our customer requirement of 8 days and our application processing cost exceeds the application fee by 18%. Customer defections represent a revenue loss of $2,500,000 per year and a cost to $165,000 for partial application processing. Current Sigma is 1.3.
Goal Statement	**Project Scope**
◆ Reduce average loan and lease cycle time to 8 days by July 1. ◆ Improve Sigma to 3.0 by July 1. ◆ Reduce processing cost by 20% by the end of the year.	◆ Loan and lease processes—begins from a call from the customer and ends with the acceptance or rejection letter sent to the customer.
Project Plan ◆ Timeline: Activity End Measure 4/26 Analyze ? Improve ? Control ? Track Benefits 10/15	**Team Selection** ◆ Team Sponsor: Howard Timmins Andy Anderson Project Manager Carrie Carson Master Black Belt Barry Bethel Black Belt Denise Davidson Customer Service Eric Edwards Sales Representative Frank Fischer Loan Department

3. **Goal Statement (or Objective)**

    The goal statement is the end point, the desired outcome you want to achieve. Both the problem and the goal statements should be quantitative in nature—that is, they should describe things with numbers and they should be specific. As a rule, goals should be SMART (specific, measurable, achievable, relevant, and timely).

4. **Scope**

    The scope is another important element of a good charter. The scope statement should clearly define the domain within which the problem and solution are bounded. The scope should define what is and is not to be addressed by the project.

5. **Financial Return (Return on Investment, ROI)**

    Return on investment is a measure of the overall value of the project. It is basically the benefits of solving the problem divided by the cost of the project. For example, if you find you are going to spend $150,000 to solve a $50,000 problem, you might want to reconsider the project. Most projects should have a minimum ROI of 1.25 or higher. I have seen some projects with ROIs of 30.0 and more. (This may be part of the business case or may be broken out separately.)

6. **Project Team**

    The project charter is a communication tool. It is used to inform stakeholders, sponsors, and other team members what the project is about. That includes the members of the team and what their roles are.

7. **Milestones/Schedule**

    Lastly, the charter should show high-level milestones. At a minimum, it should identify when one expects to complete each phase of the project.

8. **Metrics (Primary and Secondary)**

    In addition, some practitioners break out metrics into a separate section of the charter. While metrics are embedded in the business case, the problem, goal, and opportunity statements separating them can clarify how the project will be measured. Alternate measures that complement each other are recommended, since one measure alone never gives a complete picture of a problem or an outcome.

Creating a charter is an exercise in itself. If you cannot clearly complete each of these details, you may not be ready to commit resources, because some critical information about the project may be missing. If you go ahead with the project without these things, you potentially travel onto dangerous ground armed with preconceptions or assumptions. Without the logistics details, you risk compromising effective communications with your team or your sponsors.

With the earlier focus on quantitative methods, one might wonder why such a qualitative project artifact like a project charter is considered so important. The answer is ironically quite simple. Because it works. It is a widely accepted "best practice" that projects with formal charters have greater likelihood of completing and reaching their desired goals. Projects without charters often have poorly defined problems, uncertain goals or objectives, and are often vulnerable to project-wrecking scope creep or organizational ambiguity that can compromise project success. Thus, the creation of a formal project charter is one of the basic prerequisites to any Six Sigma project.

Again, the essence of a charter and a sound DMAIC project are:

▲ Clear, quantitative problem or opportunity statements
▲ Objectives and goals that are clearly defined and communicated
▲ Quantitative measure of success
▲ A clearly defined business case linked to organizational objectives
▲ Documented expectations and plans (milestones, schedules, and timelines)
▲ A formal team with assigned roles and responsibilities.

There are, however, a host of other tools, techniques, and methods that can (and often are) utilized in the Define phase. These include such activities as:

▲ Process mapping, SIPOC (Supplier-Inputs-Process-Outputs-Customer), value stream mapping, data flow diagrams, flowcharts, etc.)
▲ Stakeholder analysis
▲ Force field analysis
▲ Project or process requirements:
  ▼ Sometimes surveys or questionnaires to identify the "voice of the customer" are used to define requirements. However, serious information gathering should wait until the Measure phase.
  ▼ Kano analysis
  ▼ And more

Again, the objective of the Define phase is to define the problems, establish the projects, set objectives, limit the scope, and form the project team.

(Note: There are many sources of Six Sigma techniques that practitioners can reference for details. These range from math books on statistical methods (Hamming, 1973) to quick reference guides (e.g., *The Memory Jogger*, Brassard and Ritter, 1994).)

Measure

The next step of the DMAIC project is the Measure phase. For projects with lots of data available, this is where that data is identified, organized, and assessed. For projects with little data, this is where potential data is identified and gathered.

The objective of the Measure phase is twofold:

▲ The first is to gather data to understand the problem and its underlying nature more accurately.
▲ The second is to gather data for more detailed analysis later.

Sometimes, the collection of data is a small project in itself. In some cases, information may have to be gathered from people in the form of interviews, surveys, or questionnaires. Survey analysis and design is a complex skill and a specialization for most Black Belts. Creating a questionnaire that is accurate and valid is not as easy as it looks, nor is the analysis of the data when completed surveys are returned.

In other cases, data is present and available but it may be dispersed across many different databases or be stored in different forms and media. Pulling this data together, cleansing it, and preparing it for analysis can be difficult and time consuming.

In those situations where data does not exist, the team must implement data-gathering activities. Data definitions and data collection plans might need to be created. Data collection instruments might need to be developed (in paper-based or automated forms). Sometimes, data can be created or generated. An example of a technique that is an effective way to generate data is Orthogonal Defect Classification. When used to collect data about software defects, it combines methods of root-cause analysis and data classification to generate copious amounts of data that can reveal a great deal of information about the nature of software development processes.

The Measure phase is where the problem is clearly defined in quantitative terms and where the nature of the problem is dissected into the factors that contribute to or produce the undesirable outcomes that are to be remedied. Once again, the purpose of this phase is to identify, confirm, or discover the factors that might lie at the root of the problem and to collect data that will accurately confirm and quantify the problem and its magnitude.

A variety of different techniques and methods are commonly applied here. Some of the more common tools used in this phase include:

- ▲ More detailed process maps, e.g., value stream mapping and as-is process maps
- ▲ Information gathering (surveys, special data collection activities):
  - ▼ Operational metrics definitions
  - ▼ Data collection plans
- ▲ Survey analysis and design (psychometrics)
- ▲ Orthogonal defect classification
- ▲ Measurement systems analysis to determine what is measurable and what is not, e.g., Gage R&R studies
- ▲ Source of variation studies (SOV) to determine what factors are the greatest contributors to variation
- ▲ Factor analysis:
  - ▼ Ishikawa (fishbone) diagrams
  - ▼ Affinity diagrams
  - ▼ Cause-effect matrix
- ▲ Root-cause analysis
- ▲ Conjoint analysis
- ▲ Analytical hierarchical processing

▲ Quantitative risk assessment:
   ▼ Failure mode effects analysis (FMEA)
   ▼ Payoff tables
   ▼ Opportunity loss tables
   ▼ Decision trees
▲ Data mining
▲ Process capability or sigma performance level: Cp, Cpk, or defects per million opportunities (DPMO)

(Note: Many of these tools and techniques can also be used in the Analysis and Improve phases.)

The end result of the Measure phase is data:

▲ Data that describes the problem
▲ Data that shows the range of symptoms and results
▲ Data with details
▲ Data that is collected, or generated, or gathered, or calculated

Quite often, one primary deliverable of the Measure phase is a measurement of the problem at the beginning of the project, before corrective actions have been implemented. This is called a *process capability baseline* and is often used to measure results of the project at closure. Sometimes, the magnitude or scale of the problems is expressed by the number of defects that occur or the rate at which they occur. If converted to DPMO (defects per million opportunities), you can calculate a "sigma level" or process capability of the process. This can be compared to measures taken after the project solution has been implemented to validate whether the improvements are statistically significant.

In general, it is better to gather too much data (if it is not too costly or time consuming) rather than have too little, though either extreme can be problematic. Once data has been collected, cleansed, and organized, and the process baselined, the real work can begin.

Analyze

It is during the Analyze phase that project team members ask the most questions and where the greatest number of data-based statistical analysis techniques can be used and applied. It is beyond the scope of this chapter (or this book) to identify, enumerate, or explain them all. Indeed, it is learning about these tools and when to use them that is one of the more challenging aspects of Black Belt training.

However, it is this rigor and these statistical disciplines that direct analysts away from misleading clues and guide them toward discoveries of statistical significance. It is the nature of complex systems that many problems are hidden by complex relationships between many different factors and confounded by unimportant ones.

Much of the Analysis phase is best classified as true knowledge discovery. Experienced business analysts and data miners, Green Belt and Black Belt practitioners will systematically analyze variables, individually and in groups, to determine which factors do and do not contribute to the problems that the data describes. If sufficient rigor is applied, the influence of preconceptions, preferences, assumptions, and bias can be minimized.

In most cases, analysis consists of asking a series of questions and then applying statistical tools to determine the answers. For example:

▲ Does this factor affect that outcome?
▲ Is *this* group of data different from *that* group of data?
▲ When *this* happens, is the result different from when *that* happens?

Each of these questions can be translated into a working hypothesis and tested using simple comparative methods. The specific statistical techniques that are used to test these hypotheses depend upon the amount of data, the characteristics of the data (nominal, ordinal, cardinal, discrete, or continuous), the amount of data, and the characteristics of the data samples being compared (normal, non-normal).

It should be noted, however, that the analysis of data is not a haphazard and ad hoc endeavor. In well-organized projects, a detailed data analysis plan is created before any analysis is actually performed. Based on the types and characteristics of the data, the specific tests and analyses can be listed and planned in advance. Indeed, if one cannot write down all the tests that will be done before analysis begins, it can be argued that the analyst doesn't know what to do yet. It is not uncommon for a detailed data analysis plan to outline hundreds or thousands of tests that are to be performed. Fortunately, with computers and modern data analysis tools (and a properly organized data set) each of these tests can be performed in seconds.

Still, digging through myriad tests and searching for statistically significant results is much like panning for gold. It is slow, painstaking drudgery, but the rewards can be both surprising and rewarding.

As was mentioned earlier, the skills necessary to apply these techniques are steeped in theories of statistical analysis. Practitioners need to have more than a casual understanding of and tolerance for mathematics. These activities are not suited for the statistically uninitiated, the impatient, or the innumerate.

The purpose of analysis is to discover relationships and to discern what factors are and are not relevant to the problems being addressed. Often, things are discovered that are unexpected but are irrelevant to the problems at hand. This information should be set aside so as not to distract the team or shift the focus of the project. Sometimes, the analysis of one problem may reveal information or opportunities suitable for other projects. Unless the discoveries relate to the project's target problem or objectives, they should be noted and tabled for later action.

In addition to comparative methods, other techniques and tools are used. Correlation analysis can identify how different factors relate to one another. Regression analysis can be used to quantify those relationships. Multiple regression analysis can evaluate the interactions and statistical significance across many different factors and identify which are independent and which are not. Multivariate analysis, principal component analysis, and cluster analysis can go even further and lead practitioners to build detailed models of complex problems when data is rich and complete enough.

Some of the techniques and methods that are commonly used include:

- ▲ Comparative methods:
  - ▼ Correlation analysis
  - ▼ Analysis of variance (ANOVA), t-tests (one-sample, two-sample, paired t-tests)
  - ▼ Multiple analysis of variance (MANOVA)
- ▲ Anderson-Darling test
- ▲ Levine's test and Bartlett's test (tests of normality)
- ▲ Nonparametric tests:
  - ▼ Wilcoxon or Mann-Whitney test
  - ▼ Moody's median test
  - ▼ Kurskal-Walis
  - ▼ Sign test
  - ▼ Chi-square analysis
- ▲ Regression analysis (linear, nonlinear, binary, stepwise, polynomial)
- ▲ Multiple regression
- ▲ Cluster analysis
- ▲ Principal component analysis
- ▲ Factor analysis
- ▲ Discriminant analysis
- ▲ Sensitivity analysis
- ▲ Time series analysis
- ▲ Display techniques:
  - ▼ Pareto charts
  - ▼ Trend charts
  - ▼ Bar charts
  - ▼ Pie charts
  - ▼ Frequency distribution histograms
  - ▼ Cumulative frequency charts (ogives)
  - ▼ Box plots
  - ▼ Scatter plots

It is left to the reader to explore the references at the end of this chapter to learn more about these specific tools and techniques.

At the end of the Analyze phase, the team will typically have a good understanding of the root causes of problems and will have quantified the magnitude of the factors that may be addressed to design a solution.

Improve

It is during the Improve phase that this new understanding about the nature of problems and their causal factors can be used to design, evaluate, and select possible solutions.

Sometimes, the nature of the underlying problems suggests straightforward solutions. In other cases, there may be many different ways to address the issues and root causes. Either way, the solutions need to be thoroughly designed (and sometimes tested or piloted) before they are deployed.

When situations are complex or solutions are expensive, great care might be needed to select the best course of action and to estimate or test the impact of the solution on the systems involved.

In such cases:

▲ Teams may create robust models that can simulate the process changes to predict the efficacy of different solutions.
▲ Where insufficient details are available, it may be necessary to conduct formal Design of Experiments (DOE) to create such models.
▲ Alternatively, it may be necessary to pilot or test solutions to find out whether the solutions really work or to determine which alternative solutions are best.

It is during the Improve phase that such models are constructed or such experiments are conducted.

In most projects, solutions are rarely obvious or simple. If the problems were simple, one generally would not need to use a full DMAIC methodology to generate and analyze data. Even after having identified the root cause, there may be many different ways to alter the controllable factors to eliminate, reduce, or mitigate the problems initially identified at the beginning of the project.

It is quite common, at this point, for the team to look at the results of the Analyze phase and "brainstorm" solutions or alternatives. Many different brainstorming models can be used: taxonomy and morphology, the six hats method, round robin idea generation, combinatorial methods, idea mapping, group passing, benchmarking, best practices, synetics and bionics, and even tool-based methods to encourage ideation, discovery, and creativity, such as TRIZ.

Once potential solutions have been identified, different approaches can be used to prioritize and select the best ones to consider. Nominal group techniques and voting methods can be employed. More quantitative methods of alternative selection might be used, such as the Pugh selection process or weighted attribute analysis.

Ultimately, the chosen solutions may not be the most effective, but may involve a trade-off between more practical considerations related to operational, technical, and financial feasibility (that is, their acceptability to the end-user community who will use the modified processes, their complexity, and costs).

It should not, however, be assumed that only one of the identified solutions can be chosen. Unless the solutions are, by their nature, mutually exclusive, it is often possible, and even recommended, to implement multiple corrective actions. Rarely are individual options capable of eliminating all aspects of a problem. Sometimes, layered solutions are more appropriate strategies. For example, in trying to ensure security at airports, many different things are done: baggage is checked and x-rayed; IDs are checked multiple times; watch lists are referenced; metal detectors screen passengers; and sometimes chemical sniffers and dogs are used to detect explosives or contraband. Individually, none of these solutions are adequate deterrents; together they provide far more security than any single option alone would provide. Therefore, if multiple alternatives are identified, consider implementing more than one (again, within the constraints of reasonable cost and effort).

Once models, experiments, and tests have been performed, and after alternative solutions have been identified, prioritized, and selected, one final task needs to be completed in the Improve phase: preparation of a potential problem report.

You see, rarely are problems completely solved. More often, the problems are merely transformed. For example, before bar codes, cashiers in supermarkets hand-keyed the prices of each product into the cash register. While experienced cashiers developed impressively accurate skills, typographical errors could be made and incorrect amounts inevitably resulted. To remedy the problem of human error, bar codes and scanners were introduced.

However, rather than decreasing the error rates, the errors increased in magnitude. Where before mistakes might be limited to a few mismarked products on a single store shelf, or the incorrect price was entered on a single item in a shopper's cart, errors in the computerized list of prices would result in errors for every item of that product, on every shelf, in every store, in every state. Moreover, whereas mislabeled items before could be detected and corrected by stockboys, customers, and cashiers, errors on these automated price lists became virtually undetectable. Finally, the introduction of the new bar code technologies caused problems in the stores where they appeared. If a barcode scanner broke, someone had to be assigned and trained to fix it. Someone else was responsible for ordering and storing replacement parts. Label machines had to be maintained; special paper and toners had to be ordered and maintained on hand in adequate supply. Other people had to be trained to order these supplies, install the paper, and replace the toners. If bar codes were smeared or damaged, others had to be assigned the responsibility to print new labels on the goods in the bakery and produce areas. New procedures had to be created to report and fix mistakes that were discovered on the automated price lists, and so on.

There is a popular axiom that "every solution has a problem." And it is very true. Every time you change a process, you disrupt the surrounding organization or system. The greater the change, the greater the disruption. When processes, tools, or technologies change, roles also change. Some jobs are altered, some are eliminated, and some new jobs are created.

The critical success factors of solutions are rarely all technical in nature. The best technical solutions can fail if the impact on people, organizations, roles, and responsibilities are not recognized and addressed up front. Despite the best intentions, good solutions can fail due to administrative, logistic, operational, cultural, political, or psychological factors that were ignored or left unaddressed.

Therefore, one must examine the targeted solutions to assess the problems that these solutions will inevitably introduce. Once again, formal brainstorming can be an effective way to analyze new processes and find the problems they will create. Another effective approach is to perform a formal failure mode effects analysis (FMEA) to identify different ways that the process can go wrong. By anticipating such problems, one can modify the new process to implement additional changes to head off potential problems.

Some of the techniques and methods often used in the Improve phase include:

▲ Brainstorming techniques
▲ Simulation and modeling
▲ Monte Carlo simulation
▲ Design of Experiments
▲ Process FMEAs
▲ Weighted attribute analysis
▲ Pugh selection process
▲ Prototyping
▲ Potential problem analysis
▲ Poka-yoke (mistake proofing)

By the end of the Improve phase, you have a potential solution identified and mapped out, but the solution is not quite ready to deploy.

Control

It is during the Control phase that the final work is done prior to implementing the solution. Sadly, it also is during this phase that many DMAIC projects fail. Having identified a problem, analyzed it, and identified a solution, many teams lose momentum at this point and many good solutions simply never get fully deployed. As noted earlier, the operational and logistic aspects of implementing solutions can compromise the process changes and, if not fully institutionalized, the processes can revert to how things were done before.

The primary purpose of the Control phase is to package the solution selected in the Improve phase and add features that will ensure its enduring success.

First and foremost, the process has to be documented. In fact, it is often one aspect of the original problem that the correct way of doing things was not properly specified, communicated, or understood. As many practitioners note, "the devil is in the details," and while fully documenting new processes and procedures may be painstaking drudgery, it is essential for most changes to be implemented.

▲ Where processes are automated, specific control limits need to be defined to control all the critical factors of a process.

▲ In databases, it may be necessary to implement validation and business rules to ensure accuracy of input data and to ensure referential integrity of associated data records.

▲ For procedural processes, it will likely be necessary to prepare training materials for people who will use the improved processes before they will be implemented or used correctly.

Very often, it will fall to the DMAIC team to conduct the first round of training to ensure that the new processes are properly delivered.

It is critical to note that every process that is delivered must include appropriate process controls. At a minimum, every process must have a measurement or metrics section. If you cannot measure a process, you cannot have any confidence that it is working or that the changes you have implemented are actual improvements. Moreover, even if the process is implemented properly, without metrics, it would be impossible to detect if the process developed problems or if it stopped working. Metrics provide the essential feedback necessary to detect things when they go wrong.

The development of appropriate metrics is another one of those complex Black Belt disciplines that require a lot of training. To pick the appropriate metrics, you may need to perform a formal measurement system analysis again to discern what can reasonably be measured and controlled. Metrics also introduce behavioral changes in individuals and organizations (you affect that which you measure), and there are techniques to assist practitioners in identifying appropriate process and organizational metrics that promote desired behaviors. Victor Basili's goal-question-metric (GQM) paradigm is an example of such a technique.

In addition, some metrics controls may use statistical process control (SPC) techniques to track metrics over time. SPC techniques can be used to detect changes in processes with great accuracy and reliability, but such formal approaches require special skills to fully characterize processes, select and set up these tools, and maintain these measurement instruments.

Just as metrics were critical to the Define and Measure phases, so, too, are they vital to the Control phase. The selection of the appropriate metrics for a process at this point can make or break the success of the entire project.

As a rule of thumb, the following adage is a good summary of the principles that apply to metrics selection in the Control phase:

"If you can't measure it, you can't manage it.
If you can't measure it, you can't improve it.
If you can't measure it, you probably don't care.
If you can't influence it, don't measure it."

—SOURCE UNKNOWN

Next, any process that is to be deployed should include some way for it to be audited. If you cannot determine whether the process is being followed, then issuing it and deploying it may be a total waste of time. After all, if you issue changes to a process and no one follows the new procedures and you cannot tell this, then, sadly, it likely doesn't really matter whether the process you designed is good or bad; it is simply moot.

Fortunately, many processes are self-auditing by the nature of the very metrics that they produce. If people stop using the process, the metrics should indicate that something has stopped working. Nevertheless, periodic audit of these processes (either through self-audit or audit by external groups) can be effective controls to ensure continuing compliance after changes have been implemented. Otherwise, behaviors and practices can revert to how things were done before, and the problems that we eliminated can reappear.

Audits are important because as Watts Humphrey, the father of the capability maturity model (CMM), noted, "You cannot expect compliance without enforcement."

Lastly, you need to have instructions or procedures to direct activities when things go wrong.

In the context of SPC charts, these are called "out of control action plans" or OCAPs. If a chart goes into an alarm state due to the violation of a statistical control rule, users and operators need to know what corrective action to take. Otherwise, no action will be taken and the entire purpose of the chart is negated.

In the context of manual procedures, these corrective actions are often called response plans. For critical systems, such as IT systems, these may manifest themselves as disaster recovery plans (DRPs). However, you really need corrective action plans and response plans for all anticipated problems with processes. Otherwise, every time something goes wrong with a process, the problems may be fixed differently.

A response plan outlines what to do, who to go to, or how to react when problems occur.

In terms of formal system control theory, control is a two-stage process. The first is detection, the ability to determine that something has changed or gone wrong. The second is correction, the ability to implement changes that correct the detected error. Together they constitute feedback that is essential for stability in dynamic systems.

A process without metrics, without detection, is like driving a car with the windows painted black. You may have a steering wheel, but you are blind and powerless to correct problems that occur. A process with poor metrics might be like driving a car with thick ice across the windshield. You can make out vague shapes, but adequate control cannot be achieved and you are still at great risk.

A process with good metrics but without a response plan is like driving in the back seat of a car with nice clear windows. You can tell precisely where you are and what is happening, but you are powerless to correct errors when they occur. Once again, you are out of control.

The last formal part of the Control phase is the measurement of project benefits, the verification of the measurable project goals identified and specified during the Define phase. In some cases, financial returns for a project may be realized immediately. In other cases, it may take six months or a year to fully achieve the benefits of the process changes delivered by the DMAIC team.

Some of the tools commonly used in the Control phase include:

▲ Mistake proofing
▲ Process FMEAs
▲ GQM (goal-question-metric paradigm)
▲ Statistical process control (SPC) charts:
  ▼ U-charts
  ▼ P-charts
  ▼ C-charts
  ▼ N-charts
  ▼ Xbar-R charts
  ▼ X moving average charts
  ▼ Xbar-S charts
  ▼ Cumsum charts
  ▼ Exponentially weighted moving average (EWMA) charts
  ▼ Multivariate charts
▲ Control plans:
  ▼ Documentation plan
  ▼ Monitoring plan
  ▼ Response plan
  ▼ Training plan
▲ Rollout/deployment plan:
  ▼ Piloting
  ▼ Phase rollouts
▲ Process capability
▲ Financial verification/ROI realization

By the end of the Control phase, the team should have delivered a viable, detailed, verifiable, workable solution to the problems they were tasked to address.

However, not all problems can be solved using this approach. Sometimes during the Measure and Analyze phases, the team may discover multiple problems that need to be addressed independently of one another. In some cases, the project can be rescoped to adjust its focus and direction. In other cases, the project may be cancelled and a new charter may need to be created and a new team with different skill sets assembled.

Similarly, analysis may reveal a simple straightforward solution and the remaining steps of the DMAIC process may be unnecessary for a solution to be implemented.

In most cases, however, the DMAIC process provides a solid framework and a sound set of tools to lead practitioners and team members through the thickest and more complex problems.

The greatest risks to project success remain:

▲ A poor focus on the process and the problems
▲ A tendency to jump to conclusions and act on intuition, preference, and unconfirmed assumptions instead of fact

For the novice, these tasks may seem daunting, but to the experienced practitioner, these tools become straightforward checks and balances in a data-driven, problem-solving process that is fast and efficient. The trick is not so much knowing how to apply these specific techniques and tools, but rather knowing when and when not to use them.

In the success stories that follow, you will see examples of how these processes have been applied to real-world situations, and you will see how each of the projects are similar and how they also differ from one another.

The DMAIC process is not a rigid checklist of tasks to be performed. Rather, it is a template and a guideline, a collection of possible best practices, to consider and apply if warranted. The success stories that follow will show you the range and flexibility of DMAIC as a problem-solving approach and show the value of this methodology for your most intractable problems.

References

1. Brassard and Ritter, *The Memory Jogger*, Goal/QPC, www.goalqpc.com, 1994.
2. Crevling, Slutsky, and Antis, *Design for Six Sigma in Technology and Product Development*, Prentice Hall, 2003.
3. Hamming, R. W., N*umerical Methods for Scientists and Engineers*, Dover Publications, 1973.

CHAPTER 3

How Motorola Minimized Business Risk before Changing Business-Critical Applications

Vic Nanda

 MOTOROLA

Relevance

This chapter is relevant to organizations that want to minimize risk of disruption to business operations due to a planned process or technology change.

Examples of such changes include but are not limited to changes to key business processes, changes to existing business-critical applications, and so on.

> ## At a Glance
>
> ▲ During the redesign of a business-critical application, Motorola leveraged Six Sigma to proactively identify potential failures and quantify risk to business operations in dollars.
>
> ▲ The project team reduced Motorola's overall risk exposure by 77 percent. In addition, the risk of all high-risk failures was reduced to well below the acceptance threshold set by senior management.
>
> ▲ The project resulted in annual cost avoidance in excess of $568,000, in addition to intangible cost avoidance in millions of dollars due to potential loss of business revenue.

Executive Summary

In today's dynamic business environment, most companies face the difficult challenge of quantifying and reducing risk due to planned changes. This may include changes to a company's business processes, changes to IT applications, or changes in the overall business environment.

Without the use of a formal method, the organization-wide impact of a potential failure can't be quantified and simply becomes a subjective guesstimate, such as high, medium, or low. If a formal risk assessment exercise is conducted, only the localized impact—or the immediate consequence—of a failure is investigated, and there is no way to extrapolate the impact of a failure to an entire company.

But there is a way to see the whole picture and then act accordingly. An organization can apply an approach that enhances design failure modes effects analysis (DFMEA) to quantify and reduce its risk exposure in the face of imminent change. Motorola did so and, in the process, averted significant setbacks during an IT application upgrade and avoided spending more than a half a million dollars annually attributable to systems failures.

As part of a multimillion-dollar IT project, Motorola launched an ambitious project in 2006 to redesign and simplify the globally distributed architecture for a business-critical application. The ultimate objective was to improve operational efficiencies and cut costs by eliminating disparate IT applications that offered similar functionality for Motorola's different business units. Instead, one consolidated IT application would serve all business units.

Due to this fundamental architectural change, there was potential for post-implementation defects that would create downtime and put existing business operations at risk. During the design phase of the project in late 2007, senior management decided that the proposed architecture should be examined thoroughly for potential failures. This entailed evaluating the new architecture for design flaws that might jeopardize requirements for high availability and reliability.

A Six Sigma project was sponsored by senior management to perform this analysis and provide a prioritized list of improvement recommendations. The author of this chapter led the cross-functional Six Sigma project team comprising subject matter experts from all areas that were critical to this analysis.

The project team was chartered with the following objectives:

1. Assess and *quantify the risk to the entire company* as a result of potential failures in the proposed new IT architecture.
2. Perform risk mitigation and provide improvement recommendations so the root causes that pose the greatest risk to the company can be addressed in an orderly fashion.
3. Estimate the reduction in risk, subject to implementation of recommended improvements.

This chapter details how the project team used an innovative new approach to enhance the traditional DFMEA Six Sigma tool to satisfy the requirements for this project.

During the course of a one-week DFMEA workshop, the project team uncovered 57 potential failures (failure modes), including 27 high-risk failure modes that exceeded the acceptance threshold of risk established by the project team. All of the 27 high-risk failure modes were identified as having broad enterprise impact, with some failure modes having more extensive enterprise impact than others. All proposed improvements were evaluated for implementation risk prior to being presented to senior management (more details on this later in this chapter).

This Six Sigma project resulted in a 77 percent* reduction in overall risk to Motorola. Further, the risk of potential failure for each of the 27 identified high-risk failure modes was reduced to well below the acceptance threshold.

This project yielded a return on investment (ROI) of 6.35 times the investment in the first year alone. The estimated project cost was $77,280. Minimum projected annualized cost avoidance was $568,234 or more, depending on the types of system failures encountered. The financial benefit from this project was reviewed and approved by appropriate finance personnel in the company. Much of the more significant cost avoidance is harder, if not impossible, to quantify because it pertains to avoiding loss of business revenues (existing and future revenues from current and new customers), as detailed in the "Benefit Realization" section later in this chapter.

The graphical view of the original risk profile and projected reduction in risk profile, along with a Pareto analysis of the underlying root causes that posed the most risk to the company, served as a powerful tool for presenting project results to senior management for timely and fact-based decision making.

The results of this project are directly applicable to situations where a company wants to quantify and reduce business risks due to process or technology changes. Even though in this project the change was a technology change, the underlying Six Sigma tool of FMEA can be used to uncover potential failure modes for a process (also called process FMEA, or PFMEA) or during product design (DFMEA). Therefore, just as this project team extended the traditional DFMEA analysis to quantify risks to the entire company, likewise, PFMEA analysis may be extended to quantify organization-wide risks due to business process changes.

1. Introduction

In late 2007, Motorola's senior IT leaders that supported the business units affected by proposed changes in the IT architecture requested a Six Sigma project for risk analysis of the new architecture. The request was different from the traditional design review of a proposed new software design or architecture, because senior management wanted the risks associated with potential failures to be quantified and examined in the context of overall business impact to Motorola.

From the start of this project, senior management's sponsorship was clearly established and known within the organization. To allow for alignment of this Six Sigma project with the IT project that was to implement the architectural changes, the program manager of the parent IT project was identified as the Champion for this project. At Motorola, Six Sigma projects require a project Champion, who is a key business leader responsible for ensuring execution

---

* The remaining 23 percent should not be misunderstood as residual risk of any one failure to Motorola. Instead, it is the cumulative effect of all 27 high-risk failures taken together. In other words, it is more appropriate to consider each high-risk failure separately because the high-risk failures were not related, and thus, were unlikely to be encountered simultaneously.

of changes proposed by the project team. The Champion is the hands-on leader responsible for securing resources, maintaining sponsor support, and working closely with the Six Sigma project lead (the Green or Black Belt) assigned to the project. The request to complete this project was time-bound with a clearly defined high-level project scope that was subsequently refined during the Define phase of the project.

2. Project Background

After this project was formally authorized by the project sponsors, the project Champion requested a qualified Black Belt resource to lead the project team. Once a Black Belt resource had been assigned to the project, the project Champion worked closely with the Black Belt to communicate the business case for the project and senior management's expectations from the project. The Black Belt was then tasked with drafting the project charter.

It was determined that the project would be a traditional DMAIC project instead of a DMADV project because the goal of the project was to assess the proposed new architecture, as opposed to designing a new architecture per se. In the latter case, adoption of the DMADV approach would have been more suitable.

3. Define Phase

The Define phase of the project entailed completing the following major activities: formation of the project team and creation of the project charter and detailed project plan.

3.1 Six Sigma Project Charter

As a key requirement of its Six Sigma program, Motorola emphasizes the creation of well-defined project charters that serve as a one-page contractual agreement between the project sponsor(s) and the project team. The project charter comprises the following elements.

3.1.1 Business Case Statement

A business case statement establishes the need for the project. In other words, it describes what is also referred to as the "burning platform" to argue why the project is necessary, why it must be initiated now, which specific process or product is being targeted, and which strategic objective is tied to the project.

The key elements of the business case were:

▲ This Six Sigma project was spawned by an IT project that was one of the top five projects in the overall IT project portfolio.
▲ The shift from a distributed IT architecture for the business-critical application to a more consolidated architecture reflected a profound shift that introduced significant risks that

had to be thoroughly understood and mitigated in order to avoid any service disruption after deployment of the new architecture.

3.1.2 Opportunity Statement

An opportunity statement describes the benefits an organization can realize from executing the project. If possible at this stage of the project, the opportunity is expressed in measurable dollar terms. Otherwise, it may provide a higher-level description of savings from reduction of rework costs (costs of poor quality or COPQ), resource savings, cycle time reduction, waste reduction, customer satisfaction improvement, and so on.

The project was expected to help avoid or minimize:

▲ COPQ associated with resolving design flaws later in the project (before deployment) or, worse, the much higher COPQ if design flaws were discovered after deployment (see Figure 3.1). Motorola's data showed that rework costs after project release can be up to ten times higher than rework costs during the course of a project.

▲ Cost of business impact as a result of IT downtime to resolve defects found after deployment. Refer to the "Benefit Realization" section for further details.

Figure 3.1

Measurement tree.

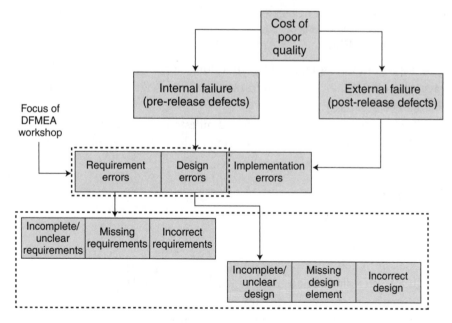

3.1.3 Goal Statement

A goal statement should be a clear and concise statement of project objectives that satisfies SMART (specific, measurable, attainable, relevant, and timely) criteria.

The project goal was to "increase the robustness of the new IT architecture by delivering measurable cost avoidance of at least $250,000* by identifying flaws in the proposed architecture by March 14, 2008."

The project due date was negotiated and agreed upon by the project team and the project sponsors to ensure it was viable, given that extensive preparation work was necessary prior to the one-week DFMEA workshop.

3.1.4 Project Scope

The project scope clearly delineates what is in scope of the project and what is not in scope of the project. A well-defined scope is vital for the project team, because it lays down clearly defined boundaries (as agreed upon with the project sponsor) in terms of what is expected to be addressed in the project and what is not expected to be addressed in the project.

The scope identified the specific architecture baseline to be analyzed and listed seven technical process flows to be evaluated to ensure their flawless execution with the new architecture. The scope also specified interfaces to other IT applications that were in the scope of the analysis. The scope statement clarified that future planned extensions to the IT architecture were out of scope of the analysis at this time, as these extensions were not yet available, and it was clarified that the expectation was that a similar DFMEA would be conducted as and when new extensions to the baseline architecture were developed.

3.1.5 Project Plan

This is a high-level sequence of key milestones for the project, typically signifying the end of each phase in the project. In addition, a more detailed project plan is also created and maintained separately.

The high-level project plan included milestone dates for the end of each of the phases in the DMAIC lifecycle of the project: Define phase, Measure phase, Analyze phase, Improve phase, and Control phase.

3.1.6 Project Team

The core project team comprised:

▲ One Black Belt
▲ One Green Belt
▲ One Master Black Belt to mentor the Black Belt and Green Belt
▲ One customer representative for the parent IT project that had spawned this project
▲ Fifteen subject matter experts drawn from various functional areas in the scope of the project

---

* Note that a minimum of $250,000 in annualized cost benefit is a Motorola requirement for Black Belt projects.

The governance team for the project comprised:

▲ Two project sponsors
▲ One project Champion
▲ One representative from the finance organization
▲ Leader of the Digital Six Sigma program in Motorola IT
▲ Immediate supervisor of the Black Belt leading the project

3.2 Detailed Project Plan

The detailed project plan drilled down into each of the phases of the project and listed start and end dates along with the responsibility for each of the underlying tasks (see Table 3.1).

Table 3.1 Detailed Six Sigma Project Plan

No.	Status	Task Description	Start Date	End Date	Responsible
1		**Define Opportunities**			
	Complete	Create Charter	11/26/07	12/20/07	Project Lead (Black belt)
	Complete	Compile list of Business Process Flows	12/17/07	12/21/07	SMEs (names purged)
	Complete	Create Project Plan	12/21/07	1/11/08	Project Lead (Black belt)
	Complete	Estimate ROI	12/26/07	1/18/08	Project Lead (Black belt)
	Complete	Submit project proposal in project management tool	1/17/08	1/17/08	Project Lead (Black belt)
	Complete	Milestone Review (Define)	1/18/08	1/18/08	Project Lead (Black belt)
2		**Measure Performance**			
	Complete	Create Measurement Plan	1/21/2008	1/25/2008	Project Lead (Black belt)
	Complete	Collect measurement data for repair costs (COPQ for Internal failure and external failure, and business impact of Post Release Defects)	1/21/2008	1/31/2008	Project Lead (Black belt)
	Complete	Milestone Review (Measure)	2/1/2008	2/1/2008	Project Lead (Black belt)
3		**Analyze Opportunity**			
	Complete	Compile list of business flows (including interface message flows)	1/7/2008	1/11/2008	SME (name purged)
	Complete	Complete Template prework for DFMEA workshop	1/14/2008	2/1/2008	Project Lead (Black belt)
	Complete	DFMEA—Team Training	2/4/2008	2/4/2008	Project Lead (Black belt)
	Complete	DFMEA workshop—Analysis	2/4/2008	2/6/2008	SSPT

(continued on next page)

Table 3.1 Detailed Six Sigma Project Plan *(continued)*

No.	Status	Task Description	Start Date	End Date	Responsible
4		**Improve Performance**			
	Complete	DFMEA workshop–brainstorm improvement recommendations	2/7/2008	2/8/2008	SSPT
	Complete	Milestone Review (Analyze and Improve)— Presentation of DFMEA results and management approval	2/11/2008	2/15/2008	SSPT
5		**Control Performance**			
	Complete	DFMEA corrective actions	2/18/2008	5/30/2008*	Project team of IT project that spawned this Six Sigma project
	Complete	Milestone Review (Final project presentation)	3/10/2008	3/14/2008	Project Lead (Black belt)
6		**Benefit Realization**			
	Complete	Calculate projected ROI based on DFMEA results	2/29/2008	3/7/2008	Project Lead (Black belt)
	Complete	Project closeout	3/14/2008	3/14/2008	Project Lead (Black belt)

Note: The implementation of the improvement recommendations from the DFMEA workshop continued beyond the end of this Six Sigma project by means of formal handoff of improvement recommendations from the Six Sigma project lead to the Program Manager of the parent IT project that had spawned this Six Sigma project. As stated previously, the Program Manager of the parent IT project was also the Champion of this Six Sigma project.

This phase ended with a formal milestone review that involved the presentation of phase results to the project governance team.

4. Measure Phase

This phase involved creating a measurement plan that listed the performance measures to be used to baseline current process performance and to calculate project ROI (see Table 3.2).

Measurement data was collected from 57 large IT projects to baseline process performance. Data showed an average of 43 pre-release defects per project—which cost an average of $132.05 per defect to resolve—and an average of nine post-release defects per project (defects reported after release of an IT application to the business users). The cost of rework for post-release defects was $1,065.85, about eight times the cost of resolving pre-release defects. This data is consistent with similar data reported by IBM.[1]

For business impact of post-release defects, data from a prior outage of a similar business-critical application in 2003 was used to estimate the impact of an outage of the new application (refer to the "Benefit Realization" section for details).

This phase ended with a formal milestone review that involved the presentation of phase results to the project governance team.

Table 3.2 Measurement Plan

Performance Measure	Operational Definition Describe Defect or Metric	Data Source and Location	Continuous or Discrete	Display Analysis Tool	Sample Size Cost? Practical?	Who Will Collect the Data?	When Will Data Be Collected?	How Will Data Be Collected?
Cost of poor quality—Internal failure	Total cost of IT rework to resolve a process defect found prior to Go-Live	Estimated based on number of defects found	Continuous	NA	100% of defects	Black Belt	Measure, Improve Phase	Computed based on agreed upon assumptions for cost of rework
Cost of poor quality—External failure	Total cost of IT rework to resolve a process defect found after Go-Live	Estimated based on number of defects found	Continuous	NA	100% of defects found	Black Belt	Measure, Improve Phase	Computed based on agreed upon assumptions for cost of rework
Business impact (of post release defect)	Impact (in $) to the business in terms of impact to bottom-line (such as cost of return freight, shipping of correct product), or top-line (lost revenue, etc.)	Business units	Continuous	NA	2007 actuals	Black Belt	Measure, Improve Phase	Manual

How Will Data Be Used?	How Will Data Be Displayed?
Will be used to calculate the projected benefit (cost avoidance) of implementing the suggested improvements	Histograms

5. Analyze Phase

This phase was divided into two subphases as follows:

5.1 DFMEA Workshop Pre-work

During this phase, the seven technical process flows identified in the project were examined and modeled. In essence, these were a higher-level abstraction of the underlying design and were meant to visually depict the sequence of events as a transaction progressed through the business-critical application and its interfaces with other applications. An example of one technical process flow is shown in Figure 3.2. The four horizontal rectangles represent four different business-critical applications involved in this particular scenario, and the multiple dotted lines from one rectangle to the other represent interface messages between the applications.

After the technical flows were developed, the DFMEA template (Table 3.3) was prepopulated with header information for each process flow separately, including the names of activities (the small boxes in Figure 3.2) and interface messages (the dotted lines in Figure 3.2) listed chronologically. This would save time during the DFMEA workshop, because the project team would have templates ready for use for each of the seven technical process flows.

5.2 DFMEA Workshop

5.2.1 DFMEA Training

The DFMEA workshop began with one-hour training for the project team on the DFMEA Six Sigma tool, covering topics such as:

▲ What is DFMEA.
▲ How to perform DFMEA.
▲ Usage of the DFMEA template (see Table 3.3).
▲ Explanation of the rubrics for scoring failure mode severity, occurrence, and detection. The team used the traditional ten-point scale for scoring occurrence and detection, with ten representing the "worst case." However, a five-point scale was used for scoring severity because it aligned with Motorola's five-point scale for scoring severity of defects reported against IT software applications. As each of these severities are well defined and known within Motorola IT, the team agreed that creation of an artificial ten-point scale for severities was unnecessary and was likely to cause confusion.

$$\text{RPN} = \text{Probability} \times \text{Detection} \times \text{Severity}$$

$$\text{Max} = 10 \qquad \text{Max} = 10 \qquad \text{Max} = 5$$

▲ Detailed discussion on Risk Priority Number (RPN), including scoring of hypothetical scenarios.

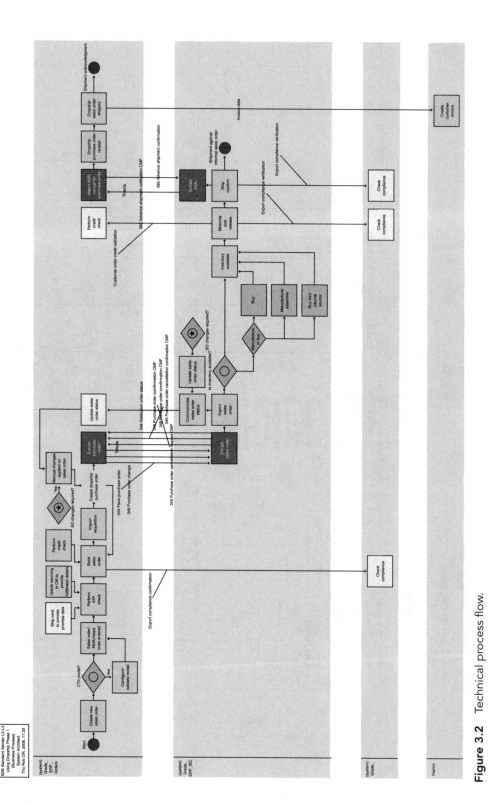

Figure 3.2 Technical process flow.

Table 3.3 DFMEA Template

Failure Modes and Effects Analysis (FMEA):

Process Product: Requirements Management

FMEA Date:

(Revised)

FMEA Team: List names here of the people who contributed to this FMEA

Black Belt:

Page: 1 of 1

Process									Actions		Results			
Item process Steps	Key process input	Potential failure mode (In what ways the key process input go wrong?)	Potential effects of failure (What is the impact on key output variables?)	How severe is the effect to the customer (1–10) 1= Most severe	Potential cause(s) of failure (What causes the key input to go wrong?)	Probability of occurrence (1–10) 10 = most probable	Current controls (What are the existing controls and procedures that prevent either the cause or the failure mode? (Should include an SOP number)	Probability of detection 10 = least probable	Risk Priority Number (RPN)	Recom- mended Action	Respons- ibility and target comple- tion date	Action Taken	Severity	Occur- rence	Detection
									0						
									0						
									0						
									0						
									0						
									0						
									0						
									0						
									0						

As a result of using a five-point severity scale, the $RPN_{Maximum}$ was:

$RPN_{Maximum}$ $= 10 \times 10 \times 5 = \mathbf{500}$

RPNr $\quad\quad$ = Reduced RPN after implementation of corrective actions

▲ The project team deemed high-risk failures as those that had an RPN > 80. This was also referred to as the acceptance threshold.

▲ Guidance on how to reduce the RPN by reducing the severity of the failure, the probability of failure, or the detection of failure or by reducing a combination of these factors.

▲ Guidance on how to categorize the risk to current project plans as a result of implementing the proposed recommendations. Refer to the "Improve Phase" section for further discussion.

5.2.2   DFMEA Analysis Approach

After the DFMEA training was completed, the Black Belt reviewed the overall DFMEA analysis approach with the project team. This approach was previously developed jointly by the Black Belt, Green Belt, and select subject matter experts for the project.

The project team was to focus on uncovering failure modes related to:

1. Planned modifications to the underlying commercial-off-the-shelf (COTS) business application. Failures in the COTS application due to inherent defects were outside the scope of the analysis because that was a risk users of the COTS application were already exposed to, and those defects were unrelated to the planned redesign of the architecture.
2. Interface messages between the business-critical application being redesigned and all other interfacing business applications.

The discovered failure modes were categorized as:

▲ Application-related failures, including application crashes, version- or patch-related problems, application errors, configuration errors, and so on

▲ Data-related failures, including data validation errors (data type and length), incomplete data, and data corruption

In the case of IT applications, there is also a category of application failures that are attributed to a failure in the IT infrastructure itself, such as network failure, memory leaks, database design errors, and capacity problems (CPU, memory, or storage). Due to the immensity of these types of infrastructure failures, it was determined that a separate DFMEA for infrastructure-related failures would be needed.

To save time, the team leveraged similarities across the technical flows during the DFMEA analysis so subsequent analysis could focus on unique differences between the flows. In other words, if part of the technical flow between flows A and B was the same, then the failure modes uncovered for that part during DFMEA analysis of flow A would be applicable to flow B.

Finally, any additional miscellaneous failures identified were included in the DFMEA analysis for risk assessment and possible hand-off to others, such as the infrastructure-related DFMEA analysis team.

5.2.3 Extending Traditional DFMEA to Assess Organization-Wide Risk

As noted earlier, a significant challenge for the project team was to devise an approach that enabled extrapolation of the results of the traditional DFMEA to get a better sense of risk exposure for all of Motorola. This entailed taking a broader view of the RPN scoring in the DFMEA exercise.

One might argue that the "severity" scale can be used for this purpose. For example, a failure with severe impact that reaches across an organization could be rated a ten on the ten-point scale, but this approach would not allow for a systematic way for computing organization-wide risk. For instance, in the case of this project, an outbound failure (from ERP 7), depending on its type, could be encountered in up to three different variants of the process (see Figure 3.3). Likewise, on the other side of the transaction, depending on the type, a failure could be encountered in up to six different ERPs (ERPs 1–6). Similarly, an inbound failure could be encountered in up to two different variants of the process. This meant that the RPN had to be multiplied by the number of process variants or ERPs, as appropriate, to get a true sense of the risk posed to the company (referred to as Enterprise-RPN, or eRPN). For example, if a risk with an RPN of 500 is only multiplied by 1, it yields an eRPN of 500. On the other hand, a risk of lower RPN, say 200, if multiplied by 7 (assuming it would be encountered in all the ERPs), yields an eRPN of 1,400! Therefore, in this example, the risk that poses greater organization-wide risk is the one with the *lower* RPN of 200! Without extrapolating the risk to the entire company by using this concept of eRPN and instead just using the traditional RPN score, the team would have been misguided in focusing on the failure with the RPN of 500!

Figure 3.3

Computing
Enterprise-RPN
(e-RPN).

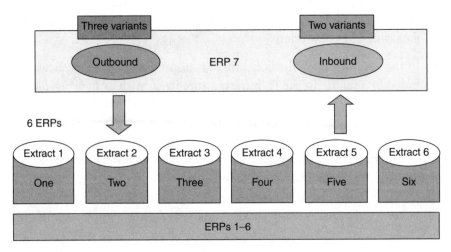

$$eRPN \quad = RPN \quad \times [\text{Number of variants } or \text{ Number of ERPs}]$$

| Max = 500 | | Max = 3 | | Max = 7 |

| | | 3 variants (outbound)/ 2 variants (inbound) | | 7 ERPs |

$$eRPN_{Maximum} = 500 \times 7 = \textbf{3500}$$

$eRPNr$ = Reduced eRPN after implementation of corrective actions

5.2.4 Identifying Root Causes that Pose the Greatest Risk

Computing eRPN is only part of the answer. Because many failure modes might share common root causes, those causes become relatively more important than others that might be responsible for one type of failure. Therefore, repetitive root causes must be carefully considered, as they are likely to pose the most risk to the organization.

By adding the eRPN of all failures that are attributed to the same root cause, you can estimate the risk posed by a specific root cause to the organization. This is referred to as cumulative eRPN. That is, cumulative eRPN is the sum of the enterprise risk of all failures that are attributed to the same or similar root cause.

5.2.5 DFMEA Analysis

The first step is brainstorming the potential failures at each step of the process. As noted earlier, prior to the workshop, for each process flow, the list of process steps and interface messages had been prepopulated in separate templates. The project team examined each process step and interface message listed in the template and brainstormed potential application- and data-related failures. In many cases, this brainstorming yielded multiple failures for the same process step or interface message. Each potential failure was assessed for cause of failure. Based on team consensus on severity, probability of occurrence, and detection, the RPN score was computed for each failure. While assigning these scores, the project team members frequently referred to the rubrics for these scales that were distributed as handouts at the start of the workshop. As one would expect, there was often some amount of discussion involved in arriving at a team consensus on the RPN score and to reconcile differences in opinion. It was stressed to the team members that minor differences in scores were not consequential, for instance, one person scoring a severity of 5 versus another scoring a 6 was not a significant difference. In such situations, the team would quickly converge on a consensus score. However, greater disparities in scoring had to be reconciled through discussion to arrive at a group consensus.

A partial snapshot of the top five high-risk failure modes (out of a total of 27), sorted in descending order by eRPN, is shown in Table 3.4. To allow for legibility, certain columns have been suppressed.

Table 3.4 Top Five High-Risk Failure Modes

Item process steps	Potential effects of failure (What is the impact on key output variables?)	Potential cause(s) of failure (What causes the key input to go wrong?)	RPN	Enterprise RPN (number of affected flows x # of affected ERPs xRPN) – eRPN	Recommended action	RPNr	eRPNr
Extract ASN—outbound	Fines from unhappy customers	Failure to provide specified info per customer in ASN	360	2160	Recommend that a detailed request for ASN data fields be specified.	24	144
Customer order credit validation	Revenue impact (shipped to customer with bad credit or exceeding available credit)	No credit check validation	300	1800	Need SOA for credit check. This may be used for compliance check as well.	10	60
Master data synchronization between two ERPs for BOM	Wrong product built	Data not synched	245	1715	This is a GEMS-specific issue (business-specific). This will create requests for Motorola PDM. Decode program should be capable for dealing with BOM changes (refer to DropShip for details).	120	840
ERP configuration differences—ERP application/process setup	Failed processes (bottomline impact)	Lack of identification of common elements of key data—ERPs perform functions differently	225	1575	Identify common elements of key data (data quality control): OM will be the system of record for the common elements.	30	210
Need common nomenclature data governance	Failed processes	Lack of identification of common elements of key data—ERPs perform functions differently	225	1575	Identify common elements of key data (data quality control): OM will be the system of record for the common elements.	30	210

ASN = abstract syntax notation; GEMS = Motorola department; BOM = bill of material; PDM = product data management; ERP = enterprise resource planning; RPN = risk priority number; eRPN = enterprise RPN; SOA = service-oriented architecture; eRPNr = enterprise RPN reduced; OM = order managaement

As explained previously, the project team was interested in determining which root causes posed the greatest risk to Motorola (that is, had the greatest cumulative eRPN). A partial snapshot of the top three root causes that posed the greatest risk to Motorola (in a total of 13 root causes) is shown in Table 3.5. All failures that were attributed to the same root cause are listed in the first column, while the associated root cause is listed in the second column.

Figure 3.4 shows the root causes organized in descending order of cumulative eRPN. The different shadings in each histogram bar correspond to the different failures that are attributed to the same root cause. For example, the first bar on the left has six different shadings, indicating that six different failures result from this root cause. The heights of these shaded portions in the first bar correspond to the magnitude of organization-wide risk (eRPN) associated with each of those failures. Clearly, root cause #1—"Errors in decode mechanism for manufacturing BOM"—poses the most organization-wide risk.

6. Improve Phase

Using the prioritized list of root causes, the project team brainstormed improvements to minimize RPN for each failure mode and compute the RPNr. A total of 15 improvement recommendations for the 13 root causes were made.

Figure 3.5 shows the original risk profile and the projected reduction in risk profile, subject to the implementation of the improvement recommendations. This figure is not intended to show a trend. It is sorted in decreasing order of eRPN. Original enterprise risk (eRPN) can be significantly reduced (eRPNr) if recommendations are implemented. This before and after visual representation of quantified risks served as a powerful way to present overall risk exposure to senior management. For four of the five failure modes that showed the smallest relative improvement, risk could be further lowered below the acceptance threshold of 80 if the strategic corrective actions for those failures were implemented. Only 1 of the 27 failure modes, with an RPN of 120, was likely to remain slightly above the acceptance threshold of 80 even after implementation of all corrective actions.

Next, each recommendation was examined for implementation impact to the IT project using the following criteria:

1. No negative financial impact (budget)
2. No schedule impact
3. No impact to business process

If these criteria were met, the recommendation was categorized as low implementation risk. If one of these criteria were not met, the recommendation was considered high implementation risk and it required further impact assessment prior to implementation.

Finally, the results of the analysis, along with the assessment of implementation risk associated with each of the recommendations, were presented to senior management.

Table 3.5 Top Three Root Causes (Sorted by Enterprise Risk)

Item process steps	Potential cause(s) of failure (What causes the key input to go wrong?	Recommended action	Responsibility and target completion date	Implementation risk (low/high)	Cumulative eRPN for the root cause
Extract SO, extract ASN—outbound, export compliance verification. 3A6 distribute order status, any interaction message	Errors in decode mechanism for manufacturing BOM	Create pre-processor logic for taking sales BOM and generating manufacturing BOM. Over long term, need general restructuring of BOMs to reflect a standard BOM configuration process.	Data purged	Data purged	4110
ERP configuration differences—ERP application/process setup, need common nomenclature data governance.	Lack of identification of common elements of key data—ERPs perform functions differently.	Identify common elements of key data (data quality control); OM will send one set of common elements.	Data purged	Data purged	3150
3A9 PO cancellation request CMF, 3A9 PO cancellation confirmation CMF, 3A8 PO change confirmation CMF, 3A8 PO change, extract PO-outbound	Change request too late in the manufacturing process, PO—SO discrepancy, order status info from ISC to OM not available, sequencing of the changes, timing of the cancellations.	Need order status synch (3A6)—near real-time—between ISC and OM-PO and SO. Design needs to incorporate this status control point. **Additional actions:** Need a process for handling SC-initiated change requests. BAM will also help detect changes. Need workflow control for L2 PO change before commit.	Data purged	Data purged	2593
Cause(s)		**Recommendations**			

BOM = bill of material; ISC = Motorola department; ASN = abstract syntax notation; OM = order management; CMF = common message format; PO = purchase order; ERP = enterprise resource planning; SC = source; eRPN = enterprise risk priority number; SO = sign off

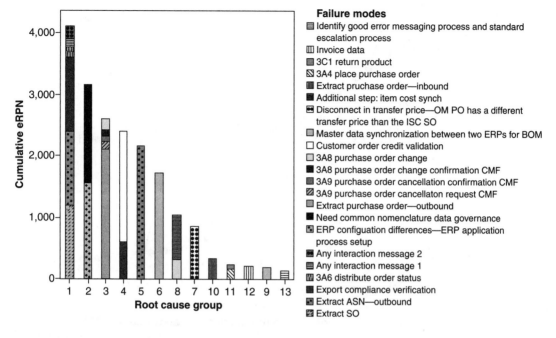

Failure modes
- ▤ Identify good error messaging process and standard escalation process
- ▥ Invoice data
- ▦ 3C1 return product
- ▧ 3A4 place purchase order
- ▥ Extract pruchase order—inbound
- ■ Additional step: item cost synch
- ▨ Disconnect in transfer price—OM PO has a different transfer price than the ISC SO
- ▢ Master data synchronization between two ERPs for BOM
- ☐ Customer order credit validation
- ▢ 3A8 purchase order change
- ■ 3A8 purchase order change confirmation CMF
- ▨ 3A9 purchase order cancellation confirmation CMF
- ▨ 3A9 purchase order cancellaton request CMF
- ▢ Extract purchase order—outbound
- ■ Need common nomenclature data governance
- ▣ ERP configuation differences—ERP application process setup
- ▦ Any interaction message 2
- ▤ Any interaction message 1
- ▨ 3A6 distribute order status
- ■ Export compliance verification
- ▣ Extract ASN—outbound
- ▨ Extract SO

Root cause group	Root cause descriptions
1	Errors in decode mechanism for manufacturing BOM.
2	Lack of identification of common elements of key data.
3	Change request too late in the manufacturing process, PO-SO discrepancy, order status information from ISC to OS not available, sequencing of the changes, timing of the cancellations.
4	No credit check validation.
5	Failure to provide specified info per customer in the ASN.
6	Data not synchronized.
7	Current process only takes the order header and lines, and does not take in adjustments.
8	Item cost not the same in both ERPs in the same entity.
9	Change management processing.
10	Change or cancellation rejection from ISC.
11	Data error—invalid or incomplete.
12	Inability to create commercial invoice holds up shipment.
13	Insufficient error handling.

Note: Errors in decode mechanism for manufacturing BOM pose the greatest risk to the enterprise.

ASN = abstract syntax notation
BOM = bill of material
CMF = common message format
ERP = enterprise resource planning
eRPN = enterprise risk priority number
ISC = Motorola department
OM = order management
PO = purchase order
SO = sign off

Figure 3.4

Root causes that pose the greatest risk to the enterprise.

Figure 3.5

Reduction in enterprise risk.

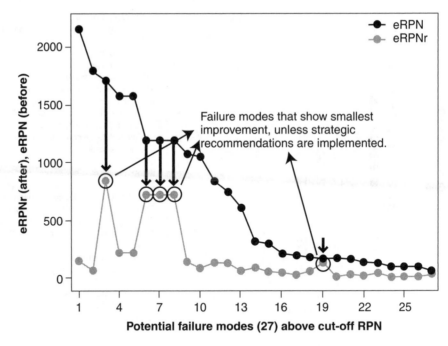

7. Control Phase

As the implementation of the corrective actions was spread over several months and phases of the parent IT project, this Six Sigma project concluded with the hand-off of all corrective actions to the program manager of the parent IT project and assignment of primary ownership and due date for each corrective action.

8. Benefit Realization

As stated earlier, the cost avoidance associated with this project was calculated in terms of effort savings resulting from avoiding COPQ associated with detecting the defects later in the project (pre-release defects) or after release (post-release defects).

Based on historical data from 57 large Motorola IT projects for pre-release and post-release defects, defect containment effectiveness (detecting pre-release defects before they turn into post-release defects), and cost of rework, the project team determined that identification of the 27 high-risk defects resulted in direct annual cost avoidance of $2,905 and $5,329 in pre-release rework and post-release COPQ, respectively.

By taking into consideration the business impact of post-release defects and extrapolating historical data from a similar outage affecting a much smaller portion of the business in 2003, the project team concluded that a similar such outage in the future, depending on the type of failure, would have an annual business impact of at least $560,000. The cost of business impact includes impact to cash conversion cycle, cost of expedited freight, manual rework of defective orders by business employees (that is, non-IT personnel), contract penalties, loss of business, and loss of export privileges.

The total minimum projected cost avoidance obtained by adding the aforementioned IT cost avoidance and business cost avoidance was $568,234.

The complete breakdown of financial projections was reviewed with the finance representative in the extended project team.

9. Conclusions and Limitations

As with any approach, there are potential limitations:

1. The limitations typically associated with FMEA analysis are applicable to this project as well. It is widely acknowledged that an FMEA cannot uncover complex failure modes, such as a sequence of smaller failures that may have a compounding effect.
2. As stated earlier, the project team did not take into consideration failure modes that were associated with failures in the standard COTS application because these were risks that the business operated with even today, and were not a consequence of migrating to a new IT architecture.
3. Finally, the scoring of failure scenarios in a FMEA analysis is always a subjective exercise because it relies on the expertise of the team to correctly detect and assess risk of potential failures, and final RPN scores are based on group consensus. Also, in this case, the DFMEA analysis was inherently more challenging due to lack of historical data regarding potential failures. This was because the new architecture had not been previously implemented in other companies or within Motorola.

Despite these limitations, the enhanced DFMEA method served as a powerful tool for quantifying and minimizing organization-wide risk due to potential failures. Results of the analysis met management expectations regarding quantification of risk at the organizational level, and the approach described in this case study can be readily adopted by other companies that face similar challenges to assess organization-level risk exposure in the face of planned changes in their business.

Acknowledgments

The author of this case study is thankful to Motorola for its permission to allow publication of this case study as a book chapter. The author would also like to thank Leslie Jones and Judy Murrah for their review comments.

10. References

1. Watts Humphrey, *A Discipline for Software Engineering*, Addison Wesley, 1995.

11. Bibliography

Pentti Haapanen and Atte Helmunen, "Failure Mode and Effects Analysis of Software-Based Automation Systems", STUK-YTO-TR, August 2002.

Andy Arkusinski, "Failure Modes and Effects Analysis during Design of Computer Software," 24th Digital Avionics Systems Conference, October 30, 2005.

G. V. Vijayaraghavan, "A Taxonomy of E-commerce Risks and Failures," Masters Thesis, Florida Institute of Technology, May 2003.

CHAPTER 4

TCS Reduces Turn-Around Time for Software Change Requests

Siddharth Kawoor*

TATA CONSULTANCY SERVICES

Relevance

This chapter is useful for companies that are interested in reducing the turn-around time (TAT) required for fixing bugs in a large and complex software system. This chapter is also relevant to organizations that are about to implement the Six Sigma methodology and want to know how to *overcome* the initial roadblocks in implementation.

At a Glance

▲ Tata Consultancy Services (TCS) successfully used Six Sigma to significantly reduce the TAT required to resolve change requests (CRs) resulting from software bugs detected during testing.

▲ The Six Sigma project reduced TAT by over 42 percent and resulted in significant improvement in the customer satisfaction index (CSI).

▲ Project resulted in cost avoidance of more than US$500,000.

Executive Summary

The process of writing code can never be foolproof. However, if you are an outsourcing company, then how soon you detect and resolve the bugs may just give your company the edge in a multivendor market. In a large and complex software system, the verification process contributes up to 20 percent of the development effort. Thus, reducing time required to fix defects will considerably reduce cost involved in developing such products.

---

* Note: At the time this project was executed, Siddharth Kawoor worked as an electronics and telecommunication engineer at Tata Consultancy Services.

In this case, resolving software defects was a complex process with excessive variation in TAT. TAT for resolving CRs varied between less than 2 days to more than 120 days. The average time required to resolve these defects was 40 percent above target. A Six Sigma Black Belt was identified to lead the initiative to reduce variation, thereby reducing TAT. This was one of the first few Six Sigma projects executed by TCS for one of its major clients. Hence, the Black Belt, in the true sense of the role, was required to work closely with senior management as well as the associates to ensure a high level of confidence and support for the project.

A business case was developed to determine the feasibility of the project, and a Six Sigma Black Belt project was authorized. Broadly described, the intentions of this project were as follows:

▲ Reduce average TAT for resolving software CRs
▲ Improve on-time delivery (OTD) of CRs
▲ Institutionalize Six Sigma deployment for this client

The initial project discussions were focused on understanding the process and its past performance. The measurement system within the CR resolution process had to be redesigned to collect accurate data. The project team could not use the traditional time and motion study to identify performance baseline. As each unit's average TAT was 28 days, this approach would have proved costly. The project team decided to focus on *reducing the variation* in the process and *centering the mean* on target. Reducing variation would make the process predictable and perform within specifications set by the client. Eliminating this variation has led to significant improvement in process performance. Process data for the past 12 months was used to study the process. Statistical tests were conducted on this data to identify significant factors causing variation. Extensive brainstorming sessions were held with the CR resolution process team to engineer solutions for the problems identified. The CR team structure was redesigned and aligned to the technical domains. Procedural changes were implemented in methods used to access technical resources, thereby eliminating bottlenecks.

The project team used a data-driven approach and advanced statistical tools to bring about a paradigm shift in the CR resolution process. Implementing these changes reduced TAT by 42 percent and improved the OTD by 51 percent, resulting in US$500,000 savings.

This project was instrumental in creating awareness and generating buy-in towards Six Sigma deployment for this client. This project initiated change in the organization's DNA and created a knowledge base and structure for aggressive Six Sigma roll-out. The result was a framework to identify future projects and ensure their smooth implementation. Appropriate training modules were developed for stakeholders involved in order to generate awareness. The project team was given training in advanced statistics and Six Sigma tools.

The project team faced many roadblocks during the course of this project—there was no measurement system in place to measure process data; CR volumes during the project were extremely low, resulting in fewer data points for analysis; and most of the associates were skeptical of the effectiveness of the Six Sigma methodology. Overcoming these roadblocks, the project team was able to make an impact.

This chapter will demonstrate how Six Sigma methodology has been used to optimize defect resolution process significantly, thereby reducing the average TAT. The Six Sigma methodology provides a host of tools to manage all aspects of such improvement projects. It provides access to tools for project management, data gathering, analysis, and more importantly driving and managing change.

The approach used to reduce TAT in this project is scalable across industries. TCS is applying lessons learned from this initiative to projects across the organization.

1. Introduction

TCS supports a leading telecommunication equipment manufacturer by providing software for its products. The relationship involves research, design, development, and support for the product software. All of these activities take place over an extremely complex set of domains evolved over years of development that require backward compatibility as well as latest feature support. OTD is of prime importance to the client, and every milestone is tracked meticulously. The customer had set a target of an average TAT of 21 days and 85 percent OTD. Before the Six Sigma project was executed, the average TAT was 28 days and the OTD metric stood at 58 percent.

With this background, the project team developed a business case to determine the feasibility of commissioning a Six Sigma project to improve the process performance, and the project was officially kicked off with management approval.

2. Project Background

The TCS support team was composed of around 35 associates involved in resolving pre-delivery and post-delivery defects called CRs. The client requirement was that 85 percent of the requests be resolved within 21 days, and 5 percent of the remainder within 28 days. The remaining 10 percent were assumed to be cases where the inherent nature of the CR would require a period of more than 28 days.

The CR resolution process itself was extremely complex. It involved designers, testers, subject matter experts, laboratory equipment, and maintenance personnel. The process also lacked a system to maintain the key metrics essential for monitoring the health of the process. This was further complicated by the distribution of resources over multiple sites and time zones and shared ownership between the TCS and client teams.

The project team had to first identify key metrics, monitor process performance using these metrics, and then take necessary corrective actions based on this data. A truly collaborative effort, along with strong support from senior management, was crucial to achieve the stated goal. Also, a cross-section of the CR team was involved to ensure all aspects of the process were covered.

MBB support was obtained from the TCS corporate quality team, and the DMAIC methodology was selected. DMAIC stands for Define, Measure, Analyze, Improve, and Control, representing the phases of the project. If an incremental improvement is desired in an existing process, then a DMAIC project is preferred.

3. Define Phase

The Define phase flowed from the business case. The objective of this phase is to clearly state the goal of the project, its scope, opportunity statement, the project team, team roles, timelines, and so on. A host of tools, such as the ARMI chart (Approvers, Resources, Members, Informed parties) and voice of the customer were used in this phase.

3.1 The Project Charter

The charter forms the face of the project. It helps focus on the core problem and provides guidelines for implementing the project. In any type of project, including non-Six Sigma projects, project teams often have to refer back to their charter before making crucial decisions. Therefore, a clear and well-defined charter is extremely important for an improvement project.

3.1.1 Business Case Statement

The TCS support team handled CRs in multiple software domains, with volumes of approximately 520 to 530 CRs annually. According to the operational process document, the team was to resolve the requests within 21 days, irrespective of internal or external factors affecting it. Not meeting these targets was resulting in considerable effort wastage. The process performance baseline before the project is summarized in Table 4.1.

Table 4.1 Performance Baseline

Timeline	Target	Baseline Performance
Within 21 days	85%	58%
Within 28 days	90%	69%

3.1.2 Opportunity Statement

Reducing the TAT would result in multiple benefits, as follows:

▲ A saving of eight person-days of effort per CR resulting in approximate cost avoidance of US$317,504

▲ Saved effort can be directed to other development activities

▲ Increased CSI

▲ Improved TAT can be leveraged to gain more business in the support function from the client

Note: TAT is defined as the total time elapsed between the start of processing the inputs to the availability of the required output. In our case, TAT is defined as the time elapsed between reporting the bug to the support team and submitting the bug-free code to the client.

3.1.3 Goal Statement

The primary goal of the project was to reduce the average TAT for resolving CRs by 20 percent for the next delivery cycle. A secondary goal of the project was to improve the OTD of the CRs to match the targets shown in Table 4.1.

3.1.4 Project Scope

The definition of project scope forms an integral part of planning for any project. The boundaries and expectations for the project team must be clearly defined in order to not overburden the team with an unrealistic goal or unmanageable scope. An in/out frame, as shown in Figure 4.1, was used to effectively communicate the scope of the project.

▲ **Out of Scope:** Certain processes, such as review procedures and code submission guidelines, were mandated by the client. These were considered to be out of scope for this project. The problem of attrition and related issues were also considered out of scope.

▲ **Partially in Scope:** The laboratory resources were located across geographic locations and time zones. Only the resources under the ownership of the TCS team were considered to

Figure 4.1

Definition of project scope.

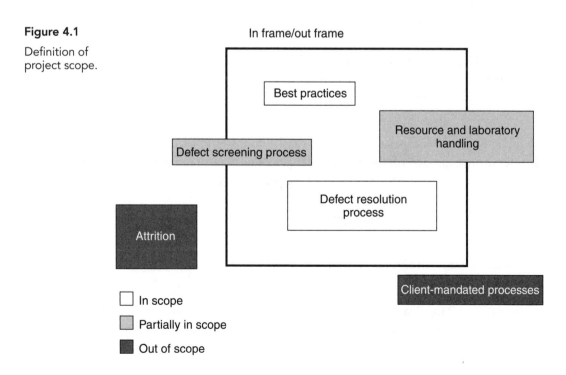

In frame/out frame

Best practices

Resource and laboratory handling

Defect screening process

Defect resolution process

Attrition

Client-mandated processes

☐ In scope

▨ Partially in scope

■ Out of scope

be in scope. Also, screening (routing the CR to the approriate domain) did not formally contribute to the TAT but was a crucial factor in determining the process performance; thus, it was considered to be partially in scope of this project.

▲ **In Scope:** The complete CR resolution process was considered to be in scope from the time the software defect is reported to the CR team until the time the new code is submitted to the client. Best practices and lessons learned from previous initiatives to improve the CR resolution process were also considered to be in scope.

3.1.5 Project Plan

The timely execution of project activities in accordance with plans ensures maximum return on investment (ROI) for improvement projects. In the case of this project, it was especially important, because future implementation of the Six Sigma methodology for this client would depend on the ROI of the first few projects. A Gantt chart was used for this purpose; it is extremely simple to use and can be implemented activity-wise or phase-wise, depending on the complexity of each phase. It clearly displays current progress and percentage completion based on a predetermined timescale (days, weeks, months, or years).

3.1.6 Project Team

An effective team plays a crucial role in a project's success. Roles and responsibilities of each team member should be clearly defined. The ARMI chart where team roles are assigned to the team members can prove useful for this purpose.

This project consisted of the following core team members:

▲ One Six Sigma Black Belt (Project Lead)
▲ Two Six Sigma Green Belt candidates
▲ One Six Sigma Master Black Belt (Coach)
▲ Project Champion (Process Owner)

Six subject matter experts (SMEs) and two laboratory and resource owners were consulted during various phases of the project. During the Measure phase, the whole CR team participated in the Six Sigma project by taking part in surveys and providing valuable data for analysis.

The governance team was composed of:

▲ Project Sponsor (Senior Management)
▲ Project Champion (Process Owner)
▲ Master Black Belt

It was agreed that team members would contribute around 20 percent of their individual bandwidth to this project so that regular operations would not be hampered.

3.2 Analyzing Strengths and Opportunities

A Strength, Weakness, Opportunities, and Threats (SWOT) analysis was performed, as shown in Figure 4.2. This helped the project team focus on the strengths and opportunities and mitigate the risk of failure due to any threats or weaknesses.

▲ **Strengths:** In the model that TCS operates with this client, ownership of resources and responsibilities is shared between TCS and client teams. Having complete ownership of the CR resolution process gave the project team tremendous freedom in making changes. Similarly, having a majority of the laboratory equipment of the CR team located at the TCS locations gave the project team the scope to implement changes in resource utilization schemes. Furthermore, plans were underway to reorganize the support teams based on technical domains. This was expected to streamline the issues of designer expertise and technical know-how. The project team planned to bank on these factors heavily during this project.

▲ **Weaknesses:** The expertise of team members in the CR team showed a large amount of variation. The team was relatively immature, with an average experience of two years. This left the senior members of the team overburdened with queries from the rest of the team. Certain bottlenecks were attributed to code submission and review procedures, but since these processes were mandated by the client, the project team could not modify them in any way. A small portion of the laboratory equipment of the CR team was under the ownership of the client, and the TCS team had no control over it.

▲ **Opportunities:** The internal tool used to store CR-related data had a wealth of information that could prove useful to the project team. The team generating the CR (the testing team) was also local to TCS. Therefore, any change required in the CR generation process could also be implemented locally.

▲ **Threats:** The current TAT was about 40 percent above target. Furthermore, there was no information about the performance of other vendors; therefore, there was a risk of loss of

Figure 4.2

SWOT analysis.

Strengths	Weaknesses
• Process owned by TCS • 80% of required laboratories owned by TCS • Experts available within TCS • Benefit from the domain structure	• Variation in expertise of team members • Client mandated processes • 20% of required laboratories owned by client • Senior staff overloaded
Opportunities	**Threats**
• Historical data from internal tools • Scope for training • Lean period for support team • Testing team is local to TCS	• Performance of other vendors is unknown • Current performance over 40% off target • Ongoing recession may affect business • Testing function may move out to another vendor

business to the competitors. The recession was another threat that could cause considerable loss of business to TCS and eventually threaten the process performance.

Understanding the SWOT analysis gave the project team a clear focus on the available resources and possible pitfalls before proceeding further.

3.3 Voice of the Customer (VOC)

Voice of the customer is a simple method to help the project team capture and translate the varied customer needs—both explicit and implicit—into actionable goals. The voice of the customer, voice of the process, voice of the employee, and voice of the business should together translate into a clear business strategy.

Simply put, it can be the verbatim comments of the customer, employees, management, or process data obtained from sources such as surveys and process metrics. The VOC should be the focal point while selecting among improvement options or alternative products or processes during the Improve phase. The one most conforming to the VOC should get preference. A summary of the VOC is shown in Table 4.2. The project team then transformed this VOC to actionable critical to quality (CTQ) parameters.

Table 4.2 Voice of Customer

Customer	Sample Comments (VOC)	(CTQs)
Client	85% CRs should be out within 21 days 90% CRs should be out within 28 days Average TAT < 21 days	TAT On-time delivery of CRs
CR project team	There's something wrong in the CR process	
CR project management	I think there's some scope for improvement, we can do better than that	

4. Measure Phase

In the Measure phase, a classic example of TAT reduction would call for a time and motion study of the process that would eventually divide the whole process into value-added time (action time) and nonvalue-added time (idle time). This way, the project team can focus on reducing the idle time. Unfortunately, due to the length of the process (average TAT was 28 days, which was too long for a time and motion study) and low churn, at the time this Six Sigma project was taken up, the cost involved in a time and motion study would have exceeded the benefits. The team had to thus explore other options to reduce the TAT.

In this phase, the project team defined and quantified the crucial factors that affected TAT and identified the ones causing the delay. As the existing process did not support such metrics, it was crucial to first determine which variables (Xs) in the process were most influential on

Figure 4.6

Test for
normality.

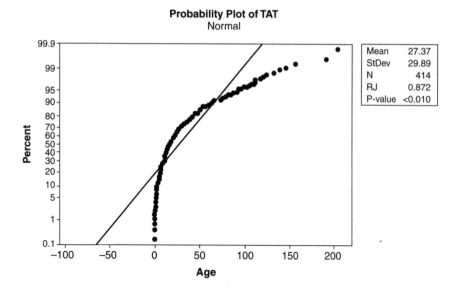

Probability Plot of TAT
Normal

Mean	27.37
StDev	29.89
N	414
RJ	0.872
P-value	<0.010

Figure 4.7

Variation and
sigma level.

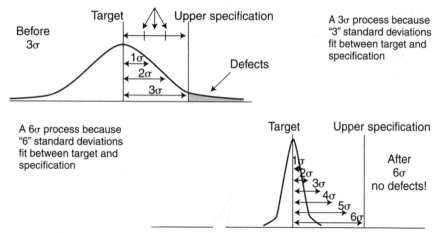

A TAT of more than 21 days had been define   ; a defect. As this TAT was achieved 58 percent of the time, we have 42 defects per 100   rtunities, thus giving us 426,500 defects in one million opportunities. This translates t   ia level of 1.69 in the short term and 0.19 in the long term. The graph in Figure 4.7 s'   ow the performance improves as we move from a lower sigma level to Six Sigma, ou'   ate goal.

4.5 Determining the Vital F

Based on the root-cause analysis   t n all the CRs over the last year with TAT above 21 days (see Figure 4.4), the pro'   the mined the few factors that contributed the most to the TAT. Figure 4.8 shows   to th lysis that highlights the factors most frequently showr

Figure 4.8

Pareto analysis.

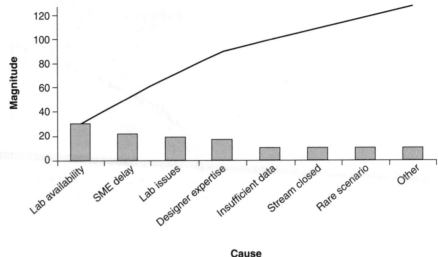

causing the delay or the ones causing the most severe delay. These factors were identified as follows:

▲ **Laboratory availability**—Using particular laboratories caused TAT to increase.
▲ **SME input**—Delay in inputs from SME caused the TAT to increase.
▲ **Laboratory issues**—Delay due to problems faced in handling laboratory resources caused the TAT to increase.
▲ **Designer's expertise**—Lack in designer's expertise level caused the TAT to increase.

Of the factors listed, SME input, laboratory issues, and designer expertise cannot be mathematically analyzed, as these are mostly qualitative factors. Factors such as designer expertise and SME inputs could be collectively analyzed by verifying the dependence of TAT on the domain to which the CR belongs. Due to this close association with critical factors, the domain was considered to be a major factor contributing to TAT.

These factors formed the basis of the analysis performed in the Analyze phase and ultimately led to the improvement actions. In addition, the project team considered the type of fix to be a factor. Type of fix is the outcome of the CR resolution process where the software bug is fixed, declared as not being a defect, declared as dependent on another CR being fixed, or one of the other such numerous outcomes. The CR follows a distinct path to reach each of these outcomes. If statistical tests proved that type of fix was a major factor, then procedural changes would be required in those paths that were causing TAT to go above target.

Thus, in the Measure phase, the following factors were identified for further analysis:

▲ Laboratory
▲ Type of fix
▲ Designer
▲ Domain

5. Analyze Phase

The project team used the data collected in the Measure phase to determine which factors were causing the variation in TAT. As the data was not normal, the usual rules for hypothesis testing did not apply to this data set. We had to either convert the data to normal form using one of many available data transformation tools such as the Box-Cox transform or use nonparametric tests instead. The Box-Cox transform, though a powerful tool in statistical analysis, does not guarantee a perfectly normal result; in fact, it does not have a test for normality. Thus, the project team decided to use the nonparametric tests for their analysis.

In order to determine which of the factors were actually contributing to the delay in defect resolution, it was necessary to find out which of these introduced variation in the output. The project team performed a detailed study to examine the variation in the output as a result of variation in one of the factors. The *student's t-test* can be used to determine whether the means of two sample groups are equal. But the student's t-test assumes that the data is normal and continuous. Also, it is limited to only two sample groups. In our case, the population was divided into multiple sample groups based on the factor being analyzed. For three or more sample groups, a similar result can be obtained using ANOVA (analysis of variance). In statistics, ANOVA is a collection of statistical models, and their associated procedures, in which the observed variance is partitioned into components due to different explanatory variables. In its simplest form, ANOVA gives a statistical test of whether the means of several groups are all equal, and therefore, generalizes the student's two-sample t-test to more than two groups. ANOVAs are extremely helpful because they possess a certain advantage over a *two-sample t-test*. Doing multiple two-sample t-tests would result in a largely increased chance of committing a type 1 error. For this reason, ANOVAs are extremely useful in comparing three or more means.

As the data was determined to be non-normal, ANOVA was not directly applicable and could result in erroneous conclusions. The nonparametric form of the student's t-test is the Mann-Whitney U test, and the nonparametric form of ANOVA is the Kruskal-Wallis test. In statistics, the Kruskal-Wallis one-way analysis of variance by ranks is a nonparametric method for testing equality of population medians among groups. It is identical to a one-way analysis of variance with the data replaced by their ranks.

The assumptions of the Kruskal-Wallis test are as follows:

▲ Data points are independent from each other.
▲ Distributions may not be normal and the variances may not be equal.
▲ There are ideally more than five data points per sample.
▲ Individuals must be selected at random from the population.
▲ Individuals must have equal chance of being selected.
▲ Sample sizes are as equal as possible, but some differences are allowed.

5.1 Hypothesis Testing

The Kruskal-Wallis test is a form of hypothesis testing with:

▲ **Null Hypothesis or Ho:** Medians of all the sample groups are equal.
▲ **Alternate Hypothesis or Ha:** Medians of all the sample groups are not equal.

5.1.1 Analyzing Laboratory

The intent of this step was to determine the effect of using different laboratories on the defect resolution process. So, if using a particular laboratory influences the time required to resolve the defect, then the reasons for the delay could be found and eliminated. The age of each defect resolved was arranged according to the laboratory used and the test was applied. The distribution of TAT across the different laboratories is shown in Figure 4.9, while the results of the test are shown in Figure 4.10.

In Figure 4.10, the p-value in the result is of importance to us. It indicates the probability of the differences between the data sets occurring by chance. As a rule of thumb, a p-value lower than 0.05 is considered sufficient to reject the null hypothesis, that is, there is 95 percent probability that the medians of each group will be different. In our case, the p-value is 0.268. Thus, the null hypothesis stands and we conclude that irrespective of the laboratory used, the delay will be uniform. Thus, dependency on a particular laboratory is not a critical factor contributing to the delay.

5.1.2 Analyzing the Type of Fix

Each of the possible outcomes of the defect resolution process follows a separate critical path. Thus, determining whether any particular outcome was responsible for causing an excessive

Figure 4.9

Distribution of TAT across labs.

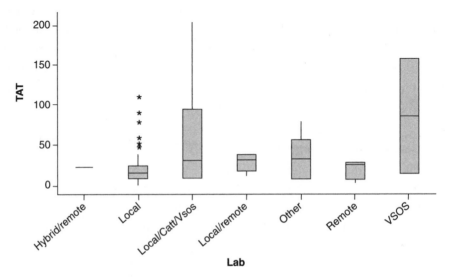

Figure 4.10

Kruskal-Wallis test for laboratory.

Kruskal-Wallis Test: TAT versus Lab

Kruskal-Wallis Test on Age2

Lab	N	Median	Ave Rank	z
Hybrid/remote	1	23.00	63.0	0.39
Local	72	15.00	46.5	-2.64
Local/VSOS	8	31.50	63.1	1.15
Local/remote	6	31.50	68.9	1.49
Other	6	32.50	60.1	0.73
Remote	7	26.00	59.4	0.73
VSOS	2	86.00	73.5	1.06
Overall	102		51.5	

```
H = 7.61   DF = 6   P = 0.268
H = 7.62   DF = 6   P = 0.267 (adjusted for ties)
```

*Note: One or more small samples

delay would help us eliminate the roadblocks in that particular path. The possible resolutions for a CR are:

▲ Completed
▲ Deferred
▲ Dependent Fix
▲ No Fix Planned
▲ No Trouble Found
▲ Not Reproducible in the Laboratory
▲ Tagged as Duplicate

We can see the distribution of TAT for each type of fix in Figure 4.11. The results of the Kruskal-Wallis test on the output variables were as shown in Figure 4.12.

From Figure 4.12, the p-value is 0.34, which is more than 0.05; thus, the null hypothesis stands and we conclude that irrespective of the outcome, the delay will be uniform. Thus, a particular resolution for the CR is not a critical factor causing the delay.

5.1.3 Analyzing the Designer

The designer handling the defect may be a major contributor in the TAT for that CR. By analyzing the dependence of the TAT on the designer, we can determine whether characteristics of a particular designer handling the CR will cause the process to have a different TAT. Figure 4.13 shows the distribution of CRs for each designer.

Figure 4.14 shows that the p-value is 0.005, indicating that there is not enough evidence to prove that the null hypothesis is valid. Thus, we reject the null hypothesis. Even though this does not prove that the alternate hypothesis is true, it is generally accepted to be. Thus, in 95 percent

Figure 4.11

Distribution of TAT across type of fix.

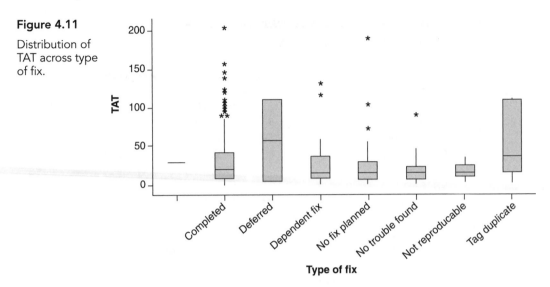

Figure 4.12

Kruskal-Wallis test for type of fix.

```
H = 19.82   DF = 18   P = 0.343
H = 19.83   DF = 18   P = 0.342 (adjusted for ties)

*Note: One or more small samples
```

of the cases, the median of each group will be different. Hence, we can conclude that different designers cause unequal delay in the process. This effect could be due to multiple causes, including all those listed on the "Man" arm in the cause and effect diagram in Figure 4.4.

The dependence of the delay on the designer indicates that the issue could lie in a few or all of the following causes:

Figure 4.13

Distribution of TAT across designers.

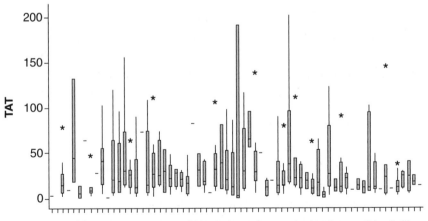

Note: Designer names were listed along the x-axis, but for confidentiality reasons, they have been suppressed.

Designer

Figure 4.14

Kruskal-Wallis test for designer.

```
H = 96.67   DF = 64   P = 0.005
H = 96.73   DF = 64   P = 0.005 (adjusted for ties)

*Note: One or more small samples
```

▲ Competency level of the designer
▲ Procedure followed to resolve the defect by the designer
▲ Coordination with the SME
▲ Designer's ability to handle the laboratory resources

5.1.4 Analyzing the Domain

This analysis would help us determine whether defects arising from particular domains would cause the TAT to increase. Figure 4.15 shows the distribution of TAT for various domains. The result of the analysis was as follows.

The p-value in Figure 4.16 indicates that there is an 8 percent chance the medians will be equal. Rule of thumb says a 5 percent chance is acceptable for a 95 percent confidence interval. Then again, this is just a rule of thumb. Considering the fact that domain in itself is representing multiple attributes, the project team decided to consider a 90 percent confidence interval sufficient for this factor. Thus, from Figure 4.16, we can see that the p-value of 0.08 means the null hypothesis is rejected, and we conclude that TAT is dependent on the domain from which the CR originates. Domain is thus considered a critical factor.

Figure 4.15

Distribution of TAT across domains.

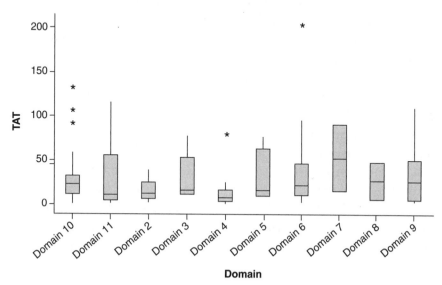

Figure 4.16

Kruskal-Wallis test for domain.

```
H = 23.02   DF = 15   P = 0.084
H = 23.04   DF = 15   P = 0.083 (adjusted for ties)

*Note: One or more small samples
```

This indicates the cause could be a few or all of the following factors:

▲ Technical complexity of the domain
▲ Designer exposure to the domain
▲ SME response time for the domain
▲ Laboratory resource requirement for the domain
▲ Training and other special needs for the domain, etc.

5.2   Inference from the Analysis

▲ Laboratory selected is not a critical factor contributing to the delay in CR resolution.
▲ Type of fix is not a critical factor contributing to the delay in CR resolution.
▲ Designer is a critical factor contributing to the delay in CR resolution.
▲ Domain is a critical factor contributing to the delay in CR resolution.
▲ Additional factor was "issues with laboratory equipment," which are consistent throughout all labs and could not be statistically analyzed but need to be resolved as well.

6. Improve Phase

In the Improve phase, the project team should devise and implement solutions for the problems identified in the Analyze phase. The team can use ways like piloting the solution and Design of Experiments to ensure that only the effective changes are implemented.

The project team compiled a list of the factors actually contributing to the delay and developed a rough draft of the possible solutions by organizing an extensive brainstorming session with the support team management and process and resource owners. All of the proposed solutions were discussed based on feasibility, cost, time to implement, overhead, and operational impact. Tools such as the effort-impact matrix as shown later in Figure 4.17 and FMEA were used to prioritize the implementation and manage risk.

The project team made more than ten strategic recommendations to the project champion and senior management. Most of these recommendations did not require piloting; however, some changes were implemented in a phased manner. Some of the major recommendations are discussed in the next section. After implementing these recommendations, the project was put on hold until sufficient data could be obtained from the new process to analyze the effectiveness of the changes. Once a positive improvement was confirmed, the Improve phase was closed and the project team officially moved on to the Control phase.

6.1   The Testing Task Force

In the Analyze phase, the project team confirmed that selecting a particular laboratory did not affect the TAT, but the CR team was facing problems across all the available laboratories and these were all causing a significant delay in the CR resolution process. Most of these problems stemmed from lack of troubleshooting experience and laboratory etiquettes while using laboratory resources. This resulted in unnecessary delay to fellow designers present in the laboratory at the time. Hardware issues, human errors, and overuse of the resources were also responsible for problems to a certain extent. At the time, there was no defined action plan for resolving these issues.

The project team recommended the formation of a testing task force—a team of experienced associates that would be the first point of contact in case of any laboratory-related issue. The team also recommended an escalation matrix with a defined timeline and protocol to follow up each issue to its resolution. This Testing Task Force (TTF) Escalation Matrix was published across the team, and all members were made aware of it.

The project team recommended including experienced associates and newly joined associates in the TTF in a 3:1 ratio on a rotational basis in order to sustain the TTF in the future and develop expertise throughout the team. This change was to be piloted for a month and revisited to decide on a full-scale implementation based on the observed results.

6.2   The Escalation Matrix

The CR lifecycle can be classified in the following six phases:

- ▲ Reproduce the scenario
- ▲ Design a fix
- ▲ Code and test the fix
- ▲ Review
- ▲ Presubmit checklist
- ▲ Submit modified code

The target TAT from receiving notification of a CR to submission was 21 days, and this was the only milestone that was tracked. This resulted in slack in the early phases and overburdened the latter part of the lifecycle. Also, the team had no defined action plan in case of delay in completing each phase (milestone slippage).

The project team devised a structured, time-bound escalation matrix to be followed by the defect resolution process. The 21-day deadline was divided into six intermediate milestones based on each of the six phases. The escalation protocol was defined in case any deadline was missed, and a person to contact at each level of escalation was identified. Tracking of each CR was to be executed under the leadership of team leads, and the roles of all team members in the new structure were clearly defined.

This change would have multifold benefits, namely:

- ▲ Problem cases would be identified early.
- ▲ Team leads could get the SME involved in the process at the right time.

▲ Designers would have a structure to refer to in case of milestone slippage.

▲ It would be possible to track which of the six phases caused the most delay and remedy the problems more effectively.

▲ The workload would be more balanced across the 21-day deadline rather than being burdened toward the last few days.

6.3 The Domain Structure

This was something the CR process team was contemplating for a long time. The team at the time had a hierarchal team structure, and each team was involved across all domains. The project team recommended a structure based on the domains to which the defects were aligned. Each domain would be staffed depending on the volume of defects observed historically. Domain leads would be the first point of contact and would act as SMEs in case of milestone slippage. This approach would have the following benefits:

▲ Designer expertise in the domain would increase.

▲ Smaller teams based on similar work would be easier to manage.

▲ Designers working in aligned fields would increase designer-SME coordination, in turn accelerating originator-designer-SME communication.

▲ The Analyze phase had provided the domains that had consistently high TAT; these problematic domains could now be tracked more efficiently in order to improve TAT.

▲ Trainings would be easier to implement on a domain basis.

The division of the 21-day deadline into intermediate milestones is shown in Table 4.3.

6.4 The First Information Report

The First Information Report (FIR) is a form that is filled out by the designer that includes the initial analysis of the defect. This document was of limited use, as the details collected were not comprehensive, and as each file was stored separately, analyzing the data was extremely difficult.

The project team first designed an FIR form that included all the key metrics that could be tracked for further analysis. Also, details were included that would aid in resource planning and risk mitigation. An online tool was developed to record and store the FIR so that reports could be generated and further analysis could be performed on the data effectively. The benefits of this change were as follows:

▲ Critical process metrics could be tracked effectively.

▲ Data storage and extraction were simplified.

▲ Resource planning would become more effective using the estimations in the FIR.

▲ Highly complex defects could be recognized and tracked effectively based on the data from the FIR.

▲ Proper estimation for laboratory resources required per CR would facilitate proper use and higher availability of laboratory resources.

Table 4.3 Milestones in the 21-Day CR Resolution Cycle

Day		Activity	
0		Incoming date	
1			
2			
3	Toll Gate	Understand defect and reproduce scenario	
4			
5			
6			
7	Toll Gate	Design fix	
8			
9			21-day CR resolution cycle
10			
11	Toll Gate	Code and test the fix	
12			
13	Toll Gate	Internal review	
14			
15			
16	Toll Gate	External review	
17			
18	Toll Gate	Checklist completion	
19			
20			
21	Final Toll Gate	Buffer for weekends and holidays	

6.5 Miscellaneous Recommendations

The project team also made the following recommendations to streamline the defect resolution process:

▲ Use internal reviews to increase expertise of designers.
▲ Create designer backups within the domain team.
▲ Increase localized availability of virtual tools to reduce laboratory dependency.
▲ Create a feedback mechanism for the screener to improve the screening process.
▲ Breaches in laboratory etiquette could be reduced by making the team members aware of the issues.
▲ Visual factory methods were prescribed for the laboratories and work sites.

6.6 The Effort-Impact Matrix

A simple way to focus on the feasible solutions is the effort-impact matrix. As shown in Figure 4.17, four quadrants are marked with the x-axis, representing the estimated effort required to implement the change, and the y-axis, representing the estimated impact the change will have on the output.

The top-left quadrant, Quick Wins, simply represents what should be your first priority, then moving to Major Projects, and then finally the Fill-Ins. The Thankless Tasks should represent the measures with a desirable impact comparable to or smaller than the effort involved in implementing them.

Figure 4.17

Effort-impact matrix.

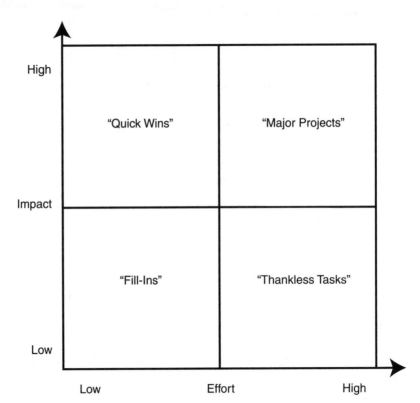

6.7 FMEA

All of these recommendations were implemented by the support team. This involved training employees on the process changes. These were then reviewed after a trial period of one month. Two detailed failure mode effect analyses were performed based on actual process failure modes encountered after the implementation of the process: one for the Escalation Matrix and the other for the functioning of the TTF. Certain critical implementation changes were made as a result of the FMEA.

6.8 Summary of Improvements

After implementing the changes and gathering sufficient data points for a consolidated analysis, the following results were obtained with a visible reduction in variation and average TAT, as seen in Table 4.4.

Table 4.4 Results Post Improvement

Trend After Implementation	Before	After
On-time delivery	58%	78%
Short-term Sigma	1.69	2.27

TAT has improved by 46%

As shown in Figure 4.18, the mean has reduced from a 28 to 15—a change of over 45 percent. Also, the variation in TAT seemed to have reduced considerably, which was further confirmed using control charts during the Control phase.

Figure 4.18

Pre and post results.

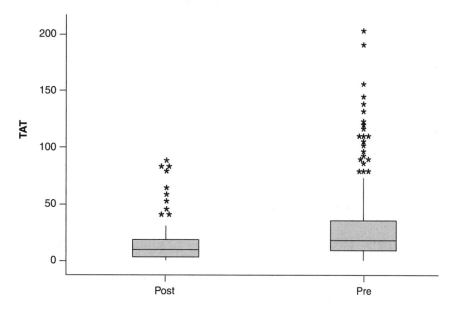

6.9 Cost-Benefit Analysis

Table 4.4 shows a considerable improvement in performance in the CR resolution process. As seen in Table 4.5, the total annual benefit of approximately US$500,000 has been in the form of cost avoidance. This implies that keeping all other factors as they are, it would have required additional resources worth $500,000 to meet the goal of 21-day TAT. The primary cost includes the cost of implementing the changes recommended by the project, and the secondary cost is the cost of running the Six Sigma project itself.

Table 4.5 Summary of Benefits

Cost Benefit Analysis			
Primary Costs	$39,000	Primary Benefits	$0
Secondary Costs	$1,545	Secondary Benefits	$500,000
Capital Costs	$0	Total Benefits	$500,000
Total Costs	$40,545	**ROI**	11.33

7. Control Phase

In the Control phase of the project, the project team has to ensure the variation in the vital X's remains in control and observe the variation in Y.

The process follows an annual cyclic pattern with somewhat predictable fluctuations. Considering the process output characteristics, the I-MR chart was selected to monitor the process stability. The I-MR chart is made up of an individual's chart and moving range chart. It allows you to track both process level and process variation at the same time and detect signs of special causes.

For accurate process monitoring, the correct control chart should be selected and the data used to populate the chart should be in the correct format. The theory of control charts is based on the assumption of normality. As our process output data has a non-normal distribution, we need to first transform it into a normal set. The Johnson transform was used for this purpose, as shown in Figure 4.19. After the transformation, the data points were plotted on the I-MR chart, as shown in Figure 4.20.

From Figure 4.20, we see that the control chart shows a healthy random variation about the central line and a contracting nature towards the end. All of these factors can be attributed to changes made in the process. The use of control charts in a non-normal data set should require some caution, as the calculation of the control limits and mean requires a fair amount of approximation.

The Control phase concluded with the official handing over of the process to the process owner. This included detailed contingency plans and protocols to be followed in case the process was found to be out of control. All of the changes implemented were standardized and published across the team. The benefit expected and achieved to date was signed off by the process owner.

8. Benefit Realization

The benefit obtained from this project was two-fold. A direct effort saving of an average 13.4 days per defect was observed. This leads to an estimated annual cost avoidance of more than US$500,000. This was US$200,000 more than the target!

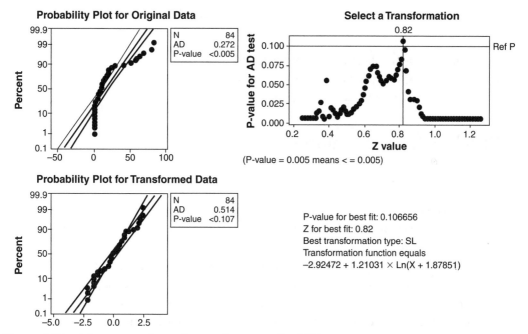

Figure 4.19 Johnson transformation used to normalize TAT.

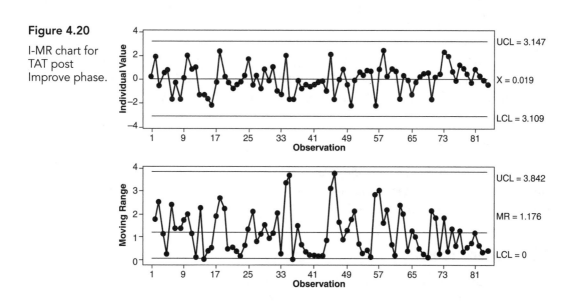

Figure 4.20

I-MR chart for TAT post Improve phase.

Secondary benefits from this Six Sigma project were as follows:

▲ Improved CSI
▲ Better measurement collecting system for further improvement projects
▲ Standardized repository for metrics and knowledge management
▲ Better accountability for laboratory resource utilization

9. Conclusions and Limitations

The project team faced a number of roadblocks in implementing this project, as such improvement projects were never taken up before. The project team had to show a lot of flexibility on account of a number of limitations such as mandatory processes and the small number of data points available for analysis. The project team had to develop training modules for the associates to make them aware of the Six Sigma framework. Special training sessions such as statistical tools and related software were also organized for the project team.

As described in the Measure phase, the CR resolution process lacked a proper system to measure and record important process metrics. The project team had to first devise a system to calculate and record these metrics so that they could be analyzed further.

It would be prudent to mention here that strong and visible support from senior management is absolutely essential for success of such improvement projects. This could be in the form of management buy-in or a derived requirement from management's priorities. Management involvement can range from approving the project cost to active involvement in the project's day-to-day execution.

Finally, each process is unique. This example should be considered as a starting point for further research and analysis for implementing similar projects. Each problem will have certain unique attributes, which makes implementing generic solutions risky. A suitable set of improvements can be engineered using proper tools and careful analysis.

Acknowledgments

The author is thankful to TCS for allowing us to publish this case study as a book chapter. The author would also like to thank the managers and MBB for their support and Vic Nanda and Jeff Robinson for the comments and suggestions.

10. References

1. Quality Council of Indiana, *CCSBB Primer*, Second Edition, 2007.
2. Forrest W. Breyfogle II, *Implementing Six Sigma*, Second Edition, John Wiley and Sons, Inc, 2003.
3. ASQ website: www.asq.org
4. iSixSigma website, www.isixsigma.com

CHAPTER 5

Defect Reduction

How TCS Helped a Major Global Bank Reduce Customer Complaints

Deepak Ramadas

TATA CONSULTANCY SERVICES

Relevance

This case study is relevant to organizations handling a high volume of transactions and experiencing high defect rate. This chapter describes how an organization can achieve significant cost savings by reducing process defects by applying the DMAIC methodology.

At a Glance

▲ A Six Sigma project was initiated for the rendition unit of a large international bank, Tata Consultancy Services—Banking and Financial Services—Business Process Outsourcing (TCS BFS BPO).

▲ The rendition unit of the bank contributed 80 to 85 percent of overall customer complaints of the bank. The unit applied the Six Sigma methodology to proactively identify defects and enhance customer experience by reducing customer complaints.

▲ The project team reduced customer complaints by 44 percent and duplicate statements by 48 percent.

▲ This Six Sigma project resulted in an annual cost saving in excess of US$225,107.

Executive Summary

One of the world's largest financial services company has outsourced end-to-end banking services to TCS BFS BPO. TCS services include providing centralized statement logistic support to the client's credit card customers that includes printing, dispatch, and delivery of customers' credit card statements.

This rendition unit is also the second-biggest cost center of the bank. In 2007 and early 2008, one of the key issues faced by the bank was customer complaints (problem incidences— PIs) related to credit card statements. These PIs significantly affected the bottom line, and also had a negative impact on customer perceptions about the bank's systems.

The rendition unit was the highest contributor of PIs. On average, the rendition unit dispatched about 1.5 million statements every month. As mentioned, the statement-related PIs contributed 80 to 85 percent of the overall customer complaints of the bank. The main reasons for these PIs were nonreceipt or delay in statement receipt by the customers. This resulted in a significant cost burden to the bank in terms of customer handling, sending duplicate statements, reversal of late payment fees, interest loss, and, of course, customer dissatisfaction.

The severity of these issues created an urgent business need to reduce the PIs and thus reduce the negative consequences to the bank and to customer satisfaction.

The key stakeholders from TCS decided to initiate a Six Sigma project to reduce the PIs. This project was perfectly aligned to the strategic imperative of the business.

The goal of the project was to reduce the PI rate (ratio of PIs to total volume of card statements) by 20 percent (from 3,828 to 3,062) and to reduce duplicate statements by 10 percent (from 30,360 to 27,320).

A DMAIC project was formed to examine the problem, to uncover the root causes of these errors, and to develop a way to eliminate or reduce them.

Root-cause analysis helped identify vital causes of the high PI rate, and statistical analysis confirmed and quantified the magnitude of these problems. They included:

▲ Poor tracing of problems and incident resolution
▲ Inadequate or delayed system updates (resulting in improperly dispatched or delayed delivery of statements)
▲ Delays in deliveries due to courier and physical statements as opposed to electronic deliveries
▲ Errors in printing due to high month-end demand and loading
▲ Errors in preparation of duplicate statements
▲ Problems related or isolated to different regions of the country

The team brainstormed potential solutions for each of the identified root causes.

Multiple overlapping solutions were subsequently implemented and effective control plans were put in place to ensure sustenance of improvements.

Ultimately, this Six Sigma project was able to provide a problem/incident (PI) reduction of 44 percent and a reduction in duplicate statements of 48 percent.

The net financial benefit of reducing these PIs (from statement errors and duplicate statements) was in excess of US$225,107 annually.

This is a classic example of how DMAIC projects can identify, analyze, and effectively correct or mitigate problems in established business processes.

1. Introduction

Each month, the credit card user is sent a statement indicating the purchases made with the card, any outstanding fees, and the total amount owed. After receiving the statement, the cardholder must pay a defined minimum proportion of the bill by a due date, or choose to pay a higher amount up to the entire amount owed. The credit issuer (bank) charges interest on the amount owed if the balance is not paid in full (typically at a much higher rate than most other forms of debt).

Interest rates can vary considerably from card to card, and the interest rate on a particular card may jump dramatically if the card user is late with a payment on that card. Therefore, it is imperative for the customer to receive the monthly statement on time in order to pay the outstanding amount before the due date. Any delay in dispatching the card statement might possibly lead to delay in payment of dues by the customer, attracting late payment fees and interest charges.

2. Project Background

Rendition is a strategic business unit of one of the leading global banks that is a client of TCS BFS BPO.

The rendition unit is an important unit in the bank. It supports printing, pouching, and dispatching of credit card statements to customers. The greatest problems faced by this unit during 2007-2008 were customer complaints (PIs) associated with individual credit card statements. Fixing these problems was costly, both in terms of time and effort to correct, but in customer satisfaction as well.

The rendition unit was the source of the greatest number of customer complaints, constituting 80 to 85 percent of all bank PIs. The most common complaints were due to late or missing statements.

3. Define Phase

The key stakeholders from TCS and the client decided to initiate a Six Sigma project to reduce the PIs. This project was perfectly aligned to the strategic objective of the business.

After obtaining approval from the project sponsor, a project Champion was identified and assigned to the project. A Champion is a key business leader responsible for ensuring execution of changes, securing resources, maintaining sponsor support, and working closely with the Six Sigma project lead. The Champion then identified a Six Sigma Belt to lead the project. The project lead then put together a draft project charter.

3.1 Six Sigma Project Charter

The project charter is a contract between the organization's leadership and the team, created with the purpose of clarifying what is expected of the team and keeping the team focused and aligned with organizational priorities.

3.1.1 Business Case Statement

The rendition unit sends statements to the card customers every month. Rendition was the highest contributor of PIs (problem incidences), and customer PIs were the key performance metric of the business; this was a chronic problem. The PIs were mostly due to nonreceipt and delay in statement receipt by customers. There was an urgent business requirement to reduce the PIs and thus reduce customer dissatisfaction.

3.1.2 Opportunity Statement

The rendition unit, on average, dispatches about 1.5 million statements every month. These statements are dispatched through regular mail (or post) and courier.

As mentioned, this unit contributed 80 to 85 percent of overall customer complaints of the bank. This resulted in a significant cost burden to the bank in terms of customer handling, sending duplicate statements, reversal of late payment fees, and interest loss.

The number of average PIs was approximately 3,828 (average of January to March 2008). The cost of each PI to the bank was significant.

Critical to quality (CTQ) is a measurable characteristic of service used to gauge if the customer is satisfied. Two CTQs were identified for this project:

▲ **CTQ 1:** PIs, that is, customer complaints
▲ **CTQ 2:** Duplicate statements

3.1.3 Goal Statement

Reduce the PI rate (ratio of PIs to total volume of card statements) by 20 percent (from 3,828 to 3,062) and reduce duplicate statements by 10 percent (from 30,360 to 27,320).

3.1.4 Project Scope

In order to better understand customers' needs and perceptions of the service, after the project kick-off meeting with the sponsor, the project team captured the voice of the customer (VOC). VOC is a useful technique used to identify key drivers of customer satisfaction, and it also helps decide where to focus improvement efforts.

Diverse sources of information about customer needs were looked into that included reactive sources like customer complaint logs, service calls, and customer notifications. Proactive sources included focus group interviews and inputs for customer service operators.

The data obtained from diverse sources was analyzed to generate a key list of customers' needs in the customers' language, and this was translated into CTQ measures, as shown in Figure 5A.1.

Figure 5A.1

Project CTQ.

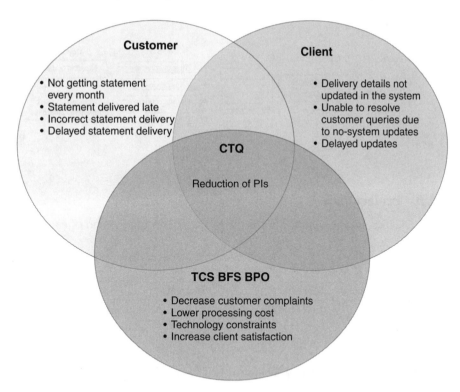

Next, an effort was made to capture the end-to-end process flow and project boundaries. A Supplier, Input, Process, Output, Customers map (SIPOC) detailing high-level steps was drawn.

SIPOC is an effective communication tool used to ensure that the project team is viewing the process in the same way. It also informed the leadership of exactly what the team is working on. This helped ensure that the current "as is" state of the process was reviewed and verified by all stakeholders. The scope included all processes under the rendition unit.

The SIPOC shown in Figure 5A.2 illustrates the flow of activities.

A detailed process flow was then mapped as shown in Figure 5A.3.

Figure 5A.2

SIPOC.

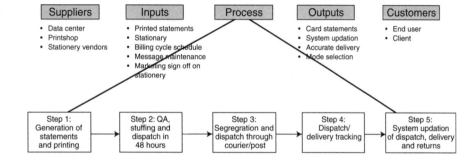

Figure 5A.3

Detailed process flow.

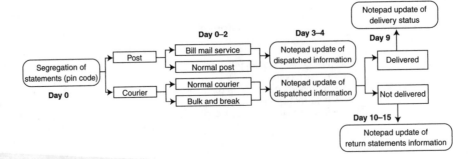

3.1.5 Project Plan

Table 5A.1 shows the milestones for the project denoting the end of each phase of the project.

Table 5A.1   Project plan

Milestone schedule	Start date	End date	Key deliverables
Define phase SIPOC VOC requirements CTQ identification	1 April 2008	30 April 2008	Project charter
Measure phase Process capability	1 May 2008	31 May 2008	As is flow
Analyze phase Defect pareto FMEA Hypothesis test Correlation analysis	1 June 2008	31 July 2008	Cause and effect analysis
Improve phase	1 August 2008	10 September 2008	Prioritize and implement improvements
Control phase Sustain improvements Business impact	11 September 2008	31 October 2008	Control phase

3.1.6 Project Team

The project team consisted of the following members from the operations and quality team:

▲ One Black Belt
▲ One Green Belt
▲ One Master Black Belt who was to mentor the Black Belt and Green Belt
▲ Six subject matter experts drawn from various functional areas in the scope of the project

The governance team for the project comprised:

▲ Project sponsor, who is the head of retail banking operations

▲ Project Champion, who is the unit head
▲ One key stakeholder from the client side
▲ Head of the Lean Six Sigma practice (process re-engineering group) at TCS BFS

4. Measure Phase

The Measure phase started with a structured data collection plan specifying what type of data was to be collected, how to ensure data consistency, and how to display the data. The important function of the Measure phase is to establish a baseline capability level.

The team realized very early the importance of a structured and robust data collection plan. One of the most important things in planning for data collection (see Table 5A.2) was to draw and label the graph that would communicate the findings properly before the collection process began. In order to be able to effectively collect data, the team had to know what they were trying to illustrate. In addition, this exercise also uncovered some issues that the project team had not considered and needed to be added to the plan.

Note: N/A in Table 5A.2 means that the operational definitions for the corresponding items were not yet fully defined when the table was prepared. Such definitions, however, are necessary before the corresponding data can be collected.

The important function of the Measure phase is to establish a baseline capability level based on existing data. Three months of data (January 2008 to March 2008) was collected as shown in Table 5A.3.

The defects per million opportunities (DPMO) method was used to calculate process sigma level. An increase in process sigma requires exponential defect reduction. The standard method of determining DPMO is to use the actual process data and count the number of defect opportunities and then scale that number up to the equivalent of a million opportunities. DPMO were 2,503 and average PI rate was 0.25 percent, yielding a process sigma level of 4.31.

The project target (20 percent improvement) was statistically validated using the one-sample t-test. This test helped determine whether the difference between μ (population mean) and μ_0 (hypothesized mean) was statistically significant or not.

$H_0: \mu = \mu_0$ versus H1: $\mu \neq \mu_0$
H_0: Null hypothesis
H_a: Alternate hypothesis

Session window output of one-sample t-test (from Minitab):

Test of mu = 3062 vs. not = 3062
Variable N Mean StDev SE Mean 95% CI T **P**
C4 3 3828.7 74.1 42.8 (3644.5, 4012.9) 17.91 **0.003**

The p-value of this test, or the probability of obtaining a more extreme value of the test statistic by chance if the null hypothesis was true, is 0.003. This is called the attained

Table 5A.2 Data Collection Plan

| Define what to measure | | | Define how to measure | Who will do it? | | Sample plan | | |
Measure	Type of Measure	Operational Definition	Data Collection Method	Person Assigned	What?	Where?	When?	How Many?
PIs	Discrete data	PIs are complaints captured by the phone banking officer/service log while interacting with the customer	Computer based	Phone banking officer/ Green Belt	No. of PIs captured	System dump	Weekly	9,000
Duplicate statements	Discrete data	Number of duplicate statements dispatched for that particular month	From MIS-control sheet generated from the system	Team member	No. of duplicate statements fired by phone banking officer	From MIS	Weekly	80,000
Physical volume	Discrete data	Total number of physical statements dispatched for the month	From MIS-control sheet generated from the system	Team member	No. of physical statements printed and dispatched	From MIS	Monthly/ daily	1,300,000
SOE volume	Discrete data	Statement on e-mail volume	From MIS-control sheet generated from the system	Team member	No. of e-mail statements dispatched	From MIS	Monthly	700,000
SLS volume	Discrete data	Volume captured in service login system	From MIS-control sheet generated from the system	Team member	No. of queries received through system	From MIS	Weekly	90,000
System dispatch	Continuous data	N/A	From system	Green Belt	Time taken to update the dispatch information of the statement dispatched	From system	Daily/ ad-hoc	5000

	Define what to measure		Define how to measure	Who will do it		Sample plan		
Measure	Type of Measure	Operational Definition	Data Collection Method	Person Assigned	What?	Where?	When?	How Many?
System delivery update time	Continuous data	N/A	From system	Green Belt	Time taken to update the delivery information of the statement delivered by courier	From system	Daily/ad-hoc	5,000
Courier delivery TAT	Continuous data	N/A	From MIS	Green Belt	Time taken by service provider to deliver the statement	From MIS	Daily/ad-hoc	10,000
Call Evaluation	N/A	N/A	From recorded calls	Green Belt	Quality of the call based on a grid prepared by GB	From recorded calls	Ad-hoc	50
Statement Return TAT	Continuous data	NA	From MIS	Green Belt	Time taken by service provider to return the statement and % of returns	From MIS	Ad-hoc	90,000
Courier feedback	N/A	N/A	From interviews	Green Belt	Difficulties faced by couriers to deliver statement	From interviews	Ad-hoc	10

Table 5A.3   Trend of PIs

Description	Jan-08	Feb-08	Mar-08
Total volume of statements dispatched	1,549,533	1,519,310	1,518,411
Duplicate statements	31,209	28,179	31,699
PIs	3,792	3,780	3,914
PI rate (PI/total volume of statements dispatched)	0.25%	0.26%	0.26%

significance level, or p-value. Therefore, H_0 is rejected, since the p-value is less than 0.05 (at 95 percent confidence level). A p-value of <0.05 indicates the target is statistically significant.

5. Analyze Phase

In this phase, the objective was to pinpoint the source of the problem as precisely as possible by building a factual understanding of existing process conditions and problems.

In the Analyze phase, the team brainstormed and generated a list of potential causes and then set out to organize those causes in order to see any potential relationship between cause and effect. All potential causes were displayed in a cause-and-effect diagram, as shown in Figure 5A.4.

The next step was to verify and validate these causes so that the improvements focused on the root cause and not on the original symptom.

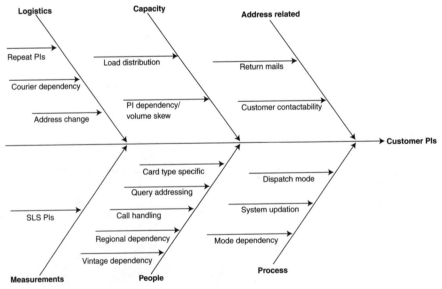

Figure 5A.4 Cause-and-effect diagram.

Detailed Analysis:

1. **Poor tracking system: System updates and tracking of dispatch, delivery, and returns.** Data collected during the Measure phase was analyzed, and it was determined that it took approximately 7 days to update the dispatch details of statements in the system and it took about 12 days to update the delivery details in the system (see Figure 5A.5).

   Whenever the customer enquired about the status of his statement, if there was no update in the system, the bank failed to provide the dispatch status of the statement; this resulted in a PI (customer complaint). Hence, it was imperative to have a reduced total actual time (TAT) for system updation. In the current process, the courier company generated an airway bill number (a reference number for tracking) after dispatch, and this airway bill number was communicated to the rendition team to update the system.

2. **Lack of system updates of dispatch and delivery.** The drill-down analysis for volume (total number of statements dispatched) pointed out that about 50 percent of the volume was not getting updated in the system (see Figure 5A.6). The unit was not able to extract any postal dispatch details, as the postal department did not provide any reference numbers or dispatch details. Without these details, the bank was unable to resolve the customer query in the first instance itself, thus resulting in a PI and subsequently leading to the issuance of duplicate statements.

3. **Post PIs vs. courier PIs.** Of the 1.5 million statements dispatched from the unit, approximately 50 percent were sent through regular mail and the other 50 percent were sent through courier. A dispatch grid was prepared based on the courier serviceability— for postal codes serviced by courier, it was the preferred method of delivery, while for postal codes not serviced by courier, statements were sent by regular mail.

   An hypothesis test using a two-sample proportion test was done to compare postal and courier efficiency. This statistical test revealed that postal efficiency was better than

Figure 5A.5

System updates TAT (dispatch and delivery).

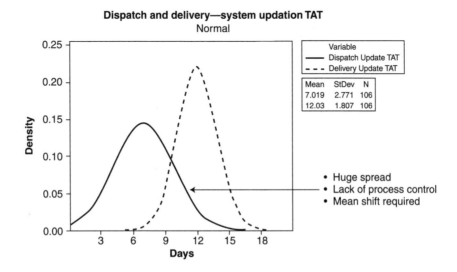

Figure 5A.6

System updates of dispatch and delivery.

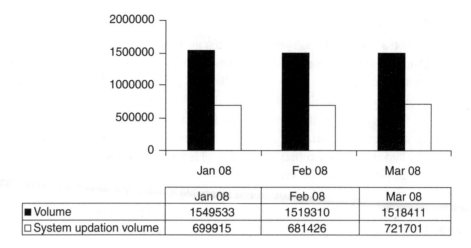

	Jan 08	Feb 08	Mar 08
■ Volume	1549533	1519310	1518411
□ System updation volume	699915	681426	721701

courier. This analysis broke the long-held belief in the organization that courier was more efficient than regular mail.

Two-sample proportion test:

▲ H$_0$: PI contribution for post and courier is the same (mc = mp)

▲ H$_a$: PI contribution for post and courier is different (mc <> mp)

Figure 5A.7 shows the dramatic differences in the PI rate between the postal and courier delivery methods. The conclusion was that regular mail resulted in fewer PIs than courier delivery.

Figure 5A.7

Box plot of postal vs. courier PI%.

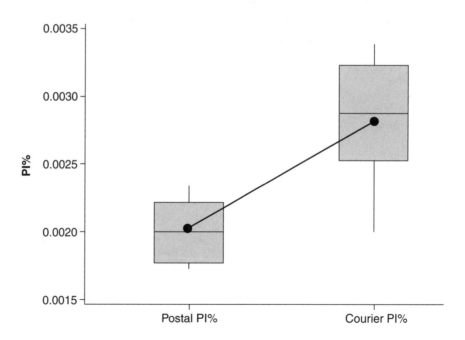

4. **Statement on e-mail (SOE) PIs vs. physical statement PIs.** Of the 1.5 million statements generated in a month, approximately 18 percent of the statements were SOE, that is, statements delivered via e-mail. Physical statements include both regular mail and courier.

   An hypothesis test using a two-sample proportion test revealed that SOE contributed far fewer PIs compared to physical statements.

   Two-sample proportion test:
   - ▲ H_0: SOE and physical statements contribute same PI% (ms = mp)
   - ▲ H_a: SOE and physical statements contribute different PI% (ms <> mp)

   Figure 5A.8 shows the dramatic differences in the PIs between physical and SOE delivery. The conclusion was that SOE contributed fewer PIs than physical statements.

5. **Load distribution analysis: Statement printing load.** The total statement volume of 1.5 million was generated in 14 different cycles (time periods) spread across the month. Each cycle had a different payment due date and statement generation and printing dates. The entire customer base was distributed across these 14 different cycles. Each cycle had varying volumes and different sets of customers. Each cycle code had different statement and billing dates. The authorization unit authorized new credit cards into different cycles based on the volume spread across these cycles. These cycle statements got generated every month on different dates starting from the 10th to the 28th of the month.

   The statement printing load at rendition was not uniformly distributed, and this resulted in a higher volume in the fourth week of every month. More than 60 percent of the volume was handled in the last eight days of the month, as shown in Figure 5A.9.

   A correlation study pointed out that PI% (number of PIs/service log volume) had a direct correlation to the dispatch volume on a day. The correlation coefficient was found to be 0.944, indicating a strong correlation.

Figure 5A.8

Box plot of physical vs. SOE PI%.

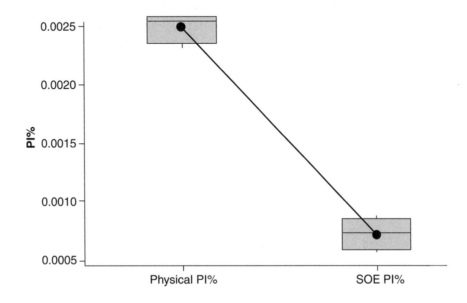

Figure 5A.9

Load
distribution.

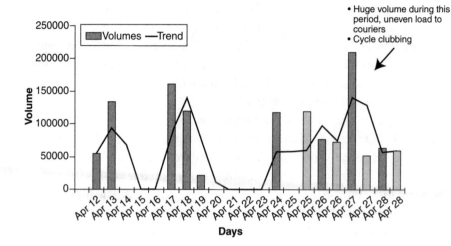

The scatter plot of PIs and volume is shown in Figure 5A.10.

The second week and third week of a month were generally less loaded for printing (see Figure 5A.9). There existed an opportunity to distribute the volume uniformly, which would lessen the courier volume in the fourth week (peak load), and this in turn would lower the PIs.

Cycles 3, 1, 9, and 11, with 46 percent volume contributed to 58 percent PIs, as shown in Figure 5A.11. Cycle 3 was the major PI contributor. The majority of the volume during this cycle went through regular mail, and the statement printing load was also very high during this cycle.

6. **Duplicate statements analysis.** The cost of sending a duplicate statement to the customer was much higher than the regular statement because of its complexity in processing and usage of premium courier service (such statements were dispatched with high priority).

Detailed analysis on duplicate statement-related queries (phone banking) was done, and a pie chart was created, as shown in Figure 5A.12.

The main reason for a duplicate statement was that the original statement was not received. On further analysis, it was found out that the address verification by phone

Figure 5A.10

Scatter plot of
PIs vs. volume.

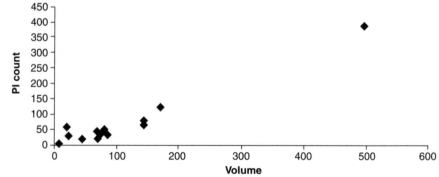

Figure 5A.11

Top PI contributors—cycle codes.

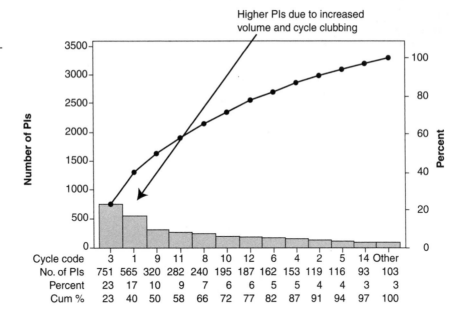

Cycle code	3	1	9	11	8	10	12	6	4	2	5	14	Other
No. of PIs	751	565	320	282	240	195	187	162	153	119	116	93	103
Percent	23	17	10	9	7	6	6	5	5	4	4	3	3
Cum %	23	40	50	58	66	72	77	82	87	91	94	97	100

Figure 5A.12

Reasons for duplicate statements.

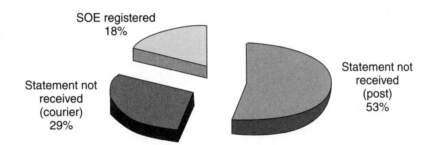

banking was not done in 80 to 90 percent of the cases and e-mail and phone verification was not done in 30 percent of the cases.

7. **Statement return analysis.** The statement return data dump was analyzed, and it was found out that most of the returns are due to address-related issues (customer moved, address incorrect, etc). The Pareto chart shown in Figure 5A.13 helped identify the top reasons contributing to return of statements. Customer address corrections in the system were initiated as a result of this analysis.

8. **PIs-region dependency.** A chi square test was done to determine whether the PIs were region dependent. It was seen that the variables are associated and performance is region dependent.

    Hypothesis:

    ▲ H_0: PIs are region independent (mn = ms)
    ▲ H_a: PIs are region dependent (mn <> ms)

Figure 5A.13

Reasons for statement returns.

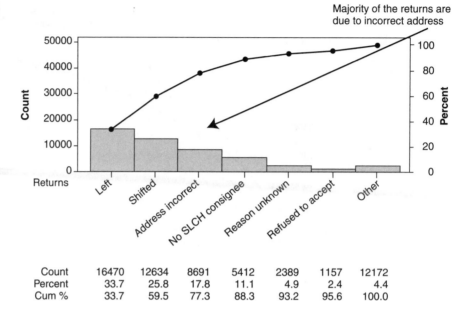

Returns	Left	Shifted	Address incorrect	No SLCH consignee	Reason unknown	Refused to accept	Other
Count	16470	12634	8691	5412	2389	1157	12172
Percent	33.7	25.8	17.8	11.1	4.9	2.4	4.4
Cum %	33.7	59.5	77.3	88.3	93.2	95.6	100.0

Figure 5A.14

Region-wise PIs.

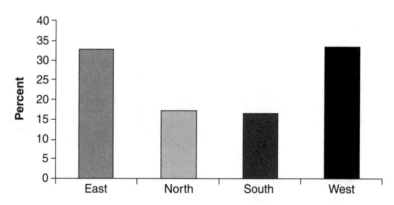

Data showed that the East and North regions seemed to contribute significantly more PIs (35 percent of the active base with 50 percent of total PIs), as shown in Figure 5A.14. The contribution of the South region was much lower than the active base.

Delivery Challenges of Couriers—Interview with Courier Vendors

To understand the difficulty faced by the couriers with delivering the statements, all courier companies were interviewed. The following issues were uncovered:

▲ Incorrect address or change in address
▲ Unable to meet customers after multiple attempts
▲ Customer's door closed at all attempts
▲ Security guards misleading them to wrong apartments

▲ Apartment security not allowing the couriers to meet the customer and delivery made at the security desk

▲ Statement delivery at corporate mail room not reaching/picked up by customer on time

▲ Customer not accepting statements

▲ Unable to meet customers at office

▲ Customer unavailable due to frequent travel

The vital causes determined from statistical validation are summarized here:

1. Poor tracking system: Statement dispatch and delivery updates on the system took around 7 to 12 days. This resulted in a PI due to lack of timely information on dispatch and delivery.
2. Lack of system updates: About 50 percent of the dispatch information was not getting updated on the system on a timely basis due to nonavailability of data.
3. Hypothesis testing (two-sample proportion test) proved that regular mail contributes fewer PIs than courier, and SOE contributes fewer PIs than physical statements.
4. Load distribution analysis pointed out that the statement printing load at the rendition unit was not uniformly distributed, creating overload in the fourth week of every month, resulting in increased PIs.
5. The primary reason for issuing duplicate statements was nonreceipt of statements (82 percent of the cases) and SOE.
6. Pareto analysis revealed that the top reasons for statement returns were that the customer had moved from the current address or the customer address was incorrect (overall 77 percent of the cases). This led to requests for duplicate statements. This occurred because phone calls were not made to verify the customer address.
7. A chi square test pointed out that the PIs were region dependent (North, South, East, and West).

6. Improve Phase

In the Improve phase, the team brainstormed potential solutions for each of the aforementioned root causes. The identified improvements were of three types:

1. Improvements that required direct client intervention
2. Improvements that required technology enhancements
3. Improvements that were perceptual and not directly measurable

The following solutions were implemented for each of the root causes after discussion with the client:

1. Introduced pre-allocated airway bill number (AWB) for speedy information updates on the system for dispatch and delivery. This number would be pasted on the statement before dispatch, while the same number would be scanned and sent for system update.

2. Created a new Excel-based macro for extracting postal dispatch information to update the system, resulting in 100 percent updation. This took into account the total volume, courier dispatch volume, and postal dispatch information.

3. Increased usage of regular mail for nonpremium card type statements. Launched SOE penetration drive by sending business reply envelope to all customers who received physical statements. The bank's website was enhanced to enable the customer to download and print the statement (matching the physical statement).

4. The authorization unit was directed to book new credit cards in the cycles with lesser volume; this would eventually help in load balancing across cycles. Also, cycle clubbing was to be avoided, wherein two cycles were clubbed during the month-end.

5. Initiated mass campaign informing and encouraging customers to check statements and outstanding balance on the bank's website. This was done through advertising in monthly statements.

6. Provided regular refresher training to phone banking officers on mandatory verifications.

7. Provided systematic feedback at regular intervals to courier companies. Phone banking was used to verify customer addresses on file. Return mail alerts were introduced and postal code corrections were performed for every customer. The onboard training module for phone banking officers was modified and now it included a module on how to perform address, e-mail, and phone number verification for all customers whenever they called phone banking.

8. Set up process of telephone calling for regions with high PIs. This was done to proactively address the customer issues.

Note: The process map developed in the Define phase was also updated to reflect these changes.

After the Implementation phase, the next step was to determine the process capability of the improved process, and it was found that the sigma level had improved from 4.31 to 4.49 (for the period January to March 2009). DPMO reduced to 1,405 (from 2,503) and average PI rate reduced to 0.139 percent (from 0.25 percent).

7. Control Phase

After implementation, an effective control plan was put in place to ensure sustenance of improvements. A management information system (MIS) tracker was created to track the volume of dispatch and delivery updates on the system. This would avoid any delay in system updates, and the unit staff could now proactively follow up with courier companies if there was any pending delivery from the courier. This MIS also helped perform a 100 percent reconciliation of system updates with the statement volume dispatched.

Part of the control plan generated in this phase included provisions to ensure that monthly analysis of PI cases was performed and resultant action plans were communicated to courier companies, along with additional feedback to ensure that identified corrective actions remained in place.

The control chart of PI rate shows a clear shift in mean value as shown in Figure 5A.15.

Figure 5A.15
Control chart—
PI rate: Before
and after.

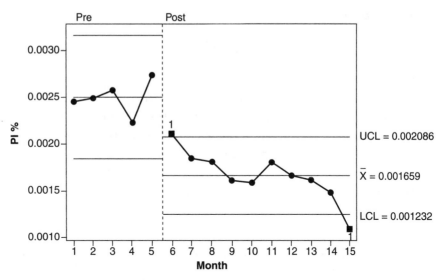

As shown in Table 5A.4, PIs were reduced by 44 percent and duplicate statements were reduced by 48 percent.

Table 5A.4 PIs and Volume Trend

Month	PIs	Duplicate	Volume	PI%
Jan-08	3792	31209	1549533	0.245%
Feb-08	3780	28179	1519310	0.249%
Mar-08	3914	31699	1518411	0.258%
Apr-08	3369	30089	1508876	0.223%
May-08	3977	34286	1449126	0.274%
Jun-08	3056	25905	1447896	0.211%
Jul-08	2632	29300	1427067	0.184%
Aug-08	2586	30201	1430675	0.181%
Sep-08	2268	25808	1409456	0.161%
Oct-08	2257	22558	1423532	0.159%
Nov-08	2543	22197	1416868	0.179%
Dec-08	2304	20278	1390295	0.166%
Jan-09	2223	17721	1375281	0.162%
Feb-09	2005	14368	1357029	0.148%
Mar-09	1464	15276	1349865	0.108%
Duplicate Stmt Reduction			**PI% Reduction**	
Jan 08 to Mar 08	30362.33		Jan 08 to Mar 08	0.250%
Jan 09 to Mar 09	15788.33		Jan 09 to Mar 09	0.139%
Reduction	**0.480003**		**Reduction**	**0.443828**

8. Benefit Realization

The actual benefits realized as a result of this project were as follows:

▲ PIs were reduced by 44 percent.

▲ Duplicate statements were reduced by 48 percent.

▲ The financial benefit by reducing PI, duplicate statements, and reversals was US$193,857.

▲ A 46 percent reduction in statement-related reversals—an approximate savings of US$31,250 in Q4 2008 (compared to Q1 2008).

▲ Dispatch update TAT decreased from seven days to two days (71 percent improvement).

▲ Delivery update TAT decreased from 12 days to 8.5 days (29 percent improvement).

▲ Query volume reduction by approximately 40 percent.

9. Conclusions and Limitations

This project illustrates how the DMAIC process can be used to:

▲ Identify and quantify problems

▲ Identify, prioritize, and verify root causes

▲ Enable the modification of processes to affect permanent corrective actions

One important element was the ready availability of data that could be used to characterize the processes involved and analyze them so that the relationships between input factors and results could be more thoroughly understood. The result of the rigorous use of Six Sigma allowed the problems to be isolated, identified, and resolved.

Improving Test Effectiveness the Six Sigma Way

José Antonio Mechaileh
Alecsandri Dias

Relevance

This chapter will be of particular interest to organizations that want to reduce defects in products by improving testing effectiveness associated with a software development process.

The high occurrence of defects can result in greater costs, lower customer satisfaction, and delay the time that it takes to get products to market.

This project is a classic example of a DMAIC project that improves a process by reducing defects through prevention and containment.

At a Glance

▲ During an initiative to reduce time to market, Motorola's Mobile Devices business unit sponsored a Six Sigma project to redesign the testing process used by test teams to evaluate and validate language-based errors on mobile devices. The goal was earlier detection of defects, resulting in reduction in translations rework later in the project and consequently reduction in the development cycle time.

▲ The project team managed to increase the detection of translation bugs in earlier test phases from 26.6 percent to 61.1 percent.

▲ The project resulted in annual cost savings of US$760,000.

Executive Summary

Motorola is a worldwide company that sells mobile devices all over the world, with user interface software in more than 50 different languages. Beginning in 2007, the Research &

Development organization of Motorola's Mobile Devices business unit realized that a large number of language-related defects present in the product software were only detected and fixed in the later phases of a product development lifecycle.

Defects such as lack of translation, wrong translations, truncated translations, and other related problems were escaping from earlier test phases and were being detected only in the later test phases, closer to the final acceptance testing by Motorola's customers. Some were not detected or reported until after product release (post-release defects).

Decades of experience in software have shown that defects detected late in the project are more expensive and difficult to resolve. They also affect the final product release date and, if discovered by customers, result in a negative customer experience with the product and the developer.

In order to address this problem, senior management sponsored a Green Belt Six Sigma project, which performed a number of important activities, including:

▲ Analysis of impact of language defects on overall product quality using procedures and techniques such as voice of the customer (VOC) analysis (thus assessing frequency and severity of different types of translation errors), including different analyses by different languages

▲ Generation of a prioritized list of improvement recommendations and subsequent implementation of these improvements (including better test methodologies, processes and tools to improve test effectiveness, and prevention of defect escapes to later test phases and customers)

▲ New development and testing methodologies to facilitate the early detection and remediation of all possible causes of language defects (defect prevention)

In order to meet Motorola's financial benefit requirements for Green Belt projects, this project was to deliver a minimum of US$50,000 annualized cost benefit.

The project utilized a standard DMAIC project lifecycle (Define, Measure, Analyze, Improve, Control).

Some of the outcomes of the project included the following:

▲ An improved understanding of escaped defects, differentiating errors by the languages involved, the most common errors, the impacts on products, and customer perceptions

▲ The development of new tools and a better test process, which focused on earlier defect detection

▲ The execution of pilot test cycles to verify the applicability and efficiency of the new tools and process

▲ Training, deployment, and follow-up on the improved test process by software development and test teams

In order to achieve the project goals, a variety of different Six Sigma tools and techniques were used, such as brainstorming, cause-effect diagrams, ranking and prioritization, comparative methods, measurement systems analysis (MSA), and a range of different graphical display techniques such as box plots, Pareto charts, etc.

This Six Sigma project helped improve earlier language defect detection *from 26.6 percent to 61.1 percent.* It was selected as the best Six Sigma project in 2008 at Motorola R&D due to the effective cost savings delivered and because it helped achieve Motorola's key management objective of reducing escaped defects.

A number of key changes were made to the overall testing process, including:

▲ The development of a new test strategy that specifically focused on language-based defects rather than just functionality
▲ The development of new testing tools and techniques
▲ The introduction of new metrics and measurements
▲ The implementation of new training for developers and testers

This project resulted in annualized total cost savings of more than US$760,000. Given the project cost of US$60,000, the annualized net cost savings was US$700,000.

1. Introduction

Motorola has a long legacy of Six Sigma usage, since 1986 when Six Sigma was created to reduce defects in manufacturing environments. Since then, Motorola has extended these methods beyond manufacturing into transactional, support, service, information technology, and engineering functions.

Six Sigma Green Belts, Black Belts, and Master Black Belts are the key to successful application of these Six Sigma methodologies. Their effective utilization reduces variation, defects, waste, and, ultimately, costs. These practitioners are the change agents that sustain and transform Motorola's business processes. The success of these approaches occurs through the combination of technical mastery of Six Sigma methods and critical leadership skills.

Motorola has a robust Six Sigma program that continuously finds, trains, certifies, and utilizes Six Sigma practitioners across its entire organization to address and solve business problems.

2. Project Background

As mentioned, this project was created to address language-based defects in IT products. Defects were being missed until late in the development process, or were escaping and were detected by customers after product release. Because of Motorola's strategic Six Sigma focus, the project was initiated as a DMAIC project, because the goal was to improve the existing test process, as opposed to designing an entirely new process per se.

Upon formal project approval by the project sponsors, the project Champion (Brazil Test Center manager), requested a qualified Black Belt resource to lead it. Once a Black Belt resource had been assigned to the project, the project Champion worked closely with the Black Belt to communicate the business case for the project and senior management's expectations from the project. The Black Belt and the project Champion then drafted the project charter, an initial step in the Define phase.

3. Define Phase

Figure 5B.1 shows this project phase and defines some of the other elements that occur in it.

During the Define phase, the improvement opportunity or problem is formally identified, based on the organization's goals and the customer's needs. During the Define phase, it is also important to document management and customer expectations and requirements. Possible defects that are considered critical to quality also need to be specifically identified. In this case, the key defects to be addressed were language-based defects, which were being discovered too late or they were being reported after product release).

Also, the specific problem to be addressed must be expressed in quantitative and measurable terms.

Finally, the project team is organized and defined, and all roles are clearly established.

3.1 Six Sigma Project Charter

The charter definition is the first step in this Six Sigma project. It is the most important part of the Define phase, because a well-defined charter is a key prerequisite for a successful project.

The project charter comprises the elements discussed in the following sections.

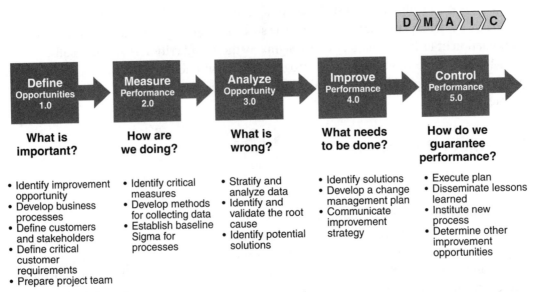

Figure 5B.1 Define phase.

3.1.1 Business Case Statement

The business case was defined as: Motorola develops and sells mobile devices in more than 50 different languages, in dozens of countries around the world. Currently, there is no systematic methodology to test the translations, which are necessary to adapt the cell phones to all different countries and markets. Many translation defects remain undetected until later test phases, where the fixes are more difficult and expensive to implement. In 2007, approximately 70 percent of all language-related defects were detected during field and customer acceptance test phases.

3.1.1.1 Definitions

In mobile devices, hundreds of text strings are displayed to the users. These include prompts, menus, error messages, pop-up messages, instructions, alarms, definitions, etc. All of these strings (which we will henceforth simply call "prompts") are initially written in English, during the earlier software development phases. However, since these devices will be sold and used in many parts of the world, these strings or prompts are then translated to the target languages during later development phases, depending on the markets where these devices will be sold.

Language defects can consist of several types of errors. Some are semantic; some are syntactic. They can include problems related to but not limited to wrong translations, lack of translations (information still in English), strings truncation, and marquee defects. Marquee defects relate to the display of information where the size of the information in one language is larger than in the original and the translated phrase or words may not fit in the space allotted.

Besides these errors, some carriers have their own requirements and directions to use when translating the strings in a phone. A given word can have different, multiple translations or meanings, depending on the carrier and the country. For instance, a word could have a translation for European Spanish, a different one for Colombian Spanish, and another one for U.S. Spanish. The same is true for French or Portuguese languages. Even in English there are different required strings, for U.K. and U.S. markets, for instance.

Another difficult situation is related to the technical terminology. Sometimes, technical instructions, used in mobile devices and other similar equipments, are not easily translated, and in some situations they are not translatable at all.

3.1.2 Opportunity Statement

The opportunity statement was:

"The development of an improved test process, focused on language tests, can increase the detection of language defects in earlier test phases, resulting in a potential net cost savings over 12 months of at least US$50,000."

The project cost (resource allocation) to complete this project (based on scope and team size) was estimated to be US$44,000.

3.1.3 Goal Statement

The goal statement was defined as:

> "To create and deploy an improved language defects detection process, applicable to all Motorola's Mobile Devices teams by December 2008, increasing the language defects detected on earlier test phases from 30 percent to 55 percent."

3.1.4 Project Scope

Project scope was defined to include most of the different development platforms and phones used within Motorola, including Symbian, P2K, and Linux-Java. The project scope was defined to cover languages accounting for 80 percent of all language defects, primarily including English (U.S. and U.K.), Portuguese (Brazilian), French (Canadian and European), Spanish (Colombian), and Chinese (traditional).

Note: While Motorola develops its phones in some different software/hardware platforms, such as Symbian, P2K, LJ, Microsoft, and iDen, this project was applicable to and focused on Symbian, P2K, and LJ platforms.

Specific platforms that were excluded and were determined to be out of scope included other platforms (such as Microsoft, iDen). Languages with minor impacts were also explicitly excluded.

3.1.5 Project Plan

The project plan included estimated dates to complete each project phase:

Phase	Due Date
Define	March 11, 2008
Measure	June 24, 2008
Analyze	Sept. 16, 2008
Improve	Oct. 17, 2008
Control	Nov. 26, 2008

3.1.6 Project Team

The project team was composed of:

- ▲ Sponsors (system test and product directors)
- ▲ Champion (Brazil Test Center manager)
- ▲ Black Belt
- ▲ Finance controller (Motorola's Brazil finance controller)
- ▲ Green Belt (GB) candidate
- ▲ Three other members with extensive technical knowledge and GB certification to help in the overall definitions and to support the technical analysis

3.2   Current Process Map (As-Is)

Another important part of the Define phase is formal definition of the process that will be modified. In this case, the team needed to develop a map of the language test process, detailing the test activities as it was currently implemented and used. The process is shown in Figure 5B.2.

Note that the initial test process consisted of five major parts:

▲   Unit tests, performed by the software developers
▲   Feature tests
▲   Interoperability tests (testing functionality and compatibility with other vendors' phones)
▲   Sanity, integration, regression, and exploratory tests, executed by Brazil Test Center
▲   System and field tests

Upon completion of these test phases, the products were delivered to the carrier for acceptance tests and then to the final customers.

In all these internal test phases, the software tests were performed without a specific focus and strategy for translations. Translations defects were typically detected by chance while testing other functional aspects of the device software.

Moreover, there was little focus on language issues in the early development phases. The software for all phones was developed in English, and translation to other target languages and markets occurred across multiple phases of the software development process. Therefore, some test phases were unable to detect translation issues because the phone's software was still in English.

Thus, most language defects were only detectable during the software's final test phases (or in the carrier's acceptance tests), where they were correspondingly more difficult and more expensive to fix.

Figure 5B.2

As-is software test process.

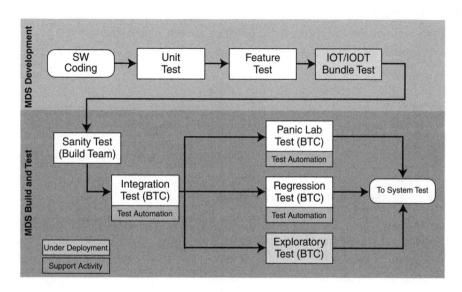

3.3 Project Phase Conclusion

With the completion of the project charter, the formalization of problem and goal statements, the formation of a project team, and the preparation of an initial project plan, the Define phase ended with a formal review, which involved presentation of the envisioned project to the project governance team, project sponsors, the project Champion, the project Black Belt, and the other team members.

4. Measure Phase

The next phase of the DMAIC process is the Measure phase. This is the phase where data will be collected that will later be used to analyze the problem and evaluate possible solutions. Figure 5B.1 shows this project phase, along with a list of key activities in this phase.

For this project, the team developed a formal "Measurement Plan" (Table 5B.1) and a "Measurement System" to identify and organize the collection of relevant data.

Table 5B.1 shows an excerpt of the Measurement Plan.

▲ Each row specifies a different piece of data or a different measurement to be collected.
▲ Each column details information about this data (where and when it will be collected; the data source and data owner; how the data will be displayed, used, or analyzed; sampling methods; etc.).

This plan shows the design of the data collection activities that were subsequently executed. Some of the data collected for review and analysis included but was not limited to:

▲ New features (software/hardware implementations) to test
▲ Languages to cover
▲ Test execution efforts
▲ Bug tracking efforts
▲ Number of test cases used per product, language, and area
▲ Number of new strings (prompts to the user) per new feature
▲ All bug-related information
▲ Languages in which the bug was detected
▲ Tools used

For each performance measure, we established the scope (software/hardware platforms), data source location, data collection method, sample sizes, and data reporting method.

4.1 Measurement System Goals

The purpose of the Measurement System was to provide a structure of systems and databases to define and support the actual collection and storage of the metrics identified in the Measurement Plan. A Measurement System enables the organization and storage of data so that it can be queried and analyzed by other tools.

Table 5B.1 Measurement Plan

Performance measure	Operational definition	Scope	Data source and location	Display analysis tool	Sample size	Who will collect the data?	When will data be collected?	How will data be reported?	How will data be collected?
New feature to test	Definitions about which new features should be tested, considering only those with new prompts	LJ, P2K, Symbian	FMD and tracking	Histogram	100%	GB candidates	Before each new test cycle	GB candidates will input data, before each new test cycle, into the tracking spreadsheet	Feature list will get from FMD, by checking the new features to be included in the product. The decision about which have prompts to test is done considering TRS/FAD/SUS/CUIS documents
Languages to cover	Definitions about which languages should be covered for the new test features, considering the target carriers	LJ, P2K, Symbian	Tracking spreadsheet	Histogram	100%	GB candidates	Before each new test cycle	GB candidates will input data, before each new test cycle, into the tracking spreadsheet	Language list per feature will be obtained from product PM
Execution effort (time)	Daily times spent (in minutes) executing language tests. This is exactly test central's data for the session: Setup (only the time to prepare the phone) Execution (actual hands on execution) Debugging (all work and investigation done after the execution in order to open a new CR)	LJ, P2K, Symbian	Test central and tracking	Histogram	100%	Testers	Daily	Testers will input data weekly into the tracking spreadsheet	Testers will enter this information in test central while executing the test case and will input data into the tracking spreadsheet weekly
CR tracking	Daily time spent (in minutes) tracking already opened CRs and giving support to development teams in order to close them	LJ, P2K, Symbian	Tracking spreadsheet	Histogram	100%	Testers	Daily	Testers will input data daily into the tracking spreadsheet	Testers will input data into the tracking spreadsheet
Amount of test cases; per cycle, per product, per language, per FA	Amount of test cases for each important indicators	LJ, P2K, Symbian	Test central and tracking spreadsheet	Histogram	100%	GB candidates	All the test cycle closing	GB candidates will input data into the tracking spreadsheet, at the test cycle closing	This data will be available in the tracking spreadsheet, and by using a pivot table, it will be possible to get this information
Amount of new prompts for each new feature	Total number of new prompts, for each new feature	LJ, P2K, Symbian	Tracking spreadsheet	Histogram	100%	GB candidates	Before each new test cycle	GB candidates will input data into the tracking spreadsheet before each test cycle	This information will be obtained from CxD team

127

Figure 5B.3

Measurement
System
overview.

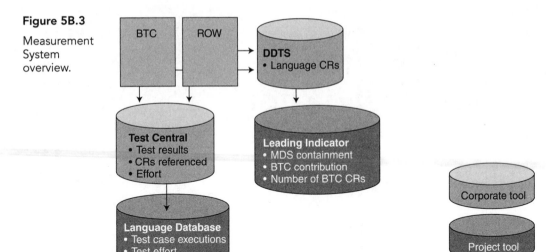

ROW = all other teams testing Motorola phones

Figure 5B.3 shows the Measurement System and its components:

▲ **BTC (Brazil Test Center):** department responsible for test planning and execution
▲ **ROW (Rest of the World):** all other test departments
▲ **DDTS:** Motorola's tool used to submit and organize bugs reported by all software and test departments
▲ **Test Central:** Motorola's tool, which serves as a repository for all test cases and test plans (groups of test cases), also used during test execution to report test results

Note: CRs (change requests) are defects or bugs with all detailed information about them (type, time and phase discovered, etc.)

All relevant data was collected from Test Central and DDTS databases.

An initial analysis of the data was then performed to establish a "baseline" of the defects. That is, the data was analyzed to determine the characteristics of the defects (their numbers, types, the languages in which they occurred, the phases in which they were detected, etc.). This baseline is necessary at the end of the project to quantitatively confirm the efficacy of any solution. That is, any acceptable solution would need to result in data that is better than this initial baseline.

4.2 Language Defects Baseline

During this phase, all language defects detected and fixed during 2007 were collected and analyzed.

Figure 5B.4 shows the result of this analysis and shows the percentage of defects detected during the different software development and test phases. Note that the primary measure of testing effectiveness illustrated here is *defect containment*. That is, the percentage of defects created in each phase that were actually caught during that phase. Thus, if ten defects were created during one phase, but only five were caught before the end of the phase, and the remaining defects were detected later, then the defect containment would be only 50 percent.

For this case, cumulative defect containment is given by:

$$\text{Cumulative containment} = \frac{\text{(Language defects caught in early tests)}}{\text{(Defects in early tests + Defects in later tests)}}$$

▲ Later test phases (field and customer acceptance tests) detected the majority of the language defects (54.1 percent of all accumulated defects during 2007).

▲ The earlier test phases (pre-field and others) detected 45.9 percent of all language defects in the year.

4.3 Measure Phase Conclusion

During this phase, essential data was defined, data collection plans were prepared, and a high-level initial analysis was performed. The phase ended with a formal milestone review with the project governance team.

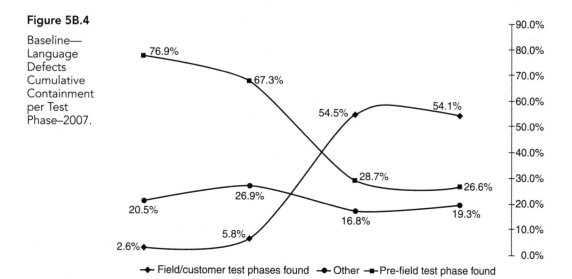

Figure 5B.4

Baseline—
Language
Defects
Cumulative
Containment
per Test
Phase–2007.

Note: This baseline improved significantly after this project was completed, as shown later in Figure 5B.14.

5. Analyze Phase

The Analyze phase is one of the most important phases of a Six Sigma project. It is during this phase that data is analyzed to discern or verify possible root causes of the defects. During the Analyze phase, the characteristics of defects are examined and relationships between different factors are quantitatively determined. Figure 5B.1 shows the relation of the Analyze phase to other phases in the DMAIC project lifecycle.

The methods and tools used in this phase included brainstorming, cause-effect diagram and prioritization, comparative methods, Pareto, MSA, and box plot.

Using these methods, defects were categorized and sorted, enumerated, and prioritized. Analysis and discovered relationships highlighted several possible root causes, and the correlations between causes and effects were examined using statistical methods to establish statistical significance. The root causes with the most impact on the process were also mapped graphically using a cause-effect diagram (also known as an Ishikawa diagram). Some other analysis tools used in the Analyze phase are described next.

5.1 Brainstorming

Identifying all possible causes for process defects requires team-based brainstorming. The team participated in a formal brainstorming session using an Ishikawa diagram, during which each team member suggested possible root causes for each type of general category (People, Methods, Machines/Tools, and Environment). After the brainstorming session, similar causes were combined, and a list of distinct possible causes was generated.

5.2 Cause-Effect Matrix and Prioritization

The list of possible defect-based root causes formed the basis of a cause-effect matrix spreadsheet. A cause-and-effect matrix is a tool that helps prioritize, sort, and display possible causes for the process defect, showing the relationship between the outcomes of the process and the causes that influence them.

The outcome of the process for this project was an enumerated, ranked list of language defects that were escaping to later test phases.

For each possible cause listed in the spreadsheet, the team members assigned a value to represent the perceived importance of the cause in the outcome of the process. In order to have a clear distinction of importance, the values 0, 10, 30, and 90 were used, in which 90 represented a very important cause and 0 represented a cause with no relationship to the outcome. The team discussed the perceived impact of each item and came to tacit consensus for the value for all causes.

Figure 5B.5 shows the resultant cause-and-effect diagram. The Y-axis represents all possible causes, and the X-axis shows the value assigned for the importance of the cause to the outcome of the process. The three causes with value 90 were considered the most critical causes to be

Figure 5B.5

Cause-and-effect matrix.

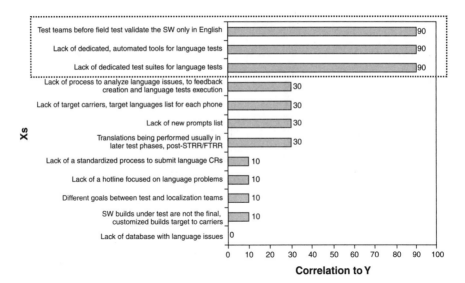

analyzed in greater detail. The team members considered those to have a greater impact on the number of language defects escaping to later test phases. This prioritization was important because of resources constraints on the project.

5.3 Root-Cause Analysis

The following issues were found to be the top three problems from the brainstorming and cause-effect analysis. Consequently, these were subjected to further analysis.

▲ **Root Cause #1:** Test teams before field test only validate the phones primarily in English and do not consider other language-related issues.

▲ **Root Cause #2:** The detection of language defects requires tools and/or people with fluency in the language.

▲ **Root Cause #3:** Test suites used are not effective to identify language defects.

5.3.1 Root Cause #1

The first root cause suggested that the exclusive focus on English in early phases caused a higher incidence of defects in other languages. If this was true, then the defect rates for English would be different from the defect rates of other languages (e.g., European, Asian, and African languages).

In fact, a Pareto chart of defects by language showed the relative proportion of language-based errors (see Figure 5B.6). More than 54 percent of language defects detected by early test phases were English defects.

However, the error rate in later testing phases was different. In later test phases, most of the defects found were in non-English languages (see Figure 5B.7). The majority of the defects detected by field tests were other European languages and Asian languages (60.6 percent).

Figure 5B.6

Pareto chart of
defects for
early test
phases.

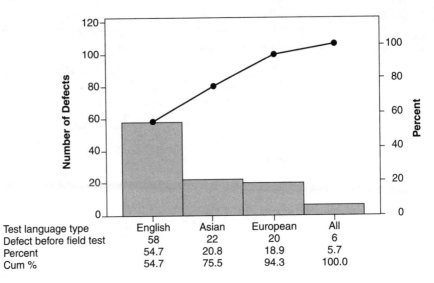

Test language type	English	Asian	European	All
Defect before field test	58	22	20	6
Percent	54.7	20.8	18.9	5.7
Cum %	54.7	75.5	94.3	100.0

Figure 5B.7

Pareto chart of
defects for field
test phase.

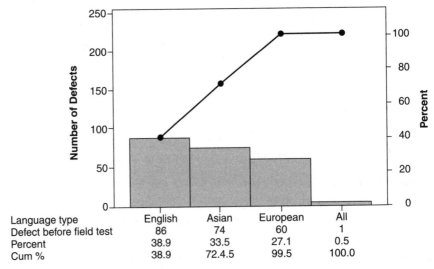

Language type	English	Asian	European	All
Defect before field test	86	74	60	1
Percent	38.9	33.5	27.1	0.5
Cum %	38.9	72.4.5	99.5	100.0

While these differences appear to be significant, the displayed Pareto charts could be an aberration and might have occurred by chance alone.

Fortunately, there is a statistical test that can discern the statistical significance of two different conditions like this. It is a two-proportion test, a comparative method that determines the likelihood that such differences could have resulted from random data gathering.

For the two-proportion test, the null hypothesis and the alternative hypothesis were:

H_o: Defect rates of early tests = defect rates of field tests
H_a: Defect rates of early tests ≠ defect rates of field tests

The results, using the Minitab tool, are shown in Figure 5B.8.

Figure 5B.8

Two-proportion test results.

```
Test and CI for Two Proportions
Sample X N Sample p
1 135 221 0.610860
2 48 106 0.452830

Difference = p (1) – p (2)
Estimate for difference: 0.158030
95% lower bound for difference: 0.0619342
Test for difference = 0 (vs > 0): Z = 2.70 p-value = 0.003
Fisher's exact test: p-value = 0.005
```

The p-value equal to 0.003 confirms that the statistical probability of these differences occurring by chance alone is very small. Therefore, we must reject the null hypothesis and can conclude that the differences in language-based defects between early test and later tests are statistically significant.

5.3.2 Root Cause #2

The second root cause suggested that many of the defects could only be detected if special tools or human "fluency" in the specific languages was used during testing. Again, the nature of defects had to be examined in more detail.

In this case, a sampling of defects was evaluated by subject matter experts who assessed whether or not fluency was required (by human or automated tool) to detect the error. The method used was called "Detectability Polling with Agreement Analysis" (using three voters).

For this study, 30 randomly chosen language defects (non-English, non-Portuguese, non-Spanish) from one of the phone platforms used in 2007 were examined and evaluated by the subject matter experts. Each voted on whether fluency in the language was essential for the detection or validation of the defect. The results of this exercise are shown in Figure 5B.9.

Figure 5B.9

Pareto of detectability analysis.

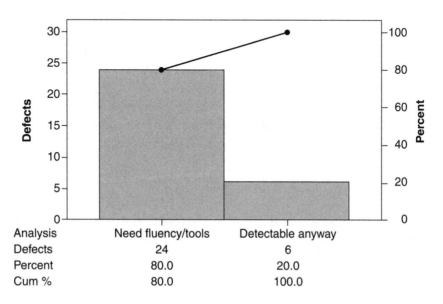

Analysis	Need fluency/tools	Detectable anyway
Defects	24	6
Percent	80.0	20.0
Cum %	80.0	100.0

Figure 5B.10

Assessment
agreement.

```
Between Appraisers

Assessment Agreement

# Inspected # Matched Percent 95% CI
  30 24 80.00 (61.43, 92.29)

# Matched: All appraisers' assessments agree with each other.

Fleiss' Kappa Statistics

Response Kappa SE Kappa Z P(vs > 0)
detectable anyway 0.666667 0.0471405 14.1421 0.0000
needs fluency/tools 0.666667 0.0471405 14.1421 0.0000
```

Three voters marked two possible values for each defect: Need Fluency/Tools or Detectable Defects was presented to the participants in random order.

The final result of the statistical analysis is shown in Figure 5B.10.

The statistical analysis suggests that, taking into account the variation across the various samples, an agreement of 80 percent is within the acceptable region (61.43 to 92.29) to suggest adequate agreement across the appraisers. Moreover, subsequent analysis showed a low p-value, once again indicating that the differences were statistically significant and were unlikely to have occurred by chance alone.

In other words, there really is a difference in defect types and most defects do, in fact, require fluency skills or tools to be detected.

This means that functional testing alone would generally be inadequate to detect these errors, thus suggesting that different tools or methods would need to be applied in early test phases to catch (or contain) these defects.

5.3.3 Root Cause #3

The third root cause analyzed suggested that the defect detection effectiveness of dedicated language test cases is greater than that for older and more dated test cases that focused only on general software tests.

In order to prove this hypothesis, a box-plot method was used (Figure 5B.11). The box plot shows that focused language tests had a much greater effectiveness (number of closed defects per executed test case) than regular functional tests.

5.4 Summary

The analysis of these three primary root causes confirmed the validity of these problems, and each one suggested different possible solutions for improving the defect detection and containment rates during the testing process.

Figure 5B.11

Box plot of test effectiveness.

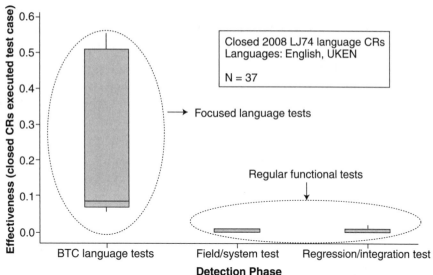

Detection Phase

5.5 Project Phase Conclusion

The Analyze phase ended with a formal milestone review with the project governance team.

6. Improve Phase

It is during the Improve phase that different potential solutions are identified and evaluated (see Figure 5B.1).

The possible solutions are defined and implemented, and the enhancements and improvements are communicated to the involved departments and test organizations. The change management plan is developed along with the communication plan.

6.1 Deliverables for the Improve Phase

The deliverables for the Improve phase constituted the changes and improvements to the test process. They included technical specifications for new testing, training and instructions, and new metrics, which would monitor the performance of language-based defects. Each of these improvements is described next.

- ▲ Test suites for all applicable software platforms: New groups of specially developed test cases focused on language defects detection
- ▲ Developed language test tools: New tools designed to help testers quickly and easily detect language defects
- ▲ Documents and procedures:
  - ▼ Language test process: Defining the new process to be used by test teams

▼ Current process map: As-is test process
▼ Should-be process map: The new proposed test process (see Section 6.2)
▼ Change management plan: How to implement the changes in the current test process
▲ New defect reporting and categorization methods (language defects submission instructions):
  ▼ This document contains instructions for the testers regarding how new language defects should be submitted and communicated to the software development teams. All defect information is described in detail in order to help software development teams quickly and effectively implement the fixes.
▲ Improvement indicators (see Section 6.3):
  ▼ Pilot results: The pilot test cycle executed in this phase is described in detail and communicated to the test and software development organizations.
  ▼ Leading indicator: A graphical way to show the enhancements and improvements achieved by this Six Sigma project.

6.2 New Process Map

At this point, it was necessary to map the new test process (Figure 5B.12), not only to document the proposed changes but to establish a new process baseline for later improvements. Some of the highlighted changes to the new test process included:

This new process is as follows:

▲ During software development, as soon the first translations are in place, the new products and their new features are analyzed, and a decision is made about which portions of the code will be subject to language-based testing (some would not).

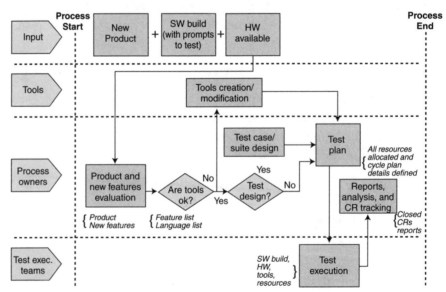

Figure 5B.12 New proposed language test process.

▲ After identifying modules for which language-based testing is appropriate, analysis is performed to define which specific tests are necessary and whether the available tools are adequate. If not, test design activities are performed, generating new test cases and adapting the current tools. After that, with test cases and tools correctly designed and implemented, a test plan (with those test cases) is organized and executed.

▲ The next steps in the process involve reporting, analysis, and bug tracking.

6.3 Improvement Indicators

Again, a key performance indicator for the effectiveness of the new process is the level of defects.

To test the new process, product testing was conducted with and without the process changes. As expected, the use of the new language-based tests and tools resulted in a dramatic increase in the number of defects detected. Figure 5B.13 shows these differences in testing at the Brazil Test Center (BTC). When focused translation testing was not conducted, fewer defects were discovered. Comparative analysis and Mann-Whitney tests showed that these differences were statistically significant.

Figure 5B.13

Hypothesis test #1.

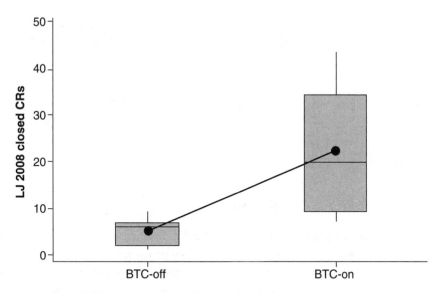

```
Mann-Whitney Test and CI: C1, C2
            N      Median
C1         11       1.000
C2         16       6.500

Point estimate for ETA1-ETA2 is -5.000
95.4 percent CI for ETA1-ETA2 is (-12.002, -0.999)
W = 85.0
Test of ETA1 = ETA2 vs ETA1 < ETA2 is significant at 0.0004
The test is significant at 0.0002 (adjusted for ties)
```

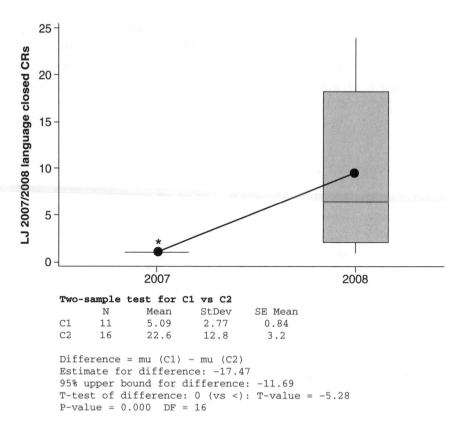

Figure 5B.14

Hypothesis test #2.

```
Two-sample test for C1 vs C2
         N      Mean     StDev    SE Mean
C1      11      5.09      2.77      0.84
C2      16      22.6      12.8      3.2

Difference = mu (C1) - mu (C2)
Estimate for difference: -17.47
95% upper bound for difference: -11.69
T-test of difference: 0 (vs <): T-value = -5.28
P-value = 0.000  DF = 16
```

Similarly, Figure 5B.14 shows the improvements in BTC defect containment metrics for different product releases in 2007 and 2008 for one of the specific platforms, LJ. Again, the increased number of defects discovered using the new tests was a statistically significant improvement.

6.4 Project Phase Conclusion

By the end of the Improve phase, the specific deficiencies identified in the Analyze phase had been addressed, with specific changes to the initial process, and the changes (including new test cases, new test tools, and new defect tracking methods) were tested for effectiveness. The phase concluded with a formal milestone review involving the project governance team.

7. Control Phase

The final phase of the DMAIC lifecycle is the Control phase (see Figure 5B.1). During this phase, controls and measures are developed to ensure that changes and improvements will be properly implemented. The lessons learned regarding language-based defects were disseminated to other teams in Motorola's R&D organization, and the new process was delivered and implemented to all test teams.

The following deliverables were created during the Control phase:

▲ A Monitor Chart to show that the improvements in the test results are consistently being sustained over time (see Section 7.1).

▲ A Lessons Learned report to put together all problems faced and all good practices during this Six Sigma project.

▲ A Process Control System to control the results after the new process is implemented. It contains critical indicators at each step, the sampling details for each of the indicators, and the actions required for possible out-of-control situations.

▲ Finance Results to show benefit realization and approval by the Finance organization.

7.1 Monitor Chart

The Monitor Chart is an easy way to graphically demonstrate that enhancements provided by the Six Sigma project are being sustained over time.

Figure 5B.15 shows that the percentage of defects detected in the latest test phases (field/customer) was 21.7 percent (before this project, it was 54.1 percent). This means that before this project most defects were caught at the end of testing.

With the new process, most defects were now detected earlier in the test process (pre-field/customer test phases) with 61.1 percent of all language defects being caught there (compared to 26.6 percent initially).

Considering these results, the improvement in language defects detection is considerable (61 percent), beyond the values expected at the beginning of this project (55 percent).

Figure 5B.15

Monitor chart.

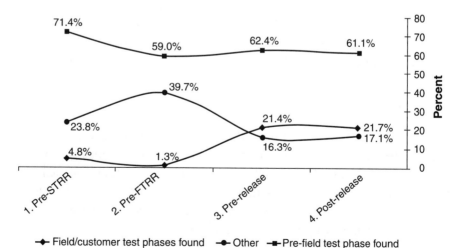

7.2 Financial Results

So how does this improvement translate into cost savings? Defects caught earlier in the software development lifecycle are cheaper to fix (with defects caught after release costing more than ten times as much to fix). Costs associated with the old and new processes are illustrated in Figure 5B.16.

In summary, this project led to a cost savings of US$760,000. Since the project costs with resources allocation and tool developments were approximately US$60,000, the net savings were US$700,000 (see Figure 5B.16).

Figure 5B.16

Final financial results.

Test phases	Initial tests	Pre-field tests	Field/ customer tests	Post-release
Baseline data (2007 CRs)				
Number of defects (all CRs)	39	13	251	76
Number of defects (found by BTC)	2	0	5	3
DSS pilot test (2008 CRs)				
Number of defects	107	168	428	176
Number of defects found with the old process (found by BTC)	5	0	9	**Old process** 0
Number of defects found with the new process (found by BTC)	27	112	231	0
New process Additional CRs contained	22	112	222	0
Costs to fix the defects (US$)	700	700	3000	10000
Soft savings (US$)	15400	78400	666000	
Total savings (US$)	759800			
Net savings (US$)	**705310**			

7.3 Project Phase Conclusion

With training and roll-out of the new process completed, and with formal financial analysis of cost savings completed, the Control phase ended with a formal milestone review that involved a presentation of phase results to the project governance team.

8. Benefit Realization

At the end of this one-year project, the net cost savings was about US$700,000, much greater than expected at the beginning of the project. The cost savings associated with this project were calculated, considering the total effort savings resulting from achieving a reduction in the rework for language translations. The earlier detection of translation defects allows Motorola to develop mobile products with reduced cycle time, delivering devices to the carriers and final customers within a shorter schedule.

9. Conclusions and Limitations

Looking at a problem in hindsight, after it has been resolved, is always so much clearer than when the problem was first identified. In retrospect, it makes sense that language-specific defects require different types of testing methods and different containment strategies than nonlanguage-related errors. However, it was the quantitative examination of test data that allowed team members to confirm these differences, and it was the rigor of the Six Sigma methodology that allowed the changes envisioned by the team to be completed and implemented in formal testing procedures so that improvements in defect detection and containment could be realized with subsequent generations of communications products.

This project provided a great opportunity to exercise Six Sigma tools, techniques, and methodologies, allowing Motorola Mobile Devices unit to deliver its products in a shorter time, with greater quality and better customer acceptance.

Moreover, the successful conclusion of this project gave additional credence to the Six Sigma approach and thus supported additional process improvement projects. The results of this project were written and published internally at Motorola, and the project team and management team received special recognition for the effectiveness of their efforts.

Finally, remember that even though this was a project to reduce language defects, the approach used and the lessons learned from this project are directly applicable (perhaps with some adaptation) to other defect reduction efforts in companies. However, different solutions might be warranted based upon the different root causes for other types of defects.

Acknowledgments

The author is thankful to Motorola for its permission to allow publication of this case study as a book chapter. The author would also like to thank Luis Bernardes from Motorola for his review comments.

CHAPTER 6

Help Desk Improvement

CHAPTER 6A

EMC: Improving Development Productivity

Rich Boucher

where information lives®

Relevance

This chapter is relevant to organizations that want to eliminate or minimize the effect of production support incidents on valuable development time. If incidents can be resolved before they reach development or eliminated altogether, more productive time can be given back to the development team.

At a Glance

▲ Beginning in 2007, EMC's Technology Solutions Group (TSG) initiated a Six Sigma project that used a DMAIC method to address the problem of defects in applications it supported.

▲ It significantly reduced the number of incidents (more than 6,800 incidents per quarter) that had to be resolved by the application development team, thus freeing up valuable application development resources.

▲ The team went further than merely transferring the responsibility for resolution of incidents from the application development team—it eliminated 47 percent of the incidents!

▲ As a result, the application development team realized significant productivity improvement and annual cost savings in excess of $600,000.

Executive Summary

EMC's largest professional services organization, the Technology Solutions Group (TSG), employs thousands of employees throughout the world and generates millions of dollars of service revenue to EMC annually. The purpose of this project was to reduce the number of production support incidents to be resolved by the application developers by either preventing them or by transferring them to the IT Global Service Desk, when appropriate. This was necessary because too much application development time was wasted in resolving application incidents rather than creating or enhancing functionality for revenue generation. Furthermore, these incidents were impeding the ability of TSG to efficiently track their time and invoice for the services they performed. Finally, too many application development resources were dedicated to resolving recurring TSG incidents.

Prior to undertaking the TSG project, in 2006, the IT organization at EMC embarked on a project to reduce the number of Collaboration (an application) production support incidents being resolved by one of EMC's application development teams. The IT Global Service Desk was resolving only 11 percent of approximately 700 Collaboration incidents logged each quarter. The remaining 89 percent were being resolved by the Collaboration application development team. This was consuming development time for new projects and delaying the resolution of the incident to the originator. A Six Sigma team was formed, and the Collaboration application development team worked with the IT Global Service Desk to identify opportunities to recognize and resolve simpler incidents without escalation to an application developer. The result of this project was that incidents resolved by the Collaboration application development team declined from 89 percent to 24 percent, which increased development capacity. This caught the attention of senior management, in particular, the VP of application development.

Given the success of the Collaboration project, in 2007, EMC formed a new team composed of employees from IT application development, IT Global Service Desk, and the IT project management group representing TSG. This was sponsored by the VP of application development, who was trained as an executive sponsor for Lean Six Sigma and saw this as an opportunity to use a structured problem-solving approach to collect data, identify the root causes of the most frequently occurring problems, and systematically resolve incidents earlier in the cycle. This project was generating 6,800 incidents each quarter, nearly ten times the size of the project completed in 2006.

The findings indicated that there were three major areas of recurring incidents. Upon analysis of these incidents, the team discovered that they could go well beyond transferring the resolution upstream to the IT Global Service Desk, the first level of resolution. The team identified the root causes of these incidents and significantly *eliminated the number of incidents from occurring in the first place*. This continued to propagate success within IT at EMC. Several other similar projects were launched and completed successfully.

This is the detailed story about one of them:

▲ The project created to address problems and incidents followed the DMAIC methodology.
▲ A charter was created and a clear problem statement and goal defined.
▲ Data was collected, and initial analysis showed that the majority of errors fell into three categories.
▲ Root-cause analysis revealed that many problems were recurring and they were being reported and resolved multiple times.

▲ By changing the routing of tickets and incidents to appropriate resolution centers, the volume of tickets was significantly reduced. Recurrence rates dropped, and the percentage of resolved problems improved.

▲ Changes to the problem resolution processes saved time, effort, and money, as well as reduced the resolution time for problems and incidents.

▲ Thus, unknown inefficiencies in the problem resolution process were corrected with major impact on developers and users.

1. Introduction

In April 2007, the VP of application development called a meeting of several members of his team, the IT Global Service Desk team, and our IT project management team representing TSG. The objective was to resolve recurring TSG incidents quickly so that TSG professional service resources could spend more time producing revenue and IT application development resources could spend more time developing new projects.

The incident management process is designed at various levels of expertise. The IT Global Service Desk (GSD) is designed to resolve routine incidents that do not require expertise in the application. This is commonly referred to as Level 1 support. The GSD uses its knowledge base, comprising books and process and product documentation (such as user guides), to engage users who report application incidents to quickly resolve their issues. Incidents beyond the expertise of the IT GSD are referred to as Level 2 or 3 production support incidents and are escalated to the IT application development team for resolution.

The expertise of the IT Global Service Desk for resolving TSG application-related incidents was inadequate due to the complexity and sensitivity of information (e.g., price information contained in the applications.) Consequently, 84 percent of the incidents (Level 2 and 3) were being resolved by TSG application development or the TSG project accounting system (PAS) teams, as depicted in Figure 6A.1.

Figure 6A.1

Incidents by level.

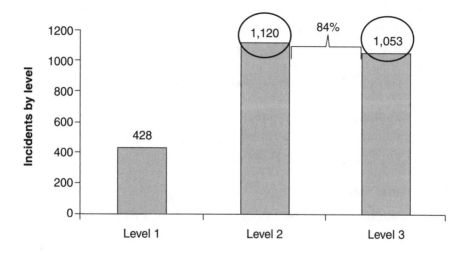

These incidents were overwhelming the IT application development team for TSG and frustrating the TSG community globally.

2. Project Background

The TSG application is complex and requires a great deal of expertise. TSG application production support issues were inundating IT, and the VP was continually receiving feedback from the TSG community to resolve these issues. The VP of application development had seen success on IT initiatives to push production support upstream to the IT support team. The VP directed his application development team and engaged the process and portfolio management group (PPMG), as well as IT support services, including the IT Global Service Desk, to address this persistent issue.

3. Define Phase

The team defined the problem as the proliferation of TSG production support incidents severely limiting time for project work and enhancements by IT application development.

3.1 Six Sigma Project Team

The executive sponsor, the VP of application development, commissioned a team consisting of the IT deployment Lean Six Sigma lead, a Black Belt who worked within the IT Service Desk, and two Green Belts, one from IT application development and an IT project manager representing TSG. A college intern was subsequently assigned to collect, analyze, and present data during the course of the project.

3.1.1 Business Case Statement

The Technical Services Group (TSG) is a global professional services organization that generates a substantial amount of service revenue for EMC and consists of several thousand employees worldwide. The EMC IT application development team was resolving too many TSG incidents pertaining to the PAS. The level of effort required to resolve these issues was adversely affecting service billing to our customers and resulting in decreased productivity by TSG, as well as impeding the IT development team from developing and enhancing applications.

3.1.2 Opportunity Statement

There was an opportunity for our IT Global Service Desk to resolve these incidents without escalating these to the application development team.

3.1.3 Goal Statement

The goal was to resolve 50 percent of the number of TSG incidents at the IT Global Service Desk globally by the end of Q1 2008.

3.1.4 Project Scope

In scope were all TSG incidents globally affecting the IT application development team. There are 12 classifications that we refer to as "affected items."

3.1.5 Project Plan

Key milestones are depicted in Table 6A.1.

Table 6A.1   Key Milestones

Phase	Deliverables	Date
Define	Project contract SIPOC Stakeholder analysis	May 2007 May 2007 May 2007
Measure	Baseline Process flow as-is	June 2007 June 2007
Analyze	Affinity diagram Pareto Improvement plan	August 2007 August 2007 September 2007
Improve	Implement improvements Measure	October 2007–March 2008 October 2007–March 2008
Control	Control plan	April 2008
Realize	Quantify the financial benefits	May 2008

Note: The as-is process flow is the "incident report management process."

3.1.6 Project Team

▲ **Sponsor:** VP, IT applications development
▲ **Black Belt Lead:** Service readiness team (IT support services)
▲ **Green Belt Leads:**
  ▼ TSG applications development
  ▼ Process and portfolio management group (PPMG), TSG
▲ **Core Team:**
  ▼ Applications development, intern
  ▼ Project accounting system operations
▲ **Key Stakeholders:**
  ▼ Director, applications development
  ▼ TSG development manager
  ▼ TSG director, process and portfolio management group (PPMG)
  ▼ IT support services director
  ▼ IT support services manager, incident and problem management
▲ **Black Belt Mentor:** IT Lean Six Sigma deployment leader

A high-level process flow was used to identify the key measurements, that is, the "Big Y," what is critical to the customer (see Table 6A.2). The key measurement in this case was the percentage of application development incidents being closed by the TSG development team.

Table 6A.2 Supplier, Input, Process, Output, Customer (SIPOC)

Suppliers	Inputs	Process	Outputs	Customers
TSG field	Incidents opened for TSE	Incident resolution process for TSG field	Critical to customer: % apps dev incidents closed by TSG dev team (PAS and TSG)	TSG field
Project coordinators for PAS	Incidents open for PAS sources: Phone, e-mail, IT online		Number apps dev incidents closed by TSG dev team (PAS)	Project coordinators for PAS
Finance TSG	IT online			Finance TSG
Business operations	Incident tracking tool			Business operations TSG
IT global service desk	Training at service desk			Portfolio and process management group (TSG)
Portfolio and process management group (TSG)	Training on tool			
Application development (TSG)	Support web sites containing documentation of affected items			

4. Measure Phase

One of the first steps was to measure the current state using the process flow for resolving an incident (see Figure 6A.2).

The team developed a process flow and *measured* the process by establishing a baseline of the current state of 6,832 incidents per quarter using data from Q4 of 2006. The team selected Q4 2006 because seasonally, it is EMC's highest revenue quarter, produces the most recurring incidents, and developers had started to look at these incidents prior to formally launching the project, In Q4 of 2006, 84 percent of incidents were being resolved by the TSG application development team, or a team within TSG called PAS operations, as a baseline. A formal measurement plan was not necessary, as the system of record captured incidents by affected item. The team ran queries against the affected items in scope to produce the information needed.

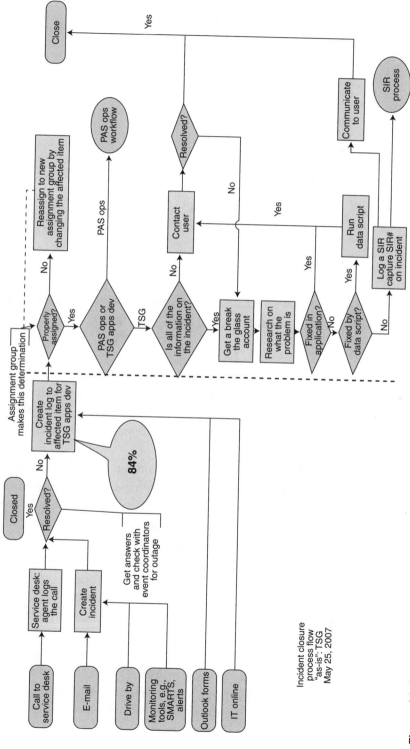

Figure 6A.2 "As-is" process map.

151

Figure 6A.3

Pareto chart of affected items.

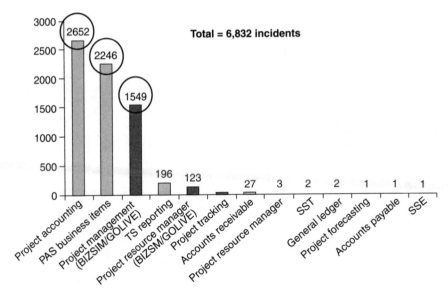

The team used a Pareto chart to classify the categories of affected items and discovered that 90 percent of the problems pertained to three affected items (see Figure 6A.3).

5. Analyze Phase

The team used the data from the Pareto chart to analyze the process. The analysis revealed that most of the problems fell into three main categories. Subsequent analysis focused on these main types of problems. The others were deferred to other follow-up projects to be addressed at a later time.

Detailed analysis of these top three items allowed the team to complete more focused root-cause analysis of these incidents and produced a Pareto chart in June by request type (Figure 6A.4).

The team discovered that business requests, 41 percent of the incidents, were being resolved by the PAS operations team within TSG with little impact to the development team. This activity was predominantly "requests," not "incidents." Requests are users in need of information, advice, or to access a service. Incidents are unplanned interruptions of the service. The team's primary focus was on "incidents," which are the defects in the process.

The team shared the data with TSG, who were conducting a TSG "optimization" initiative. Likewise, TSG shared their data analyses with IT. This led to a brainstorming "Affinity Session," consisting of the three IT groups: IT support services, application development, and PPMG, as well as TSG PAS operations.

The team developed and prioritized ideas to address the process. They decided on several areas of improvement—one of the most significant of which was to assign an error code to each incident. Beginning in Q3 2007, the team focused on identifying error codes and tagging these to TSG incidents via the IT Global Service Desk as incidents were closed.

Figure 6A.4

Pareto chart of request types.

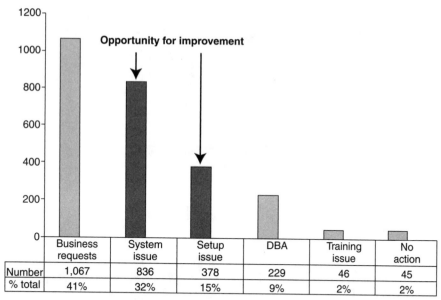

	Business requests	System issue	Setup issue	DBA	Training issue	No action
Number	1,067	836	378	229	46	45
% total	41%	32%	15%	9%	2%	2%

A college intern was brought on board, and under the direction of the Black Belt and Green Belts, analyzed the data. The analysis verified that error codes were accurately assigned to the incidents. The team reviewed the most frequently occurring error codes to develop an improvement plan.

The focus of the development team was to resolve recurring system and setup issues, which constituted a combined 47 percent of all incidents. The team created the error codes within the incident tracking system and analyzed the data (see Table 6A.3). We concentrated on error codes where we could perform corrective action and reduce the most frequently occurring incidents. Corrective action on some of the higher-frequency error codes, such as "PAS Revenue Custom," was not identified in time for the bucket release and was deferred to future releases.

We experienced a critical breakthrough. The team discovered that, unlike the project in 2006, they could prevent the incidents from recurring rather than primarily transferring incidents upstream to the IT Global Service Desk.

6. Improve Phase

The team immediately identified opportunities for project coordinators in TSG to resolve user questions without raising an incident. We utilized our point of contact for TSG PAS operations and developed a continual feedback loop to provide expertise on recurring questions. TSG PAS operations incorporated this feedback into their standard operating procedures and training.

More significantly, the team targeted application enhancement releases in December, February, and March to focus on the highest recurring incidents as a roadmap for

Table 6A.3 Error Code Analysis of Incidents

Error code	Business request	DBA	No action	Setup issue	System issue	Training issue	Total	Percentage
PAS time transfer	285	—	—	—	—	—	285	12.54
PAS OTL time transfer	—	—	—	—	252	—	252	11.09
PAS billing assignment	—	—	—	226	—	—	226	9.95
PAS partner record	155	—	—	—	—	—	155	6.82
PAS invoice/revenue reversal	123	—	—	—	—	—	123	5.41
PAS revenue custom	—	—	—	—	103	—	103	4.53
PAS employee setup	97	—	—	—	—	—	97	4.27
PAS invoice Type 6/Level 1	—	—	—	—	88	—	88	3.87
PAS time reversal	75	—	—	1	—	—	76	3.35
PAS revenue T&M	—	—	—	74	—	—	74	3.26
PAS project setup	46	—	—	—	20	—	66	2.90
PAS Miscellaneous	65	—	—	—	—	—	65	2.86
PAS ASQ issue	61	—	—	—	—	—	61	2.68
PAS redirect	—	—	—	—	59	—	59	2.60
PAS OTP/OTA problems	—	—	—	—	53	—	53	2.33
PAS multicurrency	—	—	—	48	—	—	48	2.11
PAS debook/rebook	—	—	—	—	39	—	39	1.72
PAS cost budgets	—	—	—	—	37	—	37	1.63
PAS opportunity to projects	7	—	—	—	28	—	35	1.54
PAS datascripts	—	34	—	—	—	—	34	1.50
PAS DOE issues	—	—	—	—	4	27	31	1.36
PAS invoice custom	—	—	—	—	31	—	31	1.36
PAS performance	—	—	—	—	30	—	30	1.32
PAS invoice T&M	—	—	—	—	24	—	24	1.06
PAS dashboard/stoplight	—	—	—	—	22	—	22	0.97

Error code	Business request	DBA	No action	Setup issue	System issue	Training issue	Total	Percentage
PAS ESG misc	22	—	—	—	—	—	22	0.97
PAS revenue level 1	—	—	—	22	—	—	22	0.97
PAS reporting	9	—	—	—	10	—	19	0.84
PAS data mapping	—	—	—	**18**	—	—	18	0.79
PAS forecasting	—	—	—	—	—	18	18	0.79
PAS revenue generation	—	—	—	—	16	—	16	0.70
PAS delivery org changes	11	—	—	—	—	—	11	0.48
PAS ART requests	7	—	—	—	—	—	7	0.31
PAS password	5	—	—	—	—	—	5	0.22
PAS dashboard/spotlight	—	—	—	—	4	—	4	0.18
PAS invoice generation	—	—	—	4	—	—	4	0.18
PAS contingency list	—	—	—	—	2	—	2	0.09
PAS resource management	—	—	—	—	2	—	2	0.09
PAS CS integration	—	—	1	—	—	—	1	0.04
Total	975	34	1	393	824	45	2272	100.0

Bold numbers = TSG bucket releases should reduce these incidents

improvement. The team utilized Pareto tools for prioritization and root-cause analysis to convert the error codes to actionable data. The team would take high recurring incidents, such as PAS online data entry, and determine the root cause of each incident that occurred within the preceding month. Most often, the root cause was the same for all of the incidents associated with the error code and could be rectified by enhancement releases.

The original plan was to eliminate or transfer 50 percent of the incidents from TSG application development to the IT Global Service Desk. Instead, the team took a bigger step and used this data to focus on planned proactive delivery of defect fixes in scheduled enhancement releases to improve the process by eliminating 47 percent of TSG incidents by Q1 of 2008 (Figure 6A.5) from the original baseline!

Figure 6A.5

Incident reduction trend.

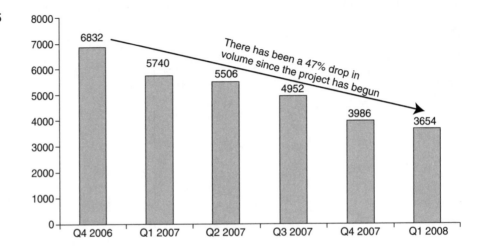

The team analyzed incidents each month and reported them to the sponsor and key stakeholders, including the director of application development, the TSG application development manager, the director of IT support services, and the director of PPMG for TSG during and several months after completion of the project. Several remarked on the outstanding success, including the executive sponsor, who said, "Many thanks to you and the team for making such great progress."

7. Control Phase

To ensure that the progress is sustained, the team instituted a process so that IT Service Desk agents included the correct error codes when they closed production support incidents. The application development team used these error codes as data in helping to determine corrective actions required in future bucket releases and implemented the control plan.

Although the project had closed, the trend continued to improve through routine analysis of error code data each month. The team measured the process at the end of Q2 2008, three months after the project completed, and incidents had decreased by 53 percent from the baseline.

The two Green Belt candidates earned their Lean Six Sigma Green Belt certifications for this project. The Black Belt earned credit for completing one of two projects in his journey to earn Lean Six Sigma Black Belt certification. The application development Green Belt presented the results of the project to an application development "all hands" meeting and the EMC quarterly rewards and recognition ceremony.

8. Benefit Realization

The team met with finance and quantified the cost savings associated with eliminating more than 12,000 incidents annually and were approved for a benefit realization of more than $600,000.

9. Conclusions and Limitations

Success breeds success. The VP of application development encouraged people to get trained in Lean Six Sigma and propagate the success of this project in support of their specific applications. Subsequently, five more successful IT projects focusing on eliminating production support issues using the same techniques were completed by following the method described in this case study. These projects reduced incidents significantly in several application areas: finance (33 percent), human resources (31 percent), shipping (40 percent), business to business (61 percent), and software licensing (64 percent).

Acknowledgments

The author of this case study is thankful to EMC for its permission to allow publication of this case study as a book chapter. The author would like to thank Tony Pagliarulo, Renu Krishnamurthy, David Minichiello, Tom Pietropola, and George Abourizk for their contributions on the project and providing content for this story. The author would also like to thank Russ Mauss, who coordinated the approvals needed within EMC to share this, and Cindy O'Reilly for helping to review and edit the content.

Infosys Helps a Global Bank Reduce Business Risk from Errors in Online Banking Applications

Prakash Viswanathan
Anshuman Tiwari*

POWERED BY INTELLECT
DRIVEN BY VALUES

Relevance

This chapter describes a project for improving uptime of a large online application running millions of transactions. Apart from uptime, the accuracy of transaction processing made this project critical to the customer. If you run a process that has lots of customers and runs an online front-end, you will find the lessons from this project useful.

At a Glance

▲ Infosys helped a global bank reduce transactional errors in an online application, which were resulting in loss of revenue and erosion of consumer goodwill, thus posing subtantial risk to business growth.

▲ The project team was successful in reducing incidents related to this online application by 18 percent (exceeding the project goal of 12 percent reduction).

▲ Timely closure of transactions resulted in a tangible benefit of about US$4 million for the bank. This enabled Infosys to grow its business relationship with the bank and become its preferred vendor.

---

* At the time of this project, Anshuman Tiwari was head of the Business Excellence Program at Infosys.

Executive Summary

This chapter presents a Six Sigma project carried out by Infosys for a large global bank with operations in more than 25 countries. Infosys is one of the main vendors of information technology services to this bank. Among the services provided are maintenance and support of critical software applications that are used by the bank's customers worldwide for day-to-day transactions. Since the entire set of applications is large for any single vendor, the bank uses a cluster of vendors to manage these applications.

The applications are legacy applications on old technology platforms and have existed for more than a decade. These applications see about a million transactions every day. Since the entire online banking division of the bank depends on online applications, any event that causes or could cause a disruption to a transaction means a serious delay in processing information and a loss of business to the bank (possibly in millions of dollars), as well as a loss of goodwill and loyalty. Such events are called "incidents." The bank wanted to minimize these incidents.

A Six Sigma project team was set up with a clear mandate of reducing errors (incidents) in online software applications. Adding to the complication of millions of transactions was the fact that the bank was using multiple vendors to manage this suite of applications. Varying documentation practices of vendors meant clear and complete documentation was not available. Keen to progressively eliminate errors in its online application, the bank set an aggressive 12 percent improvement goal for incident reduction. This improvement needed to be achieved in six months if it was to have an impact on the online banking division's business and future. The Six Sigma team set up to improve this situation included a project leader, five subject matter experts (SMEs), and a Six Sigma Black Belt.

The Six Sigma team used the Define, Measure, Analyze, Improve, and Control (DMAIC) methodology to carry out this improvement project. The key goal for the team was to reduce the number of incidents reported by the customer. This was the project's critical success factor (CSF). The team then analyzed data for all incidents for the past five months. Every transaction was treated as an opportunity, and every incident (error) was treated as a defect. The goal was to reduce the number of incidents by 12 percent from the baseline.

A detailed root-cause analysis (including brainstorming and fishbone analysis) of all the incidents revealed that a few countries contributed the most to incidents and a few causes occurred more frequently than others. These causes were traceable to inherent system design limitations, user errors, and interface errors. The team also did cluster analysis for all countries where the online applications were used. At the end of the Analyze phase, the Six Sigma team made a short list of countries and key root causes.

In the Improve phase, the team identified solutions and technological fixes during brainstorming sessions. These included workarounds, additional user training manuals, and modified data tables. An effort impact prioritization matrix was used to identify "low-hanging fruit" (low effort and high impact) items. Once implemented, these solutions resulted in a significant reduction in incident count.

In the Control phase, the team needed to ensure that the gains made during the Improve phase were sustained. These results were then shared with the customer and investment for implementing more intensive solutions was sought. Statistical tests showed that there was indeed a reduction in incident count before and after the improvements were made. The pre- and post-improvement outcomes for incident count instilled confidence in the customer about the effectiveness of the solutions. Training manuals and technological fixes for incidents were documented and shared with the customer. This ensured incident count remained in control and the risk to business was reduced.

The customer was delighted with about 18 percent reduction in number of incidents. As a result of this project, Infosys secured more work from the customer, saw a significant rise in customer satisfaction, and became a preferred vendor to this global bank. The tangible impact of the improvement project was about 4 million dollars.

1. Introduction

Information technology and software applications are part of life today, just as electricity and the phone became so more than 50 years ago. Today, we cannot imagine a bank without an error-free online service. Several other services, such as travel, shopping, utility, etc., have moved online. Most of these applications have millions of transactions a day. Managing these transactions without errors is a challenge for any software application.

Among its wide range of services, Infosys provides development and maintenance services for online applications, such as the one this global bank uses. For its customers to be satisfied, the bank requires smooth and trouble-free flow of transactions on the online applications. These large business critical applications are hosted on mainframes that have language and environment limitations.

The bank expected Infosys to reduce the incident (error) count. The customer attached high significance to incident backlog (pending and unresolved incidents), which was monitored regularly. Once the customer expressed the need to reduce incidents, senior leaders at Infosys made this project a high-priority project. The Six Sigma DMAIC approach was used to carry out this improvement project. As is common in Six Sigma teams, the team had representatives from several functions in Infosys.

2. Project Background

As explained earlier, millions of transactions take place on the online applications used by the bank. Any error in these applications not only has a huge goodwill loss, but also significant financial loss as well. Add to this the potential customer attrition, and this becomes a serious issue for any service industry company. The Six Sigma team developed a business case to estimate the return of investment and feasibility of this improvement project. The project charter detailed all roles and responsibilities of the cross-functional Six Sigma team.

By significantly reducing the errors in online applications, the Six Sigma project team could reduce business risk for the customer. This business risk included incorrect transactions and potential customer attrition.

3. Define Phase

Once the opportunity and its importance were established, the team approached the Define phase of the project. The key activities in the Define phase were:

▲ Identifying the project critical success factor (CSF)
▲ Reviewing and baselining the project charter
▲ Creating a high-level process map

The following sections provide more details on these key activities.

3.1 Identifying the Project Critical Success Factor

Identifying the CSF is necessary in the early stages of an improvement project. This helps the team remain focused on the key issues of the project. The customer expectation from this project was a reduction in incident count for online applications.

3.2 Creating the Project Charter

A useful project charter binds a cross-functional team to focus on a common improvement objective to ensure success. A typical project charter includes:

▲ Business case
▲ Problem statement
▲ Improvement goal statement
▲ Roles and responsibilities
▲ Project scope
▲ High-level project plan

The project leader created a draft project charter. This was reviewed and approved by the sponsor, who was from the senior management team at Infosys.

3.2.1 Business Case

This project had a compelling business case, since many consumers were encountering errors in the bank's online banking application. A service industry, especially financial services, cannot afford errors like this. Each error could affect the customer in multiple ways, ranging from delayed service to loss of goodwill to monetary loss.

While the business case for this project appeared obvious, the team still went ahead with documenting the same and sought approval from the sponsor. This is a useful step even if the case appears obvious. In many projects, the sponsor may provide additional inputs during charter review. In this case, the sponsor encouraged the team to aim high and include representatives from all functions that were touched by this project.

Key elements of this business case were:

▲ Software application maintenance was being executed for one of the key customers to Infosys.

▲ The software application allowed consumers to carry out transactions online. Many consumers faced errors in these applications, leading to loss of goodwill and business opportunity.

▲ It was important to reduce the incidents over a specified time, as this affected the customer's top-line.

▲ The software application used by this bank was large and complex. Furthermore, it was hosted on a mainframe and had minimal documentation. To further complicate the project, this software application was being used for millions of transactions per day.

▲ Other IT services vendors were also providing services to this customer and hence, were competing with Infosys. If Infosys could provide better value to the customer, it had a chance to secure more business from the bank.

Overall, this was a problem that needed to be fixed as quickly as possible.

3.2.2 Problem Statement

It was important to reduce the number of incidents, as they were directly linked to business revenue. An incident caused an undue delay in transaction processing, thus affecting revenue. The number of incidents was the business "Y" in this specific example.

3.2.3 Improvement Goal Statement

Infosys adopted the SMART goal principle while finalizing Six Sigma project goals. A SMART goal is specific, measureable, aggressive (yet achievable), relevant, and time-bound. In this case, the goal was to reduce the number of incidents by 12 percent within six months.

3.2.4 Project Scope

Defining the project scope is one of the most useful steps in a Six Sigma project. It allows the project team to remain focused on the project and not solve problems they don't have to solve. The authors have seen several unsuccessful Six Sigma projects early in their careers, and more often than not, the reason was improper scope definition and inability to arrest scope creep.

As is common with most global IT services contracts, multiple vendors execute parts of a workflow. In this case, Infosys managed a large portion of the online software applications suite. The scope of this Six Sigma project was thus restricted to the online applications that were managed by Infosys.

3.2.5 High-Level Project Plan

Maintaining discipline during a Six Sigma project is critical for success. At Infosys, we believe that failing to plan is like planning to fail. The Six Sigma team developed a plan to help them maintain execution discipline throughout the project (Table 6B.1).

Table 6B.1 Project Plan

S No.	DMAIC Phase	Activities	Resources	Deliverables	Planned Completion Date
1	Define phase	Define objective, scope of initiative, project charter creation	PM/SQA/ practitioner	Project charter	Apr 10
2		Sign off/initiation of project	Sponsor	Approval of project charter	May 5
3	Measure phase	Data collection and validation	SQA/PM/ practitioner		Jun 10
4		Trend analysis	SQA/PM/ practitioner	Graphical analysis	June 14
5	Analyze phase	Identify process capability	SQA/PM/ practitioner	Process capability	Jul 13
6		Identify sources for variation	SQA/PM/ practitioner/ SME	Root cause analysis	Jul 28
7	Improve phase	List out corrective actions	SQA/M/SME	Who what when plan	Aug 13
8		Prioritization of implementation actions	SQA/PM	Matrix diagram	Aug 17
9	Control phase	Executing the implementation plan	PL and team/ practitioner		Aug 27
10		Collect post improvement samples and comparing against objective set	PL and Team/ SQA/ practitioner	Hypothesis testing—pre and post improvement	Dec 8
11		Verifying benefits and full scale implementation	Sponsor, financial analyst, QM, PM, practitioner	CBA report	Dec 13

3.3 As-Is Process Map

The team then examined the process to get a better understanding of the business flow. While the objective of the as-is process map was to understand the process, it also helped in identifying bottlenecks in the process. These bottlenecks were obvious causes that led to

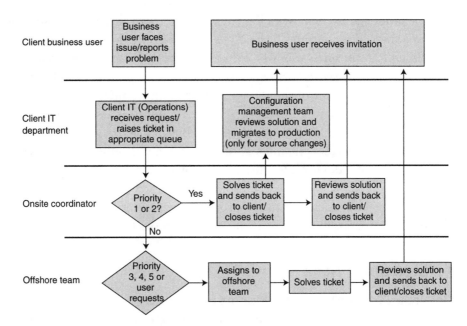

Figure 6B.1

High-level process map for incident resolution

frequent disruptions in business transactions. The high-level process map for this project is in Figure 6B.1.

There were three distinct entities involved in incident resolution. These were business users, customer IT department, and IT vendors. Business users interacted on a daily basis with the application front-end for generating reports and analyzing data. There was a customer (client) IT department who owned these applications and was responsible for ensuring uptime. There were multiple vendors who were responsible for a suite of applications. The IT department assigned a separate queue for each vendor. Infosys was one of the vendors who supported an application suite. Infosys staffed the project on the basis of the global delivery model with an optimal team spread at the customer location and led by an onsite coordinator, supported by an offshore team in India.

Every time the business user was unable to use the application front-end, whether for day-to-day report generation or data analysis for business needs, they reported this to the IT department. The IT department then looked for the appropriate queue assigned to a vendor and raised a ticket (incident) against the same. The client IT department also assigned a priority to every ticket based on the business user needs. Each priority level had a different service level agreement (SLA) between the client IT department and a business user. Infosys had an onsite coordinator who was immediately notified about the ticket. He was a technical expert who would then route the ticket depending on the impact. High-priority tickets were assigned to the Infosys onsite team, as these were to be addressed on an emergency basis. The onsite team would analyze the priority level 1 and level 2 problem tickets and suggest a solution. This would get routed to the client IT department configuration management (CM) team, who would assess the impact on other applications before making the necessary changes to the production system.

The production system was a live environment (both hardware and software), which was used to transact daily business. The ticket was closed once the solution was accepted by the IT department. The business user who initiated the problem then received notification that the problem has been solved. Similarly, if tickets were of a low priority (3 and 4), they were assigned to the offshore team. The offshore team worked on the solution and routed it to the client IT department to verify if it needed a system modification, which was then made in the production system. Once this was done, the business users were notified of the problem resolution. In a few cases, the Infosys team directly notified the business user of the solution or workaround if it did not affect the production system.

The Define phase ended with a project charter and process map in place. A Six Sigma project review panel (including the sponsor) reviewed the progress of the team at this stage and cleared their progress to the Measure phase of the project.

4. Measure Phase

One of the early questions in the Measure phase is what to measure. While many metrics can be measured, the most important one to measure is the one that needs improvement. In this project the number of errors/incidents was the key metric to be measured.

Fortunately, the number of incidents was a metric that the customer was tracking on a monthly basis. Had this not been the case, the project team would have set up a data collection mechanism, leading to a delay in the Six Sigma project. The incident tracker included fields such as incident ID, system of origin, country, severity, and date and time stamp, along with the status. This rich set of data formed the base for this improvement project.

The incident trend is illustrated in Figure 6B.2.

Figure 6B.2

Incident count—Trend analysis pre-improvement.

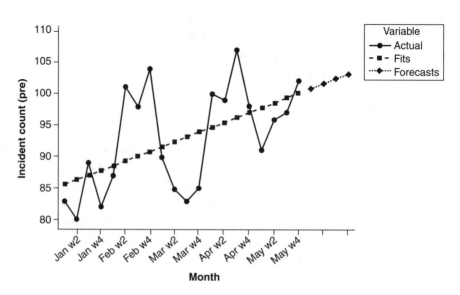

Month

The team analyzed weekly incident history data for five months. The customer noticed an increasing trend in the incident count, which was a cause for worry, as it had business (revenue) implications. After an initial internal study, the customer concluded that a 12 percent reduction in the incident count was essential to curtail revenue loss and minimize risk to business.

Key activities in the Measure phase are identifying the metric to be improved and establishing an improvement goal for the same. Having completed these two activities, the team presented its progress to the project review panel. Having secured the approval from the review panel, the project team was ready to enter the Analyze phase of the project.

5. Analyze Phase

The Analyze phase is the real "eureka" phase in a Six Sigma project. In this phase, the project team analyzes the data collected in the Measure phase and identifies key sources/causes of variation in the process.

Early in the Analyze phase, the project team reviewed data from various sources and discovered that incidents from some select countries were contributing disproportionately to the overall incident numbers.

A Pareto chart was created, which is illustrated in Figure 6B.3.

Singapore, Germany, Hong Kong, UK, India, and Taiwan contributed to over 70 percent of the total number of incidents. These countries were identified for piloting improvement solutions, which would be identified in the next phase.

Data captured through the online system had detailed root-cause information. The team performed a root-cause analysis and found three broad categories (inherent flaws, human errors, and interface errors) resulted in 99 percent of the incidents (see Figure 6B.4).

Figure 6B.3

Pareto chart of incident contribution by country

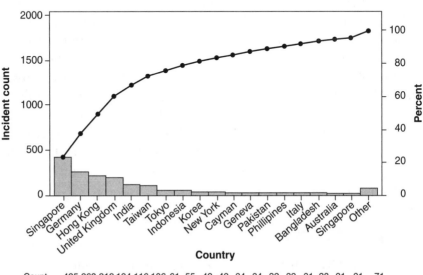

Country	Singapore	Germany	Hong Kong	United Kingdom	India	Taiwan	Tokyo	Indonesia	Korea	New York	Cayman	Geneva	Pakistan	Phillipines	Italy	Bangladesh	Australia	Singapore	Other
Count	425	263	219	194	116	106	61	55	42	42	34	34	32	32	31	28	21	21	71
Percent	23	14	12	11	6	6	3	3	2	2	2	2	2	2	2	2	1	1	4
Cum %	23	38	50	60	67	72	76	79	81	83	85	87	89	91	92	94	95	96	100

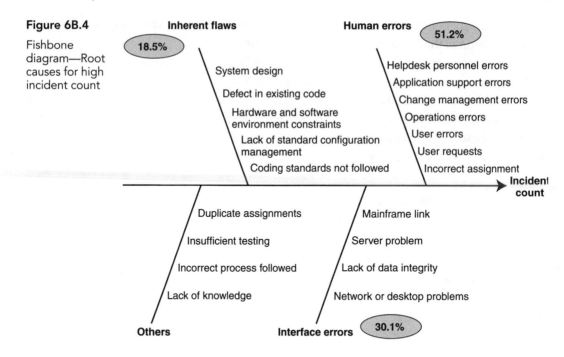

Figure 6B.4

Fishbone diagram—Root causes for high incident count

The application suite was highly interactive. In order to successfully execute a transaction, many inputs were expected to be provided by end users, for example, providing correct consumer identity, the start date of the billing cycle, and so on. Any inadvertent input could cause an incident and was classified as a human error.

The applications were maintained and enhanced over the years by multiple vendors. Furthermore, there was no proper documentation available on code fixes. There were design limitations in code, which were resulting in incidents and the application falling short of meeting consumer needs. These were categorized as inherent system flaws.

In order to successfully execute a transaction, data had to pass through various systems and interfaces involving hardware and software. Any error during the transmission or reception of data would lead to an incident. These were categorized as interface errors.

Another fishbone diagram was done for all the three of the categories—human errors, interface flaws, and interface errors—to further narrow down the specific causes.

At this stage of the project, the team had a shortlist of causes that resulted in incidents in online software applications of the customer. These were approved by the project review panel for a toll-gate review.

6. Improve Phase

In the Improve phase, after brainstorming, the team listed solutions to reduce the number of incidents. Next, these were then evaluated on two attributes: the ease of implementation and the extent of controllability, as shown in Figure 6B.5.

Figure 6B.5

Corrective
action
prioritization
matrix

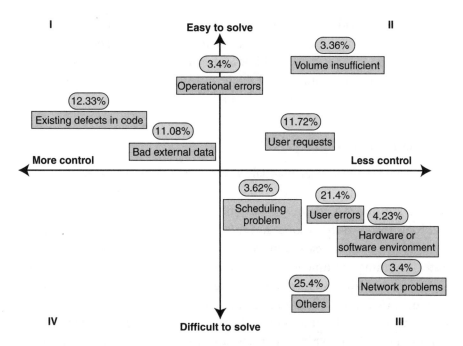

The actions in the first quadrant were relatively easy to implement and within the team's control. Existing defects in code, operational system errors, were addressed first, as they did not require customer approval. The Infosys team proactively fixed defects in code and operational system errors.

The second-quadrant actions (addressing user requests) entailed training and educating the end users on the do's and don'ts regarding incident creation. This is the concept of mistake-proofing activity. Three separate posters were created to help in training and educating end users:

▲ One poster was created for end-user identity verification. It listed the checkpoints before entering end-user details and various related attributes. The expected outcome in case the instructions were not followed was also described.

▲ A second poster for end-user rate details was created. The checkpoints were listed before entering end-user details, like calendar format, billing cycle start and end date, the base rate of interest, etc., and what the expected outcome is if these are not followed.

▲ A third poster was created to explain the details of currency conversion. This had a detailed description of attributes to be checked related to target and input currency and the correct conversion factor. The outcome if these fields were not updated was also mentioned.

These posters were distributed to various customer offices that saw a heavy influx of new employees and thus routinely had employees who were not conversant with the operational process. This was a preventive measure to reduce incident creation in the first place. Initially, the poster idea was piloted in a few regions to gauge the acceptability before being converted into a standard practice.

The third-quadrant actions involved other vendors and some upstream systems that required customer approval prior to implementation. Suggestions to optimize the scheduler were proposed to the customer. This was maintained by another vendor. The existing hardware and software had limitations, and the team suggested upgrading them to improve speed of execution.

7. Control Phase

In the Control phase, user manuals were created, with details of the error description and workaround for the end user included to prevent unnecessary incidents.

In addition, workarounds and technological solutions, which were suggested by the issues and root causes in the second and third quadrants of Figure 6B.5, were documented and formally recognized as part of the solution set. These interim solutions were useful in keeping systems running and functioning. These actions also helped reduce new incidents.

A control plan was drawn up by the team to ensure the improvement was sustainable.

The team collected data samples after piloting solutions and carried out a nonparametric statistical test on incident count (attribute data) to validate the null hypothesis. A pre- vs. post-improvement Mann-Whitney test was done to validate improvement impact on the incident count, as illustrated in Figure 6B.6.

The null hypothesis assumes no significant change in the incident count as a result of the improvement—that is, it is purely by chance that these figures are lower in the post-improvement scenario.

$$H_0: \text{Incident count}_{\text{(pre-improvement)}} = \text{Incident count}_{\text{(post-improvement)}}$$
$$H_a: \text{Incident count}_{\text{(pre-improvement)}} \neq \text{Incident count}_{\text{(post-improvement)}}$$

The p-value is 0.0041, which is less than 0.05, and hence, the null hypothesis is rejected. The solutions implemented have addressed both common and special causes, resulting in a smaller variation and mean shift. The conclusion is that the solutions implemented have been effective in incident count reduction. In fact, there was a reduction of 17.8 percent in incident count.

To ensure the initial improvements were sustainable, control charts were used to verify if there was a positive impact on incident count on an ongoing basis. This is illustrated in Figure 6B.7.

To standardize practices for incident count reduction, the team created user documentation with detailed, step-by-step instructions, along with screen shots. These were shared with the customer and integrated into end-user work instructions.

When the end users were new employees and thus not fully aware of standard operating procedures, any inadvertent error by them while performing a task could lead to an incident. Posters and user help documentation were created, which helped significantly avoid incident creation. This is a preventive measure, because in a legacy system, tracing the root cause for an incident is more challenging than the fix itself.

Figure 6B.6

Pre- and post-improvement Mann-Whitney test for incident count

Mann-Whitney Test and CI: Incident count - pre, incident count - post

```
                          N    Median
Incident count - pre      5    386.00
Incident count - post     6    317.00

Point estimate for ETA1-ETA2 is 68.50
94.6 percent CI for ETA1-ETA2 is (17, 98, 89.02)
W = 45.0
Test of ETA1 = ETA2 vs ETA1 > ETA2 is significant at 0.0041
```

Figure 6B.7

Incident count control chart

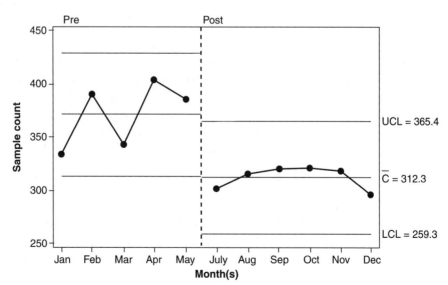

Control phase toll-gate review was conducted to verify the actual improvement in the CSF against the goal. The team then presented the project summary to an independent review panel as part of the executive review and closure.

8. Benefit Realization

The incident count and backlog volume for various transaction queues were constantly monitored by the customer's IT department. Infosys was responsible for one such queue. There was nearly an 18 percent improvement (reduction) in the incident count against the Infosys queue, reflecting a sigma level improvement from 3.5 sigma before the improvement to 4 sigma after the improvement. This significant drop in the incident volume was immediately visible to the client IT department. Consumers were delighted to see their issues resolved before they became incidents.

The solutions implemented as part of this project significantly mitigated business risk by ensuring minimal online application downtime. This helped the customer's IT department meet their SLA with consumers. Timely closure of transactions (and the related revenue) resulted in a tangible benefit of about $4 million.

9. Conclusions and Limitations

This project demonstrates that simple solutions (such as posters and user documentation), when implemented effectively, can address complex problems in legacy systems, leading to a reduction in business risk. Also, preventive maintenance is best suited to address challenges in applications with very little documentation, as it is tedious to trace the root cause and then fix it. Not for nothing is it said that prevention whereever possible is always better than cure.

The DMAIC methodology is an effective one when a disciplined approach to problem solving is required. While this project does not conform to the classical Six Sigma steps, sigma calculations, statistical tests, and so on, it demonstrates that simple and disciplined root-cause analysis is just as useful.

Although this project was successful, the same cannot be expected in each case of DMAIC implementation. Furthermore, other instances of online software applications may have varying degrees of complications, making the results of this improvement project nonrepeatable. Some of these varying degrees of complications include factors such as age of the application and platform, whether it is old or new technology, the total number of end users, and the complexity of business flow.

Acknowledgments

The authors of this case study are thankful to Infosys for its permission to allow publication of this case study as a book chapter. The authors would also like to thank Satyendra Kumar, Ramakrishnan M., Haresh Amre, Krishna Pai, and Moses R. for their review comments.

Infosys Significantly Reduces Telecom Client's Operational Expenses

Prakash Viswanathan
Anshuman Tiwari*

Relevance

This chapter describes solutions that effectively reduced operational expenses for a large telecom service provider. The challenge of operational expense reduction is common to any service provider who uses information technology (IT) software and hardware to host its services. This chapter is relevant to readers who want to reduce operational expenses. Although the project was carried out for a global telecom company, the lessons are applicable to all organizations seeking cost reductions in operational expenses.

At a Glance

▲ Infosys helped a leading telecom service provider reduce operational expenses by reducing errors in billing and delayed delivery of bills.

▲ As a direct result of improvements implemented by this project, the telecom service provider realized a 20 percent reduction in operational expenses for maintaining the customer billing application.

▲ The customer was now able to generate invoices in a timely manner, which resulted in earlier revenue realization. The annual benefit resulting from additional accrued interest on total annual revenue was estimated to be US$9 million.

---

* At the time of this project, Anshuman Tiwari was head of the Business Excellence Program at Infosys.

Executive Summary

A customer billing application is the backbone of a telecom service provider and is used to generate monthly invoices. No customer likes to receive a bill that is either beyond the payment due date or very close to it, or receive a bill with errors. Therefore, any error in billing can lead to major dissatisfaction and loss of goodwill for the company. In addition, the cost of correcting such mistakes is a cost of poor quality (COPQ). This was the predicament that drove a key customer of Infosys to drastically reduce errors in bills.

The telecom industry in the United States has been saturated for a while now. In such market conditions, companies want to improve their bottom line by reducing operational inefficiencies. Most companies have had their operations budgets reduced in the global recession. Our customer was no exception. A reduced budget had a direct impact on spending on information technology services, directly affecting Infosys.

This improvement project was carried out for a large multinational telecom company with a large, global wired and wireless network. The company has more than 100 million access lines and 30 million wireless customers. Infosys maintains the billing software for this telecom company. A key requirement of the customer was to significantly reduce any erroneous or delayed bills to subscribers, and hence, reduce operational expenses.

The objective of the improvement project was to reduce the application maintenance cost by 20 percent over a four-month timeframe. This is a critical application with a customer-specified service level agreement (SLA) for problem resolution.

A cross-functional Six Sigma team was formed. The team included a project lead, technical lead, five subject matter experts (SMEs), and a Six Sigma Black Belt. The SMEs in this team were well versed in the operational aspects of the customer billing application and business flow. The roles and responsibilities were identified in the project charter.

The project team followed the Six Sigma Define, Measure, Analyze, Improve, and Control (DMAIC) process steps. In this case, there was a clear business need to reduce the cost of maintaining the mainframe customer billing application, which is a critical success factor (CSF).

In the Define phase, the mainframe system was examined to define the improvement objective. The team discovered that the main source of errors was job abends (abnormal end). Job abends are errors that occur when multiple software applications are trying to connect to each other and exchange data or carry out a task. After system study, the team refined the problem statement.

In the Measure phase, eight months of pre-improvement abend data was collected from log files. On collating the data, the team found it tedious to track individual abends from more than 50 applications. These applications were exchanging data and information with the customer billing application. A better measure to track abends was the ratio of abends to the total number of jobs executed. This is called the abend ratio.

A detailed study of data revealed that the cost to fix an abend was directly proportional to the abend ratio. Thus, to reduce cost (operational expenses), the team needed to reduce the abend ratio. Subject matter experts suggested that reducing the abend ratio by a third would be an ambitious target. Driven by a desire to deliver benefits to the organization, the project team accepted this challenge.

In the Analysis phase, the project team established an existing process capability based on failure rate (abend ratio) against a specification limit provided by the customer. The team developed a frequency distribution by cause and found that Job Control Language (JCL) error, data exception, user abend, and file transmission errors were the key causes for the occurrence of abends. The team used the failure mode effects analysis (FMEA) tool for a detailed investigation of the causes highlighted as important in the Pareto analysis.

In the Improve phase, the team brainstormed possible solutions to address root causes for the occurrence of abends. As part of the FMEA, a Risk Priority Number (RPN) was calculated for each root cause. Key causes that had a high RPN were shortlisted for piloting improvement solutions. In order to address the key root causes, the project team proposed technical solutions, such as the setting up of test beds with exhaustive data sets and detailed test plans. The team also identified scheduling changes to be implemented while creating job books. These solutions were piloted, and early results showed that abends were reduced significantly.

In the Control phase, to sustain the gains observed during the pilot period, the team created a compendium of all minor and major changes and compiled them as a body of knowledge. The team also made some modifications to the metric report that captured additional details about abends. Statistical tests were conducted to validate the improvement to the process. This was also shared with the client.

Once implemented, the systemic solutions ensured a sustained reduction in the abend ratio, which the customer was delighted to see. Accurate customer billing invoices were sent out on time, helping in timely revenue realization. Apart from improved brand value and perception as a customer-oriented service provider, this telecom company saved over $9 million in the first year alone.

1. Introduction

Infosys provides application maintenance and support services to many customers across various industry verticals, platforms, and technologies. One such case is that of a large telecom customer providing a host of wired and wireless services to more than 130 million subscribers. As in most service companies, the customer was eager for all subscribers to receive timely and accurate bills.

More than 50 applications interfaced with the customer billing application before a subscriber's invoice could be successfully generated. The customer billing application ran as a batch job on the mainframe system. A single failure during this batch run could result in the generation of erroneous bills. These failures are called job abends. The interactions among a large number of applications resulted in a significant number of abends. Because the rectification of abends reduces errors, any reduction in abend ratio would reduce the application maintenance cost.

The client requested Infosys reduce the number of abends that were causing delays in accurate bill generation and loss of end-user goodwill. The customer attached a high significance to the number of abends and abend ratio, as this affected its business revenues.

Mainframe applications are legacy systems and pose challenges in terms of design and operational documentation. As a number of applications exchanged information with this customer's billing application, it was a huge challenge to trace the root cause for abends and find a solution. The team used the Six Sigma DMAIC approach to address this challenge of *known problem* and *unknown solution*.

2. Project Background

This telecom service provider had a business challenge to reduce its operational expenses and increase profitability. Operational expenses included dollars spent to fix abends; the more abends, the greater the operational expense. This affected the customer's margins. Thus, a reduction in abends would improve margins for the customer. A senior leader from Infosys identified this as an improvement opportunity. A business case was created to estimate the return on investment (ROI) and feasibility. A project charter was then created, and all the roles and responsibilities of the cross-functional team were detailed.

3. Define Phase

Once the opportunity was identified, the team set about to define the problem properly.
Key activities in the Define phase were:

▲ Identification of the *project critical success factor (CSF)*
▲ Creation, review, and finalization of the *project charter*
▲ Creation of a *high-level process map*

3.1 Identification of the Project Critical Success Factor

Clarity regarding a Six Sigma project's CSF helps to streamline and focus on project activities. In this case, the customer expectation was clear—reduction in abend ratio. This reduction would help reduce expenses and improve margins.

3.2 Creation of Project Charter

A project charter binds a cross-functional team to stay focused on a common improvement objective to ensure project success.
A typical project charter includes:

▲ Business case

▲ Problem statement
▲ Improvement goal statement
▲ Roles and responsibilities
▲ Project scope
▲ High-level project plan

The project team included a sponsor (part of Infosys senior management), Six Sigma Champion (Master Black Belt), project leader (PM), quality consultant (SQA), Six Sigma practitioner (Six Sigma Black Belt) and several subject matter experts (SMEs). The project leader developed the charter, which was reviewed and approved by the sponsor. This ensured that leadership was committed to this improvement project.

3.2.1 Business Case

A business case helps objectively evaluate the improvement opportunity. The sponsor provided a high-level business need, and the team prepared a detailed business case.

The nature of the project required the project team to interact with various teams within Infosys and the customer's organization. This required a dedicated cross-functional Six Sigma team to drive this improvement.

The key elements of the business case for this project were:

▲ This project was carried out for one of Infosys's key customers.
▲ The customer billing application affected the customer's margins, and it was important to ensure proper and timely billing.
▲ Application errors were related to high abend ratio.
▲ This complex business application was hosted on a mainframe and had interfaces with more than 50 other applications. It was a technological challenge for the maintenance team to ensure subscriber bills were accurate and sent on time from this critical customer billing application.

This improvement had the potential to affect the client operational cost by nearly $9 million annually.

3.2.2 Problem Statement

As abends were causing disruptions to this large business application, an abend ratio reduction was necessary. The business improvement goal (Y) was identified as operational expense, and the subimprovement goal (y) for this improvement project was abend ratio.

The benefits of reducing the abend ratio would be:

▲ The customer's end users (service subscribers) would receive their bills on time.
▲ The bills would be accurate.
▲ There would be a reduction of the customer's operational expenses for the customer billing application.

3.2.3 Improvement Goal Statement

The goal was to reduce the abend ratio by 25 percent from existing levels.

Infosys uses the SMART goal principle in its Six Sigma projects. The SMEs associated with this project agreed that 25 percent would be a stretch, yet was a possible target. The team decided on a four-month duration for the improvement project.

3.2.4 Project Roles and Responsibilities

A section in the project charter details team roles and responsibilities (see Table 6C.1).

Table 6C.1 Project Roles and Responsibilities

Role	Responsibilities
Sponsor	Provide direction and monitor the progress
Mentor (typically a Six Sigma MBB)	Mentoring and guidance on the project and Six Sigma methodology
Quality Manager (QM)	To review the progress of process implementation
	To sort out issues (if any) that are to be resolved at the senior level
	To provide guidance and directions to make it successful
	Participate in Analysis phase
SME	To provide the strategies
	Coordinate the implementation
	To clarify issues raised by the project folks in their area
Practitioner (Six Sigma BB)	Deployment
	To conduct and coordinate the initiative
	To clarify doubts and issues raised by the SQA's anchors
	To provide the status of the initiative
	To help quality anchor do the statistical analysis of data
Project Lead (PM)	To provide the solutions
	To implement the solutions
	To provide the required metrics
Software Process Consultant (SQA)	To help the improvement team in data analysis
	Pilot and deployment of improvements
	To plan and execute the project as per plan
	Conduct statistical analysis in help with practitioner
	Update project status to practitioner and in the online system

3.2.5 Project Scope

Infosys was responsible for maintaining the entire customer billing application and all interfaces with other applications hosted on a mainframe. The entire customer billing application maintained by Infosys was in scope for improvement. All interfacing applications were not in the scope of this improvement project, as Infosys did not have access to them. This boundary identification ensured that the team looked at only feasible improvement areas.

3.2.6 High-Level Project Plan

The overall start and end dates, along with the improvement stages, were included in a high-level project plan. A summarized plan is shown in Table 6C.2. Training requirements (if any) were also identified to complete the project charter.

Table 6C.2 Sample Project Plan

SN	Phase	Activities	Team	Deliverables	Estimated Completion Date
1	Define	Define objective and scope of initiative (business case, team creation, kick-off)	Project Manager/ Software Process Consultant (SQA)/ Sponsor/MBB/Six Sigma Practitioner (BB)	Project charter as-is process map	1-Mar
2	Measure	Identify the measures (data collection)	Project Manager/ Software Process Consultant (SQA)/Six Sigma Practitioner (BB)	Baseline data	21-Mar
3	Analyze	Root-cause analysis of the critical Xs for CSF	Project Manager/ Software Process Consultant (SQA)/Six Sigma Practitioner (BB)/SME	Analysis report	10-Apr
4	Improve	Identify and prioritize improvement strategies and pilot	Project Manager/ Software Process Consultant (SQA)/Six Sigma Practitioner (BB)/SME	Implementation plan	15-May
5	Control	Implement strategies (sustainability and standardization)	Project Manager/ Software Process Consultant (SQA)/Six Sigma Practitioner (BB)/SME/MBB/Sponsor	Control charts	30-Aug

3.3 As-Is Process Map

The next step in the Define phase is developing an as-is process map. Such a map helps in understanding the current and real state of the process. At times, a process map also reveals likely pitfalls. In this case, the process map helped understand the reasons for frequent failures of batch jobs by clearly identifying the interactions between different applications (including

Figure 6C.1

High-level process map for abends.

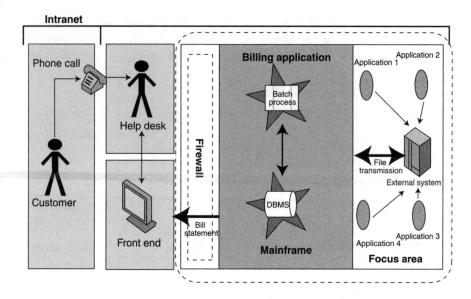

the customer billing application) and the database. The team used this map to identify bottlenecks in the process that were leading to frequent abends.

A generic flow is illustrated in Figure 6C.1.

The customer billing application is a batch job that has multiple interfaces. This application has a front-end and back-end. In order to have a reasonable scope, only the area in the dotted boundary was considered for this improvement project. The as-is map revealed an important fact—more than 50 other applications with interfaces to the customer billing application were hosted on an external server. In the event of a failure in any of the external interfaces, the batch jobs would terminate.

Armed with a charter and as-is map, the team was set for the Measure phase.

4. Measure Phase

In the Define phase, the team identified the abend ratio as the key improvement metric. The customer tracked the abend ratio on a monthly basis. Infosys maintained logs, which became the source for all previous abend history. These logs were a lucky break for the project team, as it was readily available data in Infosys's custody. Without such data, the team may have had to set up a data collection drive for about two months, further delaying the project.

The improvement team collated incident (abend cases) history data for eight months. The customer incurred a fixed cost per abend (payable to Infosys for fixing these abends), which was directly linked to the abend ratio. Any abend ratio reduction directly translated to an operational expense reduction. Although the customer was seeking 20 percent reduction, the project team established an aggressive target of 25 percent reduction in the abend ratio.

Figure 6C.2

Pre-improvement abend ratio process capability.

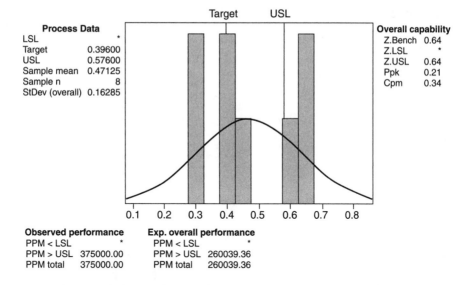

As a next step in the Measure phase, the team calculated the process capability for the abend ratio. The customer wanted 0 and 0.57 percent as the lower and upper specification limits for the abend ratio. The chart in Figure 6C.2 shows that while the sample mean was 0.47, the process was highly dispersed.

For a process to operate at Six Sigma level, Cpk should be 2. In this project, the Z score (equivalent to Cpk) was only 0.64. This indicated a significant scope for improving the process capability.

In summary, in this phase, the Six Sigma project team collected abend data, established process capability, and reviewed the improvement goal. After a thorough review by the sponsor, · the team received the go-ahead for the Analyze phase.

5. Analyze Phase

As a part of the Analyze phase, the team identified *the sources/causes for variation in the CSF*. Having analyzed the current state of the process, the team carried out a detailed root-cause analysis to identify the causes of high abend ratio using a Pareto chart, as illustrated in Figure 6C.3.

Abends due to Job Control Language (JCL), data exception, users, and file transmission errors contributed to about 70 percent of the causes. Other errors related to DB2 (mainframe database), unavailability of resources, and manual interruptions constituted the rest.

It was evident that abends were due to multiple causes. The team took up the challenge to investigate these causes and identify solutions for the same.

The team used the failure mode effects analysis (FMEA) tool to analyze each failure mode for severity (SEV), occurrence (OCC) of the cause, and detectability (DET), with current control mechanisms for the customer billing application. Weights were assigned on a scale of 1 to 10, 1 being the lowest and 10 the highest for severity and occurrence. The scale was reversed

Figure 6C.3

Pareto chart of causes for abends.

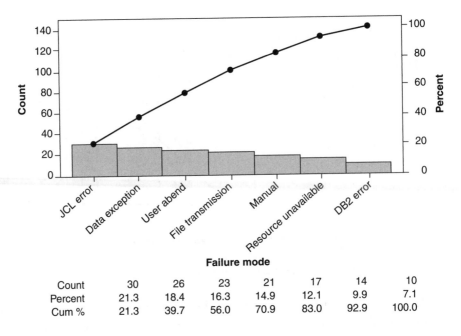

Count	30	26	23	21	17	14	10
Percent	21.3	18.4	16.3	14.9	12.1	9.9	7.1
Cum %	21.3	39.7	56.0	70.9	83.0	92.9	100.0

for detectability (10 to 1). The Risk Priority Number (RPN) was computed as the product of severity, occurrence, and detectability (see Table 6C.3).

Note: A custom rubric for the rating scale was created. A severity rating of 9 in most industries means a life-threatening situation. In this case, a 10 meant total system failure.

Table 6C.3 FMEA Sample to Investigate Potential Failure Modes

Potential Failure Mode	Potential Failure Effects	SEV	Potential Causes	OCC	Current Controls	DET	RPN
JCL Error	Abends	9	Job not refreshed after implementation	8	E-mail notification of JCL refresh	7	504
User Abend	Abends	8	Out of order sequence number	7	1. Test few override files 2. Manual review	8	448
File Transmission	Abends	7	FTP coding error	6	Code review	8	336
Resource Unavailable	Abends	7	Space problem	8	None	7	392
Manual	Abends	7	Coding error	8	Code review and independent testing	7	392
Data Exception	Abends	8	Coding error	8	Code review and independent testing	6	384
DB2 Error	Abends	7	Bindcntl not moved/ implemented	6	None	7	294

In the first iteration to prioritize failure modes, the team decided to focus on modes that had an RPN of more than 350. These modes were:

▲ **JCL Errors:** Occurred due to jobs not being refreshed after changes were made
▲ **User Abend:** Occurred due to any incorrect user input that could lead to a delay in job processing
▲ **Resource Unavailable:** Occurred due to errors resulting from insufficient space to hold data
▲ **Manual and Data Exception:** Occurred due to errors in code

In the second iteration, the team considered failure modes with RPN less than 350. These were file transmission and DB2 errors.

▲ **File transmission errors:** Occurred due to a failure in transmitting or receiving files across interfaces
▲ **DB2 error:** Occurred due to errors during code preparation

The FMEA confirmed the findings of the Pareto analysis (Figure 6C.3) done earlier in the project. Using these two analyses, the team shortlisted the following as the key causes of abends:

▲ Jobs not refreshed after modifications
▲ Space problems
▲ Coding errors

The project review panel reviewed the team's progress and approved completion of the Analyze phase. Such review is essential to ensure that the project is on the right track from a methodology perspective, as well as on course to achieve project goals and deliver anticipated benefits.

To summarize, in this phase, the team identified the most frequently occurring causes leading to abends. As this was a large batch application with multiple interfaces, a cross-functional team of experts was called in to execute a detailed FMEA to capture all possible types of failures and estimate the impact and severity of failures on the customer billing application. The team went through multiple iterations before arriving at the final FMEA and identification of the additional improvement actions.

6. Improve Phase

Armed with clarity on root causes, the Six Sigma project team entered the Improve phase.

During the Improve phase, the project team brainstormed on possible solutions to reduce the abend ratio. The following extended FMEA table (Table 6C.4) provides key actions taken to address root causes for the occurrence of abends.

As a next step, the project team identified individuals responsible for piloting solutions of key failure modes. Such identification helped the team follow up and close all action items

Table 6C.4 Corrective Action Prioritization Matrix

Failure Mode	Cause	RPN	Action Plan	Responsibility
JCL Error	Scheduler not set up correctly	**504**	1. Release instructions should include jobs to be refreshed for the release. 2. Verify jobs after implementation to ensure JCL refresh.	Module lead
User Abend	Out of order sequence number	**448**	1. Override files need to be tested before implementing in production. 2. Any last-minute modifications need to be tested again before implementing in production.	Module lead/ PM (project lead)
DB2 Error	Bindcntl not moved/ implemented	294	1. Move Bindcntl also when setting up elements in Endeavor. 2. Verify Endeavor jobs for successful bind.	Module lead
Manual	Coding error	**392**	1. Code review checklists to be strictly followed. 2. Test with more volume of data. 3. Test plan needs to ensure code coverage.	Module lead/ PM (project lead)
File Transmission	FTP coding error	336	1. Test FTP job by sending sample file to interface system.	Module lead
Data Exception	Coding error	**384**	1. Set up test bed with more volume of data. 2. Test plan needs to be exhaustive to test all code changes. 3. Final testing pointing to ACC stage in Endeavor.	Module lead

within schedule (three to four weeks). The team monitored the outcome from the pilot study and accepted all solutions for further deployment.

Having tasted success with these pilot implementations, the team was confident of moving to the Control phase, where they would have an opportunity to implement solutions in the live environment. Before implementing these solutions, however, they were shared with the customer. After due approvals from the project review panel and customer, the team was ready for the Control phase.

7. Control Phase

In the Control phase, the team implemented solutions (after successful pilots) to reduce the number of job abends.

In order to statistically verify the improvements, the team gathered new abends data after the implementation of corrective actions and recomputed the abend ratio. Abend ratio is continuous data; hence, a pre- vs. post-improvement two-sample t-test was done to quantify the improvement and validate it (see Figure 6C.4).

The null and alternate hypotheses were as follows:

H_0: Abend ratio (pre-improvement) = Abend ratio (post-improvement)
H_a: Abend ratio (pre-improvement) ≠ Abend ratio (post-improvement)

Figure 6C.4

Pre- and post-improvement two-sample t-test for abend ratio.

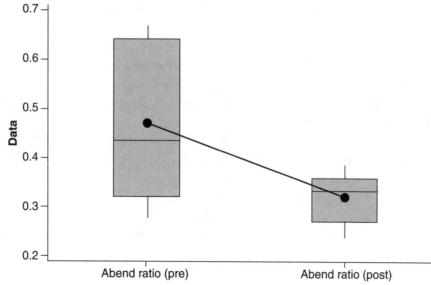

Two-sample t-test and CI: abend ratio (pre), abend ratio (post)

```
Two-sample t-test for abend ratio (pre) vs abend ratio (post)

                        N      Mean      StDev     SE Mean
Abend ratio (pre)       8     0.471      0.157      0.056
Abend ratio (post)      6     0.0542     0.022

Difference = au (abend ratio (pre)) = au (abend ratio (post))
Estimate for difference: 0.149583
95% CI for difference: (0.014293, 0.284873)
T-test of difference: 0 (vs not =): T-value = 2.50
P-value = 0.034    DF = 9
```

Figure 6C.5

Incident count control chart.

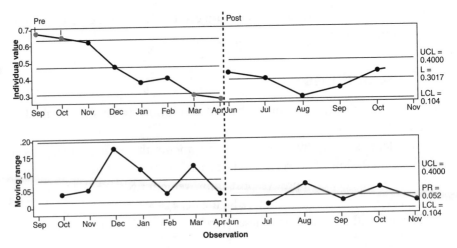

The null hypothesis is that it is purely by chance that the abend ratio is lower in the post-improvement scenario. The p-value is 0.034, which is less than 0.05, and hence, the null hypothesis is rejected. The conclusion is that due to the solutions implemented by the project team, there is a significant reduction in the abend ratio post-improvement by 31.7 percent, exceeding the goal of 25 percent.

The team tracked and monitored the abend ratio after preventive actions were implemented. Control charts were used to verify if there was a positive impact on the abend ratio. The improvements led to a shift in the abend ratio mean, and there was a reduction in the variation as well. The process control band had narrowed along with a shift in the mean. This is illustrated in Figure 6C.5.

To standardize the reduction in abend ratio, the team documented standard operating procedures that addressed the various root causes identified during the course of this project.

Next, the team verified the actual improvement in abend ratio against the goal and presented the same for review by the project review panel.

To summarize, in this phase, the Six Sigma project team validated improvements, executed pilots, and deployed solutions to prevent job abends from occurring. Of these, some solutions resulted in changes to the software application to ensure prevention.

8. Benefit Realization

The reduction in abends and operational expense was a win-win proposition for the customer and Infosys. The failure rate (abend) reduction directly affected the customer's cost of maintaining this billing application, in other words, operational expenses (reduction of 20 percent). This is because the customer was now able to generate invoices in a timely manner, which resulted in earlier revenue realization. The annual benefit resulting from additional accrued interest on total annual revenue was estimated to be US$9 million.

The reduction in abends also had a positive effect on the client's end customers (subscribers), as they began receiving accurate bills on time. As a result, this enhanced subscriber retention and avoided erosion of revenue.

By implementing changes to the customer billing application, Infosys provided a sustainable solution to address all root causes for job abends. This improvement project demonstrated Infosys's commitment to improving its customer's processes. The customer benefited from improved performance, lower errors, higher goodwill, and increased end-user retention.

9. Conclusions and Limitations

This improvement project demonstrates the power of Six Sigma in improving the processes of our customers. Typically, Six Sigma is presented as a tool to drive internal efficiencies only. At Infosys, we actively use Six Sigma to improve the processes of our customers as well. In this case, the process improvement directly affected the operational expense of the customer.

This project also reinforces the importance of Six Sigma in processes where the number of transactions is very high and service level agreements are stringent. Any improvement in such context is always beneficial to the service provider and also increases confidence in Six Sigma as a process improvement methodology.

Furthermore, such improvement projects reestablish the fact that diligent and sincere implementation of the Six Sigma methodology is more important than high-end statistical analysis. Also, the power of involving people from the floor should never be underestimated. In this project, Infosys was able to improve work culture in the team that was supporting the customer process. This was an unexpected and welcome benefit.

This project was acknowledged by the customer as useful and timely. The customer was under pressure in the marketplace, since most other large telecom players were operating with a lower operational expense. This project helped the customer reduce operational expense while ensuring that there was no adverse impact on its service level agreements (SLAs) with the subscribers.

Like all improvement projects, this case study is context-specific, and it may not be possible to achieve the same improvement in all DMAIC projects. The improvement would depend on factors such as the number of interfaces with other applications, the database structure, and the complexity of business flow.

Acknowledgments

The authors of this case study are thankful to Infosys for its permission to allow publication of this case study as a book chapter. The author would also like to thank Satyendra Kumar, Ramakrishnan M., Haresh Amre, Krishna Pai, and Moses R. for their review comments.

CHAPTER 7

Productivity Improvement

CHAPTER 7A

TCS Improves Fraud Detection for a Global Bank

Asheesh Chopra

TATA CONSULTANCY SERVICES

Relevance

This case study is relevant to organizations with a high volume of transactions and low throughput rate. This chapter describes how an organization can optimize its processes by applying Six Sigma DMAIC (Define, Measure, Analyze, Improve, and Control) methodology.

At a Glance

▲ Deposit Fraud Investigations is one of many fraud detection processes in the banking operations domain. The Deposit Fraud Investigation team of a global bank was unable to investigate approximately 25 percent of the incoming fraud alerts. These cases were purged, thereby increasing the possibility of fraud loss. Tata Consultancy Services Banking and Financial Services Business Process Outsourcing (TCS BFS BPO) applied Six Sigma to increase the process throughput, resulting in significant fraud savings.

▲ The project resulted in an increase in productivity by 27.8 percent.

▲ The project resulted in fraud avoidance saving of $3.68 million and four FTE (full-time equivalent) savings of $98,841.

Executive Summary

Deposit Fraud Investigations is one of the many fraud detection processes in banking. The responsibility to execute this process for a global bank was migrated from one vendor to TCS BFS BPO in September 2006. Fraud detection processes and resources were funded by a fraud prevention budget. The bank's fraud prevention budget, however, permitted use of only a certain number of resources for fraud detection.

Due to budget constraints, the staffing was falling short, and this led to a significant percentage of unprocessed cases. Unprocessed cases were not left pending for the next day, instead they were purged out from the queue the same day. The direct impact of not reviewing deposits was exposing the bank to severe fraud losses. With these limited resources, they were able to review/process only a portion of the total deposit alerts (transactions).

There was an urgent business need to improve the productivity of the team, thereby increasing the number of cases reviewed per person per day and consequently minimizing the bank's exposure to fraud loss.

The key stakeholders from TCS decided to initiate a Six Sigma project to increase the productivity or throughput rate. This project was perfectly aligned with the strategic imperative of the business.

The goal of the project was to increase productivity by 12 percent and reduce the referral ratio by 27 percent within four months. The project started in January 2008 and completed in May 2008.

Root-cause analysis performed during the project helped identify the most significant causes for suboptimal productivity. The team then brainstormed to determine potential solutions for each of the verified root causes.

Post-implementation, an effective control plan was put in place to ensure sustenance of improvements. Thereafter, the primary metric (productivity and referral percentage) was monitored to ensure sustenance of improvements achieved.

This Six Sigma project resulted in a productivity improvement of 27.81 percent and a referral ratio reduction of 29.48 percent. In addition, there was a four FTE saving, resulting in a net financial benefit of $98,841. Actual loss avoidance (fraud savings) amounted to $11.25 million in 2008 (the initial projection was $7.57 million), and this project success also led to a new business opportunity with the client.

1. Introduction

A door-to-door salesman sells a set of encyclopedias for $69.99. The customer pays by check, writing $69.99 to the far right on the line for the amount in figures, and the words "sixty-nine and 99/100" to the far right of the amount in the text line. The criminal uses the blank spaces on both lines to alter the check by adding "9" before the numbers line, and the words "Nine Hundred" before the text line. The $69.99 check is now a fraudulent check for $969.99, which the criminal cashes.[1]

[1] These examples were prepared by the Check Fraud Working Group, a subgroup of the interagency Bank Fraud Working Group. That working group includes representatives from the Federal Bureau of Investigation, the Department of Justice, Federal Deposit Insurance Corporation, Federal Reserve Board, Internal Revenue Service, Office of the Comptroller of the Currency, Office of Thrift Supervision, U.S. Postal Inspection Service, National Credit Union Administration, and U.S. Secret Service.

A financial institution insider identifies corporate accounts that maintain large balances, steals genuine corporate checks, counterfeits them, and returns the valid checks to the financial institution. The financial institution insider is associated with a group of criminals that distributes the counterfeit checks throughout the area and cashes them using fictitious accounts.[2]

A fraud ring provides "role players" with business checks drawn on closed accounts at a financial institution. The "role players" deposit the checks into a new account at a different financial institution through one or more ATMs operated by other financial institutions. The float time between the ATM deposits and the checks drawn on the closed accounts reaching the issuing financial institution for payment allows the criminals to withdraw funds from the new account.[3]

These events are everyday occurrences—thousands of times a day, actually. Banks in the United States process 36.7 billion checks each year with a total value of $39.3 trillion, according to the 2004 Federal Reserve Payments Study, the most recent survey available.[4]

According to the National Check Fraud Center in Charleston, South Carolina, bank fraud alone is a $10 billion a year problem. This is nearly 15 times the $65 million taken in bank robberies annually. The value or the number of fraudulent checks or deposits made is not consistent across various reporting agencies. However, the number is huge, and it is growing. The United States is most vulnerable when it comes to fraud attacks. Needless to say, with the rising losses due to fraud, it is becoming imperative for banks to invest significantly in loss detection and prevention activities.

The Concise Oxford Dictionary defines fraud as "criminal deception; the use of false representations to gain an unjust advantage." Fraud is as old as humanity itself, and can take an unlimited variety of forms. However, in recent years, the development of new technologies (which have made it easier for us to communicate and helped increase our spending power) has significantly increased the ways in which criminals may commit fraud.

As fraud attempts grow in both number and variety, financial institutions are challenged with the need for comprehensive, yet cost-effective, risk management solutions. These fraudulent or suspicious financial transactions can be identified, characterized, and red-flagged in real time providing vital information to reduce their occurrences. For example, a check deposit followed almost immediately by a cash withdrawal would be a suspicious activity and warrant a red flag to check the customer's motives.

TCS BFS BPO holds a premier position as the largest financial services BPO in India. It brings with it a strong financial services domain expertise and the ability to provide transaction processing and voice and analytics services to the complete suite of financial products in the corporate, consumer, and private banking domains.

---

[2] See note 1.
[3] See note 1.
[4] 2006 ABA Issue Summary Check Processing Facts.

The client is one of the largest financial services company in the world. It has outsourced end-to-end banking services to TCS BFS BPO.

The client process involves investigating potentially fraudulent deposits and refers (referrals) any fraudulent activity to the bank for a second-level review. The intent is to prevent loss to the bank by identifying fraudulent deposits.

2. Project Background

Fraud detection processes and resources are funded by a fraud prevention budget of the client, which permitted them to use only a certain number of resources for fraud detection. With these limited resources, they were able to review and process only a portion of the total deposit alerts (transactions).

Deposits in this case refer to all credits received by the account holder. They are of two types: check deposits and noncheck deposits (cash deposits in ATMs, salary credits, maturity of fixed deposits, any transfers, etc). TCS BFS BPO is responsible for performing a complete review of all deposits that are transferred to their workflow by the client.

In case of a check deposit, the analyst reviewing the deposit looks for suspicious red flags (suspicious activities or elements that are found in fraudulent transaction) by reviewing customer account information and similar comparable checks (the check to be reviewed is compared with a similar check of the same check series, issuer, and issuing bank). In the case of a noncheck deposit, customer account information has to be reviewed and the source of the credit has to be identified. In case red flags are identified or the source of the deposit is not found, then suitable action is taken in order to prevent the potential fraudulent transaction.

The entire workflow is split into two parts: TCS BFS BPO and the client. TCS BFS BPO works on the deposits in descending order of value. The team investigates potentially fraudulent deposits and refers any fraudulent activity to the client for a second-level review.

Due to budget constraints, the staffing was falling short, and this led to a significant percentage of unprocessed cases. These cases were not left pending for the next day, however, as they were purged from the queue the same day. The direct impact of not reviewing deposits was exposing the client to severe fraud losses.

There was an urgent business need to improve productivity of the team, thereby increasing the number of cases reviewed per person per day and consequently minimizing the exposure to fraud loss.

3. Define Phase

At the start of the Define phase, a Kano model survey was administered to the client stakeholders, and it revealed that more cases processed (increase in productivity) while reducing referrals (decrease in referral percentage) would be a definite "delighter" to the client.

The Kano model is a useful technique that helped us determine what will truly delight the customer. It divides service attributes into three categories: basic, performance, and excitement (see Figure 7A.1).

Basic attributes are the expected attributes, or "musts," of a service. Increasing the performance of these attributes provides fewer or no returns in terms of customer satisfaction; however, the absence or poor performance of these attributes results in extreme customer dissatisfaction. An example of a basic attribute would be cup holders in a car.

Performance attributes are those for which more is generally better, and will improve customer satisfaction. Conversely, an absent or weak performance attribute reduces customer satisfaction. For example, engine horsepower or acceleration (0 to 60 seconds).

Excitement attributes are unspoken and unexpected by customers but can result in high levels of customer satisfaction; however, their absence does not lead to dissatisfaction—for example, a car navigation system.

A pair of customer requirement questions in the Kano questionnaire is illustrated in Figure 7A.2.

Customer needs are evaluated based on a functional and dysfunctional approach. The aim of this approach is to closely verify the customer's feeling about the new proposal. For each need, a pair of questions is asked: a functional question and a dysfunctional question.

Customer preferences are classified into five categories:

Figure 7A.1

Kano model.

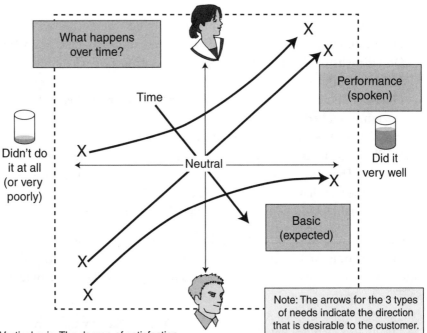

Vertical axis: The degree of satisfaction.
Horizontal axis: The degree to which the requirements has been fulfilled or achieved.
The "expanded" Kano Model is adapted from Noriaki Kano's original work.

Figure 7A.2

Functional and dysfunctional form of the question.

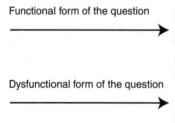

Functional form of the question

If the referral ratio was brought below, how would you feel?	1. I like it that way	X
	2. It must be that way	
	3. I am neutral	
	4. I can live with it that way	
	5. I dislike it that way	

Dysfunctional form of the question

If the referral ratio was left as it is, how would you feel?	1. I like it that way	
	2. It must be that way	
	3. I am neutral	X
	4. I can live with it that way	
	5. I dislike it that way	

If the volume worked per person per day was more, how would you feel?	1. I like it that way	X
	2. It must be that way	
	3. I am neutral	
	4. I can live with it that way	
	5. I dislike it that way	

If the volume worked per person per day was left as it is, how would you feel?	1. I like it that way	
	2. It must be that way	
	3. I am neutral	X
	4. I can live with it that way	
	5. I dislike it that way	

▲ **Attractive:** These attributes provide satisfaction when achieved fully, but do not cause dissatisfaction when not fulfilled. These are attributes that are not normally expected.

▲ **One-dimensional:** These attributes result in satisfaction when fulfilled and dissatisfaction when not fulfilled.

▲ **Must-be:** These attributes are taken for granted when fulfilled but result in dissatisfaction when not fulfilled.

▲ **Indifferent:** These attributes refer to aspects that are neither good nor bad, and they do not result in either customer satisfaction or customer dissatisfaction.

▲ **Reverse:** These attributes refer to a high degree of achievement resulting in dissatisfaction and to the fact that not all customers are alike.

▲ **Questionable:** Questionable results signifying that the question was phrased incorrectly.

Customer responses for each requirement are then tabulated as shown in Figure 7A.3.

The customer requirement was translated into specific critical to quality (CTQ) requirements. In this case, productivity, measured by the Daily Productivity Rate, was identified as CTQ.

$$\text{Daily Productivity Rate (DPR)} = \frac{\text{Cases worked by the team}}{\text{Total number of production hours}} \times 8 \text{ hours}$$

Figure 7A.3

Kano
evaluation
table.

		Dysfunctional				
	Customer requirements	1. I like it that way	2. It must be that way	3. I am neutral	4. I can live with it that way	5. I dislike it that way
Functional	1. I like it that way	Q	A	A	A	O
	2. It must be that way	R	I	I	I	M
	3. I am neutral	R	I	I	I	M
	4. I can live with it that way	R	I	I	I	M
	5. I dislike it that way	R	R	R	R	Q

A = attractive; M = must be; R = reverse; O = one dimensional; Q = questionable; I = indifferent

Process requirements					(A + O) / (A + O + M + I)	(O + M) / (A + O + M + I)
	A	M	O	I	Satisfaction	Dissatisfaction
Volume per person per day increased	1				1	0
Referral ratio lowered	1				1	0

The key stakeholders from TCS decided to initiate a Six Sigma project to increase the daily productivity rate or throughput rate. This project was perfectly aligned with the strategic imperative of the business.

After obtaining approval from the project sponsor, the project Champion was identified and assigned to the project. A Champion is a key business leader responsible for ensuring execution of changes, securing resources, maintaining sponsor support, and working closely with the Six Sigma project lead

The project Champion then identified a Six Sigma resource to lead the project.

The project lead then put together a draft project charter and a detailed project plan.

3.1 Six Sigma Project Charter

The project charter consisted of the elements discussed in the following sections.

3.1.1 Business Case Statement

The business case statement provides the broad background of the process and defines why the project was selected (from the customer's point of view), along with need for taking up the project.

The key elements of the business case for this project were the following:

1. Deposit Fraud Investigation team investigated potentially fraudulent deposits and referred any fraudulent activity to the client for a second-level review (referrals).

2. Though meeting the service level agreement (SLA) with the client (daily productivity rate as per SLA was being met), the Deposit Fraud Investigation team was unable to complete 25 percent of the fraud alerts in the queue. These cases were purged out without being reviewed in favor of more recent cases that arrived, thereby increasing the possibility of fraud loss.
3. There was an urgent business need to increase the productivity of the team (cases reviewed per person per day), thereby minimizing the exposure to potential fraud loss.

3.1.2 Opportunity Statement

The Deposit Fraud Investigation team (TCS team and client) was unable to complete approximately 25 percent of the fraud alerts in the queue. These cases are purged out every day without being reviewed, thereby increasing the possibility of fraud loss.

The CTQ parameters are identified as follows:

CTQ 1: Daily Productivity Rate (DPR) or Throughput rate

$$\text{Daily Productivity Rate (DPR)} = \frac{\text{Cases worked by the team}}{\text{Total number of production hours}} \times 8 \text{ hours}$$

Increase in throughput rate would certainly result in significant fraud avoidance savings. In addition, FTE savings would further result in internal cost savings.

CTQ 2: Referral ratio

$$\text{Referral ratio} = \frac{\text{Number of cases referred}}{\text{Total number of cases worked}}$$

Reduction in the number of referrals would allow client resources to focus on investigating high-value deposits, thereby minimizing the exposure to fraud loss.

Note: The client team investigates deposits that are of high value. Any case referred to the client for second-level review affects the client's ability to process more fresh cases (high-value deposits). Therefore, minimizing referrals would help in overall reduction of unprocessed high-value cases (see Figure 7A.4).

3.1.3 Goal Statement

To increase productivity by 12 percent (from 89 cases reviewed per person per day to 100 cases) and reduce the referral ratio from 15 percent to 11 percent (27 percent decrease) within four months.

Figure 7A.4

Case distribution.

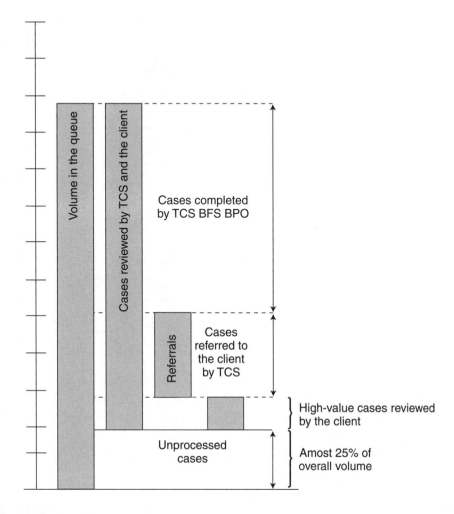

3.1.4 Project Scope

The scope included all processes under Deposit Fraud Investigations.

The SIPOC (Supplier-Input-Process-Output-Customer) shown in Figure 7A.5 illustrates the flow of activities at a macro level.

3.1.5 Project Plan

The high-level project plan included in the charter included milestone dates for the end of each of the phases in the DMAIC lifecycle of the project.

3.1.6 Project Team

The project team was composed of the following members from the operations and quality team:

▲ One Black Belt

Figure 7A.5

SIPOC.

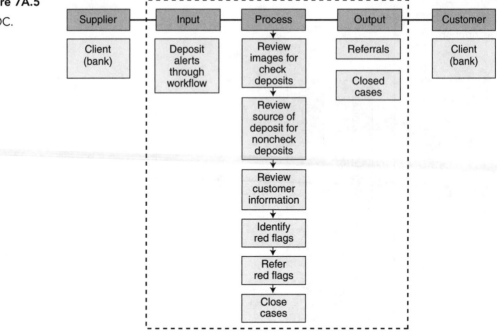

▲ One Green Belt
▲ One Master Black Belt who was to mentor the Black Belt and Green Belt
▲ Five subject matter experts drawn from various functional areas in the scope of the project
▲ The governance team for the project comprised:
▲ Project sponsor, who is the head of retail banking operations
▲ Project Champion, who is the unit head
▲ One key stakeholder from the client side
▲ Head of the Lean Six Sigma practice (process re-engineering group) at TCS BFS

4. Measure Phase

The Measure phase started with a structured data collection plan specifying what type of data was to be collected, how to ensure data consistency, and how to display the data. Three months of productivity data from October 2007 to December 2007 was collected.

At TCS, there is a strong emphasis on a structured and robust data collection plan.

One of the most important things in planning for data collection was to draw and label the graph that would communicate the findings properly before the collection process began (see Table 7A.1). In order to be able to effectively collect data, the team had to know what they were trying to illustrate. In addition, it raised questions that they may not have considered and needed to add to the plan.

Table 7A.1 Data Collection Plan

	Define to what measure			Define how to measure	Who will do it?	Sample plan			
Measure	Type of measure	Operational definition		Data collection method	Person(s) assigned	What?	Where?	When?	How Many
Daily productivity rate	Continous, process data	Cases worked by the team/total number of hours: 8 hours		Collected from MS	Black Belt	Cases worked per person per per 8 hours	Daily MS	EOD	100%
Referrals	Continuous, process data	Cases referred by the team/total number of cases worked		Collected from MS	Green Belt	Cases referred/ volume worked	Daily MS	EOD	100%
Referral feedback	Discrete, process data	Feedback received on the number of cases referred to the client		Data collected from the client	Green Belt	Feedback received from the client	Feedback tracker	Next day	100%
Case type	Discrete, process data	Type of cases: deposit types		Data collected from case tracker	Green Belt	Case type: Check or noncheck deposit	Case tracker	Oct–Dec 2007	100%

Figure 7A.6

Control chart—
Pre-
Improvement
phase—Daily
productivity
rate.

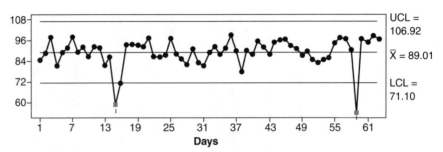

Note: The x-axis represents the days and daily productivity rate is plotted on the y-axis. The two outliers were days that were affected by system outages.

A control chart was plotted for daily productivity data to study both common-cause and special-cause variations in the process (see Figure 7A.6).

Special-cause variation means that something different has occurred at a certain time or place within the process. Common-cause variation is always present to some degree in the process. The goal is to minimize the special-cause variation and therefore, it is important to distinguish between special- and common-cause variation.

Table 7A.2 provides a quick snapshot of three months of data.

▲ Daily productivity rate averaged at 89.01 for October to December 2007.
▲ Referral ratio is not stable, but the average for October to Decembr 2007 works out to be 15.51 percent.
▲ For October to December 2007, check deposit cases were 51 percent, whereas noncheck deposit cases were 49 percent. Figure 7A.7 illustrates the distribution across the same period.

Table 7A.2 Historical Data

Month	Oct '07	Nov '07	Dec '07
(A) Number of cases reviewed	16985	16113	14332
(B) Number of referrals	2666	2753	1937
Referral ratio (A/B)	15.70%	17.09%	13.52%
Working days	22	21	20
Hours worked	1528.26	1450.39	1269.06
Daily productivity rate	88.88	88.88	90.32

In case of a check deposit, a check image comparison has to be conducted (the check to be reviewed is compared with a similar check of the same check series, issuer, and issuing bank), and if it is a noncheck deposit, then the source of the deposit has to be identified. Any red flags (suspicious elements) found in any of the activities or in customer account information would qualify the case as a referral.

Figure 7A.7

Distribution of
check and
noncheck
deposits.

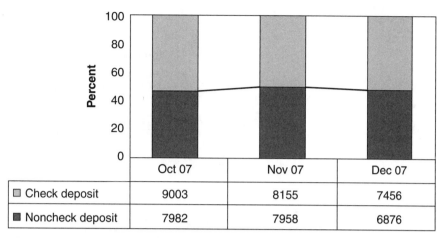

	Oct 07	Nov 07	Dec 07
☐ Check deposit	9003	8155	7456
■ Noncheck deposit	7982	7958	6876

Note: Deposits are of two types: check deposits and noncheck deposits (cash deposits in ATM, salary credit, maturity of fixed deposits, etc.).

The project goal (12 percent improvement – 100 cases) was statistically validated using a one-sample t-test.

The one-sample t-test helped determine whether the difference between μ (population mean) and μ_0 (hypothesized mean) is statistically significant or not.

H_0: μ = μ_0 versus H_1: μ ≠ μ_0
H_0 – Null Hypothesis
H_1 – Alternate Hypothesis

where μ is the population mean and μ_0 is the hypothesized mean

Session window output of the one-sample t-test (from Minitab):
Test of mu = 100 vs not = 100
Variable N Mean StDev SE Mean 95% CI T P
Daily productivity rate 95 88.608 7.450 0.764 (87.091, 90.126) –14.90 0.0000
The test statistic, T, for H_0: μ = 100 is calculated as –14.90

The p-value of this test, or the probability of obtaining more extreme values of the test statistic by chance if the null hypothesis was true, is 0.000. This is called the attained significance level, or p-value. Therefore, H_0 is rejected since the p-value is less than 0.05 (at 95 percent confidence level). P-value of zero indicates the target is statistically significant

5. Analyze Phase

In the Analyze phase, the team, led by the Black Belt, used a brainstorming technique to generate a lot of ideas to quickly identify potential causes for suboptimal productivity. This helped encourage creativity, involved everyone, and generated excitement and energy.

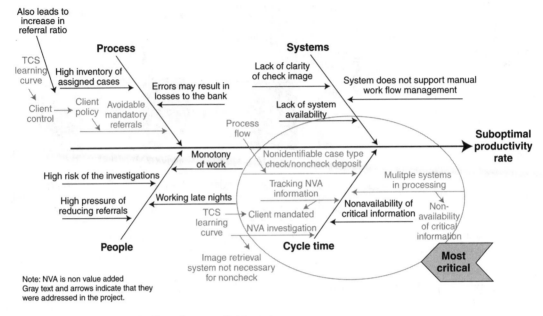

Figure 7A.8 Cause-and-effect diagram (fishbone).

The "Five Whys" analysis further helped identify the root causes. All the potential causes were displayed in a cause-and-effect diagram (see Figure 7A.8).

The next step involved collecting data and using team consensus to verify the potential causes. FMEA (failure mode and effects analysis) was done to identify and prioritize the most significant causes.

FMEA is a useful technique that can be used to help focus on data collection efforts for the input and process variables that are critical for the current process. It can be used to identify potential causes for failure and to prioritize causes as well. Therefore, it can be used in the Measure phase as well as the Improve phase, since it is a structured approach to identify, estimate, prioritize, and evaluate risk.

Table 7A.3 shows the application of FMEA to prioritize the potential root causes obtained from the fishbone diagram.

What numbers (rating) are to be placed into the boxes for Severity, Occurrence, and Detectibility are based on the Automotive Industry Action Group (AIAG) guideline that tells how to determine the correct number to be placed in each segment, predicated on certain criteria. The team used the criteria and consensually arrived at the rating for each of the potential causes.

RPN (Risk Priority Number) is obtained by multiplying the three numbers together to determine the risk of each potential cause of failure.

Table 7A.3 FMEA Table—RPN Count

Category	Potential causes	Occurrence	Severity	Detectibility	RPN
Process	High inventory of assigned cases	8	3	8	192
Process	Avoidable mandatory referrals	8	8	8	**512**
Process	Errors may result in losses to the bank	8	3	8	192
People	High risk of the investigations	5	3	8	120
People	High pressure of reducing referrals	7	5	9	315
People	Monotony of work	3	3	8	72
People	Working late nights	5	3	8	120
Systems	Lack of clarity of check image	3	3	5	45
Systems	Lack of system availability	3	3	5	45
Systems	System does not support manual work flow management	3	1	5	15
Cycle time	Nonindentifiable case types—check/noncheck deposit	9	9	9	**729**
Cycle time	NVA investigation	9	8	7	**504**
Cycle time	Multiple systems in processing	7	8		**504**
Cycle time	Nonavailability of critical information	5	8	8	320

RPN = Occurrence x severity x detection
Severity = How severe is the effect on the customer?
Occurrence = How frequently is the cause likely to occur?
Detection = How probable is detection of cause?

After assigning risk ratings to the potential causes, the top four causes that had the highest RPN count relative to others were identified.

The current process was mapped to identify opportunities for improvement.

Figure 7A.9 shows the basic process flow (as-is).

As shown in Table 7A.3, the critical X's are:

▲ **Nonidentifiable case types – check/noncheck deposit (RPN 729):** Time study showed that investigating check deposits takes 68 percent more time than noncheck deposits. Respective volumes for these were at 51 percent and 49 percent, respectively. However, the initial steps were identical for both the deposit types, thereby increasing the time taken to process noncheck deposits. The process was not capable of identifying noncheck deposits early, thus increasing the overall processing time and leading to low productivity.

▲ **Avoidable mandatory referrals (RPN 512):** Analysis of referral history showed that the significant number of cases that were referred to the client (based on the client-mandated referral criteria) as potentially fraudulent was not found to be so on review by the client. There was an opportunity to revisit the referral criteria.

▲ **Multiple systems in processing (RPN 504):** This was leading to high cycle time with low productivity. There were two image retrieval systems. Both of them were being used

Figure 7A.9

Basic process flow (pre-improvement phase).

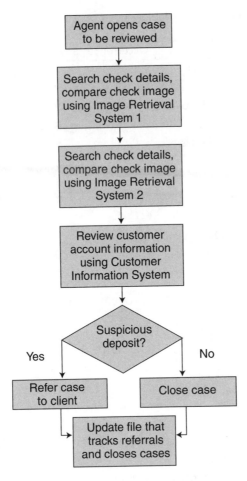

Systems/applications used:
1. Workflow management system: all cases are loaded here.
2. Two image retrieval systems: used to locate comparable check images for deposit checks.
3. Customer information system gives history of the deposit account and other linked accounts.

because there was dependency on two different sets of information available from each of the systems.

▲ **NVA Investigation (RPN 504):** In the as-is process flow, there existed an ad hoc approach with multiple steps, leading to high processing time.

▲ **Placing of comments (based on process observation):** Part of closure of each case was ad hoc and nonstandardized.

6. Improve Phase

In the Improve phase, the team brainstormed potential solutions for each of the verified root causes.

The following solutions were implemented for each of the critical X's after discussion with the client:

1. **Nonidentifiable case types—Check/noncheck deposit.** The process flow was redesigned to eliminate steps that were not required for the noncheck deposits (see Figure 7A.10). This reduced the processing time by 68 percent. The new process flow minimized possibilities of nonvalue-added investigations. This new process flow took less time, as the type of deposit could be identified with higher success rate (hence, directing the investigation in the right direction).

2. **Avoidable mandatory referrals.** Analysis was conducted on all referral criteria. Risk-based classifications were made, and it was observed that not all mandatory referrals were necessary. After several rounds of discussion between the TCS team and the client, it was

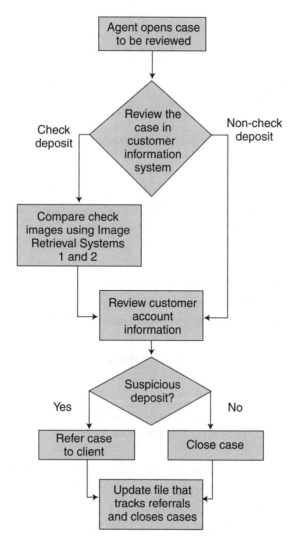

Figure 7A.10

Improved process flow (post-improvement phase).

decided that certain usual referrals need not be referred. This decision was taken while ensuring that all risk elements were adhered to. Also, an Excel-based format was created to get feedback on all referrals that the client disagreed on. These inputs were considered for the performance tracking for the agents.

3. **Multiple systems in processing.** The three different applications (workflow management system, image retrieval system, customer information system) were tested for their effectiveness in identifying the deposit type. It was found that the customer information system was the most effective application in determining the nature of the deposit.

Sampling testing for system efficiency is shown in Table 7A.4. The customer information system had the highest success rate of 96.7 percent. Hence, the combination of the workflow and customer information systems was used in the new process flow.

Table 7A.4 System Efficiency Test

System flow	Successfully determined nature of deposits	Percent
Work flow	636	63.60
Work flow + image retrieval system 1	716	71.60
Work flow + image retrieval system 2	745	74.50
Work flow + customer information system	967	96.70
Total # samples tested	1000	

4. **NVA investigation.** A gemba investigation revealed that there were several opportunities for reducing investigation time. Gemba refers to the place where all activities are actually taking place; in other words, the place where value is added.

Nonvalue-added (NVA) activities were identified and some activities were simplified. Some of them are listed here:

▲ **Linked account review:** only necessary if deposit account value is low
▲ **Updating comments in workflow:** Excel-based macro prepared, saving 15 secs/case
▲ **Observing deposit pattern in image retrieval system 1:** not required for noncheck deposits
▲ **Getting check writer account information from image retrieval system 2:** not required for noncheck deposits
▲ **Tracker updates:** no longer required
▲ **Number of check compares (the check to be reviewed is compared with a similar check of the same check series, issuer, and issuing bank):** client agreed on two compares instead of three

5. **Placing of comments (based on process observation).** This is part of the case closure. After completion of case review, every analyst had to post the observations made during the review in the comment box. A macro was created to post information in the comment box at the click of a button. This resulted in saving 15 seconds per case.

7. Control Phase

After implementation, an effective control plan was put in place to ensure sustenance of improvements. The standard practices and procedures were changed to ensure that the process changes were institutionalized. Training was provided to all the associates on the new standard operating procedures. The control chart for productivity was continuously monitored for variation (Figure 7A.11).

A two-sample t-test was conducted to check if the target (daily productivity rate) achieved was statistically significant (see Figure 7A.12).

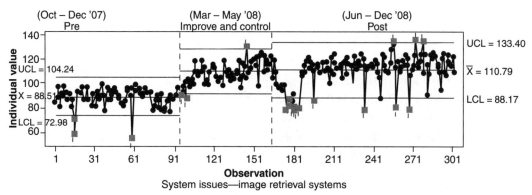

Figure 7A.11 IMR—Daily productivity rate—All phases.

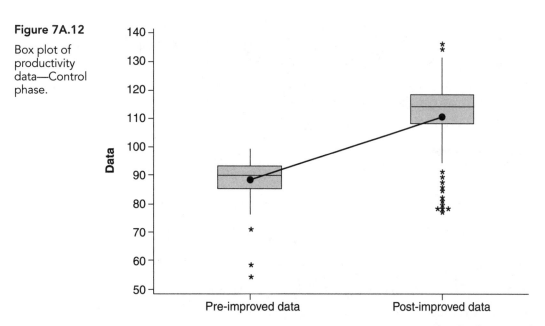

Figure 7A.12

Box plot of productivity data—Control phase.

(continued on next page)

Figure 7A.12

Box plot of productivity data—Control phase. *(continued)*

```
Two-Sample T-Test and CI: Pre-improved data, post-improved data

Two-sample t-test for pre-improved data vs post-improved data

                       N      Mean     StDev    SE Mean
Pre-improved data      95     88.61    7.45     0.76
Post-improved data     138    110.8    12.9     1.1

Difference = mu (pre-improve data) - mu (post-improved data)
Estimate for difference: -22.18
95% CI for difference: (-25.07, -19.28)
T-test of difference: 0 (vs not =): T-value = -15.08
P-value = 0.000  DF = 231
Both use pooled StDev = 11.0324
```

The p-value of zero indicates the target achieved is *statistically significant.*

As shown in Figure 7A.13, the daily productivity rate had improved by 27.81 percent and the referral ratio had dropped by 29.48 percent.

In order to have a control plan (to be executed in case of a nonconcurrence to targets), a FMEA table was created (see Table 7A.5).

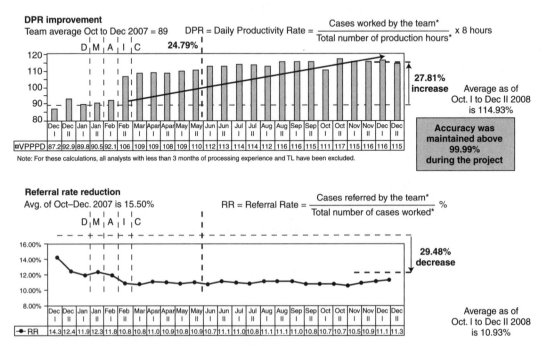

Figure 7A.13 Productivity and referral improvements.

Table 7A.5 FMEA Table

Item/function	Potential failure mode(s)	Potential effect(s) of failure	S e v	Potential cause(s)/ mechanism(s) of failure	P r o b	Current Design/Process Controls	D E T	R P N	Recommended	Responsibility and target completion date
Productivity: volume worked per person per day; where a day means 8 hours of production time	If productivity drops below 80 cases	Decrease in volume worked, impacting the overall cases left in the queue EOD	4	Process update Complacency by analyst Improper monitoring of work by TL System issues	4	Monitor production hourly Monitor DPR tracker Staff review on productivity with client Impacts staff's process updates	5	80	Identify root cause Share findings with client Escalate in case of system issues Provide training to team in case of performance points	Unit manager Target meets minimum 80 cases within 2 working days
Overall productivity of the team	If team complete less than 700 cases in a day	Decrease in volume worked, impacting the overall cases left in the queue EOD	4	Process update Complacency by analyst Improper monitoring of work by TL System issues Attrition	4	Monitor production hourly Monitor EOD volume Staff review on productivity with client Impacts staff's performance points	5	80	Identify root cause Share findings with client Escalate in case of system issues Provide training to team in case of process updates	Unit manager Target meets minimum 700 cases within 2 working days
Referral ratio: For tracking referrals made against volumes worked	If ratio exceeds 15% in a day	Impacts net volume worked; increases work load for client	2	Process update Complacency by analyst System issues	4	Monitor referrals made hourly Monitor referral tracker Impacts staff's performance points	5	40	Identify root cause Share findings with client Escalate in case of system issues Provide training to team in case of process updates	Unit manager Target to ensure less than 15% referrals within 3 working days
Referral feedback: For tracking feedback on analysts for all referrals not agreed upon	If feedback is greater than 15% in a day	Indicates poor quality; increases work load for client	2	Complacency by analyst System issues	3	Monitor referrals made hourly Monitor referral feedback tracker Impacts staff's performance points	6	30	Identify root cause Share findings with client Escalate in case of system issues Provide training to team in case of process updates	Unit manager Target to ensure less than 15% referral feedback days

211

8. Benefit Realization

The benefits realized from this project were as follows:

1. Productivity improved by 27.81 percent (114 cases per person)
2. Referral ratio decreased by 29.48 percent
3. Four FTE savings, resulting in cost savings of $98,841
4. Actual loss prevented (fraud avoidance savings) amounted to $11.25 million in 2008 (initial projection was $7.57 million), an additional savings of $3.68 million
5. New business opportunity as a result of this success

With increases in productivity, the dollar value of investigations (high-value check deposits) has also increased (from $200,000 to $500,000), and this has substantially increased the opportunity of minimizing fraud loss to the client.

This means that more fraud detection occurs successfully, thus detecting greater amounts of potential financial loss. Improved fraud detection saved money and so did the productivity improvements that reduced headcount.

The aim for the future is to ensure 100 percent completion of cases and raising the average value of reviews to greater than $500,000.

9. Future Work

As part of the plans to replicate success from this project, the client intends to implement the referral tracker concept in other fraud detection processes (fraud early warning system and electronic fraud investigation are other processes being handled by TCS). There are also plans to implement the macro concept of posting comments in similar processes across TCS. This is being addressed as part of Lean deployment in other processes.

CHAPTER 7B

Infosys Improves Software Development Productivity for a Large Multinational Bank

Prakash Viswanathan
Anshuman Tiwari*

Relevance

This chapter is relevant to readers who would like to know how productivity can be improved in software application development. Productivity improvement in software application development includes delivering additional functionality in a specified time or delivering the same features in a shorter time with fewer resources than estimated or agreed upon—simply put, delivering more with less.

A key lesson from this case is that systematic deployment of Six Sigma methodolgy almost always delivers results. Although the project was carried out for a multinational bank, it is relevant to all industries.

At a Glance

▲ Infosys provides software application development and maintenance services to a leading multinational bank with many online services. The systems offering these services are hosted on mainframe computers. The key challenge was productivity improvement in software development in the mainframe environment.

▲ Review or appraisal effort was significantly reduced, resulting in a 10 percent productivity improvement.

▲ The improvements from this project resulted in cost savings of US$0.5 million per development cycle.

---

* At the time of this project, Anshuman Tiwari was head of the Business Excellence Program at Infosys.

Executive Summary

This project was carried out for one of Infosys's large clients—a multinational bank. This bank offers the entire range of banking products to its retail and corporate customers. These include personal banking, business banking, private banking, insurance, and corporate services. The bank wanted a significant improvement in the productivity of its banking applications hosted on a mainframe. This is a challenge in mainframes where technology and platform changes are difficult. Adding to this challenge were the several development and enhancement requests from the bank's end customers. Improving productivity in such a context primarily required delivering additional functionality within agreed-upon timelines. In a similar context, productivity improvement could also mean delivering early or with fewer resources.

Improving productivity in this context was more complex than it initially appeared. Although the customer was cooperative and eager to improve, its specific requirements for changes were not always clear and complete. Technically speaking, there were limited options for improving such a legacy application. With a solution nowhere in sight, a Six Sigma team was commissioned.

A cross-functional team was created with a project lead, seven subject matter experts (SMEs), and a Six Sigma Black Belt. As is general practice in Six Sigma projects, the roles and responsibilities of these team members were defined and recorded in the project charter. The project charter also included a business case, deliverables, key risks, and mitigation plans. The team expected improvements to be institutionalized in about 13 to 15 months. Productivity improvement was possible if effort could be reduced across all software development lifecycle stages.

The Six Sigma team used the classical Define, Measure, Analyze, Improve, Control (DMAIC) approach to solving this problem. This was appropriate, since this was a "known problem – unknown solution" situation and a reasonable amount of data was available.

The key problem was long cycle times for application development.

Early in the Define phase of the project, basic data analysis showed that the appraisal or review effort was the prime contributor to long application development lead time and the critical success factor (CSF) for this project.

Appraisal effort is primarily effort invested in review. This includes effort invested in the review of various software development work products such as requirement specification document, high-level design document, detailed design document, testing specifications, etc.

The project charter was updated with this refined problem and approved by the sponsor. An as-is process map was created to understand the process flow better and focus areas for improvement.

During the Measure phase, past project data was analyzed using basic statistical methods. Analysis showed that the average appraisal or review effort was higher than expected. Every new development request is an opportunity to introduce defects in the development process and the application. Hence, the fewer the changes, the better. Each request is also an opportunity for process improvement. The project team debated and concluded that an improvement goal of 25 percent reduction in appraisal effort on the current baseline was fair. If appraisal effort could not be reduced by 25 percent or more, then such incidences were to be treated as defects. During the Analyze phase, existing process capability was computed after appraisal effort data analysis with respect to the customer specification limit. The team created a drill-down tree to isolate the appraisal effort into review and rework effort. This helped to focus on key areas of improvement that mattered most.

In the Improve phase, the team brainstormed possible solutions to address the identified root causes. Software testing tools and standards to reduce cycle time for testing were implemented. After piloting the process improvements, it was found that these solutions significantly reduced testing effort without affecting the delivered work product quality. The time to test the software application for various scenarios (business workflow) was reduced. Further rework effort in coding was also reduced by introducing customer sign-off for sample component design and unit test planning (UTP) at the detailed design stage.

In order to sustain gains observed during the piloting of improvement strategies, use of testing tools and standards was documented and made systemic during the Control phase. Post-improvement statistical tests were done to quantify improvements and validate before and after improvement impacts to the process. There was a significant reduction in appraisal and review effort by about 20 percent.

As a result of this effort reduction, productivity of the software development process for this large multinational bank improved by 10 percent. The total benefit was about $US0.5 million for each development cycle.

1. Introduction

Six Sigma is often accused of not delivering results in a software development enviroment. At Infosys, we have proven this myth to be baseless. If diligently implemented, Six Sigma can be as beneficial, if not more so, in software development as in other industries.

Among a range of other services, Infosys provides application development services to many clients across industry verticals, platforms, and technologies. One such client is a large multinational bank. The requirement was to build a large and complex application on a mainframe that hosted all the bank's critical applications. The overall objective was to build an efficient application that could enhance productivity of the bank's customers.

This was a unique challenge, because historically, mainframe applications have language and platform restrictions. The client was prestigious to Infosys, and a lot of importance was attached to the project's progress. There was an implicit expectation from Infosys management to significantly improve productivity through a structured, predictable, and sustainable approach. It was not easy to quantify any productivity improvement in tangible measure. The Six Sigma DMAIC approach was used to address this challenge. A committed cross-functional team worked with software development teams from start to finish.

2. Project Background

A senior leader at Infosys set the tone by personally monitoring the project as a sponsor. The sponsor highlighted that this project was a significant improvement opportunity due to its high impact for the customer. A business case was created to estimate the return of investment and feasibility of executing the project. A project charter was then created with all the roles and responsibilities. It was the responsibility of the project lead, along with the quality consultant (who actively supported this development project), to create the draft project charter.

As this was a large development project, it was necessary to identify the work product delivery schedule and suitably align the DMAIC phases. This facilitated effective application of preventive and corrective measures in a timely manner. The classic DMAIC approach was adopted to improve productivity.

3. Define Phase

After opportunity identification, the project moved into the first phase of DMAIC. The activities of the Define phase are identification of the project critical success factor (CSF), creation of a project charter followed by review and baselining, and creation of a high-level process map.

3.1 Project Critical Success Factor Identification

The customer's expectation from this project was a clear productivity improvement. Productivity improvement, therefore, was selected as the CSF.

3.2 Project Charter Creation

The project charter is an important Six Sigma document that binds a cross-functional team to stay focused on an improvement objective.

The project charter includes:

- ▲ Business case
- ▲ Problem statement
- ▲ Improvement goal statement
- ▲ Roles and responsibilities
- ▲ Project scope
- ▲ High-level project plan

3.2.1 Business Case

The business case evaluates the improvement opportunity and makes the case for why the opportunity must be pursued at this time. The sponsor provided a high-level business need and a business case was established. The key elements of this business case were:

▲ This development project is executed for one of the top 40 Infosys customers. It was important to deliver applications as per schedule.

▲ The application is large and complex and had to be hosted on a mainframe with limited documentation. This was a technological challenge for the development team, considering the fact that the application was to handle millions of transactions per day after release.

With an approved business case, the team could move forward and begin the step-by-step Six Sigma journey.

3.2.2 Problem Statement

The key issue in this project was to improve productivity. As explained earlier, this was essential to retain a large and key client. The sponsor's expectation was a clear productivity improvement, which is also known as Y (output variable) in the Six Sigma world.

Productivity improvement in this context was a bit more difficult than usual, due to the twin challenges of business complexity and technology and platform limitation. Additional issues included:

▲ Lines of code could not be increased without the customer's approval.

▲ It was important that any effort reduction did not adversely affect delivered work product quality.

The team decided to explore effort reduction alternatives across software development lifecycle (SDLC) stages for productivity improvement.

3.2.3 Improvement Goal Statement

Infosys uses the SMART goal articulation principle in its Six Sigma projects. This encourages teams to set specific, measurable, attainable, relevant, and time-bound goals.

A productivity improvement goal of 10 percent (Y) was set to be accomplished in a 13- to 15-month timeframe. This timeframe coincides with the development time for the software application covering requirement gathering to integration testing lifecycle activities. A 25 percent effort (x) reduction was expected in this timeframe.

This goal was specific to the development project SDLC activities; the unit of measure was function points built per person hour. Based on historical data, this improvement goal was achievable and realistic with a clear time span.

3.2.4 Project Roles and Responsibilities

Team roles and responsibilities were clearly documented in a section of the project charter.

3.2.5 Project Scope

Once the project team was identified, it was important to define the project scope. There were tasks that were clearly in scope for this improvement, and there were others that were out of scope. This boundary identification ensures proper measurement and helps avoid scope creep (unplanned increase in scope) as the project progresses.

Although Infosys provided a wide range of services to this bank, only the mainframe development projects were in scope for this improvement.

3.2.6 High-Level Project Plan

The overall start and end dates, along with the improvement stages, were captured as a part of the high-level project plan, as shown in Table 7B.1. Training requirements (if any) and risks were also identified at the start of the project.

Table 7B.1 Sample Project Plan

S No.	DMAIC Phase	Activities	Resources	Deliverables	Planned Completion Date
1	Opportunity identification	Define scope and objective of initiative	PM, SQA, Practitioner	Project charter	9 Oct
2	Define phase	Business case	PM, SQA, Practitioner		
3		Formation of team/ project charter that defines roles and responsibilities	PM, SQA, Practitioner		
4		Sign-off/initiation of project	Sponsor	Approval of project charter	12 Nov
5		Kick-off meeting	PM, SME, SQA, QM, Practitioner, Mentor, Sponsor	Meeting minutes	21 Nov
6		Obtaining details of the current process	PM, SQA	Process mapping	16 Nov
7		Define phase gating review	PM, QM, SQA, Practitioner, Mentor, other stakeholders	Approvals	21 Nov
8	Measure phase	Identify the critical success factors	PM, SQA, Practitioner	QFD	12 Nov
9		Define performance standards	PM, SQA, Practitioner		12 Nov
10		Collect the data and validate the accuracy and adequacy of data	PM, SQA, Practitioner	Gauge R & R	16 Nov

S No.	DMAIC Phase	Activities	Resources	Deliverables	Planned Completion Date
11		Identify the current process capability	PM, SQA, Practitioner	Process capability	27 Dec
12		Measure phase gating review	PM, QM, SQA, Practitioner, Mentor, other stakeholders	Approvals	10 Jan
13	Analyze phase	Set performance objective	PM, SQA, QM, Practitioner	Benchmarking	15 Jan
14		Identify sources of variation	PM, SQA, SME, Practitioner	Root-cause analysis	4 Mar
15		Analyze phase gating review	PM, QM, SQA, Practitioner, Mentor, other stakeholders	Approvals	6 Mar
16	Improve phase	List out corrective actions	PM, SQA, SME, Practitioner	Who What When plan	12 Apr
17		Prioritization of implementation actions	PM, SQA	Matrix diagram/effort impact matrix	25 Apr
18		Improve phase gating review	PM, QM, SQA, Practitioner, Mentor, other stakeholders	Approvals	29 Apr
19	Control phase	Executing the implementation plan	PM, SME, Project Team	Progress tracker	5 May
20		Collect post-improvement samples and compare against objective set	PM, SME, SQA, Practitioner	Process capability report, post-improvement/ hypothesis testing, pre- and post-improvement	15 Sep
21		Verifying benefits	Sponsor, Financial Analyst, PM, SQA, Practitioner, Mentor	CBA report	21 Sep
22		Full-scale deployment of the learnings	PM, SQA, Project Team	Control plan to sustain the improvement	28 Sep
23		Control phase gating review and closure of the project	PM, QM, SQA, Practitioner, Mentor, Sponsor	SOP creation, approvals/hand-off	30 Sep

3.2.7 Project Team

The team for this project consisted of:

- ▲ Sponsor (senior management)
- ▲ Six Sigma Champion (Master Black Belt)
- ▲ Project leader/manager (PM)
- ▲ Quality consultant (software quality advisor – SQA)
- ▲ Six Sigma practitioner (Six Sigma Black Belt)
- ▲ Several subject matter experts (SMEs)

The project charter was developed by the project leader.

Apart from the team carrying out the Six Sigma project, the project review panel also plays a critical role in ensuring a successful Six Sigma project. This review panel typically includes a senior delivery manager, quality head of the involved business unit, Six Sigma MBB, and select subject matter experts (when required). An independent executive review was done at the conclusion of the project, after the Control phase. This review is to ascertain whether the improvement project objectives as set by the sponsor are met in a timely and sustainable manner.

This panel diligently reviews a project from start to end, and their approval at closure of each phase of DMAIC is mandatory.

3.3 As-Is Process Map

To identify the process bottlenecks affecting productivity, a cross-functional team was created. The team developed an as-is map to understand the current process state. An as-is process map was built to identify improvement areas. A generic SDLC is illustrated in Figure 7B.1.

Figure 7B.1

High-level process map.

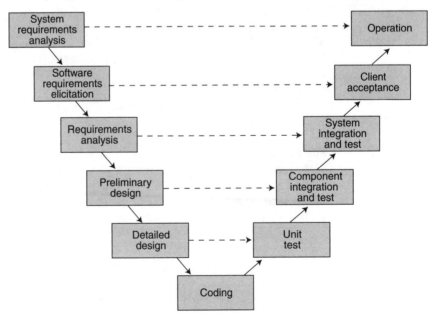

From historical data, it was observed that detail design and coding software lifecycle stages are most effort intensive.

The Define phase was reviewed by the project review panel prior to progressing to the next project phase.

4. Measure Phase

After defining and scoping the improvement project, the next step is to identify the metrics for measuring improvement in the critical success factor (CSF). Steps in the Measure phase include translating customer requirements to functional requirements, defining the performance standard, data collection, measurement system analysis, and establishing capability baseline.

Productivity is defined as the ratio of size in function points (FP), which is a metric to measure the size of the software, to the total project effort invested (person months) across all SDLC stages. Function points were agreed upon during estimation, requirements, and design. The other variant that significantly affected productivity was effort spent during SDLC stages. In order to improve productivity, it was necessary to reduce effort spent. For this project, effort and size data were captured from in-house measurement systems.

The team created a drilldown tree to segment the effort data into various SDLC stages. The effort spent in design, coding, review, rework, and testing-related activities was examined. The team discovered that review effort was disproportionate to the effort spent in the remaining lifecycle stages.

To be able to accurately (as much as possible) assess the current capability of the process, the team needed to collect and baseline data from the past year. Past productivity data for 11 development projects was used establish the current baseline. Productivity is expressed as FP/Person Month.

The minimum desired productivity target was 13 FP/Person Month, which is the Lower Specification Limit (LSL). The as-is process capability computed using the classical Z formula is about 0.3, as shown in Figure 7B.2.

A significant number of data points were below the LSL. This indicates the need for a mean shift to improve process capability.

The appraisal or review effort was higher than the expected value and a stretch target of 25 percent improvement was chosen. All development/enhancement requests for mainframes were considered as improvement opportunities. An inability to reduce appraisal or review effort by 25 percent is a defect.

After establishing the improvement goal and identifying the defect/opportunity, a Measure phase gate review was carried out. Armed with an approval from the review team, the project moved into the Analyze phase.

Figure 7B.2

Capability pre-improvement (sample).

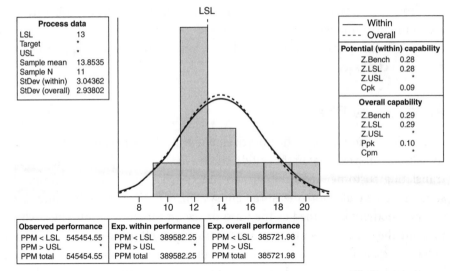

5. Analyze Phase

To study the causes for data variation, a detailed drill-down tree was created to segregate the appraisal effort into review and rework effort, as shown in Figure 7B.3. Task effort for coding

Figure 7B.3

SDLC effort distribution (sample).

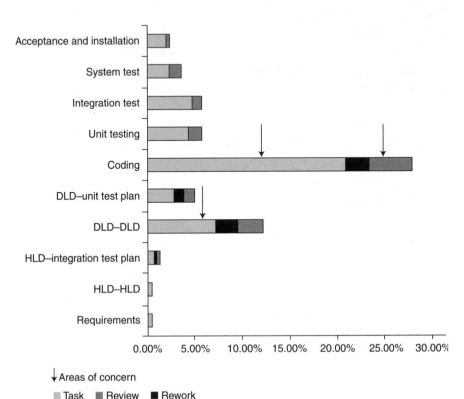

and detailed design stages was contributing significantly to the overall project execution effort. This has a negative impact on productivity. Rework effort in coding also had a negative impact on productivity. Any reduction in appraisal and rework effort significantly improves productivity. Thus, it was important for the team to implement actions in this direction.

After isolating task and rework effort, the team brainstormed key areas of improvement. A fishbone diagram was created, as shown in Figure 7B.4, to drill down into various root causes leading to high task and rework effort. Two separate fishbone diagrams were built to examine all factors leading to high task and rework effort.

Client dependency, data and environment variables, people dependencies, and other miscellaneous factors were identified as causes for low productivity. Key process input variables (KPIV) were identified and categorized. As a next step, the team segregated the vital few from the trivial many. Unclear requirements from the customer, complex database structure, low business application knowledge, and inadequate tools usage are some of the significant causes affecting productivity.

Another fishbone diagram was created to identify the factors affecting engineering and project management effort. From Figure 7B.3, it is evident that the coding and detail design task effort are areas in need of improvement.

Data showed that significant effort was spent by the project managers in client communication. Artifact reviews and approvals were cited as one of the primary factors leading to an increase in project management effort.

After establishing the current as-is process capability and identification of root causes, the team brainstormed on innovative solutions to reduce variation in the KPIVs. This concluded with an Analysis phase gate review by empanelled experts and senior management.

The key root causes identified at the end of the Analyze phase were:

1. Tools not implemented adequately
2. Insufficient usage of reusable components
3. Resources not trained adequately on the system
4. Inclusion of multiple test cycles in the projects

In the next phase, the team worked to either eliminate or reduce the impact of the identified root causes.

6. Improve Phase

After identifying the key root causes, the cross-functional team brainstormed three times on ideas to minimize the key root causes: effort-intensive tasks, and appraisal and rework effort. The brainstorming addressed the root causes mentioned in the Analyze phase—for example, on how to select tools and improve code reuse to reduce review and test effort. More than 25 ideas were shortlisted, and cost-benefit analysis was performed.

Based on a cost-benefit analysis and prioritization matrix, certain corrective actions were shortlisted for piloting (see Table 7B.2).

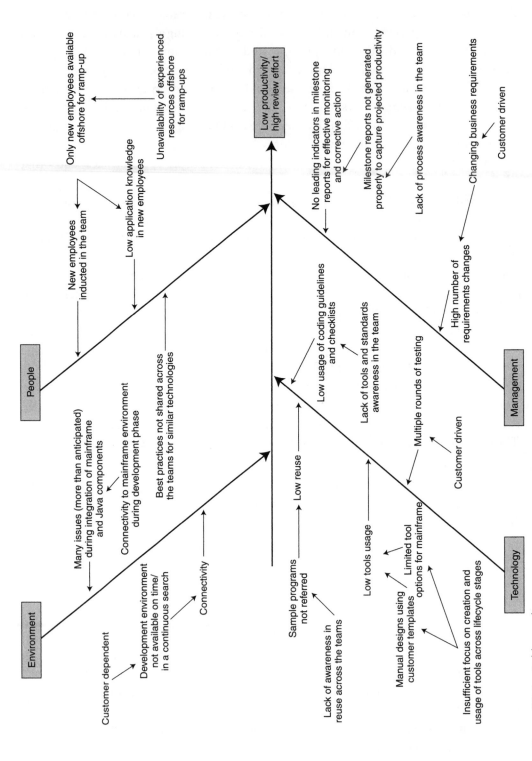

Figure 7B.4 Fishbone diagram.

Table 7B.2 Corrective Action Prioritization Matrix (Sample)

S. No.	Root cause	Sources of variation	Corrective actions to be implemented	Effort (high/ med/ low)	Impact (high/ med/ low)	Priority (high/ med/ low)
1	Inadequate tools usage	Inter- and intraproject knowledge sharing mechanism	a) New tools in mainframe and Java will be shared across all the projects (tools to create production and test JCLs) b) Anchors to train and help projects in other locations to drive tool usage	Med	High	High
2	Insufficient reuse implemen-tation	Fresher not using the reuse components effectively because of lack of knowledge, awareness/exposure	a) Program repository to be built (e.g., data routines, e-mail components, etc.); will be part of account repository b) Training and awareness sessions on available components, their benefits and sharing of location-specific case studies	Med	High	High
3	Team members not trained adequately on the systems	Project-specific induction training, system	a) System appreciation documents to be developed for projects b) SMEs to be identified to conduct these sessions c) Lessons learned for each project to be documented and shared periodically	Med	Med	Med
4	Inclusion of multiple test cycles in projects resulted in low productivity	Dependencies on the client internal terms	PMs to proactively communicate dependencies and dates to client at the time of planning	Low	High	Low

The areas of improvement identified in the Improve phase were listed and solutions to address them were evaluated by the team using the effort impact (prioritization) matrix.

Software tool usage was one of the areas of improvement identified. The team selected tools in the mainframe platform, and experts were identified to conduct training sessions. This had a positive impact on reducing SDLC review and testing effort. For example, the COBOL code review tool, code alignment tool, standard checker tool, and parser tool were customized and used to reduce the manual effort spent on reviews.

The practice of code reuse was another area of improvement that was addressed through the creation of a project repository of components. Developers were trained on the use of components to help in reducing manual coding effort. Good practices were shared by SMEs with senior developers, module leads, and project leads once a month.

The project had a mix of new and moderately experienced team members who were not well conversant with the system environment. The team developed system appreciation documents and trained personnel through SMEs. Further lessons learned from previous project execution were held. Collectively, these actions reduced review and rework effort. For example, sessions on standards and checklists are made a part of new team member induction and project kick-off meetings.

Multiple rounds of testing were effort-intensive, and the team chose to proactively communicate the impact of the same on the overall project schedule to the customer. For example, customer sign-off is obtained for a sample component design and UTP at the start of the detailed design stage, which results in rework effort reduction.

These corrective actions were implemented as part of the project planning stage for all new projects initiated after the completion of this improvement project. Furthermore, as part of compliance audits of new projects, it was verified that these corrective actions had been implemented.

7. Control Phase

In the Control phase, during the implementation of improvement actions, information on standards and use of testing tools was documented and made systemic for sustenance.

Post-improvement data samples were collected and, in order to study the characteristic of the mean shift (due to implementation of the improvement actions), a post-improvement Z score was calculated, as shown in Figure 7B.5.

To verify the impact of the change statistically, a pre- vs. post-improvement paired t-test was done as illustrated in Figure 7B.6.

The null hypothesis was that there is no significant change in the productivity levels before and after improvement. It is incidental that these figures are higher in the post-improvement scenario. The p-value is 0.042, which is less than 0.05; hence, the null hypothesis is rejected. Therefore, it can be concluded that due to the corrective actions implemented to reduce engineering task and review effort, there has been a positive impact on productivity.

Appraisal effort was significantly reduced by 20 percent and productivity improved by 10 percent. Productivity was tracked using control charts after the implementation of improvement actions. The improvements have led to a positive shift in the mean and a reduction in variance. The process is now controlled, and the variation band has narrowed along with a positive mean shift. This is illustrated in Figure 7B.7.

To ensure the gains are sustainable in the long term, the project adopted a control plan that documented the selective interventions and monitoring mechanisms for project leaders (see Table 7B.3).

Figure 7B.5

Capability
post-
improvement
(sample).

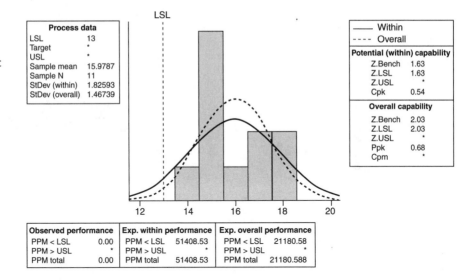

Figure 7B.6

Paired t-test.

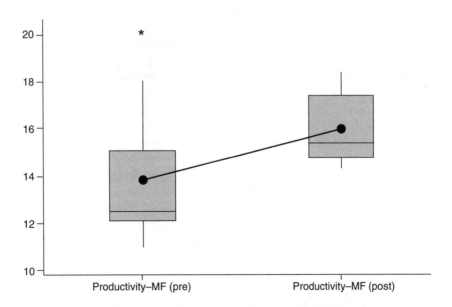

Paired T-Test and CI: Productivity–MF (pre), productivity–MF (post)

```
Paired T for productivity - MF (pre), productivity MF (post)

                   N      Mean     StDev    SE Mean
Productivity - M   11   13.8535    2.8656    0.8640
Productivity - M   11   15.9787    1.4312    0.4315
Difference         11   -2.12527   3.67688   1.10862

95% upper bound for mean differnce: -0.11594
T-Test of mean difference = 0 (vs < 0)
T-value = -1.92   P-value = 0.042
```

Figure 7B.7

Productivity
control chart
(I-MR chart).

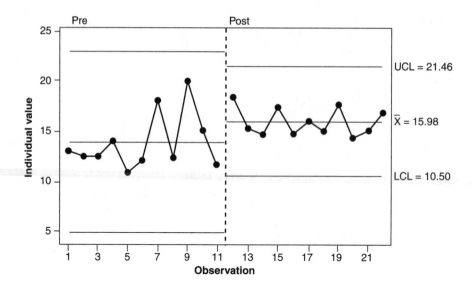

Table 7B.3 Control Plan

SI #	Improvement plan	Process step	Parameter to be monitored	Monitoring mechanism frequency	Frequency	Responsibility
1	Tools usage	Planning	Tools usage index	Checklist	During project initiation	PM and SQA
		Coding/ testing	Tools usage and saves	Project tracking tool	Monthly	PM and SQA
			Tools usage and saves	Process to check, verify, and monitor usage of tools	Once every two weeks	PM and SQA
2	Reuse	Planning	Reuse planning and saves	Checklist	During project initiation	PM and SQA
		Coding/ testing	Reuse actual effort	Project tracking tool	Monthly	PM and SQA
3	Training	Planning	Training at the portfolio level for reuse, tools and system appreciation	Project tracking tool	During project initiation and then on an as-needed basis	PM and SQA
4	Knowledge transfer	Planning	Knowledge transfer training for new lateral entries in the account	Project tracking tool	During project initiation and then on an as-needed basis	PM and SQA

Using software tools and code reuse helped reduce task and rework effort. These are now integral parts of the project planning. In addition, checklists were created to ensure appraisal and rework effort was minimized. All the changes were documented in order to institutionalize the improvements and to minimize any risk of human oversight.

Control phase toll-gate review was performed at the end of this phase to verify the actual improvement in the CSF against the goal.

Finally, the project team presented the project summary to an independent review panel selected from senior management of business units as part of the executive review and closure.

8. Benefit Realization

As a direct outcome of the productivity improvements due to this project, the customer was able to roll out changes as planned. These changes were primarily in the software application and were originally proposed by end business users. In some cases, these changes were possible before schedule, due to the reduced cycle time of software development. The desired business functionality is available to end users. This had a significant impact on the bank's business, improving end-user experience with online transactions.

In order to arrive at the net effort savings due to this project, the effort expended by the cross-functional team working on the project was deducted from the overall effort saving. The dollar impact is the product of net effort saved and standard rate for each role in the development project. The productivity improved by 10 percent, which is significant for a large development project. This translates to approximately US$0.5 million, which was validated by the finance department.

9. Conclusions and Limitations

A delicate balance had to be maintained while dealing with appraisal and rework effort reduction in a software development lifecycle. While improving productivity, the team had to ensure that work product quality was maintained as well. A key metric that must not be ignored is work product quality. Any reduction in delivered product quality would have had a major impact on Infosys's relationship with this customer.

There is no right measure of what the proportion of appraisal and rework effort must be, as this depends on various factors like technology, platform, language level, and type of application.

The magnitude of improvement resulting from this project cannot be expected in every single development project, due to uncontrollable external variables, as mentioned earlier. However, the root causes and corrective actions, if identified correctly, can significantly improve engineering productivity.

Acknowledgments

The author of this case study is thankful to Infosys for its permission to allow publication of this case study as a book chapter. The author would also like to thank Satyendra Kumar, Ramakrishnan M., Haresh Amre, Krishna Pai, and Moses R. for their review comments.

CHAPTER 8

DMAIC Conclusions
and Lessons Learned

The nine chapters just presented represent a good mix of DMAIC Six Sigma projects for the high-tech industry. They all focus on software or IT processes and apply the same rigor and disciplines found in traditional production environments.

The nine success stories addressed a wide range of problems:

- ▲ One sought to reduce business risk.
- ▲ Another focused on reducing cycle time of software development processes.
- ▲ Two addressed defect reduction.
- ▲ Three targeted help desk improvements for IT and service industries.
- ▲ Two strove to achieve productivity improvements.

These are the types of classic process improvement goals typical of DMAIC projects. Such projects focus on a single measurable problem and use data to drill down to root causes until substantive permanent corrective actions can be identified. These stories also provide excellent illustration of use of Six Sigma tools for software and IT process and quality improvement. For a list of which Six Sigma tools that are highlighted in each chapter, refer to Appendix A.

The focus areas of defect reduction, cycle time reduction, productivity improvement, and problem resolution effectiveness are perennial themes in these Six Sigma stories. After all, these are the factors by which most business processes are measured and evaluated. Since the performance metrics of cycle time, cost (which includes productivity), and quality predominate how core business processes are monitored, it should not be surprising that problems with these factors are first recognized and addressed.

All of these projects have this common focus on problems and processes. That is one reason that they all utilized the DMAIC lifecycle model. They all developed charters with clear problem statements and clearly defined goals. They followed the same phases: Define, Measure, Analyze, Improve, and Control. They all prepared process maps and changed these processes, although in different ways, to achieve their objectives.

Still, the nature of the specific problem typically varied from one project to the next, along with the complexity of the process and associated systems. Therefore, different problems warranted the use of different tools and techniques.

In these nine projects, a wide variety of tools was used:

▲ Most performed some form of measurement systems analysis and formal data collection.

▲ Those with multiple root causes or complex problems used FMEAs to prioritize or characterize those problems.

▲ All of them used some statistical tools to search for or verify relationships between different factors. Whether they used descriptive statistics, correlation analysis, ANOVA, or other comparative methods, all sought to objectively identify and measure problems before solutions were implemented.

▲ Some piloted their solutions before they implemented them, but all of them established some type of ongoing control(s) to ensure that any changes that were made were substantive, real, and supported so that solutions would remain in place and effective over time.

This is what Six Sigma DMAIC projects are all about—addressing problems systematically, with formal methods, in such a way that solutions are clear, effective, and enduring. Although not all projects in a company address process improvements, those that do are often well suited to the DMAIC approach, and all can reap the financial benefits of effective problem resolution when properly applied.

In the end, the project teams were able to apply the DMAIC approach to solve their problems and achieve quite remarkable financial savings for their companies.

PART TWO

Lean Six Sigma Projects

PART TWO OF THIS BOOK SHIFTS ITS FOCUS from traditional DMAIC projects to Lean Six Sigma endeavors. As explained in the primer that follows, Lean Six Sigma is a blend of two different process improvement approaches that developed independently of one another. Like DMAIC projects, Lean Six Sigma addresses problems or issues with existing processes. Both methodologies use formal methods and practices and statistical methods for analysis. Lean Six Sigma, however, tends to focus more on waste and cycle time-related inefficiencies, though most process problems can be effectively addressed using either project type.

Revealing its Japanese origins, Lean Six Sigma also introduces new fast-paced projects (Kaizen events or Kaizen blitzes), as well as new methods of analysis (Kano analysis) and value stream mapping, a greater awareness of waste (muda), and new solutions like mistake-proofing (poka-yoke), line-balancing (kanban), and general cleanliness and organization (5S).

While the terms and methods differ between Lean Six Sigma and DMAIC, the focus on fixing existing processes and the robust use of data remains the same. Both methods are effective, and the main differences are often a matter of style and tool preference, given their appropriateness for use as explained in the first chapter.

CHAPTER 9

Lean Primer

Jeffrey A. Robinson, Ph.D.

Lean Six Sigma is a combination of two different approaches: *Lean* and *Six Sigma*. Both are focused on improving processes, but they use different methods and techniques. They also start at different points to achieve essentially similar, but not identical goals.

History of Lean

The roots of *Lean* can be traced back as far as Henry Ford, who built on the scientific management work of Fredrick Taylor and Frank Gilbreth's philosophies, which in turn later established the foundations of industrial engineering.

Gilbreth was the visionary who analyzed how people worked and introduced "time and motion" studies. In one anecdotal success, he analyzed the motions of bricklayers and reduced the number of operations and actions to lay a brick from 18 to 4.5. His detailed methods for analyzing processes and activities remain to this day accepted practices to streamline, simplify, and optimize processes. The core philosophy of Lean focuses on the elimination of waste, with the objective of improving efficiency, reducing effort and cycle time, and lowering costs. When combined with other innovative practices that were developed and popularized throughout the twentieth century, Lean forms a strong and effective approach for process improvement, particularly for manufacturing, though its principles can be readily applied to almost any process.

Many of the more common practices of Lean have come from Japanese approaches to quality, practices that predate the more recent Six Sigma methods.

History of Six Sigma

The origins of Six Sigma are more recent, dating from the mid-1980s. Six Sigma uses statistical analysis and quantitative methods to identify and eliminate sources of variability. Born in the revolution in quality that shifted attention from the "quality of a product" to the "quality of the processes" that underlie production, Six Sigma has become a popular improvement methodology, more recently in the high-tech industry, where automation is common and a wealth of data is available for analysis and study.

Commonalities and Differences

Both Lean and Six Sigma approaches seek to improve processes. Lean, by eliminating waste and errors; Six Sigma, by identifying and eliminating sources of variation (and errors).

However, Six Sigma is born of formalisms and mathematical techniques, methods of detecting, measuring, and evaluating hypotheses and making decisions based on statistical analysis, whereas Lean is more of a collection of methods that have proven effective over decades of use.

Combining Two Winning Approaches

The combination of Lean and Six Sigma produces a simple yet robust set of tools that are ideal for process problems of all sizes. Lean approaches are often best suited for companies who are just starting efforts to fix or improve processes or who are just learning about Six Sigma.

This marriage results in a practical and robust suite of techniques that are more practical for those learning about formal process improvement methodologies. The combination of approaches complements one another well. Lean provides simple tools and techniques that are easy to understand and implement. Six Sigma adds quantitative rigor to these solutions and assures long-term success through monitoring of quantitative measures and controls.

Lean Six Sigma fills the gap between "quick-wins"—just-do-it projects where the problem and solutions are well understood—and full Six Sigma projects where the detailed analysis of data may be the only way to effectively solve complex problems with high risk or high cost. Indeed, one of the greatest problems with programs is when all projects are forced to be Six Sigma projects. Lean Six Sigma may be considered a variant of DMAIC, DMADV, and DFSS that focuses on simplification and elimination of waste, rather than improvement, optimization, or defect prevention. It is no less effective and is often faster.

Lean Six Sigma Strategies

The overall sequence of activities in Lean Six Sigma closely models the DMAIC lifecycle model. It is only common sense, after all, to first formalize the problem you are seeking to address (if only to avoid unwanted scope creep in a project), then to collect and analyze information or data related to the problems before identifying, prioritizing, and selecting solutions so they can be effectively implemented.

The more rigid adherence to phases in DMAIC are not common to Lean Six Sigma projects, since Lean projects focus on advancing from problem to solution with minimal delays. Six Sigma projects (using DMAIC or DFSS) often take longer, but then they are also used more predominantly where there is greater complexity and where there are higher costs and risks with associated problems and their solutions.

Lean projects tend to involve fewer people and are typically smaller; these teams are more focused and often faster than standard Six Sigma projects.

Some Core Lean Concepts

One of the core ideas underlying Lean is the concept of waste. The Japanese term *muda* is commonly used to describe the types of waste that Lean seeks to minimize or remove. There are eight types of waste. They include:

▲ **Waiting:** Whenever goods are not in transport between steps or being processed, they are waiting. In traditional processes, a large part of an individual product's life is spent waiting to be worked on.

▲ **Overproduction:** This happens each time you engage more resources than needed to deliver to your customer. For instance, large batch production because of long change over time exceeds the strict quantity ordered by the customer. Producing parts that aren't needed can result in wasted units that must eventually be scrapped. It also means that effort and time is wasted too.

▲ **Rework:** This is scrap or defects, as well as repair. Whenever defects occur, extra costs are incurred in reworking the part, rescheduling production, etc.

▲ **Motion:** This refers to the producer, worker, or equipment. Wasted motion means wasted time. It can also relate to opportunities for damage, and wear and safety. It also includes the fixed assets and expenses incurred in the production process.

▲ **Transportation:** Each time a product is moved, it stands the risk of being damaged, lost, or delayed as well as being a cost item with no added value. In other words, transportation does not make any transformation to the product that the consumer is supposed to pay for. Combining steps or making them closer can reduce transportation effort and time.

▲ **Processing:** Typically, this means unnecessary processing or overprocessing. Overprocessing occurs any time more work is done on a piece than what is required by the customer. This also includes using tools that are more precise, complex, or expensive than absolutely required.

▲ **Inventory:** Inventory, be it in the form of raw materials, works in progress (WIP), or finished goods, represents a capital outlay that has not yet produced an income. Any of these three items not being actively processed to add value is waste.

▲ **Intellect:** Wasted intellect is associated with people whose ideas are not valued and whose recommendations are not listened to. Operators in the factory often know more about problems and potential solutions than consultants or managers. Sometimes, however, they are not heard, and this is a waste of ideas and people.

Each of these categories represents real classes of loss. Lean practitioners systematically examine these different wasteful practices and activities and strive to eliminate waste in all its different forms, thus optimizing lost times, lost material, unnecessary effort, and lost opportunities.

Another key concept from Lean is yet another practice from the Japanese—the application of the "5 S's" (pronounced the five ess-ez). This is a mnemonic for the principles of:

▲ Sort (seiri)
▲ Store (seiton)
▲ Shine (seiso)
▲ Standardize (seiketsu)
▲ Sustain (shitsuke)

This five-step approach basically means that you need to clean up your workstation and manufacturing lines, your workspace, and work areas. Order and cleanliness is a prerequisite discipline that needs to be developed and maintained for other more rigorous practices to succeed.

▲ **Sorting (seiri):** This means organizing a workstation so that it is well arranged and logical; a place for everything, and everything in its place. A tool rack with outlines for every tool is an example. With such a simple system, each tool is easily and quickly located and missing tools are immediately apparent.

▲ **Storing (seiton):** This is a strategy that eliminates piles of parts (incoming or outgoing) and assigns specific places for items to be stored. Just as seiri organizes things like tools and equipment, seiton creates bin or storage locations for other types of materials that pass through the workstation. For office workers, this may be a place to file forms, incoming and outgoing documents, pencils, and supplies material. This organization not only makes work more efficient, it eventually relates to a matter of pride for the individual workers.

▲ **Shine (seiso):** This element of 5-S is basic cleanliness. This means the work areas are clean. Cleanliness is a measure of pride in work and reflects the disciplines that accompany good workmanship. At the end of each shift, clean the work area and be sure everything is restored to its place. People who are sloppy with their work areas may also be sloppy with their work effort.

▲ **Standardize (seiketsu):** Work practices should be consistent and standardized. Everyone should know exactly what his or her responsibilities are for adhering to the first three S's.

▲ **Sustain (shitsuke):** Maintain and review standards. Once the previous four S's have been established, they become the new way to operate. Maintain focus on this new way and do not allow a gradual decline back to the old ways. While thinking about the new way, also think about better ways.

It is important that the first step in implementing a Lean Six Sigma program be to start with a 5-S initiative and to literally clean house before the real work of process improvement begins.

"Kaizen" is another term frequently used in association with Lean and Lean Six Sigma. Kaizen, which is Japanese for "improvement" or "change for the better," is a philosophy or practice similar to Lean that focuses on continuous improvement of processes in manufacturing, engineering, supporting business processes, and management, though it has also been applied to other industries like IT, healthcare, banking, etc. In most cases, when used in the business sense and applied to the workplace, kaizen refers to activities that incrementally improve all functions and involve all employees, from the CEO to the assembly line workers. Kaizen was first implemented in several Japanese businesses after the Second World War, influenced by visiting American business and quality management experts. It has since spread throughout the world.

In modern usage, a focused kaizen is designed to address a particular issue over the course of a few days or a week and is often referred to as a "kaizen blitz" or "kaizen event." These are limited in scope, and issues that arise from them are typically used in later blitzes. This high focus can compress traditional projects and achieve substantial results in much shorter periods.

Kaizen events are team-based, multiday activities that are often used to kick off projects. Having multiple subject matter experts working together to do a value stream map or to participate in brainstorming sessions can greatly accelerate the analysis of processes and the identification of appropriate corrective actions.

Similarly, a *kaizen blitz* is an approach that compresses a project to a minimal length by dedicating teams full-time to a problem until it is solved. This approach can compress a project that might have taken a month down to a few days.

Both kaizen events and blitzes require experienced team leaders to maintain the correct focus and pace.

As with other principles of Lean, kaizen is a recursive process similar to the Shewhart cycle, Deming cycle, or PDCA(s): you look for things that can go wrong; you look for opportunities to improve your processes. You act and implement fixes and improvements. And then you start all over again.

A final core principle of Lean that we will highlight is the concept of mistake-proofing. "Poka-yoke" is based on the adage "the best way to eliminate mistakes is to prevent them." The Japanese call this approach to process improvement *poka-yoke* (pronounced "po-ka yok," where poka means an "inadvertent mistake" and yoke means "prevent"). This roughly corresponds to the Western axiom of Murphy's law that anything that can go wrong, will. It seeks to implement features in processes that make it difficult or impossible to make mistakes; examples are check-letters on part numbers (error-detecting codes that catch typos when they occur), or interlocks on elevator doors so they cannot be opened except when a car is present, or asymmetrical parts (like plugs or USB ports) that prevent you from inserting plugs upside down or backwards. The philosophy is sound and eminently practical.

Starting Projects

One of the first things to note about Lean Six Sigma projects is that, like kaizen, they are recursive. You look for things that can go wrong; you look for opportunities to improve your processes. You act and implement fixes and improvements. And then you start all over again.

Lean Six Sigma has a popular anecdote about lowering the water in a lake. As the waterline gets lower and lower, rocks begin to protrude above the water. These are the biggest problems that you can solve through process improvements. You can remove the rocks, but as more improvements occur (or as the waterline continues to drop), more problems are exposed and become identifiable. As you progress, you eventually find smaller and smaller rocks.

Lean Six Sigma's recursive approach to problem solving addresses major problems first, but as you fix them, sometimes smaller problems and opportunities become visible and may be addressable.

The second thing to note is that *all process improvement activities are not Lean Six Sigma projects.* Sometimes, you uncover problems that are easily recognizable, where the problem is clear and the solution obvious. These are "quick wins" or "just-do-it" activities. They do not require additional analysis or Lean Six Sigma projects. If such solutions to problems are found, do not delay the fixes by creating a project for them. Just implement the solutions at hand.

If the problems are more substantive, or if some analysis seems warranted to confirm or verify root causes, or if there are multiple methods of solving the problem, a Lean Six Sigma project may be called for. However, there may be many different problems competing for attention or for limited resources, so such projects need to be prioritized and addressed in order of criticality and effort.

Finally, if the problems are complex, or the solutions are risky or expensive, then a full Six Sigma project may be warranted.

One important lesson about Lean Six Sigma is that one size (one methodology) does not fit all (problems or projects). This is one of the important premises of Lean Six Sigma that make it effective. The methodology strives for practicality and utility, and it actively avoids falling in love with itself.

Unfortunately, the focus on fixing problems is often one born of *crisis*, when problems arise that persist, problems that defy immediate correction and can no longer be ignored. It is when problems become abundant that people stop and realize that something needs to be done. This is the ideal environment for introducing Lean Six Sigma, primarily because people become highly motivated and there are a lot of problems to address and lots of opportunities to improve.

The true measure of success of a Lean Six Sigma program, however, is whether this recursive approach of looking for, finding, and fixing problems remains in place after the immediate crises pass.

Identifying the Problems

For any given process, there is often a wealth of opportunities for improvement. The trick is to look at processes in a way that can highlight and identify these opportunities. There are many ways to do this.

Value Stream Mapping

One such tool is value stream mapping (VSM). VSM is a method of documenting a process, breaking it down into component steps, and analyzing it. The method is simple, relatively fast, and replete with opportunities to quantify the characteristics of the process so that improvements can be measured, ranked, and evaluated systematically.

Steps are identified, named, and numbered, and details like "how long the process takes" and "how long a product waits at each step" and "how long queues are at each step" and "how much movement or transportation time occurs between the steps" are examined and recorded (see Figure 9.1). Each step in turn is assessed and classified as a value-added (VA) or nonvalue-added (NVA) activity. Eliminating steps that are repetitive, that take too long, or that add no real value to the product being created are prime candidates for elimination.

Subsequent analysis looks for opportunities to reduce movement and transportation activities, sometimes by combining steps or changing the locations where work is done. Sometimes, steps can be eliminated or shortened. The VSM methodology allows the rapid development of a hypothetical new process model with clear measures of improvement in effort and cycle time. Simplifying processes that have over time grown increasingly complex is one of the major ways that processes can be improved.

Because of the complexity of many processes and the fact that many people are more familiar with some parts of the process than others, VSM is often a team-building activity. Full VSMs of complex manufacturing facilities can take two or three days, but upon completion of a value stream map, many problems and potential solutions are commonly identified. Some solutions may be immediately implementable; others may require fundamental changes to manufacturing or processing activities to fully achieve identified improvement opportunities.

VSMs can be used in traditional manufacturing processes, but can be applied to any defined process, from financial processes (processing invoices or orders or loan applications) to IT activities, like the procedures for processing help desk tickets or dispositioning or escalating trouble tickets.

Figure 9.1

Value stream map.

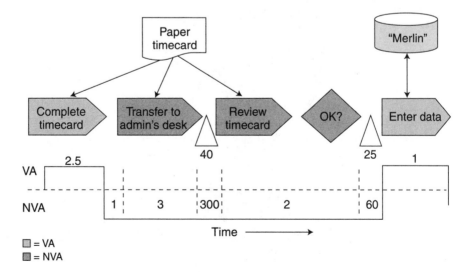

SIPOC

Another simple approach to process modeling is the SIPOC method (see Figure 9.2). SIPOC is an acronym that stands for Supplier, Input, Process, Output, Customer. It is based on the model that every process step has *inputs* (information or material) and *outputs* (final product or service) that result from the active *processing* of those inputs. Moreover, these inputs come from a specific source (a *supplier*), and outputs for the process step go to other specific destinations (customers).

The SIPOC approach is a simple and effective method for documenting a process. It can be used in conjunction with VSM and other process models.

Spaghetti Diagrams

A spaghetti diagram (Figure 9.3) shows the flow of a process (lots or manufactured units, piece parts, works in process, order forms, etc.) over its lifetime. Often, you will find that work crosses

Figure 9.2

SIPOC for preparing a burger meal.

Suppliers	Inputs		Outputs	Customers
Store customer	Order		Burger	Store customer
			Fries	
Food services	Ingredients	Process	Drink	
Employees	Labor			

Figure 9.3

A spaghetti diagram.

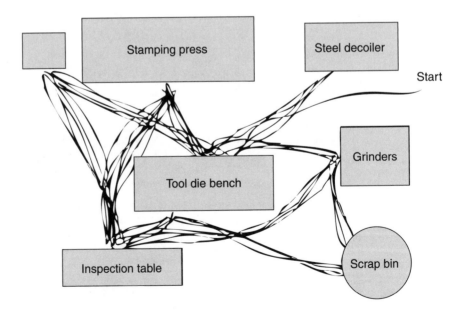

its own path over and over again. An example is an invoice that is processed and comes back to the same inbox multiple times over the course of it being completed or a manufacturing operation where an operator travels repeatedly over the same path again and again.

The spaghetti diagram highlights wasted motion and effort and often suggests opportunities to eliminate this and reduce overall cycle time. Rearranging workstation locations, for example, can reduce travel and wasted motion and time.

Muda

Another approach that can be used in conjunction with any of the process modeling activities is to examine processes and attempt to identify all the types of waste that occur—wasted motion, wasted waiting time, rework, nonvalue-added labor, overproduction, excess inventories, and so on.

Each type of waste identified can become an opportunity for improvement and may spawn separate Lean Six Sigma projects to study the problem and engineer solutions.

Big Y – Vital X

This concept from Six Sigma is based on the idea that an outcome (Y) results from different factors (X's)

$$Y = f(x1, x2, x3, x4, ...)$$

To ensure that projects are properly aligned with realistic business goals, key objectives or outcomes need to be identified and listed (identifying the Big Y's). Then each Y is examined to list the different factors that affect, influence, cause, or contribute to or negatively affect that outcome.

This approach can identify the factors (X's) that can be measured or controlled, and by improving the control of those factors, the corresponding outcomes become more achievable and predictable.

This approach also helps maintain a focus on important goals and can head off the distraction of making many little improvements that don't really matter while critical business objectives go unaddressed.

Requirements

Another important element of crafting Lean Six Sigma processes is developing a clearer understanding of what the processes really need to do. Such requirements, often called voice of the customer (VOC), are often essential to understand before processes are changed or altered; otherwise, improvements that are not needed or even desirable can result. A clear definition of requirements can avoid distractions of unnecessary improvement opportunities that in themselves represent a new kind of muda or waste.

Kano Analysis

Where multiple requirements are identified, it is important to note that not all requirements are of equal importance and that many requirements are only implied. Kano analysis (named after the noted Japanese quality expert Professor Noriaki Kano) is an approach to discover new requirements and to classify them so that they can be effectively prioritized.

Prioritizing Projects

After the use of the aforementioned methods, one may now have many different potential projects. Since resources to work on projects are often limited, some method of prioritizing is often essential.

Different approaches can be used to rank, order, or prioritize projects. They can be sorted by cost and effort (project size), or impact and importance (potential cost savings or impact), or the ratio of the two (benefits/effort = return on investment or ROI). You can also use other methods to sort and compare projects so that you address them in the most efficient manner.

Another approach is to characterize projects by their effort (high, medium, or low) and importance of impact (high, medium, or low) using an effort-impact matrix. This divides projects into nine sections or groups that make projects more recognizable. The most attractive projects are those that have high impact and low effort. The projects to avoid are those that have low impact and high effort.

Such analyses can be readily displayed using a technique called bubble chart (see Figure 9.4), when more than two factors are considered (such as cost, risk, or complexity and urgency).

In order to properly prioritize projects, it is important to create charters for each. Not all projects can be started at the same time. They probably should not be. Multitasking is another form of waste (see Chapter 11). Charters can form a backlog of opportunities, and a few of them can be initiated while the others wait. Those not yet begun can be started in later "waves," and as each finishes, others can be started in turn.

Projects will fall into several categories:

▲ First, there are the "quick wins" (sometimes called "low-hanging fruit"). These are the simpler problems, the ones that do not require a lot of analysis or verification. These are the projects whose problems are well understood and whose solutions are relatively clear. They are often called "just-do-it" projects, because to implement the solutions, all you generally have to do is get permission and fix the problems. They tend to be small, fast, and effective. They do not even require charters.

▲ The second class of projects requires formal Lean Six Sigma. Using the charter as a starting point, formal data collection and analysis can begin and a thorough explanation of the problem can proceed. These are the projects described in this chapter. Solutions are identified; most are tested or piloted before being implemented in production.

Figure 9.4
Bubble chart.

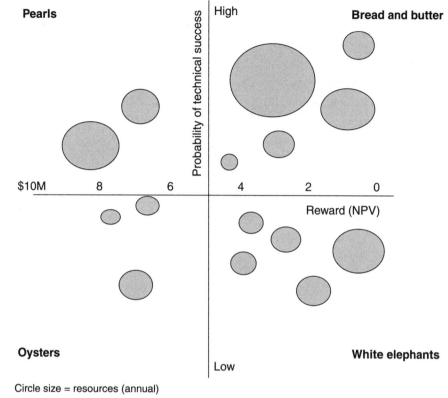

Circle size = resources (annual)

▲ The third class of projects includes those that involve greater complexity. This may be suited to other Six Sigma lifecycles, like DMAIC or DFSS. This third group of projects is typically not Lean. They may not be solving problems or eliminating waste. Different tools, techniques, and approaches should be considered for these endeavors.

Understanding the Data

Once a project has been identified, approved, and initiated, it may be appropriate to revisit the activities discussed earlier. The VSM or process mapping performed earlier may be revisited with a focus just on specific steps that are the focus of the project. It is at this point that additional data normally needs to be collected for analysis.

Measurement Systems Analysis

Measurement systems analysis (MSA) is a methodology that examines the viability of data for a specific process. It asks questions regarding what data can be collected, what data should be collected, how reliable and accurate that data is, what needs to be measured to ensure the project goals are achieved, and so on.

With the relative abundance of data available on most processes, with the vast number of possible things that could be measured, it is easy to collect far too much data and to be overwhelmed with the daunting task of analyzing it all.

Since one of the precepts of Lean is to "keep things simple," it makes sense to be diligent in only collecting the minimum amount of information that is necessary. Anything more would be a waste of time and effort. Accordingly, there is a popular axiom that summarizes this approach:

> If you can't measure it, you can't manage it.
> If you can't measure it, you can't improve it.
> If you can't measure it, you probably don't care.
> If you can't influence it, don't measure it.
>
> —SOURCE UNKNOWN

MSA examines actual and prospective data in a number of different ways. Data is evaluated on factors such as accuracy, precision, stability, linearity, reproducibility, and repeatability. Gage reliability and reproducibility studies (*Gage R&R*) can be used to identify which data is reliable and accurate and which is not and suggest approaches to improving data quality through the recalibration of tools, the acquisition of new measurement instruments, or the training of people performing the measurements.

Other Graphical Analysis Tools

There are, however, simpler tools to organize information for analysis or display.

- ▲ *Affinity diagrams* allow different topics or issues to be mapped, often hierarchically. Groups and patterns and categories can be rapidly defined or discovered.
- ▲ Another effective technique for organizing factors or root causes is an *Ishikawa diagram* or a *fishbone chart*. Like the affinity diagram, different levels of interactions can be captured for analysis or display.
- ▲ A *Pareto chart* is another graphical technique that can summarize problems and rank them by frequency of occurrence. A Pareto chart is essentially a sorted frequency distribution histogram. The problems that occur most often are showed with less frequently occurring issues, problems, or categories shown side by side in descending order.
- ▲ Other methods of displaying data include trend charts, run charts, stacked area charts, and contour plots for showing more complex numerical data.

Graphic displays of information like this make it easier to visually characterize patterns in data for initial analysis or for reporting.

Analyze the Process

Before a process can be improved, a significant understanding of that process is required. Just as important, before claiming improvement, you have to measure performance improvement from the baseline.

Baselining a process is a critical step in process improvement. It documents the conditions before improvements and quantifies the performance. If you do not measure process performance (in terms of throughput, or cycle time, or defect rate) before changes occur, then the effects of any improvements are not measurable or provable. Claims of improvement without measurement are opinions, not facts.

Characterizing a process normally involves some degree of MSA. Characterizing a process means measuring enough of it so that you can characterize it quantitatively. Specific questions are asked, such as:

▲ How long does each step in the process take (mean, median, mode)?
▲ How far do parts or materials physically move between process steps?
▲ What is the actual time and effort spent at each step?
▲ How much time is spent waiting (before or after this effort actually occurs)?
▲ How much material is consumed in the processing of each step?
▲ How much scrap or waste occurs?
▲ What tools are used?
▲ What procedures are in place?

Sometimes, this data is collected using time and motion studies. Some of this information is collected at the end of each step. Often, data collected is logged in statistical process control (SPC) charts that act as a mechanism to both store and display data.

All of this data and information are essential to thoroughly understand a process before actions are taken and changes are made.

One outcome of process characterization is a measurement known as *process capability* (or *sigma score*). This is the measure of quality or the measure of defects that occur over time. This is a measure of how well the process is able to meet the requirements or criteria expected of it. For instance, if you are making ball bearings, this would be the measure of what percentage of ball bearings produced meet success criteria (size, roundness, shape, density, etc.). Process capability is commonly expressed in terms of *defects per million opportunities* (DPMO), and this is sometimes converted to a sigma score, which represents the overall measure of a process's ability to meet required outcomes.

A sigma score of 3.0 means that the variation of the process is such that acceptable outcomes fall within three standard deviations of the desired target.

▲ A sigma score of 3.0 corresponds to 66,807 DPMO.

▲ A sigma score of 4.0 (four standard deviations) would correspond to fewer defects or about 6,209 DPMO.

▲ A sigma score of 6.0 (six standard deviations) would correspond to 3.4 DPMO.

In any case, the quantitative measurement of quality and capability is essential before the project so that the effects of changes can be measured again after changes have been made. Otherwise, the impact of improvements cannot be confirmed or proved.

Other Techniques

There are many other techniques that can be used to analyze processes. These include such things as:

▲ **Line balancing:** This method examines the throughput of each step. If different steps have different productivity rates, the line can become imbalanced and bottlenecks and backlogs will appear at the steps that are inherently slower than those before or after them.

▲ **Takt time:** This is a measure of each process step. It is the maximum time per unit allowed to produce a product in order to meet demand. It is derived from the German word *taktzeit*, which translates to "cycle time." Takt time sets the pace for industrial manufacturing lines.

▲ **Eliminating waste, simplification:** This is a direct application of the principles of muda described earlier. The process is examined in detail for each type of waste so that corrective actions can be identified or problems addressed.

In addition to these Lean tools, traditional Six Sigma methods are commonly used. These include, of course, tools like:

▲ FMEA (failure mode effects analysis)
▲ Root-cause analysis (Ishikawa or fishbone)
▲ Cause-effect matrix
▲ Mathematical modeling
▲ Queuing
▲ Brainstorming

Implementing Solutions

Prioritizing Solutions

Finally, after the processes have been examined and studied in detail, problems have been identified, and potential courses of action have been identified, then the team may have more potential solutions than they can implement. Normally, when there are multiple possible solutions, one or more need to be selected. The challenge is to find the best solution or the best combination of solutions (don't limit yourself to just one if others complement each other; sometimes, more is better).

Solutions thus need to be prioritized, similar to the way initial projects were earlier. Again, the solution set can be sorted by cost and effort (project size), or impact and importance (potential cost savings or impact), or the ratio of the two (benefits/effort = ROI). You can also use other methods to sort and compare projects so that you address them in the most efficient manner. (Such techniques include the *Pugh matrix* and *weighted attribute analysis*; see Table 9.1).

Piloting Solutions

Once solutions have been identified, they need to be detailed (documented, new process instructions and training materials developed), but sometimes it is important to pilot the changes first. Not all ideas work out as planned, and sometimes new solutions also create new problems. Therefore, it is common practice to test the solutions before formally implementing them. Pilots tend to be small (just on a few products) or for a short period. These trials evaluate the fixes, and this gives the team the opportunity to test things out to find if the solutions work or if the solutions create other unanticipated problems.

When changes are substantial, there can be hidden resistance to these changes. Piloting can demonstrate success and overcome this resistance. Pilots can improve the likelihood of success and give teams the opportunities to avoid changes that don't work as expected.

Some Common Solutions

In addition to the solutions mentioned earlier such as 5-S and poka-yoke, there are some other popular Lean solutions.

Kanban is a method of running a process manufacturing line that is faster and has less waste. It does require considerable planning and reorganization. Rather than each step proceeding independently, each operator waits until the next step needs a part before they begin. It is often called a "pull" system as opposed to a traditional "push" system. This eliminates queues between steps and, if problems occur, minimizes the amount of product that is adversely affected. It is a different concept of manufacturing and is often difficult for new practitioners to accept.

SPC charts are data collection mechanisms that make it easy to track performance along a process. An integral part of control plan SPC charts are effective methods of detecting when processes change. Things like processing time, defect rates, and quality can be measured and tracked continuously. SPC charts use statistical methods to detect patterns in data that reveal subtle but important shifts in processes and are, therefore, popular in stabilizing process changes that have been implemented. There are many types of SPC charts, and some training is required to apply them properly (see Figure 9.5).

Of all the traditional Six Sigma tools, SPC charts are the most commonly implemented statistical tools in Lean projects because they monitor and maintain the process changes.

Table 9.1 A Pugh Matrix

Requirements category	ID	Requirement	Weight (1 Minor, 2 Major, 3 Critical)	Product 1 Score (0 Fails reqrt. 1 Partially met 2 Meets reqrt. 3 Best in class)	Product 1 Element score (Weight x score)	Product 2 Score (0 Fails reqrt. 1 partially met 2 Meets reqrt. 3 Best in class)	Product 2 Element score (Weight x score)	Product 3 Score (0 Fails reqrt. 1 Partially met 2 Meets reqrt. 3 Best in class)	Product 3 Element score (Weight x score)
Functional requirements	FUN01	<Text of FUN01 requirement>	3	2	6	2	6	0	0
	FUN02	<Text of FUN02 requirement>	2	3	6	2	4	0	0
	FUN03	<Text of FUN03 requirement>	2	2	4	2	4	2	4
	FUN04	<Text of FUN04 requirement>	1	1	1	3	3	1	1
	FUN05	<Text of FUN05 requirement>	3	2	6	2	6	2	6
		Requirements category total			23		23		11
Performance requirements	PER01	<Text of PER01 requirement>	3	3	9	2	6	1	3
	PER02	<Text of PER02 requirement>	3	3	9	2	6	1	3
	PER03	<Text of PER03 requirement>	2	2	4	2	4	2	4
		Requirements category total			22		16		10
Other requirements	OTH01	<Text of OTH01 requirement>	3	2	6	2	6	1	3
	OTH02	<Text of OTH02 requirement>	3	2	6	3	9	2	6
		Requirements category total			12		15		9
		Overall rating (Max. possible score: 75, min. required score: 50)			57		54		30

Figure 9.5

Sample SPC chart.

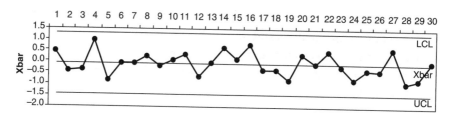

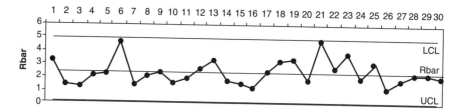

Control Plans

As with traditional Six Sigma projects, Lean Six Sigma projects finish up by establishing control plans to ensure that changes remain in place and remain effective. This includes the preparation of:

▲ Implementation plans
▲ Training plans
▲ Monitoring plans
▲ Control plans

Summary

The essence of Lean Six Sigma is a methodical approach to problem solving:

▲ Establish a philosophy to look for and address problems of all sorts, beginning with waste and inefficiency.
▲ Search for problems to be fixed, document them, and prioritize them.
▲ Form teams to work on these problems; avoid multitasking.
▲ Develop project charters.
▲ Use Lean tools in the context of a Lean Six Sigma methodology (typically DMAIC) based on the types of problems:
   ▼ Use quick wins or just-do-it projects where problems are well understood and solutions known.
   ▼ Use kaizen events for faster team-based solutions.
   ▼ Use blitzes for faster analysis and solutions.
   ▼ Use Lean projects for classic process problems.

▼ Note: Use full Six Sigma methodologies (DMAIC or DFSS) for complex problems that require more analysis and more robust methods.

▲ Deploy the improvements, and monitor them over time.

▲ Start over and do it again; develop a project-tracking system and strive to keep those good ideas coming.

CHAPTER 10

Leaning Six Sigma Projects: How to Run a DMAIC Project in Five Days

Jeffrey A. Robinson, Ph.D.

Relevance

This chapter challenges the misconception that Six Sigma projects are complex and take a long time to complete. The project summarized here shows how a DMAIC project can be initiated and completed in a mere five days. For those who want their Six Sigma projects to be "Lean," this is where to find tips and tricks to make your projects fast and effective.

At a Glance

▲ DMAIC projects have a reputation for taking a long time (six months or more) and involving large teams.
▲ The project here started and completed a full DMAIC lifecycle in five days.
▲ The problem was a classic scenario of severe defects in the field that were about to initiate a full product recall of all products issued.
▲ The reasons that the project was atypical of most DMAIC projects include factors such as high management commitment, 100 percent dedication of team members until the project was complete, the immediate availability of data, and an experienced MBB who led the team and tailored the process to focus only on what was essential to achieving project goals.

Executive Summary

In 2001, a manufacturing facility made telemetry units that utilized cell phone technology. Complex hardware was designed, built, and sold to agricultural companies to facilitate the monitoring of large tracts of farmland. Each unit monitored a variety of factors, including temperature, humidity, wind direction and speed, soil water content, cumulative rainfall, and more. All of this data was transmitted back to a base station, where it was compiled and used to determine the status of crops on large tracts of lands (some spanning more than 100 or 200 square miles).

The problem was that since initial delivery, the hardware had been plagued with a series of failures that, despite efforts by the engineering team, seemed to be growing.

Finally, after a large increase in field failures, the management team was faced with a total recall of all the units produced over the last 18 months and with the possible closure of the facility.

Before taking such drastic action, however, management called in a Master Black Belt (MBB) to address the issue.

The problem turned out to be both technical and perceptual in nature. There were real defects that needed to be addressed, but equally severe was the fact that the engineering team had mismanaged data reporting with inaccurate, incomplete, and often conflicting reports so that management no longer had confidence in what they said was or was not the problem.

This project needed to achieve two primary goals:

▲ To identify the roots causes of the field failures and to implement appropriate permanent corrective actions to the manufacturing processes
▲ To report and organize information in such a way that the confidence in the engineering staff was restored

A formal DMAIC project was initiated in which one day was devoted to each phase:

▲ **Monday:** The Define phase was completed.
▲ **Tuesday:** The Measure Phase was performed.
▲ **Wednesday:** The Analyze Phase was conducted.
▲ **Thursday:** Improve phase activities were done.
▲ **Friday:** Control plans and implementation plans were created and findings presented to management.

In technical terms, problems were found, identified, quantified, and analyzed. The elements of the manufacturing process responsible for these errors were noted and new process procedures were identified to address activities at these steps. Again, this is classic DMAIC problem analysis and follow-through.

Actually, this type of project is sometimes called a "kaizen blitz," which compresses a project into a short amount of time. It is an approach frequently applied to Lean Six Sigma projects. It nevertheless illustrates how a normal DMAIC project can be completed in far less time than normal.

In logistical terms, what made this project unusual was the short time in which it was completed. What made this project special was:

▲ Complete commitment by the management staff

▲ One hundred percent dedication of team members for the duration of the project
▲ Immediate availability of engineering and manufacturing data
▲ Access to robust data analysis tools
▲ Tailoring of the generic DMAIC lifecycle to focus only on those steps that were absolutely essential to project completion
▲ Facilitation by an experienced MBB with a strong manufacturing background

In the end, the recall was avoided. The plant continued operations. The engineering team learned new skills and vindicated themselves. New processes were implemented to forestall future problems of this kind. And a difficult problem was resolved in an amazingly short time.

1. Introduction

In 2001, a manufacturing facility of a prestigious Fortune 500 company made telemetry units that utilized cell phone technology. This organization had a long history in Six Sigma. Complex hardware was designed, built, and sold to agricultural companies to facilitate the monitoring of large tracts of farmland.

Each unit monitored a variety of factors, including temperature, humidity, wind direction and speed, soil water content, cumulative rainfall, and more. The hardware was both complex and diverse. It consisted of power supplies, a wide range of different types of sensors and transducers, and internal CPUs with software to interpret and store data, as well as cellular technology to call, connect to, and transmit this data to a remote base station.

The units had to literally weather the worst conditions (heat and cold, rain and wind, hail and snow) and continue to operate without human intervention or assistance. Each unit was a stand-alone unit, and large agricultural enterprises planned to have one unit positioned roughly every square mile so that a farm of 100 square miles or more could be monitored without requiring human beings to take a half-day to travel to the far end of the property and back.

The units were quite popular when installed, but over time, some failed, for a variety of different reasons. Initial versions of hardware had used the most inexpensive parts possible, but this financial strategy turned out to be problematic. For instance, the first CPU boards were only rated for operations between 0°C and 50°C. Not surprisingly, in the winter, the units in Wisconsin stopped transmitting, as did other units when the temperatures grew too hot inside the metal units that stored the hardware. In another failure mode, some motherboards corroded due to wet conditions, rain, and condensation, and more expensive hardware needed to be acquired where the electronics had additional conformal coating to keep the water out of the electronic circuits.

Still different problems cropped up and continued to occur, vexing both management and the engineering teams responsible for producing replacement units.

Finally, when matters couldn't seem to get worse, a report was issued that reported a massive spike in failures and management had to decide whether to recall all products or shut the facility down. It was at this time that the general manager decided to call in a Six Sigma expert to address the problem.

Though the unit at which this project was performed was later sold off and subsequently shut down, it stands as a model of how Six Sigma projects can be run.

2. Project Background

The call for Six Sigma support and the assignment of an MBB to travel to the manufacturing facility occurred on a Friday. The MBB travelled on Sunday and met with the general manager of the manufacturing facility on Monday shortly before being introduced to the engineering team.

The problem immediately became twofold.

First, there was the technical problem of defects in the field. These problems were real. They had persisted since the first release of the product, and they seemed systemic to the design or manufacturing of the units because they kept cropping up and would not go away.

Second, there was the obvious hostility of the engineering team. These engineers had been performing double duties—designing, maintaining, and improving the telemetry units—while managing the manufacturing activities and handling any problems that arose. As is typical with start-up ventures, this last task of problem solving seemed to consume most of their time. Now, with all their other problems, an outsider had arrived—a Six Sigma expert—to "solve" their problems. They didn't know how this Six Sigma nonsense was going to fix everything. All they knew was that they were going to be pulled into meetings and be forced to spend valuable time away from the critical problems that demanded their attention.

Thus, the problem turned out to be both technical and perceptual in nature.

There were real defects that needed to be addressed, but equally severe was the fact that the engineering team had mismanaged data reporting with inaccurate, incomplete, and often conflicting reports so that management no longer had confidence in what they said was or was not the problem.

This project needed to achieve two primary goals. It needed:

▲ To identify the root causes of the field failures and implement appropriate permanent corrective actions to the manufacturing processes
▲ To report and organize information in such a way that the confidence in the engineering staff was restored

But first, the engineering team need to be appeased so that they would be willing to participate in these Six Sigma activities when no one even knew what Six Sigma was.

3. Define Phase

A formal DMAIC project was initiated to address the defects. The engineering team was small (consisting only of engineers), and the MBB asked to have all of them on the team. The general manager protested that pulling all of the engineers into the project would shut the manufacturing line down. The MBB noted that the line was already shut down.

When asked for how long the engineers would be needed, the MBB suggested one week. The assessment was that anything longer than that would likely result in the permanent closure of the facility anyway.

Note: This was initially a guess by the MBB. He assumed that given a week, they could identify the problem and take other actions later to mitigate it. To the surprise of the MBB and the rest of the team, all phases of the DMAIC were completed in that week.

As noted, the time before the initial meeting was awkward. The engineering team did not want an interloper added to the team in their moment of crisis. All they really wanted was for this intruder to leave so they could get back to work and fix the problems that were now at crisis proportions.

The MBB explained his role was not to take over engineering operations or to solve the problems. The MBB reassured the engineering staff that they knew far more about the nature of the problems, the history, the details of the manufacturing processes, and hardware and software of the product line than the MBB would ever learn in a long time. The role of the MBB, it was explained, was to help them manage, organize, analyze, interpret, and package the data that the team already had so that *they* could solve their own problems.

With these reassurances, the engineering team decided to accept this help, albeit with some trepidation and reluctance.

The first item of action was to conduct a short course on Six Sigma and DMAIC. None of the engineering team had ever taken any Six Sigma courses; none were certified Green Belts or Black Belts; none had ever worked as a member of a Six Sigma project team.

An extemporaneous course on Six Sigma was presented to the engineering team, focusing on the quantitative nature of the Six Sigma methods and techniques and the phased structure of the DMAIC methodology.

Between the initial meetings with management, the engineering team, and the overview on DMAIC, the day was already half gone.

The Define phase of the project began in earnest after lunch.

3.1 Project Team

The project team consisted of the MBB and the entire engineering team:

▲ General manager (sponsor/Champion)
▲ Master Black Belt
▲ Engineering manager (project leader)
▲ Three other engineers

3.2 Project Charter

The charter was documented in a PowerPoint slide for presentation to management at a later time (see Figure 10.1).

Figure 10.1

Draft charter from day 1.

Business Case	Opportunity Statement
Problems are occurring of an unknown magnitude and frequency. Management fear they are approaching crisis levels, but data is scarce and reports are anecdotal in nature.	Due to a paucity of data on the matter, confidence of customers and management is rapidly fading. The problems need to be analyzed and addressed. Confidence in engineering staff needs to be reestablished.
Goal Statement Reduce field failures to zero, or develop plans to mitigate them and eliminate them.	**Project Scope** Only post-release problems will be addressed. Faults and errors in software development will not be considered. Only those associated with assembly or failures after ordering and delivery will be considered.

Project Plan (timeline)		Team Selection	
Activity	End	Team characteristics/composition:	
Define	9/28	General manager	Sponsor/champion
Measure	TBD	Jeff Robinson	Master Black Belt
Analyze	TBD	Engineering Manager	Project leader
Improve	TBD	Three other engineers	
Control	TBD		
Track benefits	11/15		

3.2.1 Business Case Statement

The problem statement developed clarified the focus of the endeavor:

> "Problems are occurring of an unknown magnitude and frequency. Management fears they are approaching crisis levels, but data is scarce and reports are anecdotal in nature. Lack of data is impacting business and a product recall or factory shutdown may occur if the problem is not resolved."

3.2.2 Opportunity Statement

A complementary problem or opportunity statement was drafted:

> "Due to a paucity of data on the matter, confidence of customers and management is rapidly fading. The problems need to be analyzed and addressed. Confidence in engineering staff needs to be re-established."

3.2.3 Goal Statement

The team was unanimous in the formulation of the goal statement:

> "Reduce field failures to zero, or develop plans to mitigate them and eliminate them."

3.2.4 Project Scope

The scope was limited to post-release defects (not errors within the manufacturing line):

> "Only post release problems will be addressed. Faults and errors in software development will not be considered. Only those associated with assembly or failures after ordering and delivery will be considered."

3.2.5 Project Plan

In project team discussions, it was revealed that all the field failure data was on hand (albeit in various formats and not centralized into an organized database). Still, the ready availability of data suggested that rapid analysis might be possible; therefore, an aggressive schedule was established.

It was decided to strive for an unusual DMAIC phase schedule. One day would be devoted to each phase, as follows:

▲ Monday: Define
▲ Tuesday: Measure
▲ Wednesday: Analyze
▲ Thursday: Improve
▲ Friday: Control

4. Measure Phase

This phase involved several activities. First, the team had to develop a process map to document its understanding of the current design, manufacturing, testing, and installation activities. Something in one or more of these processes was going wrong for all these defects to be occurring. Second, the team had to gather and assemble all the data that was available on the field failures so that it could be analyzed.

The MBB split the team into two groups:

▲ One group was assigned to develop a process map, even though no formal, written documented process existed.
▲ The other team was charged with collating all field defect data from all available sources.

The first group spent the morning drafting and drawing, revising and redrawing their view of the current process. Even with a team of only a half-dozen engineers, there was already specialization and different engineers "owned" different pieces of the overall process.

Figure 10.2 illustrates the "as-is" process. Even though there was no formal written process, this is the process that had "evolved" and was the de facto method of building and delivering units to customers. This was the third major rewrite of the process, and some team members

Figure 10.2

De facto
process map.

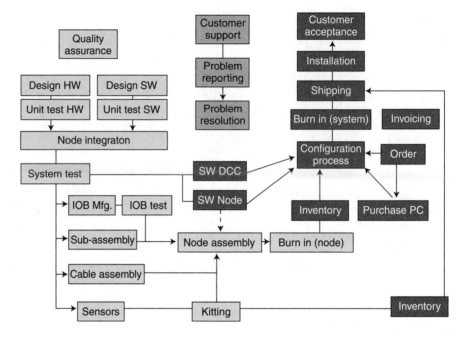

were surprised at how much detail there really was in their development, manufacturing, and installation activities.

The second team spent the morning digging through different documentation on the field defects that had occurred. They had some reports they had issued. There were e-mails and trip reports. There were written letters and replies regarding incidents and field failures. The data existed, but it was scattered. It had not been gathered into a single place or database. Because there was no formal defect reporting process, data on defects was often anecdotal and based on written complaints, reports by technicians who went to customer sites to diagnose and repair the units, and recollections by the entire team.

After lunch, the two teams met and shared the information they had gathered with each other. The teams then began organizing the data they had gathered. To do this they did several things:

▲ First, process steps were numbered and named.
▲ Next, a defect database (in the form of an Excel spreadsheet) was created and every incident that had occurred was entered as a separate row in this simple tabular database.
▲ Data attributes were assigned to each column of the defect database and each row (or defect) had data entered regarding the occurrence:
  ▼ Customer
  ▼ Date and time
  ▼ Location
  ▼ Product version
  ▼ Problem description

At this time, it was discovered that there was no categorization method for identifying defect types. There was no list of defect types that could be used to further classify the field failures. The team, therefore, started to develop such a classification scheme.

First, they went through the defects they had entered into the Excel spreadsheet and started creating codes. Within a few minutes, they had a dozen different classifications and they started organizing the defects into higher-level categories (hardware, software, customer errors, setup or installation errors, etc.).

When they were done, they had a nice list of codes, but the MBB asked "Are these all the things that can go wrong? Are these all the defects that could possibly occur? What other customer, hardware, software, or setup errors could occur?"

The synergy was immediate, and the team identified many more items not on their original list of defect codes. In a surprisingly short time, they developed a nice defect rubric for failures of virtually all the things that could go wrong.

This type of brainstorming about possible failures was a preface to a more formal failure analysis process called failure mode effects analysis (FMEA), which the MBB promised would be demonstrated further during the Analysis phase.

Table 10.1 shows an excerpt of the defect code rubric the team created.

Table 10.1 New Defect Code Table

Computer	Customer	Component	Setup
G0–Other	C0–Other	N0–Other	P0–Other
G1–Field node failure	C1–Customer input	N1–Modem	P1–Documentation error
G2–Communication error	C2–Customer shutdown data collection window	N2–CPU	P2–Bad node names
G3–Host PC failure	C3–Disabled feature	N3–IOB	P3–Missing license
G4–Host node failure		N4–Sensor	P4–Install wrong software version
G5–Sensor failure		N5–Cable	P5–Wrong configuration ordered
G6–Lost data		N6–Power supply	P6–Wrong configuration shipped
G7–Erroneous data		N7–Solar panel	P7–Wrong configuration installed
G8–Other		N8–Green radio	P8–Installation error
		N9–Black radio	P9–Missing cable
		N10–Antenna	P10–Wrong cable
		N11–Bad calibration/gauge	

The new defect code table was use to categorize all the incidents that had occurred. Some incidents involved more than one failure code (in other words, when the technicians fixed the problem, they determined that more than one thing had gone wrong).

Once the initial data on defects had been compiled, the final activity for the day involved calculating the field defect rate.

As they counted up the failures, the team was surprised that the percentage of defects was far higher than they had realized. Indeed, nearly 30 percent of all units sent to the field had experienced some type of failure. (The actual percentage appeared higher because some units had failed multiple times for different reasons.)

Because field failures are attribute data, process capability was calculated using defects per million opportunities (DPMO):

▲ Forty-two incidents (out of the 111 installations)
▲ Seventy-nine field nodes and 29 base station nodes
▲ About 37.84 percent of installed items had failed
▲ Sigma value of 1.81 (DPMO = 378,378)

By the end of the day, the team was overwhelmed with the information that they hadn't even known they possessed and they left for the day, proud of the fact that they now had a chance to dig into that information to get a grasp on the root causes of the problems.

Thus ended the Measurement phase.

5. Analyze Phase

Wednesday began with the team eager to analyze the data. They all had ideas about what to look for, but most were seeking to confirm their current perceptions. The MBB took a more systematic approach.

First, the MBB reviewed the different types of analyses that might be of value and introduced them to some new analysis tools. Specifically, he showed them tools built into Excel and introduced the team to a new tool called Minitab. The first hour of the day involved a cursory show-and-tell of the various statistical analysis capabilities within each tool. Then the analysis of their new defect database began in earnest.

First, the team looked for overall patterns. Pareto charts were created (by defect type) to show which defects were the most frequent (Figure 10.3).

The MBB then introduced the team to the concept of time series analysis to look at errors over time. Did errors occur after a certain time in the field? What was the mean time between failures, or the time between installation and failure? Did errors occur most often during different seasons? (Accompanying mini-lectures introduced concepts of periodicity and techniques to look for cyclic patterns using autocorrelation.) Unfortunately, there did not seem to be any substantive temporal patterns to the occurrence of defects. Indeed, most defects seemed to occur almost immediately after installation.

Figure 10.3

Defect Pareto charts.

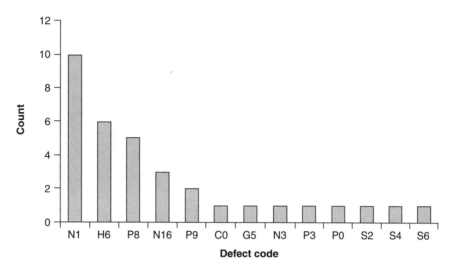

Next the MBB taught the team how to do basic cluster analysis to determine if different defects tended to occur together. But, again, no significant patterns were found.

Other tests were also run: correlations and defect occurrence rates. There seemed little opportunity to perform regression analysis, and some simple nonparametric statistics did not confirm differences between different regions or countries.

At this time, the MBB interrupted the analysis and revisited the FMEA that had begun the previous day. The list of defects was embellished and some new columns were added to the spreadsheet. First, the different defects were scored by severity and detectability. The defect database provided scoring details regarding frequency and occurrence rates. The three numbers were multiplied together (using formulae in Excel), and the resultant prioritized defect list showed which defects were the most important by the nature of their high severity or frequency, or both.

The team then identified the symptoms of each failure, and a likely root cause was assigned to each defect type.

This new focus on root cause generated some ideas, and the process maps from the previous day were reviewed. Examining the process steps, each defect type was studied to determine what step in the design, assembly, or installation processes were responsible for the generation of the defect or allowed the defect to slip past for delivery to the customers.

Some interesting patterns appeared. The same process steps seemed to show up as problem points over and over again. Indeed, as each defect was associated with origins within these processes, it became apparent that some process steps were far more problematic and more critical than others. The Pareto chart of defects was converted to a Pareto chart of culprit process steps where these defects originated.

However, mere identification of the problematic steps within these processes was not sufficient to suggest a remedy.

Additional analysis was started to examine the nature of these different process steps. Some of the activities at individual process steps were documented; most were not. Some were deemed robust; some were ad hoc and performed almost intuitively. Some were measured or were measurable (like testing); others weren't. Some were periodically checked or audited.

A three-point scale was created, and each process step was evaluated and rated according to these criteria: existence of formal documentation, robustness, metrics, auditability, and the number of problems or defects that had been attributed to each.

Table 10.2 shows an excerpt of the initial mapping of defects to process steps and their associated ratings as assigned by the engineering team.

Table 10.2 Initial Mapping of Defects to Process Steps (Unsorted)

	Documented/ format	Robustness	Metrics	Audited	Attributable problems
	0–3	0–3	0–3	0–3	
Board manufacturing	2	2	2	0	
Burn-in (node)	2	2	2	1	
Burn-in (aystem)	2	2	2	1	
Cable assembly	2	1	0	0	3
Configuration	1.5	2	2	0	5
Customer acceptance	0	2	1	0	
Customer support	1	1	1	0	

By late afternoon, the team was amazed at how much analysis their small stash of data had allowed. Several important conclusions were made by the team:

▲ First, it was discovered that most of the defects that had technical root causes had already been identified and fixed in previous product revisions. However, until this analysis had been done, it was not clear if the changes in product design and manufacturing had really been effective.

▲ The second conclusion was more disturbing. Far more frequent than technical failures were process errors, mistakes that simply should not have been made. These included building units with the wrong (often old or outdated) components, or components that did not work or function together. Sometimes, these were kitting errors. That is, sometimes the wrong units were simply used and there was no process to double-check the assemblies until the units were delivered to the field. Sometimes, parts were shipped with missing cables, or the wrong connectors. Sometimes, they were lacking power supplies or necessary sensors or transducers.

▲ Worst of all, the majority of these process-related human errors were limited to a small number of the total process steps, and the analyses indicated that these were the ones that were the least documented and the least robust.

▲ Overall, the lack of formal process documentation (particularly for these most critical steps) was attributed to be the root cause for most field failures. With no formal processes, there was no formal training, no measurability, and no ability to audit or check that these steps had been performed properly.

▲ Finally, it was noted that until this project, there had been no formal process to log, track, resolve, or report defects. Each failure was handled as a unique incident. When several occurred together, action would often be swift since a crisis was discerned; but if single failures occurred, they were often not fixed in a timely manner. Depending on the severity of the problem and the individual assigned to resolve it, the problem might be fixed quickly or not. Reporting seemed to be as error ridden as manufacturing. Indeed, the event that had precipitated the current crisis was determined to be a reporting error. Defects that had occurred during unit testing before shipment over the previous month had mistakenly been reported as field failures. This appeared to be a massive increase in field failures alarming management and alerting them to a sudden change in failure rates that had actually not occurred. The entire crisis in confidence with engineering had actually been a "reporting error" due to sloppy handling of defect-related data.

By the end of the day, a plan for fixing the factory's problems was emerging, the Analysis phase was completed, and everyone went home.

6. Improve Phase

By the time the team reconvened the next morning, everyone on the team had suggestions on how to fix the problems. But before detailing and prioritizing these solutions, the MBB had the team review the findings from the previous day and brainstorm solutions.

Brainstorming on possible solutions revealed that most technical problems had already been solved, even though defects were still occurring. Most of the technical defects that were still occurring were noted to be "latent defects" from older versions of units that were still in the field that had not been upgraded or replaced. In fact, it was suggested that, with other historical data, it would be possible to calculate how many units associated with these potential field failure modes were still waiting to occur. One of the possible solutions suggested was to identify these units and fix them or upgrade them proactively, before they actually failed. This would be a limited type of recall that would fix failures before they actually occurred.

The MBB noted that most of the remaining problems were pure process errors, as opposed to technical or component-related failures. These were related to deficiencies in the design, assembly, test, and installation processes that allowed errors to be made or allowed them to escape out into the field. The MBB suggested that the solution to these problems could be fixed by fixing these "culprit" process steps identified from the previous day.

The question for the team was "What was the best way to fix these problematic points in their manufacturing process?"

The MBB suggested the development of formal process documentation that would establish specific practices and procedures to be performed so that these activities could be performed consistently. Moreover, the creation of these documents would facilitate training of personnel and enable the measurement of compliance, performance, or quality at each step, and allow these steps to be audited later on to make sure that problems did not reappear at those locations.

Unfortunately, the team was uncertain how to document processes in such a manner. No one on the team had documented a process or written process procedures before. The MBB proceeded to pull out a copy of a process document he had and the document was stripped down to become a template for process development.

The MBB suggested that the team start writing procedures beginning with one subprocess that the team had been working on all week—the process of documenting and logging field defects.

Since the team was now confident about how to process defect-related data, all they had to do was document procedures that described what to do when failures occurred—how to capture, record, track, resolve, and report defects properly.

It took the team the entire morning, but by noon they had their first formal process document. It specified:

- ▲ The existence of their new defect database.
- ▲ The procedures for entering and recording defects.
- ▲ The methods for classifying the defects and determining the root cause.
- ▲ Specific reports (to be run monthly) and responsibilities for creating reports for the team and for management.
- ▲ It even included a formal control plan with proposed statistical process control charts and audit schedules.

The team had now formalized their role in managing defects and defect data, and established procedures to ensure that future problems would be identified, reported, and resolved efficiently. Sample reports from the week's work were included and folders were created on a shared server to store the new procedures, the new defect tracking database, and reports for future use.

The team was not experienced in writing a process document, but they had a lengthy list of other process steps that needed to be fixed in a similar manner. Moreover, the team realized that after these process documents were developed, everyone in manufacturing would need to be trained on these new procedures.

It was going to be a lot of work.

The MBB split the team into two groups once more. The first group worked to prepare a report for management on their findings. The second group worked to create a project plan for the creation of the remaining process documents. They estimated the time and effort that it would take to write these other procedures and they created a work breakdown structure scheduling this work over the next few months. They finalized their planning with a formal request for a technical writer so the engineering team could get back to their manufacturing and engineering activities.

By the end of the day, the team had documented their plan. They had a report for management about the nature of defects, accurate data and statistics that would reassure management about the nature of the defect problems, the magnitude of their risks, and associated root causes and solutions.

They had a plan with a newly documented procedure as an example, and a project schedule with staffing requirements to address the remaining process problems.

With solutions in hand, the team adjourned for the day and a meeting with management was scheduled for the next morning.

7. Control Phase

The next morning the team convened with management to make their final report. The MBB started the meeting; explained the crisis, the formation of the team, and the nature of the DMAIC project; and concluded with the charter.

The engineering manager and his staff presented the rest of the slides, explaining the problems, their analyses, the root causes, and their proposed solutions. They showed their slides and explained their conclusions. They explained the new metrics and reporting procedures. Finally, they offered their recommendations about changes to processes and procedures.

Management listened and was relieved. The magnitude of the problems was not as great as feared. The nature and causes of the errors were clear and the team's recommendations were well received. The planned solutions were accepted and the project plans, schedules, and estimates for staffing were accepted.

The decision was made not to initiate a product recall and to restart the manufacturing line while expediting the corrective actions to the process that had been recommended by the team.

After the meeting, the engineers were exuberant. Not only were they more confident in their understanding of the myriad problems they had been wrestling with, they were pleased that they had re-established their credibility with their management. Each of the team members had learned valuable new skills, and the team now had new tools and methods to help them in their daily jobs. They had a new appreciation for how data should be handled and for how Six Sigma techniques could be used to identify and remedy problems.

The MBB worked with the team to formalize "lessons learned" over the week and reinforced the importance of following through on the recommendations and plans they had developed. The MBB then met with the management team and re-emphasized the same points. Management was quite pleased with the results and was impressed with the volume and specificity of the engineering team's report that morning.

The project was officially closed and congratulations were offered to the team.

8. Benefit Realization

ROI was not formally calculated, but a rough estimate is provided here.

▲ Costs:
  ▼ Not including MBB travel, totaled about 25 staff-days (5 people × 5 days); estimated $10,000
▲ Benefits:
  ▼ Avoidance of a prolonged factory shutdown (which would have cost more than $150,000 per week) or a total product recall (which could have cost more than $1 million)

The ROI is thus estimated to range from 15:1 to 100:1, depending on the outcome that would have occurred without the DMAIC project.

9. Conclusions and Limitations

In technical terms, problems were found, identified, quantified, and analyzed. The elements of the manufacturing process responsible for these errors were noted and new procedures were identified to address activities at these steps. This was a classic example of DMAIC problem analysis and follow-through.

In logistic terms, what made this project unusual was the short time in which it was completed. Several factors made this short project duration possible. These factors included:

▲ Complete commitment by the management staff
▲ One hundred percent dedication of team members for the duration of the project
▲ Immediate availability of engineering and manufacturing data
▲ Access to robust data analysis tools
▲ Tailoring of the generic DMAIC lifecycle to focus only on those steps that were absolutely essential to project completion
▲ Facilitation by an experienced Master Black Belt with a strong manufacturing background

In the end, the recall was avoided. The plant continued operations. The engineering team learned new skills and vindicated themselves. New processes were implemented to forestall future problems of this kind. And a difficult problem was resolved in a surprisingly short time.

10. Lessons Learned

So precisely how was this project different from a normal DMAIC? Several factors distinguished this project from the typical Six Sigma project. First, the team was small. There was no committee or large body of stakeholders. Indeed, after project initiation, only a small core team was involved. Moreover, all the team members were subject matter experts and practitioners.

Next, the team was 100 percent devoted to the project for its entire duration, including four engineers (and the MBB), 100 percent of the time for one week!

Third, a smaller set of activities was used than on a typical DMAIC. Some of the normally performed activities were not relevant and were omitted, for example, gathering requirements, gathering voice of the customer (VOC) or voice of the business (VOB) requirements, extensive data gathering activities. Resultant work was specific to the project and more focused.

Last, data related to the problem was relatively easy to acquire. While not already gathered, it was generated quickly for a variety of sources, thus avoiding the delays associated with special metrics-gathering activities.

The fact that the problem was perceived to be a crisis was probably instrumental in getting the team assembled and participating.

Let's review some of these factors, albeit not in the sequence listed earlier.

10.1 DMAIC Process Tailoring

First, despite what you have learned in Six Sigma classes, it is *not* necessary to complete *all* tasks on a DMAIC checklist. The DMAIC techniques are tools for consideration, but no project will use all of these tools.

Six Sigma techniques are tools and the DMAIC process can be tailored to individual tasks. Six Sigma is like a hardware store. No project will require you to use all the tools in the store.

Practitioners need to know *how* to use specific techniques and *when* to use them (or not use them). Unfortunately, novice BBs often follow DMAIC checklists too rigorously (and often they do too many things). It typically takes a year to train a BB on how to follow the formal DMAIC process and two years to teach them how to deviate from it.

10.2 Percent of Time on Project (the Dangers of Multitasking)

The main reason that DMAIC projects take too long is that most of the time, people are simply not working on the project. Consider that if a project takes 40 hours of work for each team member but each team member only works 2 hours per week, then the project will take 20 weeks to complete. (If team members only devote 5 percent of their time, projects will take 20 times as long as they would if they were dedicated to the project.) As a rule, if the team cannot commit at least 20 percent of their time, then cancel the project. Ideal participation level is 50 percent to 100 percent.

Unfortunately, multitasking is a way of life in most organizations. People think it improves productivity. However, it should be avoided at all costs, especially for Six Sigma projects.

To understand the real dangers of multitasking, consider the following example of multitasking three projects (A, B, and C) where each project will take 40 hours or five days to complete.

In the first scenario, you do pure multitasking. On Day 1, you work on one project. On Day 2, you work on another; on Day 3, the third. Then you repeat this process (Figure 10.4).

Figure 10.4

Scenario 1:
Time sharing
(project
hopping).

Day

1	2	3	4	5	6	7	8	9	10	11	12	13	14	15
A	B	C	A	B	C	A	B	C	A	B	C	A	B	C

A = 13, B = 14, C = 15

Average completion = 14 days

Figure 10.5

Scenario 2:
Dedicated/
contiguous
projects.

Day

1	2	3	4	5	6	7	8	9	10	11	12	13	14	15
A	A	A	A	A	B	B	B	B	B	C	C	C	C	C

A = 5, B = 10, C = 15

Average completion = 10 days

With this scenario (taking turns to equally distribute work across all the projects), the first project finishes on Day 13, the second on Day 14, and the third on Day 15. Thus, the average project finishes in 14 days.

In the second scenario, we will work on each project until it is finished before moving to the next one (Figure 10.5).

In this case, project A finishes on Day 5, project B finishes on Day 10, and project C again finishes on day 15. While the last project still takes 15 days to finish, the other two projects are finished much sooner and overall, projects finish sooner with the same effort. The average project now finishes in ten days, a 39 percent reduction.

In reality, however, the improvements achieved by *not* multitasking are even better. This is because you do not have to take the time to stop and switch projects or take extra effort to pick up where you left off from where you stopped before. Because you spend less time switching back and forth, labor is more than 20 percent more efficient. Rather than projects taking 40 hours, they actually take less time—in the range of 32 hours. Thus, scenario 3 illustrates what can really happen if you decide not to multitask.

In this case, project A finishes on Day 4, project B on Day 8, and project C on Day 12, for an average project completion of eight days (a 43 percent improvement over our initial scenario).

In summary, multitasking is *bad*. It interrupts efficient work and delays all projects involved. Projects are forced to take much longer than they should.

So why do people multitask? It is because they honestly believe it will keep projects moving, when just the opposite occurs. The inability to prioritize projects motivates participants to try to give equal attention to all of the projects involved. Unfortunately, the real reasons projects take too long is that there is insufficient management commitment for any of the projects.

11. Final Lessons

Here are some do's and don'ts for projects in general:

▲ Don'ts:
  ▼ DMAIC projects don't have to take a long time.
  ▼ You *don't* have to do everything on the DMAIC checklist (keep your projects simple).
  ▼ You *don't* have to do everything in order.
  ▼ Avoid scope creep. Create a clear charter and stick to the problem at hand. Don't get distracted by other problems you discover. If they are substantive, you can always address them in other short projects.
  ▼ Don't multitask (keep commitment high and durations low).
  ▼ Don't involve people on the team who cannot immediately contribute (unless you want the project to turn into a training activity).

▲ Do's:
  ▼ As a rule, avoid starting DMAIC projects unless team members can dedicate at least 50 percent of their time to project activities.
  ▼ Keep teams small.
  ▼ Plan on highly dedicated resources (>50 percent).
  ▼ Plan on tailoring standard DMAIC activities (work smarter, not harder).
  ▼ Have a team leader who is an experienced Six Sigma practitioner.
  ▼ Finally, you *do* need adequate commitment to a project or you will waste everyone's time and delay the results you want and need.

DMAIC projects don't have to take a long time. They should ideally be fast and effective.

CHAPTER 11

How IBM Reduced Help Desk Escalations and Overhead Activities

Timothy Clancy

IBM

Relevance

This chapter is relevant to companies that want to reduce help desk ticket escalations by systematically identifying and eliminating problem root causes.

At a Glance

▲ IBM leveraged Six Sigma to reduce escalation of unresolved help desk tickets reported by one of its major outsourcing customers.

▲ The project achieved a 47 percent reduction in escalations and an 87 percent reduction in the cause that was the biggest contributor to escalations.

▲ By significantly improving a strategic pain-point with the client's help desk operation, IBM demonstrated the utility of Six Sigma and gained management and employee buy-in for further Six Sigma work.

▲ Financial savings were low, approximately four team lead hours per day, due to reduced escalations needing to be resolved, or $12,000 a month in labor among the four team leads working at the client site, but the techniques were easily replicated to other projects.

Executive Summary

The case involves IBM Strategic Outsourcing (SO) help desk support of a major New York investment firm. The help desk received, on average, between 700 and 1,000 calls daily, resulting in hundreds of tickets. Customers dissatisfied by open tickets could file escalations and over time, the percentage of all tickets that were escalations was creating strain on the support organization.

A project team was formed consisting of client and IBM participants with an initial goal of improving the escalation process. The challenge of a goal like this is it seeks to optimize the waste organization processing escalation rather than attacking the root cause of why tickets were escalated to begin with. A revised problem statement clarified the problem simply to be "reduce escalations," which allowed greater focus on the root problem.

The project team measured escalations for 30 days, placing them into segmented categories—both why the tickets had been escalated by the customer and where the originating tickets came from (help desk queues). A convergence among different segments emerged, indicating that a majority of all escalations were either directly or indirectly related to the wireless provisioning process because that process took too long and when completed, was inaccurate.

The project team held a *value stream mapping kaizen* to look at the wireless provisioning process and find ways to reduce cycle time and improve quality. Opportunities identified in that effort indicated that removing extra or unneeded steps might reduce the cycle time from 25 days to 9 days. The reduction in complexity, when combined with mistake-proofing, would also reduce the errors, which reduced the quality of the final delivery.

As implementation of the kaizen ideas proceeded, control charts were used to compare future state to baseline, both for wireless provisioning as a percentage of all escalated calls and total escalations as a percentage of all tickets. The control charts showed a dramatic (87 percent) reduction in escalations due to wireless provisioning, which corresponded to a 47 percent reduction in overall escalations and financial benefit of $US12,000.

The success of the project proved that a joint partnership utilizing the Lean Six Sigma methods could produce meaningful results for client pain-points. IBM was invited to present the project both to the client's CIO and IT directors, which built momentum and enthusiasm among key stakeholders for future cooperation between the client and IBM on the Lean Six Sigma deployment and projects.

Editors' Note: It is important to note that not all Six Sigma projects have monumental financial returns. It is a common expectation that most Green Belt projects should have at least $50,000 in net financial returns and Black Belt projects should be higher. However, some projects have priority due to factors beyond mere financial payback. Some projects have great importance for safety or security reasons; some are vital to maintaining brand name or image. Often, as in this case, projects are tests of Six Sigma methodologies and, as noted in our concluding chapters, Six Sigma programs often start small, and longer-term viability of programs as a whole is often built upon the success of smaller initial projects.

Financial returns are only one measure by which the success of projects can be measured. Not all benefits are quantifiable; some are intangible and more difficult to document. Success depends on many factors, and sometimes the value and impact of projects exceeds mere monetary gain.

1. Introduction

The IBM SO account had only recently begun deploying Lean Six Sigma projects into their environment. Unfortunately, as is the case of most initial deployments, early projects were stymied by a lack of full-time project resources, the overwhelming number of potential tools to select from, lack of a Master Black Belt (MBB) coach for the Black Belt (BB), and a general hesitancy over whether any of this would result in a real change. As is also often the case in early deployment, both the client and IBM Executive Champions were committed to deploying Lean Six Sigma, but the workers remained skeptical that this method could produce meaningful results or be anything more than a flavor of the month.

To help build momentum for the deployment and show results, a Black Belt candidate was selected to lead a joint IBM-client team on the specific pain-point of the relationship—help desk escalations—and a Master Black Belt was brought in to coach the project. IBM and the client both recognized that the project might not generate significant financial savings. However, both agreed that demonstrating the utility of Six Sigma to solve recognized pain-points presented a strategic opportunity to convince middle-management and employees to support additional Six Sigma projects. The Six Sigma project team consisted of the IBM MBB and BB, four to five subject matter experts (SMEs) pulled from both the client and IBM, and the two Executive Champions. A successful project with participation from both partners could help build enthusiasm and momentum for follow-up efforts by delivering meaningful results.

2. Project Background

As part of the relationship, IBM's help desk created tickets based on client technical inquiries. If a client wasn't happy with the progress or status of a ticket, they could "escalate" the ticket. At the beginning of the project, the reasons for the escalations were unknown, but what was known was that the escalation process created a lot of overhead and friction. A new type of ticket, escalation, was created and routed to team leads to investigate. The investigations of escalation tickets were accumulated and a database created to store the escalations with a new role defined as a "resolver" who receives the escalation and acts on it and a database to store the end results of the investigation. Although a small number of tickets each day were escalations, each escalation took quite a bit of time to resolve, about 53 minutes. Since escalations were resolved by team leads, the significant amount of time required for all resolutions affected their ability to manage their teams. Even though the financial cost of the time was not great—an estimated $12,000 per month—all agreed that the lost time of the team leads and client frustration due to escalations justified the expense of the project. Furthermore, if this problem could be solved with Six Sigma, it could increase interest and willingness in using Six Sigma in other areas within both the client and IBM staffs.

Once the escalation process was formalized, it was observed that more and more of all the tickets worked by help desk were actually related to escalations. This was beginning to consume a significant amount of their time spent.

Because the cause of the escalations was unknown, this problem lent itself to Six Sigma; however, the goals of the project were initially unclear and were to be clarified in the Define phase.

3. Define Phase

Dr. Ohno[1] refers to Japanese folklore where farmers spend half the year sleeping, only to awake and rush to harvest their fields in the last part of the year. Six Sigma projects can be much the same way—too much time spent up front, constantly revising and updating the charter and define documents, leading to rushing through the value-added portions of DMAIC, which most often occur in the Measure, Analyze, and Improve phases.

Delays in the Define phase occur when there is no clear focus in the problem statement and too much time is spent trying to get every detail of the Define toll-gate right without really knowing what is actually going on. In a way, this is natural—Six Sigma projects are about finding solutions where the cause of the problem is unknown. The key in the Define phase is to document what you know, admit what you don't, and move on. Everything you think you know in the Define phase may be called into question or invalidated when you get into the Measure phase, so why spend time documenting information that may or may not prove valid?

The proposed schedule of the project on escalations was to run January through August, with about 1.5 months per phase, but the Define phase ran clear through April, when the MBB joined the team.

The original problem statement indicated "user escalation tickets represented a time consuming process for helpdesk team leads (TL), the Install-Move-Add-Change (IMAC) coordination team, and the End User Support (EUSS) team to research, document, and follow-up on each escalation." A difficulty with this problem statement is that it is not clear what the goal of the project is and can easily lead a Lean Six Sigma (LSS) effort down the wrong path. Valid solutions to the problem might be to find solutions that decrease the time it takes to resolve an escalation, or the cycle time in which answers are returned to their customer. There's also a lack of focus—what are you supposed to measure: cycle time of resolution, cycle time for answers, customer dissatisfaction? Just as the project was to enter the Define toll-gate, the problem statement was rewritten to: "Significantly reduce the daily percentage of escalations (0.957 percent April/May) of all tickets from users to helpdesk, IMAC, and executive support team by August." Put simply—reduce the number of escalations.

3.1 Understanding the Core Defect

To understand why the original problem statement was flawed and the rewritten one was better, it is important to first understand what the core defect is and what are waste organizations. This is important, because many organizations utilizing Six Sigma often try to optimize waste organizations rather than fixing and eliminating core defects.

A core defect of unknown type creates the need for overhead activities to manage the symptoms or consequences of the defect rather than address the defect itself. These waste organizations can be self-perpetuating—beginning with establishing a mechanism to capture customer complaints, then a database to manage the complaints, mechanism for tracking and reporting status of customer complaints, and staffing of specialist roles to handle the complaint management system.

All of these are forms of overhead because none of them solve the defect themselves— rather, they just manage the waste that the defect produces, which is why they are called waste organizations.

Often companies will start a Six Sigma project aimed at incrementally improving performance in an overhead activity managing the defects. But, this is just an optimization of the waste, rather than the elimination of it. Instead, effort should be applied to identifying the core defect and eliminating it so that none (or very minimal) of the overhead activities are needed.

The project team revised the problem statement to "reduce the number of escalations" provided that level of focus on the core defect. This clarity would then lead the team through the rest of the DMAIC phases with a renewed focus: find what caused the core defect, which caused the volume of escalations, eliminate it, and keep doing that until escalations were reduced to a manageable level.

4. Measure Phase

4.1 Measure Symptoms Rather Than Causes

With the new problem statement in hand, the project team entered the Measure phase with a clear understanding of what they wanted to measure. What they found was that even though the client and IBM agreed on the problem of escalations, no one really knew why customers were escalating, or from where in the help desk processes those escalations originated. In other words, the waste organization was more focused on processing the *symptoms* of the core defect, rather than the causes. Even though plenty of data was available from the escalation database, it was the wrong kind of data. The project team had to create a segmented measurement system based on causes rather than symptoms in the escalation tickets. Once contributing causes were known, they would be tracked so that a Pareto analysis could be performed to identify what causes contributed the most to escalations.

4.2 Segmentation of the Measure:
Identifying Where and Why for the Escalations

The escalated help desk ticket already contained information regarding the originating queue. The "why" was determined by following up on previous escalations to find out why customers escalated tickets. The project team came to the following potential causes of why a ticket might be escalated:

1. **Expedite:** Any escalation that is received in either fewer than two days or in less than the service level objective (SLO) target.
2. **User Expectations:** Any escalation received in either fewer than two days or in less than the SLO target and that that does not fit the expedited subcategories listed.
3. **VIP:** VIP expedites will be highlighted as an acceptable expedite.
4. **Can't Work:** Those tickets that are escalated due to the user not being able to perform critical work.
5. **Work Flow:** Any escalation due to a ticket not being performed efficiently enough to satisfy the user.
6. **SLO Missed:** Any escalation by a customer on a ticket that has already missed the SLO or is older than two days.
7. **Process Break:** Any escalation where the ticket type has a defined procedure, but the procedure was not followed.
8. **Process Enhancement:** Any escalation where the ticket type either does not have any procedure defined or the existing procedure should be enhanced to improve efficiency and/or expediency.

These eight categories can be divided into two broad divisions: Expedite and Workflow. Expedites were when customers would escalate even though the SLO was being met because performance wasn't meeting their expectations. Workflows, on the other hand, were circumstances where the escalation indicated a failure to meet the established service level, as depicted in Figure 11.1.

It was tempting to eliminate Expedite escalations altogether, because those represent escalations where the customer is asking more than what was agreed to in the SLO. However, doing so is wrong on two fronts. First, the entire point of Lean Six Sigma is to understand

Figure 11.1

Segmentation of escalation call types.

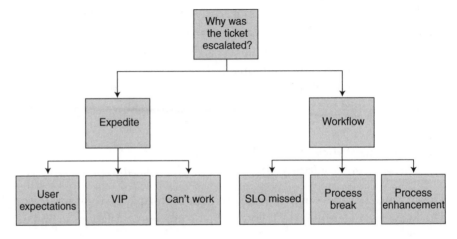

what customers want and how to deliver it, both effectively (from the perspective of the customer) and efficiently (from the perspective of the provider). Second is that you never want to artificially limit your data sets in the Measure phase based on assumptions (customers are asking for what wasn't agreed to) because you might miss key clues down the road in finding the root cause of a problem.

This segmentation provided enough granularity on two axes that when calls were tracked against these subcategories, a Pareto chart could be created to determine what was truly driving the majority of escalated calls. Working with her team, the Black Belt tracked all escalations for a 30-day period, manually categorizing every call within the segmentation. Escalations during the Measure period were also tracked as a percentage of all tickets, since this was the key performance indicator that the project was seeking to improve.

4.3 Use of Control Charts to Set the Baseline

The project team tracked escalations as a percentage of total calls for the control chart. For a 30-day period, every escalation was tracked as a percentage of all calls. Figure 11.2 shows the baseline measurement, taken from mid-April through mid-May.

Figure 11.2

Escalations as a percentage of total calls—All escalations baseline.

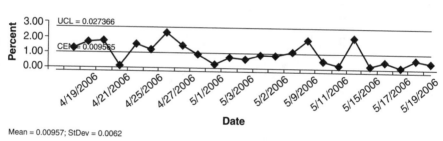

The mean of the process was that, on average, 0.9565 percent of all calls were escalated. The range of variation (lying between the UCL and LCL) was between 0 percent and 2.7366 percent of all escalations. These measurements were key—changes to the mean (lowering the bar) or shrinking of the variation (narrowing the gap between the UCL and LCL lines) over time would give an indication if the project was having any meaningful impact on the amount of escalations the help desk was receiving.

5. Analyze Phase

5.1 Identifying the Core Defect Through Pareto Analysis

After the 30-day data collection period that ran from April to May, the project team began segmenting the escalation calls using Pareto analysis.[2]

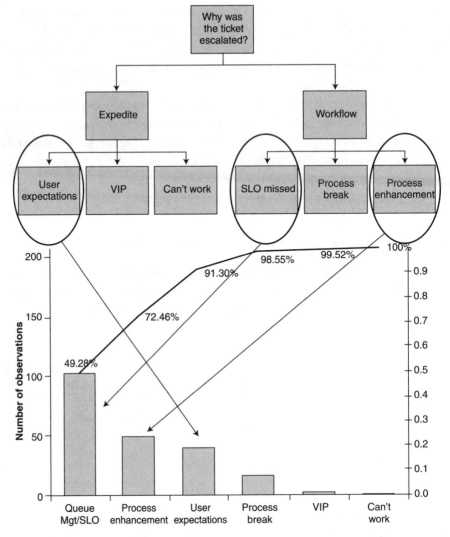

Figure 11.3

First-level
Pareto analysis:
Segmentation
causes.

The approach was to use the Pareto analysis to understand first *why* the escalation occurred (Figure 11.1) and then *where* it occurred based on the queue of the original ticket escalated. Note in Figure 11.3 how only a few of the subcategories in the Why segmentation diagram drive a majority of the causes.

The analysis now had three specific categories of "why" to focus on. They decided to drop process enhancements, because on further investigation, they had no commonalities that were measurable, leaving the team with missed SLO and user expectations. Running a second-level Pareto analysis, the team evaluated the "where" in terms of which queue originated each escalation, as applied to the top two categories of "why." The largest queues from which escalated tickets emerged were wireless (48.04 percent), messaging ops (18.63 percent), end-user support (14.7 percent), and procurement (8.83 percent); collectively, these four queues accounted for 90.2 percent of the escalations. Delving another level down into those four queues and running more Pareto charts yielded further interesting results. Within the wireless queue, the most significant root causes for all escalations were obtaining either a new or replacement device (51.85 and 11.11 percent, respectively). Also, some inquiries showed that e-mail/calendar not working (11.11 percent), transfer of liability (11.12 percent), and synching problems (3.7 percent) were likely to be encountered when operating a new device.

Looking back at the first-level Pareto analysis in Figure 11.3, the third largest source of causes was user expectations. The second-level Pareto analysis of user expectations by origination queue is depicted in Figure 11.4.

Note that the second cause, messaging ops, also has to do with wireless. But what about procurement? Taken on its own, it's somewhat vague, so the project team conducted a third level of Pareto analysis specifically on procurement, as seen in Figure 11.5.

Figure 11.4

Second-level Pareto analysis: User expectation escalations by originating queue.

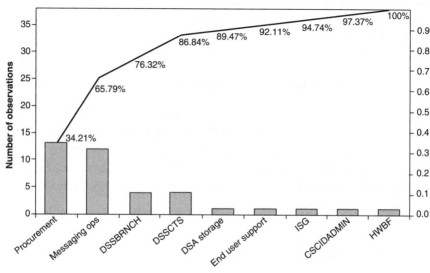

Figure 11.5

Third-level Pareto analysis: Procurement escalations by type.

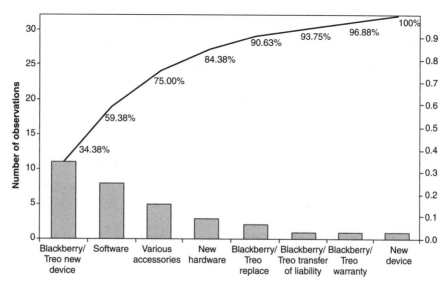

Note that many of the causes of procurement escalations, which is a primary driver of user escalations, are related to the provisioning of wireless devices, which was also the primary cause of escalations caused by queue management/SLO. Although not exclusively, many of the other queue management/SLO escalations could be partially attributed to the provisioning of new devices.

All of these Pareto analyses began to converge around a common core defect: Provisioning of new or replacement devices takes too long and/or is inaccurate.

This convergence revealed something interesting: customers were complaining about the process of obtaining wireless devices primarily because of how long it took, and even after many delays, the devices were still not set up correctly. It became apparent that the true core defect of escalations lay not in the escalation process itself, but in a wireless provisioning process that took too long and was inaccurate when delivered. Creating a positive change here might move the dial sufficiently on all escalations to achieve the project goal. From symptoms, the team had moved to cause.

5.2   Applying Value Stream Mapping (VSM) Kaizen to the Core Defect

As the convergence emerged from the Pareto analyses around a core defect, the team decided to create a value stream map (VSM) of the wireless provisioning process in a kaizen (for more on constructing VSMs, refer to Chapter 9). The goal of the kaizen was to map out the VSM of the wireless provisioning in the current state, identify areas of opportunity to reduce cycle time while improving quality, and begin to implement those areas within a few days. The current-state VSM is depicted in Figure 11.6.

5.3   Order and Delivery Current State

The VSM showed that the cycle time of wireless delivery led to customer escalations—and that the primary factors of cycle time were complexity, number of hand-offs to redundant functions that added no value, and the high rework on wireless delivery forms that had to be filled out with each ticket. Incorrect information passed throughout the process resulted in the errors users experienced when they received their device.

The VSM metrics of the current state of wireless provisioning are shown in Table 11.1.

Table 11.1   Current-State Metrics

VSM Metric	Current State
Complexity	7 groups, 28 steps, 9 hand-offs
Cycle time	26.8 days (routine), 25.7 days (expedited)
Value-added (VA) time	5.05 days
Process cycle efficiency (PCE)	19%

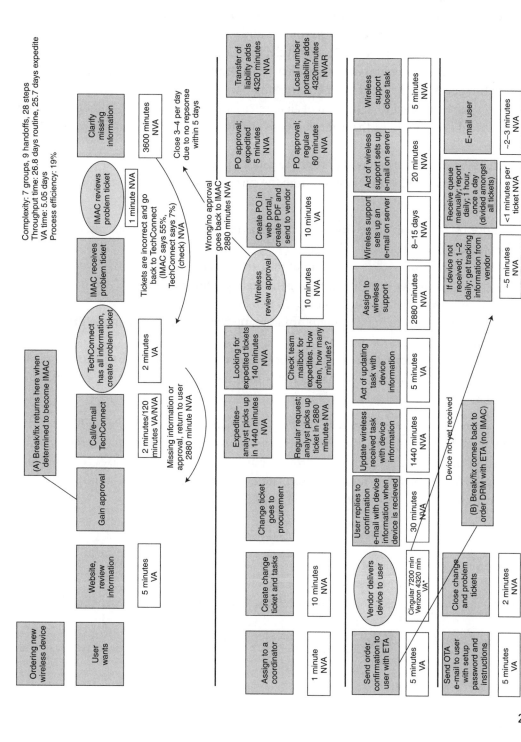

Figure 11.6 Value stream map, current state.

Complexity: 7 groups, 9 handoffs, 28 steps
Throughput time: 26.8 days routine, 25.7 days expedite
VA time: 5.05 days
Process efficiency: 19%

283

Complexity speaks to the opportunity for rework, mistakes, and extra time, as well as the loss of customer-provided information at the beginning of the process that leads to wireless devices being sent out incorrectly.

Every hand-off creates an opportunity for lost time and incorrect information to enter the process. Also, the difference between routine and expedited orders was enlightening. VIP expediting only gained 1 day on a 25-day cycle, but each VIP expedite rippled through the process, requiring everyone to stop what they were doing and work on the "rushed" order before getting back to what they had been working on.

The VA time indicated what portion of all the steps in VSM were actually adding value, and PCE indicated the amount of that time compared to the total cycle time, in this case, only 19 percent.

Each of these measures allowed the team to gauge a potential future state to see if there had been improvement. Notable observations from the current state and the improvements decided on to fix them were:

▲ Remove IMAC coordinators from the process, as the IMAC team did not add any value to the process. *Time savings: 4-8 days*
▲ Mistake-proof (poka-yoke) front-end of the process. *Time savings: 3-5 days*
▲ Remove searching for expedites (looking through queues to find tickets that needed to be rushed through). *Time savings: 2 hours, 20 minutes*
▲ Change tracking procedure to gain information up front rather than waiting. *Time savings if eliminated: 1 day, 35 minutes*
▲ Obtain tracking information from the vendor upon shipping. This information would be forwarded to the user in an e-mail communicating the expected time of arrival. *Time savings: 1 hour*

The future-state value stream with the proposed changes is shown in Figure 11.7.

The team then compared the VSM metrics from the current state to the future state proposed. If the core defect is that the provisioning of wireless systems takes too long and quality is too low, then the current and future-state VSM metrics should show improvements in both complexity and cycle time (see Table 11.2).

Table 11.2 Current vs. Proposed Future State

VSM Metric	Current State	Proposed Future State
Complexity	7 groups, 28 steps, 9 hand-offs	5 groups, 17 steps, 5 hand-offs
Cycle time	26.8 days (routine), 25.7 days (expedited)	7.77 days
VA time	5.05 days	5.05 days
Process cycle efficiency	19%	65%

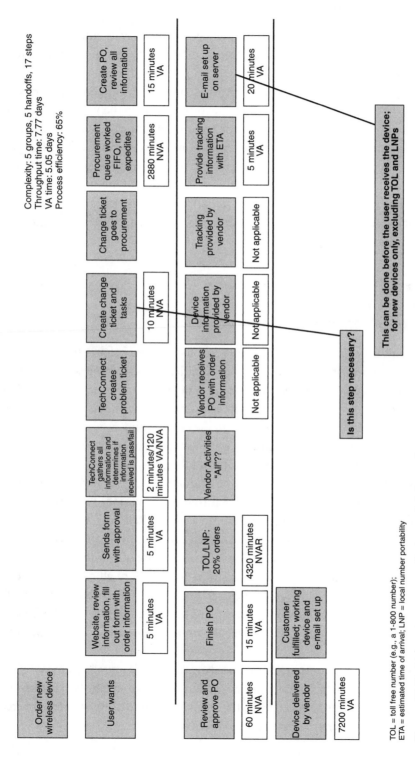

Complexity: 5 groups, 5 handoffs, 17 steps
Throughput time: 7.77 days
VA time: 5.05 days
Process efficiency: 65%

Order new wireless device						Create PO, review all information
						15 minutes VA
User wants	Website, review information, fill out form with order information	Sends form with approval	TechConnect gathers all information and determines if information received is pass/fail	TechConnect creates problem ticket	Create change ticket and tasks	Change ticket goes to procurement	Procurement queue worked FIFO, no expedites	E-mail set up on server
	5 minutes VA	5 minutes VA	2 minutes/120 minutes VA/NVA		10 minutes NVA		2880 minutes NVA	20 minutes VA
Review and approve PO	Finish PO	TOL/LNP: 20% orders	Vendor Activities "All"??	Vendor receives PO with order information	Device information provided by vendor	Tracking provided by vendor	Provide tracking information with ETA	
60 minutes NVA	15 minutes VA	4320 minutes NVAR		Not applicable	Not applicable	Not applicable	5 minutes VA	
Device delivered by vendor	Customer fulfilled; working device and e-mail set up							
7200 minutes VA								

Is this step necessary?

This can be done before the user receives the device; for new devices only, excluding TOL and LNPs

TOL = toll free number (e.g., a 1-800 number);
ETA = estimated time of arrival; LNP = local number portability

Figure 11.7 Value stream map: Order and delivery, future state.

285

6. Improve Phase

Both Executive Champions approved implementing improvements identified within the kaizen, and enthusiasm and motivation were again high, both within the project team and the IBM SO Account and client teams.

Not all the improvements would be implemented simultaneously, however. Most interesting is the finding that expedites didn't significantly help the VIPs they were entered for, but caused a great deal of process chaos. The team wanted to eliminate expedites altogether, but the client wasn't comfortable doing so until the average cycle time of wireless provisioning could be reduced from 25 days to 15 days while increasing accuracy. Implementing a combination of the improvements during the course of May, June, and early August reduced the provisioning time from 25 days to 15 days, and then, when expedites were removed and all requests worked first-in, first-out (FIFO), the cycle time plunged to 9 days, as depicted in the control charts in the next section.

7. Control Phase

7.1 Confirm Reduction of Core Defect

As the team put the improvements in place, they used control charts to understand whether the core defect, wireless provisioning, was significantly changing as a percentage of all escalation calls. Since the team had no baseline data, they started measuring from when they first discovered the convergence in mid-May; as kaizen improvements were implemented, they were marked on the control chart. If the fixes to the provisioning process significantly decreased the cycle time, in theory, there should be fewer escalations related to wireless and that would show. Figure 11.8 shows control charts indicating the progress from one month prior to the VSM kaizen to two months after, showing the dramatic change in wireless escalations.

Note there is some overlap from the 7/18 through 8/18 period. There were several out-of-control points in 7/18 and 8/22, which, when checked, resulted from an increase in escalations due to the process changes. (Initially, VIPs weren't too happy about having their expedites removed for those already in process.) So as a normal-cause of variation, the team kept them in. However, the mean consistently dropped from 24 percent of all calls in May to 18 percent in July to a negligible amount in late August and September, an overall reduction of nearly 87 percent. The variation also narrows substantially, meaning that the team's efforts improved both consistency and performance in terms of eliminating wireless calls as a percentage of all escalated calls.

Had the team's success meaningfully reduced escalation calls overall? They continued tracking from the original baseline throughout the kaizen and all improvements (see Figure 11.9).

Note that in the bottom two charts, the scale has been decreased from 0 percent to 3.5 percent and from 0 percent to 1.8 percent to increase clarity.

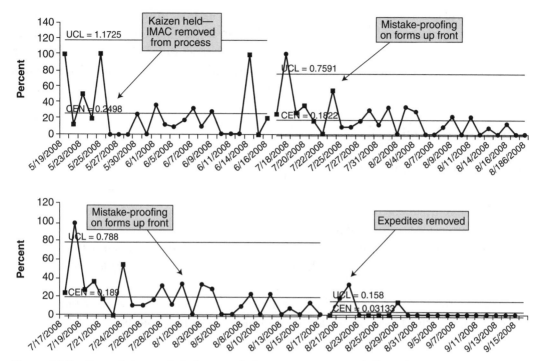

Figure 11.8 Control charts of wireless escalations as percentage of all escalations.

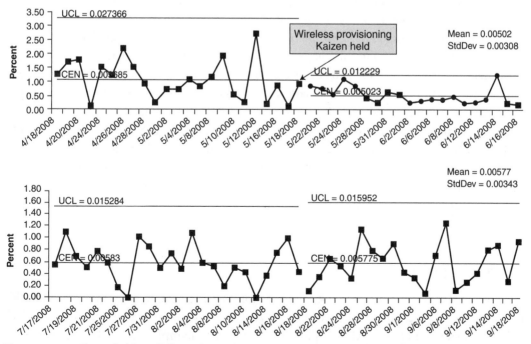

Figure 11.9 Control chart of all escalations as percentage of all calls May–September.

What these control charts clearly show is that the significant improvement in wireless provisioning escalations resulting from the kaizen in Figure 11.9 contributed to a sharp decrease in overall escalations as tracked as a percentage of all calls throughout July through September, a decrease of nearly 47.5 percent in all escalations and a decrease of 53 percent in variation.

8. Benefit Realization

The following benefits were realized as a result of this project:

▲ Forty-seven percent decrease in escalated tickets (tracked via control charts)
▲ Eighty-seven percent decrease in wireless escalated calls (primary contributor of all escalations prior to project start) (tracked via Pareto analysis)
▲ Savings of approximately four team lead hours per day due to reduced escalations needing to be resolved, or $12,000 a month in labor among the four team leads working at the client site
▲ Significantly increased client and IBM team enthusiasm for additional Six Sigma projects
▲ Techniques formed basis for replication on future Six Sigma projects

9. Conclusions and Limitations

This case illustrates the importance of fixing the core defect rather than the symptoms of the defect—reducing the cause of escalations rather than making them takes less time. Too many Six Sigma projects mistakenly aim to incrementally improve the waste organization.

The approach outlined in this chapter is easy and quick to replicate in many areas where a core defect is driving the expansion of a waste organization. By repeating these simple projects sequentially (in this case, find the next largest cause and reduce that) or in parallel (same type of project in multiple areas), new Black Belts can gain confidence in using these tools while building momentum and enthusiasm for the organization that LSS actually creates results, and gain buy-in from the client for future cooperative efforts on more complex problems.

The simplified set of tools, however, limits this approach. For super high-performing areas (already at or near five or six sigma), where the mere act of implementing a change might cause as much harm as good, more sophisticated modeling tools should be used to understand the impacts of change before changes are made.

Acknowledgments

The author of this case study is thankful to IBM for its permission to allow publication of this case study as a book chapter. The author would also like to thank Jessica Rosado, IBM SO Black Belt Candidate, the other IBM SMEs and the Executive Champion, as well as the client for their participation both as Executive Champion and providing SMEs. LSS can rarely be successful without a partnership between the provider and customer.

10. References

1. Ohno, T., Dr. (1988). *Toyota Production System: Beyond Large-Scale Production.* New York: Productivity Press.
2. Pyzdek, T. (2003). *The Six Sigma Handbook: Revised & Expanded.* New York: McGraw-Hill.

CHAPTER 12

Motorola Realizes Significant Cost Avoidance by Streamlining Project Documentation

Vic Nanda

 **MOTOROLA**

Relevance

Most companies today are beset with one common challenge—too much documentation. This chapter is relevant to organizations that want to reduce process or project documentation and make it "just right." The chapter describes a step-by-step method developed at Motorola that can be used to identify and eliminate *nonvalue-added* document content, thus realizing significant resource savings.

At a Glance

▲ Motorola IT launched a Lean Six Sigma project to reduce project documentation.

▲ The project team reduced project documentation by 35 percent and significantly reduced the effort required to complete project documentation.

▲ The project resulted in annual cost avoidance of US$1.08 million and a return of investment of 10.57.

Executive Summary

A fundamental challenge for many software development and support organizations today is that of excessive documentation—either process documentation that practitioners have to follow or documents that they have to create during the course of their everyday work. Formal research has also shown that most software documentation is not updated consistently and the need is for powerful yet simple documentation strategies and formats (templates) that software engineers will likely maintain.[1] This challenge is further compounded by the fact that, due to nonvalue-added activities, there are almost always inherent inefficiencies in an organization's processes. This problem often goes unnoticed or is acknowledged but remains unaddressed due to lack of organizational expertise or an effective method to tackle this problem. This results in precious resources being wasted in performing nonvalue-added work, and it limits the organization's ability to respond with the *speed* that today's customers and markets demand.

Like most organizations, Motorola IT faced this challenge. In the 2008 Digital Six Sigma (DSS) jumpstart for one of Motorola's large IT organizations, senior management, led by Motorola's CIO, sponsored a Six Sigma project to streamline the common development process (CDP) used to design, develop, and deliver large IT software development projects to Motorola's business units (these were projects with a budget greater than US$100,000).

The CDP is a suite of common processes, standards, and tools for software and service delivery based on capability maturity model integration (CMMI) and support processes based on the Information Technology Infrastructure Library (ITIL) framework.[2,3]

It was to be a rapid improvement blitz effort that was to be completed within six months. During the jumpstart meeting, senior management identified a Black Belt (the author of this case study) to lead this effort and authorized the creation of a cross-functional Six Sigma project team (hereafter referred to as *project team*).

Due to the vastness of CDP, the project team and sponsor agreed that a voice of the customer (VoC) survey would be conducted during the Define phase of the project to help define the specific scope of this project. The VoC survey was administered to a sample size of approximately 500 IT employees worldwide, and had an approximately 20 percent response rate. Overall, results of the survey revealed that there was 65 percent dissatisfaction with the CDP, with 60 percent of those being *slightly dissatisfied*. Likewise, there was 60 percent dissatisfaction with the project documentation associated with large IT projects, with 60 percent of those being *slightly dissatisfied*. Statistical analysis of survey results showed significant correlation between dissatisfaction with CDP process and dissatisfaction with large project documentation. The project team and project sponsor thus agreed upon streamlining of large project documentation as the specific scope of the project.

The project team leveraged the traditional concept of process value stream mapping, including value-added and nonvalue-added activities, and developed an innovative analysis method for streamlining documentation. The chapter describes this method and provides detailed guidance on its application with one real-world example from Motorola. Similar to the pictorial representation of an organization's value stream in traditional value stream maps,[4] this method provides an equivalent way to pictorially represent value-added and nonvalue-added content in a document by use of a *document value map* (DVM). During the course of a one-week kaizen blitz workshop, the project team applied this method and analyzed 549 sections and subsections in project document templates and deemed 190 of them to be nonvalue-added (including redundant sections), *an improvement of 35 percent over the baseline*. In addition, 47 of the 142 medium and high effort value-added sections were simplified to require less effort to complete, *an improvement of 33 percent over the baseline*.

Thirty-two requests for changes were submitted as a result of the workshop, with a minimum annual cost savings of US$1.08 million. The project yielded a return on investment (ROI) of 10.57 times the cost.

Although the analysis method devised as part of this project was applied to streamlining of project documentation, it is readily applicable to streamlining of any documentation, such as process documentation. Also, it was our experience that use of DVMs provided an excellent means to visually highlight elements in each document that needed to be streamlined, enabled easy comparison of document value before and after streamlining, and was thus an effective tool to systematically conduct documentation streamlining, communicate streamlining results, and calculate ROI due to the improvements.

1. Introduction

In 2008, Motorola's CIO and senior management from the IT organization participated in an annual DSS jumpstart for the IT organization of one major business unit of Motorola: Business-2-Business IT, or B2B IT. The purpose of this annual event is to launch Six Sigma projects that are directly tied to key initiatives and objectives on management scorecards. This event was a half-day exercise, and it was jointly facilitated by the leader of Motorola IT's Six Sigma program and the author of this chapter.

Senior IT leaders began by listing what they considered to be their top-five metrics (as applicable to their role) from the set of metrics listed in the B2B IT annual scorecard. After all the metrics had been collated, each leader was asked to vote for their choice of the top three metrics in the consolidated list. In order to avoid inadvertent CIO influence on the voting by the IT leaders, the CIO voted last. All the votes were tallied and yielded the final set of top-five metrics.

For the top five metrics, each IT leader was asked to write down a maximum of five initiatives that would favorably affect the metrics. Some flexibility was allowed in this step. For example, an IT leader could choose to list two initiatives for one metric and none for another. Further, for each suggested initiative, the respective IT leader was asked to rate the anticipated impact of that initiative (low, medium, or high) and the complexity of that initiative (low, medium, high). After all the initiatives had been listed, each IT leader was asked to vote for their choice of the top three initiatives in the consolidated list that included initiatives proposed by all IT leaders. Again, the CIO voted last in order to avoid any inadvertent influence on the voting. All the votes were tallied and yielded the final set of top initiatives, prioritized in order of number of votes received. At the end of this exercise, the proposed initiative to streamline the CDP received the second-highest number of votes.

The jumpstart concluded with assignment of senior IT leaders as project sponsors for the top initiatives, including this project. The CIO specifically requested that the project to streamline the CDP be a quick turnaround project, delivering streamlining recommendations for the organization in less than six months. The author of this chapter was assigned as the

Black Belt to lead this Lean Six Sigma project. A preliminary identification of key project team members was also completed during the jumpstart.

2. Project Background

Motorola IT's CDP is one common suite of processes and tools comprising processes for:

▲ Designing, developing, and enhancing IT applications, which are referred to as *Build* processes and are founded on CMMI

▲ Maintaining the capability of the IT infrastructure and sustaining IT applications after deployment, including defects and ongoing maintenance (excluding functional enhancements), which are referred to as *Run* processes and are founded on ITIL.

The CDP also comprises process assets that include *standardized* procedures, work instructions, templates, forms, checklists, methods, and tools for use.

Just as with any business process, formal and informal practitioner feedback from using the CDP over the years uncovered opportunities for process improvement, primarily around improving process efficiency by eliminating nonvalue-added activities in the process.

3. Define Phase

The Define phase of the project entailed completing the following major activities: formation of the project team and creation of the project charter and detailed project plan.

3.1 Project Charter

After the jumpstart, the Black Belt for the project started working with the project team to draft the project charter for review with the governance team and final approval by the project sponsor.

3.1.1 Business Case Statement

The key elements of the business case were:

▲ In the annual B2B IT DSS jumpstart, senior management requested that the CDP be streamlined to improve IT operational efficiencies.

▲ Results of an earlier internal IT survey (unrelated to this project) indicated a need to streamline the process and requests for practitioner involvement in the effort.

▲ There was a similar CDP streamlining request from the jumpstart in another IT organization within Motorola in 2007.

3.1.2 Opportunity Statement

The expected benefits from the project were the following:

▲ Deliver higher business value:

   ▼ A significant amount of hours were expended to deliver large IT projects. Removing activities that were nonvalue-added (NVA), that is, activities with limited or no customer or business value, Motorola IT could realize valuable savings in resource effort expended on IT projects.

   ▼ Assigning resources to other additional IT efforts would deliver greater value to the business.

▲ Decreased costs:

   ▼ Elimination of NVA activities from the CDP would help reduce the cost of IT projects.

   ▼ Lower operating expense (cost avoidance) would be delivered to Motorola, estimated to be US$576,000 (refer to the "Benefit Realization" section).

3.1.3 Goal Statement

The project goal was to "streamline the common development process (CDP) for large IT projects by reducing the percentage of non-value added activities* by at least 5%[†] by Dec. 15, 2008."

The project due date was reviewed with the project team to ensure its acceptability, given that extensive preparation work was necessary prior to the one-week kaizen blitz workshop.

3.1.4 Project Scope

The scope of this project included:

1. Reducing NVA activities in CDP (for *Build* processes) for large IT projects across Motorola IT. Note: Large projects are defined as greater than US$100,000 in total cost, including internal and external expenditure.

2. Streamlining interfaces to the *Run* processes of release management and change management because they directly interface with the *Build* processes.

The scope of this project excluded:

1. Reducing NVA in the CDP for small projects to develop and deliver defect fixes and fulfill maintenance requests after release of an IT application.

2. This project would not examine process flows in the Run process of release management and change management that are not invoked during delivery of IT projects, but are for the purpose of managing change requests received *after* deployment of an IT application.

---

* After conducting a VoC survey in the Define phase of the project, the project team identified CDP project document template streamlining as the specific focus of this project.

† This percentage matches the actual improvement realized from a Six Sigma project earlier in the year to streamline the front-end portfolio management process for IT projects. Further, due to business reasons, the CIO placed a constraint that there be no radical changes to the CDP process at this time. Therefore, while more improvements to the CDP may be possible by undertaking radical redesign (refer to Chapter 17), a goal of more than 5 percent improvement would have been unrealistic, given the comparison to results of a similar Six Sigma project and the constraints for this project.

3. This project scope was limited to the CDP only and thus excluded streamlining of documentation in IT organizations that had recently been acquired through business acquisitions by Motorola and had not yet deployed the CDP.

3.1.5 Project Plan

The high-level project plan included in the charter contained milestone dates for the end of each of the phases in the DMAIC lifecycle of the project: Define, Measure, Analyze, Improve, and Control.

3.1.6 Project Team

The project team comprised:

▲ One Black Belt
▲ Nine practitioners covering all processes in scope of the streamlining and representing all IT organizations
▲ Four process owners, including overall CDP owner and key process owners who reported to the overall CDP owner
▲ One Master Black Belt
▲ One kaizen coach

The governance team for the project comprised:

▲ One project sponsor
▲ One project Champion
▲ One representative from the finance organization
▲ Immediate supervisor of the Black Belt leading the project
▲ Twelve stakeholders responsible for Build and Run processes, including the overall Sarbanes-Oxley (SOX) lead for IT

3.1.7 Voice of the Customer Survey

Because the Build processes in CDP span 18 different process areas (of which 17 are key process areas in CMMI) and more than 200 process assets, a goal to streamline all the processes would have been far too ambitious and unrealistic for a single Six Sigma project. The project team concluded that conducting a VoC survey with the IT employees would be the best way to obtain practitioner feedback on which process areas should be the subject of this streamlining effort.

A short ten-minute VoC survey was created and administered via the intranet to approximately 500 IT employees worldwide from all functional areas and IT organizations. Subsequent analysis of 91 survey results (including 83 free-text comments) provided valuable insight into streamlining opportunities as perceived by the process users, including elements of the CDP with highest user dissatisfaction.

Survey results showed that 6.6 percent of the respondents were very dissatisfied with the CDP, 19.8 percent were dissatisfied, and 38.5 percent were slightly dissatisfied, resulting in overall dissatifaction of 64.9 percent. The remaining 18.7 percent were neutral and 16.4 percent were satisfied. Similarly, 60.5 percent of the respondents were dissatisfied with the project documentation associated with large projects (5.5 percent were very dissatisfied, 18.7 percent were dissatisfied, and 36.3 percent were slightly dissatisfied). The remaining 26.4 percent were neutral and 13.1 percent were satisfied.

A side-by-side comparison of these two key satisfaction measures is shown in Figure 12.1. The scale on the Y-axis is as follows: 1 = very dissatisfied, 2 = dissatisfied, 3 = slightly dissatisfied, 4 = neutral, and 5 = satisfied. The median satisfaction with CDP and project documentation was 3 (slightly dissatisfied).

Extrapolation of the VoC survey results to all of Motorola IT indicated that one could say with 95 percent confidence that satisfaction with large project documentation was between 3.00 and 3.45 (see Figure 12.2), likewise the satisfaction with CDP was between 2.95 and 3.42.

Because the CDP dictates the quantity and content of documentation required for large projects, and the level of dissatisfaction with the process and documentation were similar, a logical question arose: "Is overall dissatisfaction of IT employees with CDP *correlated* with their dissatisfaction with CDP large project documentation?"

A correlation study was done, and the Pearson coefficient of 0.52 established a moderate correlation (a value of 0 would have implied no correlation, while a value of 1 would have implied strong correlation). Further, the p-value of 0.000 was less than 0.05 and thus the null hypothesis that there was no correlation between the two factors was rejected and the alternate hypothesis was accepted. As a general rule, correlation does not naturally imply causation, due

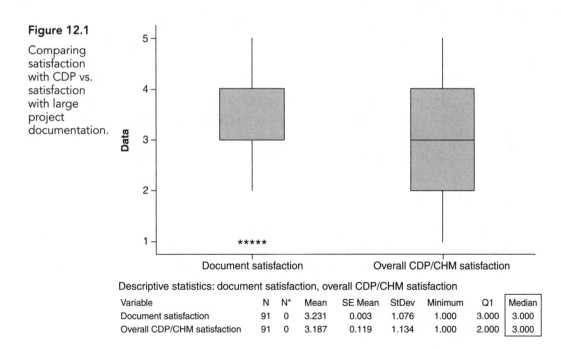

Figure 12.1

Comparing satisfaction with CDP vs. satisfaction with large project documentation.

Descriptive statistics: document satisfaction, overall CDP/CHM satisfaction

Variable	N	N*	Mean	SE Mean	StDev	Minimum	Q1	Median
Document satisfaction	91	0	3.231	0.003	1.076	1.000	3.000	3.000
Overall CDP/CHM satisfaction	91	0	3.187	0.119	1.134	1.000	2.000	3.000

Figure 12.2

Extrapolating
VoC survey
results for
project
document
satisfaction.

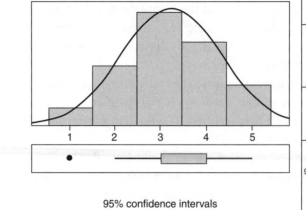

Anderson-Darling Normality Test	
A-squared	3.27
P-value <	0.005
Mean	3.2308
StDev	1.0758
Variance	1.1573
Skewness	−0.093308
Kurtosis	−0.540554
N	91
Minimum	1.0000
1st quartile	3.0000
Median	3.0000
3rd quartile	4.0000
Maximum	5.0000
95% Confidence Interval for Mean	
3.0067	3.4548
95% Confidence Interval for Median	
3.0000	3.3006
95% Confidence Interval for StDev	
0.9390	1.2596

to the aforementioned relationship between the two (quantity and content of large project documentation being dictated by CDP), in this case, correlation implied causation.

Additional survey results provided data on the number of streamlining requests for each project document template. In order to prioritize the list of templates for streamlining, the streamlining requests were tallied and sorted in descending order.

Finally, the project team conducted affinity analysis on the 83 survey comments, and the resulting Pareto diagram is shown in Figure 12.3. Results showed that the 27 comments were for *reducing documentation* and *improving template usability* (23 and 24 suggestions, respectively), and 32 requests were for simplifying the process* (see requests for "simplify process," "reviews and approvals," and "right-size process to organization/projects").

Based on the VoC survey results, the project team recommended to the governance team that reducing documentation be the specific focus of this project, and that there be a separate Six Sigma project to streamline the process. This recommendation was accepted, and the team started prework tasks in preparation for a one-week kaizen blitz workshop.

4. Measure Phase

This phase involved creation of a measurement plan (Table 12.1). This provided pertinent information about the measurements that were to be collected to baseline current process performance and to calculate ROI from the project (see the "Benefit Realization" section).

---

* Due to the much larger scope of this request, there was a separate Six Sigma project to simplify the process, and this is described in Chapter 17.

Figure 12.3

Pareto diagram of VoC comments organized by affinity analysis.

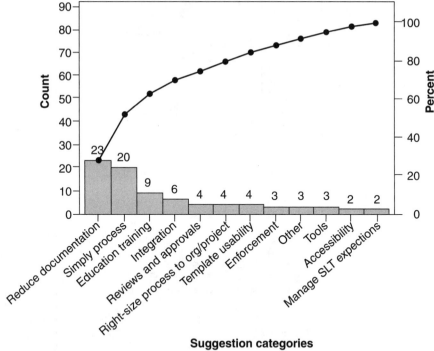

In preparation for the one-week kaizen blitz workshop, the project Black Belt and kaizen coach started developing an approach to adapt the traditional process value stream mapping method for use during documentation streamlining.

The proposed method entailed creation of a DVM as shown in Table 12.2. The DVM provided a one-page visual summary of the document sections in terms of:

1. **Value-added sections:** These are categorized as:
   - ▲ Customer value added (CVA): Document content that is deemed of value by the customer. That is, it specifies requirements or implementation to change the requirements into the final product.
   - ▲ Operational/business value added (OVA): This is document content whose removal would jeopardize the organization's ability to deliver the product or service within cost and schedule, and per agreed scope and predefined quality criteria. Information that is subsequently used in the process (or future projects), or is necessary to support the product or service after release (for example, design documentation), or is otherwise critical to the flow of the process is also OVA. Finally, information that is required by senior management at Motorola (in IT or in the business units) is also considered OVA.
   - ▲ Required for Sarbanes-Oxley (SOX) compliance.

   Also, the effort involved with completing these sections was classified using color-coded cells to represent Low Effort (green), Medium Effort (yellow), or High Effort (red).

Table 12.1 Measurement Plan

Performance Measure	Operational Definition Describe Defect or Metric	Data Source and Location	Continuous or Discrete	Display Analysis Tool	Sample Size Cost? Practical?	Who Will Collect the Data?	When Will Data Be Collected?	How Will Data Be Collected?
Count of non-value added document sections and sub-sections	Any information that customer or business is not willing to pay for to be documented, managed, and maintained and results in wasted effort (resources) Considerations: Overproduction (document information subsequently not used) Inventory (too much)	CDP document templates	Discrete	Spreadsheet	All CDP templates that are in scope for the workshop, as determined from the VOC/VOB survey	Black Belt and individual Track leads	Analyze phase	Calculated based on team consensus on which sections and subsections are nonvalue-added per the streamlining criteria provided to the team

How Will Data Be Used?
Data will be used to form initial baseline for value map of the document and implement streamlining improvements

How Will Data Be Displayed?
Spreadsheet

2. Nonvalue-added (NVA) sections

3. Redundant sections: Information that is duplicated elsewhere in the same document, in other documents, or in tools. As per Lean Six Sigma concepts, redundant information is considered nonvalue-added. However, during this project, it was considered in a separate category because the project team explicitly wanted metrics on how much information in CDP documents was clearly redundant.

Note that "sections" refers to document sections, down to the lowest-level subsections in the document. For example, in a spreadsheet template, each field, row, or column of data may be considered an individual section. This is because during document streamlining, streamlining opportunities are most likely to be found in lower-level subsections as opposed to higher-level sections. Say a document section X contains three subsections x.1, x.2., and x.3. In this case, while section X might be value-added, if we do a value analysis on its subsections, only subsection x.1 might be value-added while subsections x.2 and x.3 might be nonvalue-added or redundant! Therefore, during any document streamlining exercise, it is important that the streamlining team examine the lowest-level subsections in the document. All such subsections are expected to be listed in the first column of the DVM shown in Table 12.2.

Table 12.2 Document Value Map (DVM)

Document Section Heading	Candidates to Eliminate or Reduce Note: Place Check Mark		Candidates for Streamlining (Red or Yellow Cells)			Notes
	NVA	Redundant	CVA	OVA	SOX	
Header sections	X					
Section 1		X				
Section 2	X					
Section 3				Low effort		
Section 4		X				
Section 5			Medium effort			
Section 6				Low effort		
Section 7					Low effort	
Section 8			High effort			
Section 9		X				
Section 10	X					

Save this worksheet and create a copy or separate version before streamlining the document

↑ Focus on these first   **↑ Focus on these next**

At the start of the kaizen blitz workshop, a DVM was completed for each document and saved separately. The DVM provided a succinct visual representation of potential improvement opportunities in the document. Subsequently, during the kaizen blitz workshop, the project team agreed on recommendations to:

▲ Maximize elimination or reduction of *nonvalue-added* and *redundant* document sections (this was the *primary goal* of the workshop)

▲ Reduce effort involved with completing sections that were deemed value-added but *Medium Effort (ME)* or *High Effort (HE)* (this was the *secondary goal* of the workshop)

At the end of streamlining, each document's DVM was updated to reflect changes in the value and effort classification of each section. For example, several nonvalue-added and redundant sections were marked for deletion, and several Medium Effort sections were now Low Effort sections, while several High Effort sections were now either Medium Effort or Low Effort sections. This enabled comparison with the DVM of the document before streamlining (also referred to as the *baseline*). Besides serving as a powerful tool to depict improvements as a result of streamlining, it enabled calculation of improvement metrics. Table 12.3 shows data collected from the baselined DVMs as number of document "sections before" streamlining, number of sections that were "NVA before" streamlining, number of sections that were "redundant before" streamlining, number of value-added sections that required ME before streamlining, and number of value-added sections that required HE before streamlining.

5. Analyze and Improve Phases

This phase was divided into three subphases to execute a kaizen blitz workshop. Kaizen is a Japanese word typically translated to mean "continuous improvement." Traditionally, kaizen was used to refer to gradual improvements made over time.[5] However, in recent years, the methodology has evolved to include the *kaizen blitz* (also referred to as a kaizen event)—a rapid improvement project that delivers significant improvements within a short period, typically as little as one week.[6, 7]

5.1 Kaizen Blitz Kick-Off

Two weeks before the kaizen blitz workshop, the project's kaizen coach trained the project team on Lean Six Sigma and kaizen blitz concepts. This training also covered topics such as:

▲ Value stream mapping

▲ Definition of value-added and nonvalue-added

▲ Seven types of waste typically encountered in business processes (overproduction, inventory, transporting, waiting, processing, unnecessary motion, and defects)

▲ Kaizen blitz methodology to deliver rapid improvements in the shortest possible time

Table 12.3 Baseline Data from Document Value Maps (Before Streamlining)

	Sections Before	NVA Before	Redundant Before	ME Before	HE Before
Configuration Item List Template	6	3	1	0	2
Customer Requirements Template	35	14	5	10	4
DAR Template	18	4	1	4	1
Development Requirements Template	59	18	9	25	6
Gap Analysis Template	13	6	0	4	0
Peer Review Template	29	16	6	1	0
Process and Project Waiver Template	17	1	0	0	0
CM Plan Template	32	11	2	3	2
QAP Template	20	6	5	2	0
Release Plan Template	17	7	4	3	1
Release Support Notification Template	20	5	0	8	0
Requirements Traceability Matrix	13	7	0	6	0
Support Sustenance Plan Template	31	8	4	5	12
Technical Design Template	21	7	1	11	0
Test Report Template	31	7	1	9	0
Universal Test Plan Template	51	21	13	6	1
Communications Plan Template	40	11	13	0	0
Lessons Learned Template	10	6	2	2	0
Test Strategy Template	13	8	0	2	0
Project Kickoff Template	28	15	11	1	0
PM Plan Template	39	14	5	9	0
RASCI Matrix Template	6	4	0	0	2
Total Document Sections	**549**	**199**	**83**	**111**	**31**

ME: medium effort; HE: high effort

The training was followed by a presentation of:

▲ Detailed agenda for the kaizen blitz workshop.
▲ Assignment of each core team member to one of three tracks to discuss streamlining suggestions (each track was a logical grouping of templates by type, that is, engineering templates, project management templates, and support templates).
▲ Detailed guidance on what constituted customer value-added content, operational (business) value-added content, SOX required content, redundant content, and nonvalue-added content.

▲ Definition and quantification of Low Effort, Medium Effort, and High Effort. For example, a company may choose to define Low Effort as 0.25 person hours, Medium Effort as 1 person hour, and High Effort as 2 person hours.

▲ Each of the three tracks was assigned a track lead to moderate the discussions, and they were provided with a suggested order for reviewing documents for streamlining opportunities. This list was sorted by the number of streamlining requests received from the practitioners for each document (see the section "Voice of the Customer Survey").

5.2 Kaizen Blitz Prework

As part of the mandatory prework for the kaizen blitz workshop, each core team member was required to review all templates assigned to their track, looking for streamlining opportunities as per the guidance provided during the kaizen blitz kickoff. This prework spanned two weeks. Reviewers were required to note all their improvement recommendations for discussion within their respective track meetings during the kaizen blitz workshop.

As part of this prework, each track lead generally sampled five completed documents for each of the templates to take into account how the templates were actually being completed in projects. This provided valuable insight into what document sections were inconsistently completed across projects (possibly due to unclear instructions); were skipped altogether because they were possibly deemed nonvalue-added; or generally took an excessive amount of time to complete, which indicated a need for reducing the effort needed, where possible, for completing those sections.

This prework was the key to the success of the workshop, because it enabled the project team to be thoroughly prepared for the workshop. Consequently, during the workshop, the team was able to cover far more templates than it might have otherwise been able to cover.

5.3 Kaizen Blitz Workshop

During a one-week kaizen blitz workshop, each track lead led their team in reviews of the templates assigned to their respective tracks:

▲ First, the track lead, working with their team and by consensus, completed the DVM to capture document value prior to streamlining.

▲ Second, using the DVM as a constant reference, the team reviewed each document in detail, discussing improvement recommendations they had identified from the prework to eliminate nonvalue-added and redundant sections, and to reduce effort involved with Medium and High Effort sections.

▲ Third, the team updated the DVM, again using team consensus, to agree on document value after streamlining.

▲ Finally, for each document, using the baselined DVM and the DVM after improvements, the team computed improvement metrics as shown in Table 12.4. Note that in this example, 12 of 14 nonvalue-added sections were eliminated and 3 of 5 redundant sections

Table 12.4 Actual Example of Improvement Metrics for a Document

	Baseline (Before Streamlining)	After Streamlining	Improvement
Number of document sections	39	20	19
Number of NVA sections	14	2	12
Number of redundant sections	5	2	3
Number of high-effort sections	0	0	0
Number of medium-effort sections	9	4	5

were eliminated, which is a net removal of 15 sections for the 39 sections originally in the document. In addition, some other sections were consolidated with other sections, resulting in 20 sections in the revised template.

At the end of the workshop, improvement metrics for all documents were consolidated. Reductions in document sections (including nonvalue-added and redundant sections) and reduction in effort required to complete value-added sections were shown using bar charts as shown in Figure 12.4. The number of NVA sections was reduced from 199 to 108—a reduction of 45.73 percent; the number of redundant sections was reduced from 83 to 11—a reduction of 86.75 percent; the number of value-added medium effort sections was reduced from 111 to 85—a reduction of 32.44 percent; and the number of value-added high-effort sections was reduced from 31 to 10—a reduction of 67.75 percent.

Figure 12.4

Overall reduction in document sections.

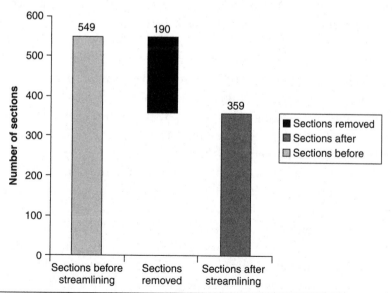

	Sections before	Sections after	Sections removed
Total document sections	549	359	190
Percentage improvement			34.61

6. Control Phase

As a result of the kaizen blitz workshop, 32 formal requests for change (RFCs) were submitted against the templates. Besides the reduction in document sections and reduction in effort involved with completing value-added sections, the kaizen blitz workshop recommended retirement of two templates and the merger of four templates with other templates in the CDP. Effectively, this resulted in the elimination of six different templates.

Each request for change included several improvement recommendations, which were documented in the Notes column of the DVM for that document (Table 12.2). All the RFCs were reviewed and approved by Motorola IT's Change Advisory Board for CDP assets. Each RFC was assigned for implementation to the next planned release of CDP.

7. Benefit Realization

The benefit realization for this project was computed twice:

▲ The *estimated benefit* was computed during the Measure phase. It was a rough estimate of expected benefit based on certain broad assumptions.
▲ The *actual benefit* was computed at the end of the project and was far more accurate.

The estimated benefit was computed as follows: Based on actual project data from 86 large projects in B2B IT prior to this project, the project team calculated that mean effort expended on large projects in B2B IT was 2,232 person hours. Extrapolating the data for the full year showed that 191,952 person hours were spent in B2B IT on project activities (including documentation). Estimated resource cost (including contractors) was $11,517,120.00. Assuming equal distribution of effort across project activities and with the assumption that most of the work performed at organizations is nonvalue-added,[8] a 5 percent reduction in the estimated effort for the projects would result in annualized savings of US$576,000 for B2B IT alone. Because the benefits of this project would be realized by all of Motorola IT, the actual benefit would be much higher—more likely two or three times this amount. Subtracting the project cost of $93,480 yields an estimated net benefit for the first year of $482,376—an ROI of 5.16 times the investment.

This project resulted in 35 percent reduction in the number of document sections prior to the start of the project, including 46 percent reduction in nonvalue-added sections and 87 percent reduction in redundant sections. Note that in Lean Six Sigma, it is accepted that it is not realistic to completely eliminate nonvalue-added content because several indispensable activities, such as document reviews, approvals, and testing, are considered nonvalue-added as per Lean Six Sigma definitions of value-added and nonvalue-added activities. Also, although most of the redundancy in documents can and should be eliminated, in some cases, redundancy is necessary to enable quick access to important information.

The actual benefit was computed as follows:

First, the actual effort expended in creating project documentation using the templates *before streamlining* was collected by interviewing several practitioners. The team was interested in finding out how much time it took a person to record information in a project document template if the person had done the necessary background work to gather the information required to create a document. This is because the scope of this project did not include examining and streamlining background process activities, such as requirements gathering, but the recording of collected information in project documents. For example, in the case of requirements definition, once the requirements had been gathered, the project team wanted to know the mean time it took practitioners to simply record all the information in the requirements template. Effort data gathered for all the project documents is shown in Table 12.5.

Table 12.5 Actual Effort Data (in hours) for Project Templates (After Streamlining)

Project Document	Best Case	Most Likely (Avg)	Worst Case
Customer Requirements	2	12.67	20
Project Kickoff Presentation	2.5	4	6
Project Management Plan (PMP)	3.5	5.2	7
Development Requirements	6	21.4	40
Communications Plan	0.5	1.58	4
...			
DAR	10	10	10
QAP	10	10	10
RASCI	1	3.34	6
Total Duration	**156**	**280.48**	**448**

Next, a Monte Carlo simulation with 2,000 runs was executed, and results showed that the project team could say with 95 percent confidence that the mean amount of effort required for completing all project documentation in a large project before streamlining was at least 228 person hours, rounded from 227.83 hours (see Figure 12.5).

Again, this was simply the time needed to record the information and does not include time spent to execute the process activity, such as, for instance, brainstorming, discussing, and establishing optimum design for a new software application in meetings, which is a separate and much more time-consuming task, compared to simply recording the software design in a document. The effort estimate of 228 hours was conservative because it only assumed one requirements and one design document for a large project when, in fact, large projects typically have one product requirements document and several underlying software requirements specifications, as well as several software design documents for each of those software requirements specifications. Therefore, the actual effort expended was most likely even higher than the 252.50 hours upper limit shown in the Monte Carlo simulation.

Figure 12.5

Monte Carlo simulation for actual effort spent on project documentation before streamlining.

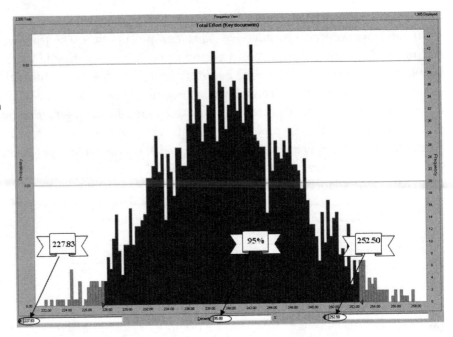

The effort data (228 hours) was multiplied by the number of large projects for the year and hourly resource rate, and showed that the B2B IT organization alone was minimally spending US$1,764,720 annually to complete project documentation.

After template streamlining, some anecdotal information had started to come in, with practitioners reporting significantly less time to create project documents, but because large projects take considerably longer to complete, the project team would have to wait for several months to again interview practitioners to gather effort data for completing project documentation after streamlining. Therefore, the team brainstormed an alternate way to compute actual benefit realization by quantifying the effort savings due to removal of document sections and the effort savings by reducing certain High Effort sections to either Medium or Low Effort sections, and by reducing certain Medium Effort sections to Low Effort sections.

This was done as follows:

▲ One hundred sixty-two of the 190 sections eliminated were deemed either nonvalue-added or redundant, while the rest were either High Effort or Medium Effort (redundant with other sections). After examining several nonvalue-added and redundant sections, it was observed that the vast majority were Low Effort sections. Therefore, by multiplying the number of Low Effort sections removed by effort savings for each Low Effort section (assumed to be 0.25 hours for the purpose of this project) and the number of large projects in the year and hourly resource rate, the actual effort savings was US$638,550.

▲ Likewise, effort savings due to 36 Medium Effort sections (assuming Medium Effort section takes one person hour) becoming Low Effort was: 36 sections multiplied by 0.75

person hours (reduction in effort), multiplied by number of projects and hourly resource rate yielded savings of US$208,980.

▲ The reduction in effort due to 10 High Effort sections (assuming each High Effort section takes two person hours) becoming Medium Effort sections and 11 High Effort sections becoming Low Effort sections yielded savings of US$234,135.

Adding these savings yielded a net savings of US$1,081,665, a 62 percent reduction in resource effort (compared to the project goal of US$575,856).

8. Conclusions and Limitations

The hidden waste in an organization is not only the inherent inefficiencies and waste in the organization's business processes, but also the documentation produced by the process. Because software project documentation consumes significant time and effort in software development projects, in order to reduce waste and save resources, a company must streamline the documentation as well.

This chapter described an innovative method to extend the application of the traditional concepts of Lean Six Sigma and value stream mapping to streamline a company's documentation. The results produced by this method are dependent heavily on having the right subject matter experts scrutinize each document carefully using specific criteria to assess if the content is truly value-added, and to assess the amount of effort typically involved with completing the value-added sections. Having a common set of criteria for use by all reviewers enables quicker consensus on decisions regarding desired document sections and their structuring so as to minimize nonvalue-added and redundant content in a document. Also, having a clear set of definitions regarding Low Effort, Medium Effort, and High Effort sections facilitates evaluation of effort involved in completing value-added sections and exploration of ideas to reduce effort.

Acknowledgments

The author of this case study is thankful to Motorola for its permission to allow publication of this case study as a book chapter. The author would also like to thank Leslie Jones and Judy Murrah for their review comments.

9. References

1. Timothy C. Lethbridge, Janice Singer, and Andrew Forward, *How Software Engineers Use Documentation: The State of the Practice*, IEEE Software, 2003.
2. Capability Maturity Model Integration (CMMI), http://www.sei.cmu.edu/cmmi, Software Engineering Institute, Carnegie Mellon University, Pittsburgh, USA.
3. Information Technology Infrastructure Library (ITIL), http://www.itil-officialsite.com, OGC, UK.

4. Tony Manos, "Value Stream Mapping—An Introduction," *Quality Progress*, June 2006.
5. Masaaki Imai, *Kaizen: The Key to Japan's Competitive Success*, McGraw-Hill, 1986.
6. Anthony C. Laraia, Robert W. Hall, and Patricia G. Moody, *The Kaizen Blitz*, Wiley and Sons, 1999.
7. Anthony Manos, "The Benefits of Kaizen and Kaizen Events," *Quality Progress*, February 2007.
8. David P. Spencer, "Lean on IT: Applying Lean Manufacturing Principles Across the IT Organization," White Paper, Infosys, May 2007.

CHAPTER 13

Boiling the Ocean with Value Streams, Kaizens, and Kanbans

Timothy Clancy

IBM

Relevance

This chapter is relevant to practioners of Lean Six Sigma (LSS) confronted with the need to drive broad improvements across large complex organizations—"boil the ocean" efforts. Such large improvement efforts often have a negative connotation because they involve too much complexity to manage through a DMAIC methodology. However, this chapter demonstrates how high-level value stream kaizen events and use of kanbans and heijunkas can help drive broad sea-change efforts across diverse functions of an enterprise.

This chapter assumes an advanced level of knowledge in LSS, but references are made where appropriate for additional reading that can help inform on terms found within.

At a Glance

▲ The project team had a goal of dramatically decreasing the 280-day cycle time for end-to-end server deployment while achieving other benefits.

▲ Value stream maps, kaizens, kanbans, and heijunka boards were used in an eight-month effort (April–November 2007) spanning three different companies and involving 27 functional teams (24 of which were geographically dispersed) along with more than 100 subject matter experts.

▲ Cycle time reduced by 20 percent to from 280 days 226 days.

▲ Financial benefits included 40 percent reduction in financial penalties, 95 percent reduction in inventory, and freeing up of more than one hundred workers, resulting in savings of several million dollars.

Executive Summary

This case involves outsourced IT support to IBM for a major credit card company's data center operations. One of the client's core business activities is to run marketing and promotion campaigns for customers and vendors. With the development of Web 2.0, the technical nature of these marketing campaigns became ever more complex—frequently requiring sophisticated databases, web-based applications, and numerous interactions with the existing infrastructure of loyalty programs, card management tools, and other systems. The client's data center, located in the southwest of the United States, was one of three global locations where the servers running this infrastructure were housed, containing by some estimates close to 10,000 systems. One of IBM's responsibilities was to respond to client requests for service (RFSs) for new server installations and provide end-to-end deployment services.

The project office at IBM was frustrated with service delivery, as was the customer—server deployment took too long; processes were variable and subject to heroics to complete, and once completed, were often wrong. The resources were also frustrated, as they wanted to do a better job but could not improve the interdepartmental coordination to reduce waste. The cycle time for delivery of complex servers was nearly 300 days, and the client expressed a desire to reduce this cycle time to 13 days, a reduction of nearly 97 percent!

Executive Champions from the client's senior management and IBM agreed to sponsor a project utilizing Lean tools adapted for IT, referred to as Kaizen-IT. First, a high-level value stream map (VSM) identified the entire end-to-end process from request for a new server to successful provisioning of the new server. This identified the extent of the ocean that required boiling. Second, a series of sequential and cascading kaizens were conducted in the core, enabling, and dependent activities to set the ocean to boil. Finally, 21-day and 91-day checkpoints after each kaizen kept the ocean boiling. The introduction of kanbans and heijunkas to control workflow and encourage continuous improvement within each area ensured that processes would not backslide.

Over the course of eight months—from April 2007 through November 2007—a total of eight major kaizens were conducted in different areas and hundreds of employees were trained in the Kaizen-IT method. This chapter highlights the work at the high-level VSM, which helped choreograph efforts across eight kaizens, in particular, two of those eight kaizens: RFS of servers and server build kaizens. Together, these two kaizens reduced the overall cycle time by 20 percent in addition to other financial benefits.

1. Project Background

Prior to the start of this project, the client had experienced a controversial top-down, corporate-directed Lean deployment focused on reduction of costs rather than process improvement. The IBM executive responsible for service delivery to the client described the previous effort as:

> "What didn't work was a top-down Lean effort that had no consideration for the customer's needs or the resources performing the tasks. This dictatorship approach

undermined the momentum of the customer-focused account approach ... with the employees and the customer."

The project team consisted of two IBM Master Black Belts, including the author, and two IBM Black Belts. Two Executive Champions, a delivery process executive from IBM, and a vice president of the client helped provide sponsorship on both sides of the fence. Over the course of the effort, several hundred subject matter experts (SMEs) either directly participated in kaizens or as part of changeover efforts.

Because of the previous unsuccessful Lean effort, the new project team could not discuss their methods as Lean, else they would not be accepted by employees or the client. Instead, the methods were named "Kaizen-IT." Throughout the chapter, references to Kaizen-IT can be interpreted as modified Lean and Six Sigma specifically for the transactional environment. Both Master Black Belts had worked together previously but they had different backgrounds in Lean—the author in technology and transactional environments, and the other in manufacturing. The diversity of viewpoints based on different experiences and perspectives was a key success factor in tackling the problem.

The contrast between successful application of Lean concepts to manufacturing and transactional environments led to the development of Kaizen-IT methods. Transactional environments found in technology, healthcare, financial, and public sectors are characterized by the intersection of highly variable, if well-intentioned, human work and imperfect information. These environments compound both structural complexity (number of moving parts) and interactive complexity (how interrelated to one another those moving parts are). With more moving parts and the interactions between those parts constantly varying, a holistic view—systems of systems—must be taken. So, kaizens grow in size from a few participants to 20 to 40, just to include the appropriate number of experts to understand the system in question. It's also hard to see the environment or make any physical measurements of reliability because they are virtualized—processes conducted solely through a computer (and these days that could be a desktop, laptop, or smart phone) and running on a corporate, home, or cloud computing network. These processes receive, manipulate, modify and supply digital information streams. Geographically dispersed workers execute many of these virtualized processes, so there is no one location where a process can be physically observed. Instead, process activities are functionally subdivided and dispersed into hundreds of small tasks managed through elaborate ticketing systems. This means that creating visibility and transparency, hallmarks of Lean, must be approached in fundamentally different ways. To be clear, the concepts and underlying theories of Lean are still applicable, but it is the methods that must be modified, a point recognized in the Preface to the English edition of Dr. Ohno's seminal work.[1]

To reflect this modified approach, the major sections of this chapter have been renamed from DMAIC as follows:

1. Mapping the ocean: High-level value stream map (Define and Measure)

2. **Setting the boil:** Sequential and cascading kaizens (Analyze and Improve)
3. **Keeping the rapid boil:** 21-day and 90-day checkpoints, kanbans, and heijunkas (Control)

Since this deployment, elements of Kaizen-IT methods have been deployed successfully by IBM at several different companies and public-sector agencies with similar transactional complexity.

2. Mapping the Ocean—High-Level Value Stream Map

The first step was to interview SMEs to understand the nature of the problem. It was like the proverbial handling of the elephant. Every SME knew their specific task but no one understood the whole process. This functional view of the elephant, split across two dozen or so technical teams, is depicted in Figure 13.1.

As the project team continued asking questions, it learned about the business needs that drove client server deployment—typically, new marketing campaigns connected with Web 2.0 applications required server deployment to support the applications and databases for the new campaigns. The last thing the client wanted to worry about was whether the servers were there or having to push the marketing out even a few days due to a server-related delays.

The Executive Champions were loud and clear about client expectations. In Figure 13.2, where "stuff happened," cycle time had to be reduced from about 280 days to a stretch goal of 13 days—a 97 percent reduction. The project team began building out from the "stuff happens." The high-level VSM was decomposed into four broad sections, referred to as core activities.

The project team now had a high-level understanding of how work flowed. It then took the documented "as-is" processes and laid them out underneath these four core activities to create timelines and identify key opportunities.

From this simplified version of the wall chart, one can begin to understand the benefit of the high-level VSM: depicting visual flow of work, beginning with customer desire and ending with customer fulfillment.

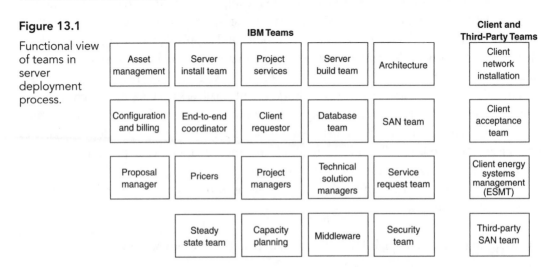

Figure 13.1

Functional view of teams in server deployment process.

Figure 13.2

Expansion of high-level VSM into four core activities.

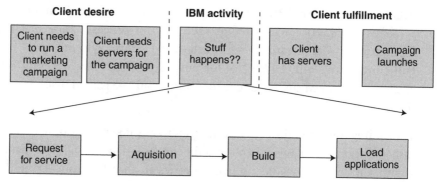

Editors' Note: The VSM for this project is not shown in its entirety. Only some of the key metrics and the results are presented for the readers. It is important to recognize that a VSM is much more than a simple process map that shows process steps and their sequence. The VSM process collects a lot of data that fully characterizes the process and includes details like processing time, waiting time, queue lengths, scrap and rework rates, and more; details that indicate, not only the means but the variation of these measures as well. For more details on VSMs, refer to Chapter 9, "Lean Primer."

The project team knew that the high-level VSM wasn't completely accurate, however, it gave an indication of where to begin. It helped alter the view with which everyone perceived the problem—from one of functional teams (see Figure 13.1) to the high-level value stream itself (see Figure 13.2).

Acquisition was not included in the initial efforts because it was not a large contributor to cycle time, and as a function of IBM Corporate, it was outside the initial scope of influence.

2.1 High-Level VSM Metrics

"One of the things we did that I haven't seen very much of in high-level VSMs before is the use of different strategic metrics in addition to cycle time and throughput … you can also look at dollars, manpower expenditures, quality metrics, work in process (WIP), revenue in process (RIP)—identify whatever is your strategic objective to change and lay that out on the VSM to visually see where those metrics accumulate, disperse, or stagnate in the flow of value."

—IBM LSS MBB

For this effort, the project team identified the following strategic VSM metrics:

▲ **Cycle Time:** The total cycle time of the process.
▲ **Work in process (WIP):** The dollar cost of the servers IBM was paying for but hadn't been delivered to customers and therefore couldn't be offset by revenue.
▲ **Revenue in process (RIP):** The potential revenue waiting to be realized when the client begins paying for successfully completed work.

▲ **Penalties in process (PIP):** Even though penalties charged for missing service level agreements (SLAs) are assessed at the end of a process, those penalties are accrued throughout a work stream wherever delays and lost time occur.

▲ **Full-time equivalent (FTE):** Not for purposes of cost removal, but to help identify if efficiencies between current and future states could be gained by improving quality while reducing cycle time.

The findings on the high-level VSM metrics were somewhat notional because there was no established tracking mechanism in place at the time (see Table 13.1). Note that this chapter only deals with kaizens that would improve metrics in the RFS and Server Build categories. Additional kaizens briefly described here and shown in Figure 13.4 targeted other areas (Acquisition and Load Application) to achieve management's goal for an overall cycle time reduction to 13 days.

Table 13.1 High-Level VSM Metrics (Current State)

VSM Metric Current State	RFS	Acquisition	Server Build	Load Application	Current State
Cycle time	~68d	~10d	~48d	~156d	~280d
RIP	Confidential	Confidential	Confidential	None	High
WIP	None	None	Unknown	None	Very High
PIP	Confidential	Confidential	Confidential	Confidential	Confidential
FTE	Several hundred across all activities				300–400

Although some of the numbers in Table 13.1 cannot be disclosed, they were quite large, easily numbering over eight figures. Because the project team had shifted from a functional to a value stream view of work, it was also hard to distribute findings into specific core activities and accurately vet them.

Revenue was only earned from the client at the completion of Server Build, leading to RIP in RFS, Acquisition, and Server Build. However, even that number was not exact, because, as the project team learned in the RFS kaizen, the process took so long that many clients would put in a request for a server, implying RIP, to "get their spot in line" and only submit the real order as they closed in on the needed date for the marketing campaign.

For WIP, IBM only had to pay for the servers between when they were acquired and the application transition, where client developers could begin accessing them and thus began paying. This is why WIP only shows accumulation in the Server Build phase.

FTE measurements also proved problematic. A single FTE from the storage area network (SAN) team might participate in the RFS to provide solution advice, work with the server build team when configuring and later deploying the SAN, and finally be consulted in the Load Application phase. Was that a partial FTE, one FTE, or three FTEs? The project team didn't know the answer, but it knew that taken as a whole, FTEs were in the hundreds.

3. Setting the Boil—Sequential and Cascading Kaizens

If you're going to burn down a forest, you need to start a lot of fires. Likewise, to drive fundamental change in a large, diverse, and complex company, you're going to have to work multiple efforts. The key improvement vehicle of Kaizen-IT is the use of kaizen events to drive a rhythm of change.

> "The kaizen approach was the magic. Value stream mapping of the current morass of processes, and the wall walks of the mess of the customer and the account team leadership helped them see how the situation was not any one department's problem, but it was everyone's responsibility … including the customer."
>
> —IBM PROJECT EXECUTIVE CHAMPION

Figure 13.3 lays out a notional three-month lifecycle of a kaizen event.

3.1 Conducting Kaizens in a Transactional Environment

As discussed in the project background, kaizens in transactional environments can look a lot different from those in the manufacturing environments. This is not to say that other elements of traditional kaizens are absent—the focus remains on identifying what activities are value-added versus muda (waste), as well as making changes by the end of the kaizen week.

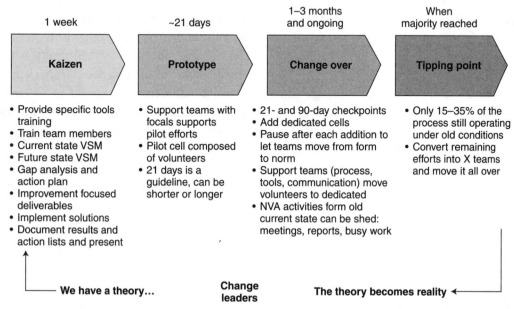

Figure 13.3 Three-month notional kaizen plan.

Figure 13.4

Sequential and cascading kaizens aligned to high-level VSM.

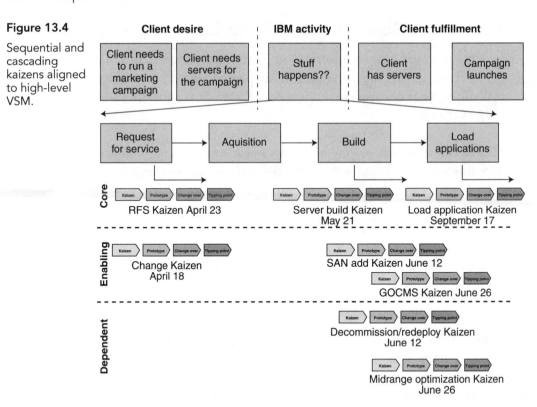

Figure 13.4 depicts how the project team went about starting the fires that would bring the ocean to a boil—core activities from left to right in the high-level VSM would be targeted with *sequential* kaizens, then *cascading* kaizens would target enabling and dependent activities. The approach was that each kaizen should be nominally independent of one another to allow each area to improve before linking up with others. The project team provided continuity as the facilitators of all kaizens, and as it turned out, many members of the functional teams would participate in more than one kaizen, since each functional team participated in all areas, bringing an additional level of continuity. The high-level VSM metrics also helped frame the context for all kaizens, that is, the overall goal of reducing cycle time, PIP, WIP, and RIP.

This approach allowed successful kaizens to spark momentum across the value-chain, lighting another fire to boil the ocean as it were, while a less successful kaizen wouldn't jeopardize all the progress. Kaizens were scheduled so as to be frequent enough to maintain momentum.

3.2 Steps to Run a Kaizen

The following approach was used for all kaizens:

1. Just-in-time training
2. Create the current-state VSM
3. Value-added (VA) analysis

4. Muda analysis
5. Rework analysis
6. Calculation of current-state VSM metrics
7. Record current-state observations
8. Create future-state VSM
9. Compare current- and future-state VSM metrics

For additional readings on any of these areas, refer to Ohno;[1] Womack, Jones & Roos;[2] Womack & Jones,[3,4] Pyzdeck,[5] and Balle & Balle.[6] Keep in mind that most traditional literature puts kaizens and Lean concepts in the context of a manufacturing environment and these may require translation or modification to fit the transactional environment.

3.2.1 Just-in-Time Training

The project team only trained the SMEs at the beginning of each kaizen on what they needed to know for that kaizen. One of the Master Black Belts explained the approach as:

"We didn't waste much time in bringing people up to speed; we did one to two hours of training only in what they needed, which kept things moving and kept people excited. Everyone got talking about the process."

3.2.2 Create the Current-State VSM

Using sticky notes on a wall, begin documenting the current-state VSM. Using sticky notes is important; process complexity almost guarantees that no two people in the room will perceive the process the same way. Processes in transactional environments start out as complex—your VSM, when completed, should not be a picture of perfection, but it should reflect accurately the reality of what work the workers do.

3.2.2A Customer Desire and Fulfillment

Book-end your VSM with customer desire and customer fulfillment, the true start and end of a process, these reflect the true value stream as it flows end to end.

3.2.2B Group Legend

On the lower left, under customer desire, there should be a legend of what the different-colored sticky notes represent in terms of groups. Any other key symbols, for example, your rework, value-added, and muda markers, can go here.

3.2.2C Document Process Steps from Left to Right

The value stream is built by placing sticky notes representing activities progressing from left to right across the wall, starting with customer desire. The main flow represents the "correct path" with activities in the VSM proceeding from left (desire) to right (fulfillment). Rework loops diverge up or down away from the proper path and backwards "upstream" from right to left. For each activity, identify the duration of work in the lower left (calendar time) and the effort in the lower right (people time).

Always ask where mistakes occur, and on any step where a mistake occurs more than 5 percent of the time, document a rework loop with a red sticky indicating what causes the rework (e.g., incorrect specification or missing information), the percentage of time that rework occurs, and how long it takes to resolve the issue. Keep in mind, in some key steps, you'll find many rework loops, each occurring more than 5 percent of the time. These are key pain points to resolve down the road.

When documenting process steps, be specific as to what activity is actually going on. Take an e-mail exchange as an illustrative example. One might say it takes a day to get a response on an e-mail, but it's not really a full day to get the response. It's actually several hours of the e-mail sitting in the recipient's inbox, a few minutes for them to respond, and then several more hours of the e-mail waiting for us to read it. The "response" takes only a few minutes; the "wait" many hours. A good rule of thumb is to pursue granularity to the next smallest measure of time, one unit more than that used to measure the VSM cycle time. If VSM is measured in months, pursue individual actions down to days; if measured in days, break it down to hours. For example, if on a VSM that takes six months a worker says that "Design Document" takes one week to create, be sure to ask questions about how that time might be subdivided into component steps. A more accurate reflection of reality might be that the week is really three steps: "Wait for All Information to be Finalized" (2 days), "Create Design Document" (1 day), and "Fix Problems in Document" (2 days). The one activity of designing the document is actually three, some of which may be VA and most of which is probably nonvalue-added (NVA).

If many different software applications are used in a process step then list them under each process step. Having to move back and forth between different applications to complete a step may introduce the muda of transportation for the information being entered and muda of motion for the worker who has to access multiple applications.

3.2.3   VA Analysis

Once the process is mapped, go back and identify which steps add value. Determining this is definitely more of an art than a science because it's most likely to lead to fisticuffs and riots. But this is important. Conflict is important—conflict is useful. Conflict raises the reality of disagreement that has been embedded in the process and may represent hidden perceptions that are critical to capture. If you can't determine what truly is value-added to the client, bring a client in the room—they won't hesitate to tell you. Just keep in mind that circumstances matter with clients as well—a transactional user may have a different view of what is value-added than a CFO; both perceptions are important as long as VA is a reflection of client perception.

3.2.4   Muda Analysis

Once the VA steps are identified, everything else, by default, is NVA, and all NVA is a form of waste. That doesn't mean that the future state will eliminate all forms of waste—sometimes, we have to continue with activities we know clients aren't interested in because we can't run the business without them; or they represent legal requirements. In Lean for manufacturing, there is a standard set of mudas to consider—most of these hold true for Kaizen-IT, with some important clarifications or additions to account for the kinds of waste found in virtual work environments where the "product" might be information and how a worker utilizes it. For Kaizen-IT, the project team used eight mudas, listed In Table 13.2.

Table 13.2   Eight Types of Muda and Examples

Muda type and description	Examples
Overprocessing: Output of products or service beyond what is needed for immediate use	Extra information collected that isn't needed and isn't used Excessive paging or alert notifications that don't produce an action Reports that do not produce a decision or action
Transportation of goods or information: Unnecessary movement of materials, products, or information	Hand-offs of information between groups that use different applications/databases Escalations Expedites
Motion of workers or customers: Excessive movement, either physical or virtual, to accomplish a task	Multiple software applications to complete a process step Customers forced to navigate complex websites, call centers, phone trees
Inventory: Physical or virtual areas where work idles accumulate	Databases with old or inaccurate records Ticket queues Excessive documentation
Wait: Any delay between when one process step ends and the next step begins	Waiting for approvals Events are handled in sequence rather than in parallel Interruptions due to expedites or escalates
Defects: Any aspect of the service that does not conform to customer needs	Incorrect contact information that leads to "tracking down" the right contact Multiple tickets opened for same request Incorrect or insufficient information supplied to complete a request
Improper utilization: The waste of not having the right person do the right job with the right tools	Expert staff performing routine work, or routine staff performing expert work Untrained staff working on areas that require expertise Attending meetings or filing reports, that do not produce an actionable deliverable to the attendee
Mind: The waste incurred by not having all heads involved in solving a problem	Punishing workers who identify problems and recommend fixes Isolated functional silos that don't communicate horizontally Automation efforts that dull the mind with repetition of rote tasks

Take time to discuss what kinds of waste the other steps represent and whether it's feasible to fix them up front—chances are they'll be one of the eight versions of muda in Kaizen-IT. Look for where the mudas besides rework (discussed next) are clustered, as opposed to the VA steps. In most transactional processes, the first two-thirds of a process are variations of muda; the last one-third is where most of the VA occurs, but this can vary from process to process.

3.2.5 Rework Analysis

Go back through the process again and where rework occurs (identified by a rework loop in red stickies), assign it a number, and then ask the team to identify where they believe that rework loop *originates* in the process with another red sticky note with the same number on it. Most defects originate upstream in the process from where they're discovered, and focusing on fixing waste where it is found is like cleaning pollution from a river once it has traveled several hundred miles downstream—it's much better to identify the source and fix it there. Also, by identifying where each rework loop starts, you'll soon notice patterns of clustered defect originations in certain process stages early on; these areas will be critical to drive poka-yoke (mistake-proofing) activities, as the benefits will flow across the entire process.

Figure 13.5 shows two depictions of the current-state process as documented in the RFS kaizen in April. The first image is the actual VSM itself, showing all the steps and hand-offs, shrunk to fit on a page. Even though you can't read the small boxes in the first image, you can begin to appreciate the factors of complexity (groups, steps, and hand-offs). The second image is a consolidated view of the process for clarity, where each labeled box represents many of the steps in the first image grouped together. The abbreviation REQS stands for the technical requirements of the RFS solution—the blueprint of the servers to be built.

3.2.6 Calculate Current-State VSM Metrics

Just as our high-level VSM had VSM metrics to help us understand what was going on, each kaizen VSM will have its own metrics. Five common metrics used frequently in Kaizen-IT were:

▲ **Volume:** The number of transactions begun each day, month, quarter, or year, as appropriate.
▲ **Cycle time:** Add up the duration cycle times (lower left) of all sticky notes in the *correct* path (left to right). Then multiply each rework loop's cycle time by the percentage of time it occurs, adding that to the correct path. This correctly aggregates the time a normal transaction takes to navigate the VSM, with defects. Cycle time represents the calendar duration a transaction takes between desire and fulfillment (*as perceived by the client*).
▲ **FTE time:** Add up the FTE (lower right) of all sticky notes in a similar manner as cycle time. This represents how much labor time is expended on each transaction. Multiplying FTE by volume gives an indication of approximate labor consumed in a VSM.

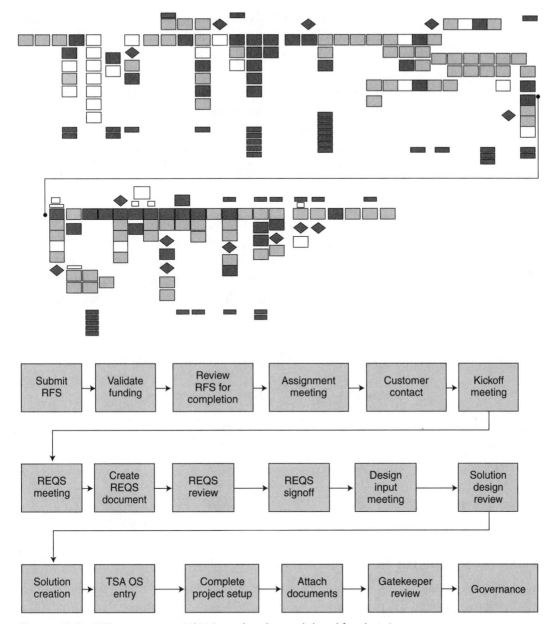

Figure 13.5   RFS current-state VSM (actual and consolidated for clarity).

▲ **VA time:** From the sticky notes identified as VA only, add up all of the *FTE* times—these are the parts of the processes that your team works on that the client finds valuable. You do not add together the duration times, because it may include wait time—only the FTE time should be considered VA.

▲ **Process cycle efficiency (PCE):** PCE is calculated by dividing VA into cycle time and is expressed as a percentage. This is the percentage of how much of the time the teams spent doing something was actually valued by the client. Don't worry if your PCE is low, even in the single digits. If your PCE is very high right out of the gate, you're probably being overly generous in what is being counted as VA or not specific enough in the activities described in 3.2.2.c.

▲ **Complexity:** Complexity is measured in three parts: groups, steps, and hand-offs. It reflects both the structural and interactive elements of complexity that combine in transactional processes. You already know the number of groups from your legend on the VSM created in 3.2.2.b; just count the number of different-colored stickies representing groups. Then count the number of stickies in the correct flow (not including rework loops) from customer desire to customer fulfillment—this is the number of steps. Finally, since the stickies are color-coded to the groups, any color change within the process is indicative of a hand-off.

In the RFS kaizen, the current-state VSM metrics are shown in Table 13.3.

Table 13.3 RFS VSM Current-State Metrics

Kaizen-IT Metrics	Current-State RFS Metric
Cycle time	68 days
FTE time	181 hours
VA time	2 days
Process cycle efficiency (PCE)	2.9%
Complexity	9 groups/120 steps/40 hand-offs

3.2.7 Record Current-State Observations

In addition to calculating the VSM metrics, ask participants what they saw in the process as it was being mapped and discuss what kind of muda it might represent. For RFS, some of the observations were:

▲ Gathering requirements was a major challenge. The process took so long that the client entered RFS before they knew what they wanted (mudas of wait, improper utilization, and mind).

▲ There were no standing teams of experts to respond to RFSs. This resulted in the requirements for each RFS being reviewed in the project kick-off meeting. Then managers would find available SMEs to assign to the RFS. During the SME assignment process, 20 percent of the RFSs ended up in limbo without the proper resources ever being assigned. Most SME teams had never worked together before as a team (mudas of mind and improper utilization).

▲ More than 31 different software applications had to be used to complete an RFS, many required re-entry of data or information (mudas of overprocessing, movement, and motion).

▲ One hundred percent of the project kick-off meetings had to be held twice because not enough information was provided up front for managers to understand which SMEs would be needed (mudas of defects and mind).

▲ There was a lot of rework and cancellations at the end of the process—40 percent were cancelled before being approved (muda of defects).

▲ Inconsistencies in terminology and process variations by country and project manager (muda of mind).

▲ Lack of partnership between IBM and client made it confusing to understand what the client really wanted, resulting in rework of solutions (mudas of defects and mind).

▲ Process complexity drove a steep learning curve (mudas of overprocessing and mind).

▲ Only two days of effort in the whole process were value-added to the client—solution design and agreement with the client, communication of schedule to the client advising them when the request would be complete, and the client's governance board approval of funding for the solution.

▲ A large amount of staff time (approximately 13 percent) was spent creating reports and attending meetings to update the status of RFSs.

3.2.8   Create Future-State VSM

With a full knowledge of the current state—what is VA, what is waste, where and how rework loops originate—it is now possible to begin constructing the future state. The future-state VSM is not a blank sheet of "tell me what you want it to look like" from a traditional business process redesign (BPR). Instead, the future-state VSM should be constructed with the knowledge of the actual problems being encountered in the current state, as well as what the client finds to be VA. Indeed, one practice is to take the VA sticky notes from the current state and place them on the sheet for the future state, the challenge being to develop a method of connecting those VA steps in as little time, effort, and with highest quality as possible. Also, keep in mind it's not necessary to try and eliminate every NVA step uncovered in the current state. The goal is to identify the key pain areas, the root causes of the process defects as exposed by the VSM, and target fixes to those. Sometimes, a group discussion and consensus on improvement ideas, or even breaking into individual teams to try different approaches, are useful. Five good ideas are better than two dozen marginal ones.

Themes from future-state discussions for RFS kaizen included:

▲ Eliminate a silo-functional approach to work and combine all resources necessary into a single "event" rather than repeated hand-offs as ticket tasks.

▲ Form joint teams into a work cell, similar in purpose to a u-cell. In Lean manufacturing, a u-cell is a work area where all the tools and parts necessary to complete an activity are within easy reach of the worker. In Kaizen-IT, a work cell applies the same concept to information and software applications. Work cell teams have all the necessary information at their fingertips: software applications, skill sets, and, most importantly, authority, to complete an activity that would traditionally be subdivided and passed sequentially to many members in a process chain. This approach substantially reduces the hand-offs, waiting, and opportunity for defects and ensures the best use of mind through the collaboration of the team across the work cell. In the RFS future state, these cells would include client representation and attempt to complete most of the VA activities in one "event" lasting between several hours to one to two days.

▲ Gather requirements in real-time, that is, face-to-face with the client and all technical members of the team.

▲ Greater up-front consulting with the right technical people between the client and IBM.

▲ Leverage standards and create standard work packages that could be leveraged to build solutions by component, rather than from scratch each time. If RFSs used those preapproved standard work components, they could proceed on a streamlined course with less custom solution development time.

▲ Reduce project management involvement in the beginning of the process and engage them later.

▲ Create standing teams focused on delivering RFS solutions so that they can become experts on the process and create solutions customers desire.

▲ Develop a kanban pull system to manage workflow in a new "RFS factory" and leverage a heijunka (see the description of kanbans and heijunkas later) to reduce meeting times for updating and reporting status.

For the RFS kaizen, once the future state was completed it looked very different from the current state (see Figure 13.6).

3.2.9 Compare Current- and Future-State VSM Metrics

Once the future VSM is complete, its VSM metrics can be calculated and compared to the current state to see if real change has been achieved (see Table 13.4).

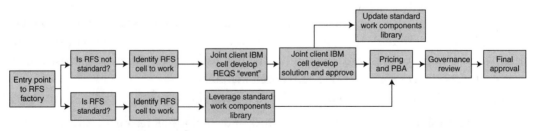

Figure 13.6 RFS future-state VSM.

Table 13.4 Comparison of Current and Future VSM Metrics

Kaizen-IT Metrics	Current-State RFS Metrics	Future-State RFS Metrics
Cycle time	68 days	28 days
FTE time	181 hours	80 hours
VA time	2 days	1.5 days
PCE	2.9%	5.3%
Complexity	9 groups/120 steps/40 hand-offs	3 groups/7 steps/3 hand-offs

In the RFS kaizen, a significant change was to collapse multiple functional groups into three event groups. These event groups would act as a cross-functional team working with the client to create the requirements and other necessary documents that started server deployment in a short, intense event. This contrasted with the typical approach of work being handed off as tickets. The future state, it was believed, would greatly reduce the number of steps and hand-offs, halving the potential cycle time. The future-state metrics were estimates, and would be validated via a pilot.

3.3 Server Build Kaizen Results

The server build kaizen, held from May 21–24, 2007, followed the same steps as earlier (Table 13.5). Figure 13.7 shows a consolidated view of the current-state VSM.

Table 13.5 Server Build VSM Current-State Metrics

Kaizen-IT Metrics	Current-State Server Build Metrics
Cycle time	48 days
VA time	27 hours
PCE	7%
Complexity	19 groups/79 steps/17 hand-offs

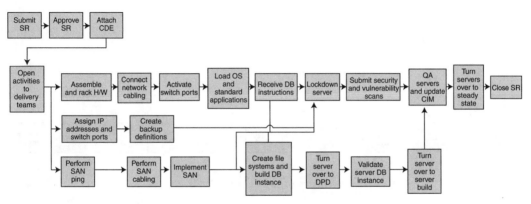

Figure 13.7 Server build current-state VSM (consolidated for clarity).

3.3.1 Current State Observations

▲ Confusion around "what the process is" between different groups

▲ Perception was "we're fine" because output was good, but the reality was a process in chaos

▲ More rework than there is work—46 rework loops—the majority of which were clustered at the front of the process where incorrect information was received

▲ IBM control of process was split in the middle by insertion of client and third-party teams

▲ Reaction to next task at hand rather than planned process steps

▲ All groups were running independently, only coming together at the end

As with the RFS kaizen, the teams then worked on creating a future-state VSM, depicted in Figure 13.8.

3.3.2 Future-State Common Themes

▲ Design new "server build factory" that streamlines throughput by reducing hand-offs and flows work in a logical rather than a functional sequence

▲ Reduce hand-offs and unnecessary steps

▲ In order to reduce rework loops, create an up-front cell with full representation to inspect all work coming into "factory" before any work begins (quality inspection at beginning rather than end)

▲ Expanding team functions—reduce silos and increase flexibility across IBM, the client, and other contractors employed by the client

▲ Deploy heijunkas/kanbans for pull-based workflow across all three companies rather than separate ticketing methods

3.3.3 Comparison of Current- and Future-State VSMs for Server Build

Comparison of current- and future-state VSM metrics is shown in Table 13.6.

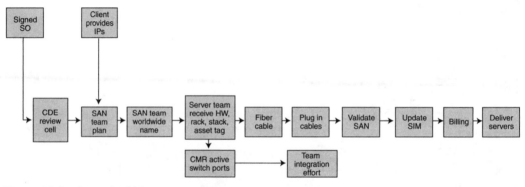

Figure 13.8 Server build future-state VSM.

Table 13.6 Current vs. Future VSM Metrics (Server Build)

Kaizen-IT Metrics	Current-State Server Build Metrics	Future-State Server Build Metrics
Cycle time	48 days	12 days
VA time	27 hours	27 hours
PCE	7%	28%
Complexity	19 groups/79 steps/17 hand-offs	9 groups/13 steps/10 hand-offs

"Get the experts in the day-to-day processes involved and keep the executives out of the meetings or quiet. The ... improvement was viral, and all involved became extremely excited about the progress made in process improvements."

—IBM EXECUTIVE CHAMPION

4. Keeping the Rapid Boil—Checkpoints, Kanbans, and Heijunkas

Some forest fires go out before their time. To guard against this, it's important when boiling the ocean to keep regular checkpoints on the progress made after the initial kaizen. For our effort, the team used 21-day and 90-day checkpoints for each kaizen team, as well as a system of kanbans and heijunkas to regulate workflow in the future state. Kanbans are both a visual workflow management system and a means through which to accomplish a just-in-time (JIT) pull-based production system. Heijunkas are the control boards wherein kanbans are managed to give an enterprise view of the workflow managed by the kanbans. In rough analogy, kanbans are like records in an enterprise resource planning (ERP) system, while the heijunka is the ERP system within which all the tickets exist.

As shown in Figure 13.3, the first checkpoint is the 21-day checkpoint to complete prototyping of the proposed changes. During this time, the future ideas identified in the future-state VSM were validated, and lessons learned were incorporated into rapid modifications. Following this checkpoint, the kaizen team had to report to senior leadership, and more importantly to themselves, where they were at. The Executive Champion should always delegate responsibility for conducting the work but retain accountability. The 21-day checkpoint is often as much a reflection of executive sponsorship as it is of kaizen enthusiasm.

The second checkpoint occurs 90 days after the initial kaizen. Unlike the 21-day checkpoint, which focuses on prototyping, this second checkpoint focuses on the progress being made to make the changes permanent. The push here is to reach the *tipping point* where a vast majority of work is being conducted in the future state rather than the current state. Some kaizens may be "all-over," switching 100 percent from current to future processes at one point. Others may deploy by phases or segments. During this 90-day period, there is also work on the organizational changes needed, such as redefining job roles, documenting standard work components, and training staff in the new future state.

Figure 13.9

90-day RFS checkpoint.

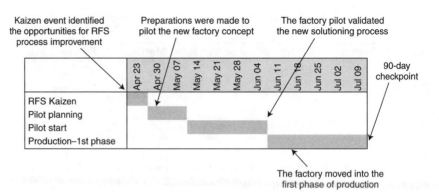

In Kaizen-IT, the project team strove to move from the functional view of work to the process view, and one way of doing that was to begin labeling the core activities as "factories." Even though there were 18 different teams in the Server Build core activity, it was one factory that produced built servers—no one team could declare success unless the whole factory was successful in producing built servers. This reinforced the upstream and downstream importance of work outside the functional team's immediate area. Figure 13.9 shows the activities performed after the RFS kaizen and the 90-day checkpoint.

4.1 21-Day Checkpoint: Prototyping

In order to avoid downstream rework, each critical design elements (CDE) form was reviewed by the server build CDE cell before any server build work began. This work cell consisted of representatives from all groups who had the collective expertise and knowledge to identify missing or incorrect information. This effort improved CDE accuracy once it got past the cell into the server build factory to 100 percent accuracy, as opposed to 99 percent inaccurate previously. Also, the CDE cell identified the root causes of erroneous information created upstream and worked with groups to correct them. The process was piloted across 32 servers representing seven different customer RFSs.

Server build cycle time reduced on average 74 percent (from approximately 48 days to 12.48 days) during the pilot. However, this average appeared to be influenced by a disproportionate number of virtual servers in the pilot compared to normal conditions, and virtual servers in the new process were completed in a cycle time of roughly 3 days per build. Non-virtual server build times during the pilot decreased 40 percent (from about 48 days to 30 days). Since in normal conditions non-virtual servers would significantly outnumber virtual servers, based on these results the pilot team set a near-term target of 25-day cycle time for all servers.

4.2 90-Day Checkpoint: Changeover

First prototype was successfully completed in a cycle time of 21 days, 7 less than estimated. Cells of work teams were developed—the first cell was able to absorb 20 percent of all RFSs even though they were less than 5 percent of the available workforce. Factory production of RFS began on June 11, three months after kaizen. Factory full staffing was 22 FTEs—about 10 percent of the current FTEs assigned to the RFS process.

As the RFS kaizen was the first one we conducted, we learned quite a bit:

"You've got to have a coach and mentor available to guide the 21-day and 90-day team if they aren't experienced (in kaizen) to walk them through the early stages and be available to keep things moving …It could've benefited from iterative work, things got up and running, then leveled off. Do your improvement in waves; schedule your second kaizen when you're already running the first to keep momentum going."

—IBM LSS MBB

4.3 Kanbans and Heijunkas

Once each kaizen effort reached a level of maturity between the 21-day and 90-day checkpoints and began thinking of themselves as factories, the project team led them in implementing kanban (a method of workflow) and heijunka boards (where kanbans are tracked) to control workflow. Kanbans have a long history in manufacturing (see Ohno, Womack et al. and Balle et al.). But pull-based systems have been slow to find their way into many transactional environments where ticketing systems attempt to subdivide workflow into hundreds of individual tasks, distributing them to geographically dispersed workers who have no upstream or downstream visibility. As each worker completes their task in isolation, they push the ticket downstream to someone else they don't know, as someone upstream they don't know pushes another new task into their queue. It's similar to Womack's critique of the Ford assembly line (1990).

However, the process of *producing* what the clients desired—configured servers delivered on time and accurate so that the marketing campaign could begin—is a form of manufacturing. Therefore, the transformation in how workers understood their work from a push-based, task-oriented ticket system to a pull-based workflow kanban system was critical to boiling the ocean.

As they determined they were ready, each kaizen team implemented its own kanban system. This allowed the teams to understand how kanbans worked in their area first before linking the kanbans between work areas. Refer to Balle (2005) for full descriptions of how kanbans work, but you can simply think of them in layman's terms as a visual method of controlling a pull-based production system. Imagine a fast food restaurant with a burger slide behind the counter. When a customer orders a burger, the cashier turns and takes a burger out of the slide, causing the ones above it to slide down and creating an empty space at the top. This empty space in turn signals the cook in the kitchen to create a new burger. In this way, burgers are "pulled" as they are consumed even if a cashier pulls several burgers at once, the cook can fill up the "spaces" until there is no more room, ensuring that there isn't too much inventory accumulated at any one time. For our server build factory, we agreed upon a set of kanban rules that would govern our production methods as follows:

1. No work starts until all work can begin.
2. No work is delivered until work can be received.
3. No pre-work is done before the kanban is received.
4. Only "pull" work when work is needed.
5. Only do the work authorized by the kanban (quantity, type).
6. Work the kanbans FIFO (first in – first out).
7. Do not do work outside of the kanban.
8. Do not reorder the kanbans.
9. Deliver perfect quality for the work being performed; don't cut corners.

The server team spent quite a bit of time looking at their future-state VSM. They wanted to create a heijunka that would hold all kanbans that would visually depict to workers all the work in the server build factory, a far cry from the task-limited view of ticketing systems. A snapshot in time of that heijunka board is included in Figure 13.10.

Keep in mind that this figure presents the entirety of the server build factory operations at the moment it was taken. Although the figure is condensed, it provides many visual clues to help read it. Figure 13.11 is a blown-out portion that makes easier reading, and represents just the upper-left corner of the overall heijunka.

In Figure 13.11 you can see the group (CDE QA Cell) on top and the three process steps (New Hardware, Redeployed Hardware, and Virtual OS) they are responsible for. In the heijunka depicted in Figure 13.10, these sections are magnified from the first three columns across the far left, and they represent the key groups and process steps in the new future-state VSM created in the kaizen. If you would follow them left to right, they represent all the steps and hand-offs work needs to flow through to complete production of a server build.

Returning to Figure 13.11, each square beneath the process steps represents a kanban for that work area, and although the picture is grayscale, the original color of the square represents the state of the kanban:

▲ **Red (dark gray):** The activities of the kanban in this process step are being worked by the team of the process step.
▲ **Green (light gray):** The work of the kanban in this process step has been fully completed, and is ready to be pulled to the next process step.
▲ **White:** There is capacity for a kanban to be pulled in from the previous process step.

By looking at Figure 13.11, one can tell at the time this snapshot was created where the active work was ongoing (the two dark gray kanbans) and what kanbans had been completed and were ready to pull (light gray kanban). Looking at the factory heijunka in Figure 13.10, you can get the same sense for the entire server build factory: many red squares indicating kanbans in process are in the final process step on the far right of the heijunka. Because the downstream process step is

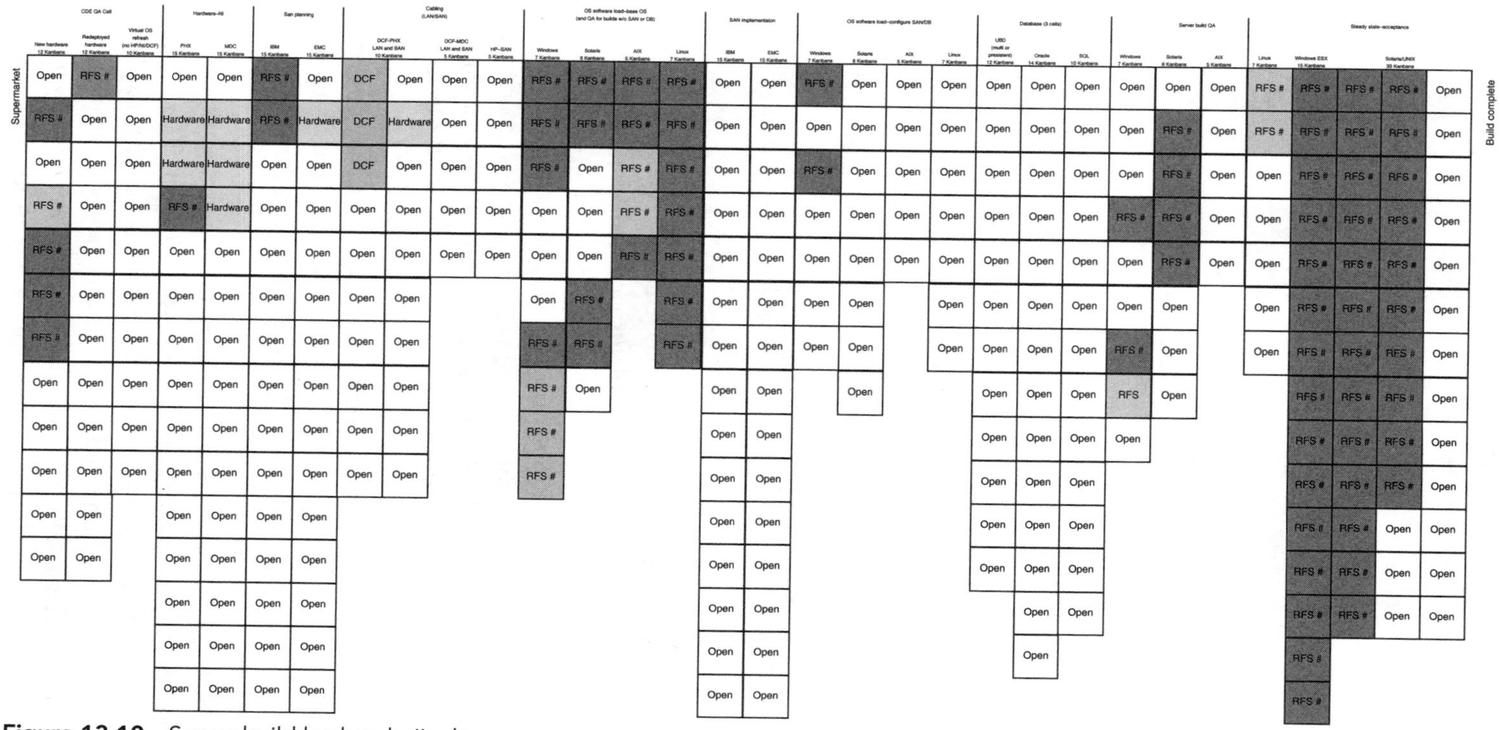

Figure 13.10 Server build kanban heijunka.

333

Figure 13.11

A blowup of heijunka.

CDE QA cell		
New hardware 12 Kanbans	Redeployed hardware 12 Kanbans	Virtual or OS refresh (no HP/NI/DCF) 10 Kanbans
Open	RFS SR #	Open
RFS # US P-WXSX	Open	Open
Open	Open	Open
RFS SR #	Open	Open

the one that pulls available work, a green square may remain for a limited time until it is "pulled," in which case a square in the next process step will turn red (indicating work in progress) and the one in the previous step will revert to white (capacity to do more work).

Were you to look at either the snapshot or the blown-out portion in a few hours, the squares currently marked green would probably have converted to white (capacity to do new work), with the work they had represented being marked as red in the downstream step. At the time this snapshot was taken, there were lots of white cells, indicating that the factory was not running at capacity and indeed had, at the time, excess kanbans. In the blow-up (Figure 13.11), you can see that each kanban had identifying information (blacked out here for confidentiality purposes) that indicated the order number the kanban was associated with and codes to describe the kind of work necessary to complete that server build at each step. These codes represented the "kanban type" referred to in Rule #5 in the list of kanban rules: each process group would know from that code what they were supposed to do based on standard work instructions.

The system works on a FIFO basis so that the work completed earliest gets pulled first if given multiple choices to pull from. If a work area filled up with red kanbans (which is occurring on the right side of the heijunka), that means they are reaching their limit of being able to take on additional work. Because the kanban spaces were finitely limited, this allowed the server build factory to quickly and visually spot where problems were occurring *before* a large accumulation of work orders backed up. Any worker at any time could look at the whole factory and understand, at a glance, what areas were accumulating work (red) and those completing it (green). Likewise, a project manager looking for a status update on a specific order could quickly see where their order was by what process step it was located beneath, and whether it was being worked or ready to pull. This limited the interruptions of the production teams to provide status updates. Keep in mind Rule #6: even though project managers could see the status, they could not expedite or escalate it, as all work was completed FIFO.

Kanbans do not create workflow problems; they reveal them. Heijunka workflow systems in complex environments are adaptive and evolutionary in real time by identifying, via the kanbans, where workflow problems are occurring. When no more kanbans are available in a certain area, the factory team begins 5-Whys root-cause analysis until they understand why the red kanbans accumulated in a certain area and corrective action is taken.

Heijunka boards with their kanbans keep the ocean boiling once it's been set. If a work area or factory has a consistent level of "white kanbans" over time, begin removing some a few at a time. In the previous example, if, over time, the many white squares stayed white, it would be time to begin removing kanbans and then see what changes. If kanban backlogs occur at the lower level per root-cause analysis and fix it. If no backlog of red kanbans occur, remove some more. It's a self-regulating method of identifying workflow problems.

The "white kanban" means that the entire "kanban inventory" is not needed; this is, basically, the size of the input queue. Reducing white kanbans reduces WIP inventory possibilities. It also indicates that the process is under control and not too variable.

If two adjoining work areas on the heijunka board both have very low kanbans, then see if that work can be merged into a single function (as the server build team did many times in its initial deployment). When work crosses core activities, separate heijunkas can be set initially for each factory. When kanbans in two adjoining heijunkas reach a low level, look to see if you can consolidate core activities by conducting a VSM kaizen that incorporates both core activities as if they were one.

5. Benefit Realization

So how did the cycle times actually improve for the kaizens as they progressed? Figure 13.12 is a control chart that indicates the improvement of the RFS kaizen before and after results. Keep in mind from the 90-day checkpoint on, the kaizen went into factory production beginning in June 2007.

Note how wide the variation (space between the control limits) is prior to the kaizen being held. This shows a wide variation—the inconsistencies produced by the complex process. Also note how the mean remains constant at about 68 days; this indicates the long cycle time of the delivery. However, once the kaizen improvements are implemented, notice how both variation reduces (control limit lines come closer) and the mean drops almost in half. The slight tick up at the very end of the process represents when the RFS factory "tipped" over and brought in all outstanding RFSs, including some very old ones that were finished out, but due to the age distorted the numbers slightly.

The server build team showed similar improvements (see Figure 13.13). You can see that even though volume varies, the average cycle time to build a server steadily begins reducing after deployment of the server build factory to about half—48 days to 25 days.

By including our results from just these two kaizens in the high-level VSM metrics, we got a good comparison of the impact of the changes between current and future states (see Table 13.7).

Figure 13.12

Before and after control chart of cycle times for RFS.

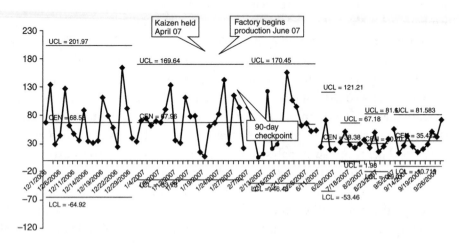

Figure 13.13

Results of server build kaizen— Average # of days per server build and total servers built per month.

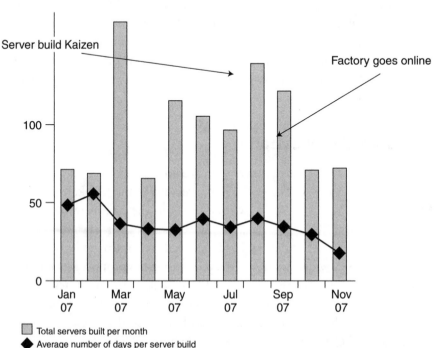

Table 13.7 Revised High-Level VSM Metrics After Two Kaizens

VSM Metric Current vs. Future	RFS	Acquisition	Server Build	Load Application	Change from Current State
Cycle time	~68d down to ~35d	~10d	~48d down to ~25d	~156d	~280d to ~226
RIP	50% reduction	Confidential	50% reduction	None	About 40% less
WIP	None	None	95% reduction	None	95% less
PIP	Confidential	Confidential	Confidential	Confidential	Confidential
FTE	30	NA	30	NA	100+ FTE reduction

In both areas we were able to achieve close to a 50 percent reduction in cycle time, leading to corresponding reductions in PIP (approximately 40 percent), inventory (95 percent reduction), and FTEs (more than 100 reduction). This resulted in several US$1 million a year in direct savings. Additionally the improvement in cycle time reduced the amount of Revenue in Process (RIP) accelerating the recognition of that revenue. Of course, these were the results from only two of the eight kaizens; we still had a long way to go to meet the stretch goal of 13 days (for which separate kaizens were initiated per Figure 13.4). But we also knew for sure that we only needed about 30 workers in each of the two core activities focused on by the first kaizens. That doesn't mean the other hundreds of workers went away over night, but they were freed up to focus on other areas of the organization.

6. Conclusions and Limitations

The approach illustrated in this chapter is best attempted with the help of a Master Black Belt familiar with the underlying theories of VSM, VA, muda, kanbans, and heijunkas who can facilitate and mentor the knowledge base in the workers. Strong executive championship is critical to any attempt to boil the ocean. Tuning the number and pace of kaizens, running fewer, or allowing more time between each can customize the pace of change to each organization's ability to absorb it. In addition, even though the technology sector can often be considered transactional, not all IT operations are. For example, for certain customer-facing applications or servers where downtime can be no more than mere minutes per month of operation, more traditional Six Sigma tools may be applicable.

Also note that you may not get to the client-stated goal in the first pass. Lean and Kaizen-IT are about the continuous application of process improvement methods. Once results have stabilized and improvements are showing, go back and continue the exercise again. Look for new refinements and optimizations that can be made to decrease cycle time and effort or increase value-added time.

Acknowledgments

The author of this chapter is thankful to IBM for granting publication permission. The author would also like to thank Don Woodward, IBM Deployment Executive and Executive Champion for the project; David Brechesein, LSS MBB; Kimberly Austin-French, LSS BB; and Jamie Martin, LSS BB, as well as all the SMEs who gave of their time and enthusiasm to make this effort what it turned out to be.

Diagrams created with Visio, Office, and PowerPoint. Statistical charts created with SPC XL.

7. References

1. Ohno, T., Dr. (1988) *Toyota Production System: Beyond Large-Scale Production.* New York: Productivity Press.
2. Womack, J., Jones, D. & Roos, D. (1990) *The Machine That Changed the World: The Story of Lean Production.* New York: Free Press.
3. Womack, J. & Jones, D. (1996) *Lean Thinking: Banish Waste & Create Wealth.* London: Simon & Schuster.
4. Womack, J. & Jones, D. (2005) *Lean Solutions: How Customers & Companies Can Create Value and Wealth Together.* New York: Simon & Schuster.
5. Pyzdeck, T. (2003) *The Six Sigma Handbook,* Revised & Expanded. New York: McGraw-Hill.
6. Balle, M. & Balle, F. (2005) *The Gold Mine.* Cambridge: Lean Enterprise Institute.

CHAPTER 14

How a Global Retailer Improved the Reliability of Software Development and Test Environments

Sanjay Dua
Ambuli Nambi Kothandaraman

Relevance

This chapter is relevant to organizations that have been plagued with nonproduction environment configuration challenges during the development and testing phases of the software development lifecycle (SDLC). A nonproduction environment can be defined as an environment in which code is developed, tested, and certified before it is released into production. Environment configuration can be defined as setting up a "production-like" environment that meets the needs of a given project. Project cycle time and budget are dependent on having access to stable and certified environments (e.g., sandbox, application development, QA, and performance) that are used throughout the SDLC. Learn how one global retailier applied Lean Six Sigma to reduce test defects due to environment-related defects during software development.

At a Glance

▲ TCS, in partnership with a global retailer, launched a Six Sigma project to improve the stability and configuration of nonproduction environments used in application development and QA testing.

▲ The initiative resulted in greater than 50 percent reduction in high-severity environment defects identified during testing, as well as a 39 percent cycle time reduction in the fulfillment of nonproduction environment requests.

▲ The reduction in project overruns resulted in an annual cost savings of approximately US$840,000.

Executive Summary

IT development plays a major role in the success of one of TCS's major clients—a large international retailer. As a major company in the specialty retail market, this global retailer spans four countries in North America and Asia. Yearly revenue places this company in the Fortune 50 of U.S. businesses. The information technology (IT) function is recognized as a main catalyst that facilitates this company's success.

TCS began its technology process outsourcing (TPO) relationship with this global retailer in 2007. As part of the transition, about 70 percent of the testing and quality assurance (TQA) function was outsourced.

Within IT, retail systems are the life blood that enables customer transactions to flow through the business. As part of the retail system's TQA organization, the configuration of nonproduction environments is a major function. Many teams are dependent on a stable and reliable environment to complete development of code and QA testing. Without stable environments, the testing effort can become extended, and in some situations, the quality of code released to production could be suspect, consequently leading to production outages.

The environment readiness (ER) team under TQA is responsible for fulfilling formal requests to set up nonproduction environments (e.g., application development [AD], QA, and so on) for their respective customers. Part of the team's process is to apply the:

▲ Correct version of code within these nonproduction environments
▲ Management and provisioning of devices used in these environments
▲ Application of the appropriate data sets for testing

Customers of this team could be defined as development teams, as well as QA testers, who are internal to the TQA department. The TQA organization measures defects, which are categorized during the QA testing process. The purpose of categorization is to determine the root cause for each defect, and to use that information for continuous improvement of the development and testing processes.

An ER defect is defined as a defect that is caused by the way in which the environment is configured. An example would be if a handheld point of sale (POS) device was provided for QA testing, but it was not provisioned with the appropriate software. Ultimately, a test case (or set of test cases) that requires the use of this device would fail, and the failure would be attributed to an "environment" defect.

A severity level ranging from one through five is assigned to testing defects to establish the priority and the impact the defect has on testing activities. Severity 1 and 2 defects are the highest-priority issues, as they either cause a longer-than-normal turnaround time due to a major code fix, or they cause a blockage, which prohibits the completion of testing subsequent and dependent test cases.

The TQA department had a major problem. During a three-month study, an average of 41 percent of Severity 1 and 2 test defects were attributed to environment-related issues. This caused an increase in testing cycle time (CT), a significant amount of rework, and many dissatisfied customers (internal QA testers and development). In addition, the rework often required the unplanned time and support of cross-functional teams to help troubleshoot the nonproduction environment (in other words, to determine why the nonproduction environment was unstable and unreliable for development or testing).

The scope of this project was to focus on stabilizing and driving consistency in the configuration of nonproduction environments.

The approach used was a blend of Lean and Six Sigma. A project team comprising members of the TQA organization was formed. In addition, an extended cross-functional team was identified to participate as SMEs during interviews and kaizen events, as well as to implement identified corrective actions. A kaizen event in this context refers to a five-day workout session used to pinpoint waste, pain-points, and ultimately solutions to improve the process. This term will be used at various points within this chapter.

This project led to a 50 percent improvement in the first-time pass rate (i.e., the number of test cases that passed divided by total test cases executed), reducing the number of defects attributed to environment-related issues. In addition, the cycle time for fulfilling nonproduction environment requests made by the development teams and QA testing was reduced by 39 percent.

1. Introduction

With an annual IT capital budget of US$100 million, this global retailer has multiple IT projects spanning retail systems, supply chain, and merchandising, as well as enterprise resource planning (ERP) technologies.

The IT retail systems organization plays a pivotal part in ensuring the most innovative and latest technologies are implemented, maintained, and operating smoothly across the enterprise. The retail systems TQA department provides the safety net before new technology is released into production. TQA provides the following services (work streams):

1. **QA testing:** Testing that completely validates the functionality of a business process from beginning to end, including validation of operational and technical components.
2. **ER:** Provision of nonproduction environment configuration for development and QA testing.
3. **Performance and load testing:** Testing conducted to validate various aspects of the system for response, workload, scalability, reliability, and resource usage.
4. **Automation of test scripts:** Used to improve the efficiency and cycle time for a given testing cycle.

TQA has a "customer first" focus. The mission of this organization is to "ensure that high-quality software and technologies are delivered that meet the needs of our customers." In July 2008, TQA embarked on a Lean Six Sigma initiative focused on improving the ER process. The average cycle time from initial request of an environment to provision of the requested environment was 7.7 days. In addition, the stability and completeness (or lack thereof) of the environment typically resulted in defects that affected test execution. The four-month average of Severity 1 and 2 testing defects attributed to environments was 41 percent (April–July 2008).

A number of upstream processes and dependencies fed into the TQA environment readiness process. As an IT organization, the process for configuring a nonproduction environment was not transparent. Cross-functional teams who contributed to the process were not aware of when their support was required and how long it should take them to provide their time and resources. In addition, when key process inputs were received, there was lack of clarity on what the operational definition of "complete" was. This lack of process knowledge had an unfavorable impact on the completeness and on time delivery of test environments.

2. Project Background

To understand the full dynamics behind this project, it is important to note that this global retailer had previously established a formal Six Sigma program. The program resulted in over US$300 million in cost reduction over a four-year span and helped strengthen operational controls across the enterprise. However, in an effort to bolster customer service in stores by dramatically increasing associate staffing, the organization identified ways to streamline corporate overhead costs to offset this human capital investment. As a by-product, the decision was made to cut the Six Sigma program and the associated staff. The views of the corporate staff were mixed, as the dissemination of Six Sigma and its philosophy was selective. In some instances, recipients of a Six Sigma solution may not have been fully involved and immersed in the process, thus not fully understanding, and in some cases undervaluing, the approach and rigor of the program.

In spite of this, many leaders across the business still valued the results that Six Sigma methodology brought to process improvement. In order to effectively gain support and buy-in throughout this project, formal Six Sigma terminology was used sparingly. In addition, the typical formality and rigor that exemplifies the Define, Measure, Analyze, Improve, and Control (DMAIC) method was blended with tools and techniques of Lean Enterprise.

3. Define Phase

The scoping of this project required the succinct definition of the problem in quantifiable terms. The scope, goals, and initial timeline for the project were mapped out. A Supplier, Inputs, Process, Output, Customer (SIPOC) map was completed to provide a high-level view of the key inputs, outputs, and customers of the process. In addition, primary project team members from the TQA organization were identified, along with secondary cross-functional team members who would later play a pivotal part in the kaizen event.

The following sections describe the activities and deliverables that were created as a result of this phase.

3.1 Business Case Statement

The development of the business case for this project included two perspectives:

▲ **Internal drivers:** Those pain-points and opportunities experienced within the TQA organization
▲ **External drivers:** Those forces outside of TQA that required the organization to dive deeper into the end-to-end process and the respective dependencies

3.1.1 Internal Drivers

▲ On average, 41 percent of high-severity defects detected during test execution were attributed to "environment-related" issues (see Figure 14.1).
▲ Average cycle time for fulfilling a nonproduction environment request was 7.7 days.
▲ Improvement in efficiencies was key. The ER team's workload (workload based on new project requests as well as rework required to troubleshoot environment-related test defects) kept increasing as funding was decreasing.

3.1.2 External Drivers

▲ There are a number of upstream processes and dependencies that feed into the TQA environment setup process.
▲ More than eight "suppliers" provide critical inputs to the environment setup process.
▲ Rework requiring participation from other teams became an ongoing challenge. Due to obligations of other teams, it was a challenge to get resources to help troubleshoot problems experienced in these nonproduction environments.

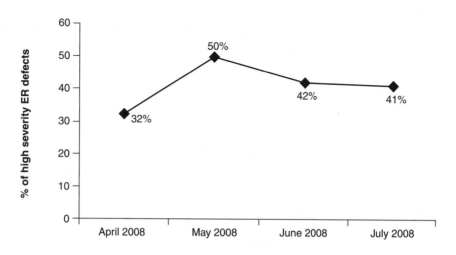

Figure 14.1

Environment readiness severity 1 and 2 defects trend.

3.2 Opportunity Statement

By executing this project, the customer expected to realize two key benefits:

▲ Reduction in cycle time for testing as well as resource savings due to reduced rework (or cost of poor quality) associated with resolving nonproduction environment defects.
▲ Reduced environment cycle time due to improved "first-time pass rate" for provision of nonproduction environments. The reduction in environment cycle time could be used to more exhaustively test the product or deliver the product to customers faster.

3.3 Goal Statement

The project team, in collaboration with the project sponsor, established the following goals for the project:

▲ Reduce Severity 1 and 2 environmental defects detected during test execution by 50 percent (baseline was ~41 percent).
▲ Reduce average environment cycle time by ~50 percent (from baseline of 7.7 days to goal of 4 days).

3.4 Project Scope

The project scope included providing reliable environments for parallel project testing for more than 190 annual projects that spanned 50 nonproduction environments, maintained across four geographically dispersed locations. The various test labs were housed with equipment that emulated the actual store environment (e.g., POS registers, scanners, handheld devices for inventory management, printers, etc.). In addition, to ensure time and energy was spent on the biggest "pains," the focus was on Severity 1 and Severity 2 environment defects, which posed the greatest risk to test execution and on-time delivery of code to production.

3.5 Project Team

The core Lean Six Sigma project team comprised:

▲ Executive Project Sponsor—VP of Retail Systems IT
▲ Cross-functional Leadership Council—VP Infrastructure; Director of Retail Systems Development; Senior Director/Chief Architect
▲ Project Sponsor—Director IT Testing & Quality Assurance
▲ ER Resource Manager
▲ TQA Black Belt—Project Manager
▲ TCS Project Manager
▲ TCS Quality Consultant—Master Black Belt

- ▲ Four QA Test Leads
- ▲ One Performance Test Lead
- ▲ Two ER Systems Engineers
- ▲ Two TQA Architects

In addition to the core Lean Six Sigma team, a secondary group of 19 cross-functional team members (SMEs) were identified to help solve problems in the end-to-end process. They spanned the following IT departments: network engineering, application architecture security, application framework security, development, networking, software provisioning, deployment, and device engineering.

3.6 Detailed Project Plan

In the spirit of providing layman terminology for the method used to communicate and execute this project, the following terminologies were used interchangeably with the DMAIC lifecycle:

1. Leadership scoping (Define)
2. Baseline "as is" processes (Measure)
3. Analyze process complexity and capability (Analyze)
4. Develop and implement proposed solution (Improve)
5. Transition project to process owner for control (Control)

Given the Lean nature of this project, quick-win solutions (i.e., low-hanging fruit around waste reduction) identified earlier in the project (e.g., Measure and Analyze phases) were implemented immediately instead of waiting until the Improve phase.

3.6.1   Leadership Scoping (Define)

During this phase, the plan was to form a team, define roles and responsibilities, and define the process and data analysis approach.

A stakeholder analysis was conducted to ensure the right level of buy-in could be obtained at the appropriate levels of the organization. The project sponsor recognized the criticality of this step in socializing this effort with key senior leaders. Without assessing their level of support, determining the context that led them to request this project, and executing the appropriate influencing strategy to gain their buy-in, the success of this cross-functional initiative would have been in jeopardy from the beginning.

3.6.2   Baseline As-Is Processes (Measure)

During this phase, the goal was to develop a high-level view of the current nonproduction ER process. The tools to be used for this purpose included SIPOC, and detailed process maps were built in swim lanes to illustrate the touch points across cross-functional teams. In addition a voice of the customer (VOC) survey was used to help gauge the perceived pains and quality expectations from a customer point of view.

3.6.3 Analyze Process Complexity and Capability (Analyze)

The Analyze phase would focus on the root causes affecting the ER process. Tools such as failure mode effects analysis (FMEA), along with kaizen events, would help identify the opportunities for improvement related to cycle time and defects.

3.6.4 Develop and Implement Proposed Solution (Improve)

As solutions were identified, the thoughtful rationalization of variety, scope, and complexity of the solutions was to be performed. Tools such as impact effort analysis, along with training and organizational change readiness plans, were deployed to ensure any changes that were implemented were sustainable in the long term.

3.6.5 Transition Project to Process Owner for Control (Control)

After implementation of solutions, the control activities to monitor and reinforce process performance were validated prior to transition to the process owner.

4. Measure Phase

During the Measure phase, the core project team was divided into two subteams.

▲ **Process Analysis Subteam:** This team's primary focus was to map out the "as-is" ER process. The team met twice weekly to map out the workflow of the process. A detailed process map illustrating the high-level work flow was documented. As the process was mapped out, certain steps were further decomposed as a drill-down into the process. This is known as level mapping, where Level 0 is the highest level of a process and Level 3 is a further breakdown of tasks within a process (see Figure 14.2). These processes were developed in a swim lane format. As a complement to the process mapping activity, the team estimated the actual time it took (processing time) to perform tasks and activities within the process and the overall cycle time (processing time plus amount of wait time when a product or service stood idle in the ER engineer's input queue). This helped the team visualize where potential works in process (WIP) could build up, and was later used in developing a comprehensive value stream map (to be discussed and illustrated during the Analyze phase.

▲ **Data Analysis Subteam:** This team was formed to identify the key metrics required to measure outputs of the ER process and the in-process measures that required the establishment of metrics that may not have currently existed, or that were measured but may not have been illustrated from a trending perspective.

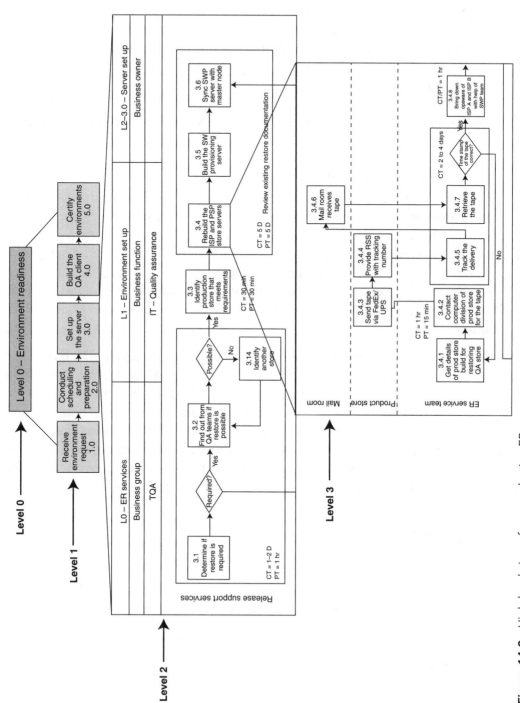

Figure 14.2 High-level view of nonproduction ER process.

347

The data collection plan leveraged raw data from two main sources:

1. **Environment request database:** This database was the central repository in which customers would place an environment request for a project.

    This data provided cycle time information, as well as the capability to segment data based on request type. This helped determine what the average cycle time was for specific requests, where the longest cycle time was, and where efforts needed to be concentrated for cycle time improvement.

    The following were typical ER requests:

    ▼ **Environment restore:** Total rebuild of a nonproduction server to fully emulate a production environment taken from a tape at an actual store

    ▼ **Clients:** The peripheral attachments that are part of a store environment (e.g., handheld devices, self-checkout lanes, scan guns, etc.)

    ▼ **Printers:** Separate client used to print bar codes, reports, and other pertinent material

    ▼ **Deployment:** The promotion of a new version of code to a nonproduction environment

2. **HP Quality Center:** This application was used as the central database by the development and project teams to track defects found during QA testing, including defect category, severity of defect, defect owner, and defect frequency. This system of record was pivotal in filtering out and pinpointing the environment-related defects across, between, and within all projects.

In addition to quantitative analysis, the team created and administered a VOC survey. The VOC was administered across 400 IT development partners as well as internal TQA testing associates. The survey provided insight into the pains experienced across segments of the population, as well as shed light on the expectations of the customer in terms of timeliness and quality of the environments delivered. Free text comments from the survey were grouped into common themes and proved to be invaluable during the subsequent kaizen event. The survey results were used to assure the end-user community who participated in the survey that their feedback would be taken into consideration to improve the existing process.

5. Analyze Phase

The Analyze phase of this project included the following activities:

▲ FMEA analysis

▲ High-level process validation with cross-functional stakeholders

▲ Kaizen preparation

▲ Kaizen stakeholder kick-off meeting

▲ The kaizen event

5.1 FMEA Analysis

After the high-level process was mapped out, the team conducted a FMEA as an early launching pad for the kaizen workout session (see Table 14.1). FMEA is a tool used to identify and mitigate potential failures for a process or system. Risk can be mitigated by prioritizing actions based on the probability and severity of failures.

During the FMEA, as failures and related root causes were identified, the team, in some instances, was able to take immediate action to control the situation. For example, when performing a full rebuild of a store environment, the team requested the retail store to send a production backup tape via courier to the IT headquarters. The tape would contain the latest version of code and software, per the requirements of the ER request. Half of the time, the tape would either be defective or would not have the right versions of software required to restore the nonproduction environment. Ultimately, this would tack on an additional two days of cycle time, due to the re-request of an adequate production tape. The team quickly brainstormed ways to minimize this type of rework. The solution was to partner with the store computer room associate by virtually logging on to the actual store environment to view the content and structure of the tape. As a result, the ER team could immediately pinpoint if it had the correct version and the correct snapshot in time to be used for the restore of the nonproduction environment. Furthermore, they could run the appropriate systems check to ensure the tape was not defective. The team was able to immediately implement this solution.

The remaining information captured during the FMEA exercise was used in preparation for the kaizen event, which will be described in the subsequent sections.

5.2 High-Level Process Validation with Cross-Functional Stakeholders

After completing the as-is process mapping, the team validated the activities that had multiple touch-points across boundary teams. Adjustments were made to the documented activities and the sequence of the flow, which helped prepare for building out a value stream map for the end-to-end process.

5.3 Kaizen Preparation

One of the most critical activities performed throughout this process is the kaizen workout session. Kaizen is a Japanese term for stressing continuous incremental improvements to achieve the goal of creating value and perfection in the organization. The term kaizen in the context of this project describes a session in which the ER value stream was validated and adjusted, and the pain-points and opportunities were linked to specific activities or work products in the process.

As participants' time was scarce, preparation for this event required meticulous planning. A creative yet effective plan was employed to time-box team members based on the relevant parts of the value stream.

Table 14.1 FMEA for ER Workflow Process

Process or product name		Environment set up – L1 receive environment request			Prepared by		Page _____ of _____				
Responsible							(rev)				
					FMEA date (orig) _____						

Process step	Key process input	Potential failure mode	Potential failure effects	S E V	Potential causes	O C C	Current controls	E E T	R P N	Actions recommended	Resp.	Actions taken
3.4.1 Get details of prod store build for restoring QA store	Person requesting the restore	Delivering the wrong store type	Delay testing	7	Inadequate information provided; incorrect information	3	None	10	210			
3.4.1 Get details of prod store build for restoring QA store	Person requesting the restore	Delivering the wrong store type	Rework required	7	Inadequate information provided; incorrect information	3	None	10	210			
3.4.2 Contact computer room operator of prod store for the tape	Request for recent (1 week's prior) back up tape	Connecting with (and getting buy in to send) the appropriate associate, in a timely manner	Delay in completing the restore, requires follow up calls	6	Inability to get hold of someone. Person doesn't have authority to release. Skepticism about ER authority to request it. Language barriers. Lack of on-hand tapes to be able to give…	10	None	2	120			

The kaizen workout event was scheduled for five days. It was a challenge to get this amount of time, but based on the anticipated value and payback of the kaizen, the project sponsor was able to negotiate with her peers to get commitment of their teams' time for their participation.

A detailed and well-structured agenda was created, listing the sequence and timing of events, as well as the appropriate participants required during each segment of the workout.

Process maps created earlier in the project (Figure 14.2) were integrated and streamlined into an end-to-end value stream of the ER process. Activities were grouped into clusters, but the details were made available to spark discussion around the process. Cycle time and processing time were also illustrated on the value stream map (see Figure 14.3).

5.4 Kaizen Stakeholder Kick-Off Meeting

Two weeks prior to the kaizen workout event, the project sponsor held a kick-off meeting to educate the project team members and key stakeholders on the forthcoming event. The audience included the sponsor's peer group as well as the executive sponsor of the program.

During the kick-off meeting, the following items were covered:

▲ A recap of why the initiative was undertaken (in order to establish a sense of urgency for the initiative)
▲ The baseline "as-is" process
▲ Current process metrics
▲ Overview of how the event would be conducted and expectations from the project team during and after the event

5.5 Conducting the Kaizen Event

The five-day kaizen workout event was strategically planned to ensure the right amount of participation, by the right team members, at the right time.

The first two hours of the first day included everyone who would participate in some part of the session. This was to ensure that everyone received the same orientation and had the same opportunity to walk through the value stream and suggest any necessary changes or additions prior to diving deeper into the event.

There were two main facilitators during the event. One focused primarily on walking the team through the techniques to brainstorm pains and solutions, while the other focused on documenting and capturing verbal discussions of the process.

During walkthrough of different segments of the process, various SMEs led the discussion of the process and clarified why information flowed a certain way into their work stream.

Various brainstorming techniques were used to determine the pains and opportunities within the process; for example:

▲ One technique used was as follows: After a segment of a process was discussed, the participants would break out into teams of four and silently write down on Post-it Notes as many pains they could think of that occurred during the process. At the end of the silent brainstorming, they would discuss and narrow down the top five or seven, and then post it over the segment of the process in the value stream.

▲ Another approach used was more of a free-form brainstorming. That is, the participants were instructed to identify the pains on their own and then post it on the value stream map. At this time, the facilitator also included the pains discovered during the FMEA activity, previously described in the "FMEA Analysis" section, and posted them on the value stream.

After each brainstorming session, the facilitator would review the items on the Post-it Notes. Common themes would be grouped together, and then the participants who documented the pains would discuss their perspective and experience. A good amount of discussion and debate was used to help narrow down and clarify the actual problems. The Post-it Notes were then placed on the value stream map (VSM) located on the flip chart, directly above the corresponding process activities.

The "5 Whys" technique was used on numerous occasions to get to the heart of what caused a problem. After the problems were fully identified, the team was instructed to do a similar silent brainstorming session to pinpoint opportunities (solutions) that would reduce or eliminate the pains. The facilitator would again group common opportunities to avoid redundancy. The opportunities were then placed on the VSM, under the corresponding process activities (see Figure 14.3).

At the end of each day, the facilitators would capture and document the pain-points and opportunities identified during the session.

As a capstone for the event, the entire cross-functional team of participants came together to review the pains and opportunities, and to prioritize the proposed solutions.

Figure 14.3

ER value stream map.

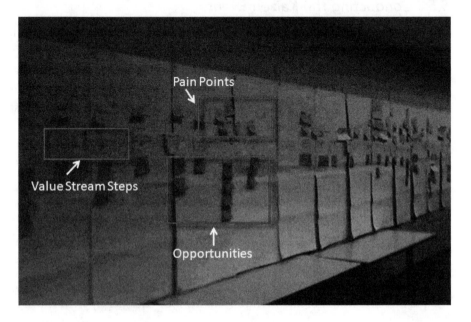

The two key criteria that were used to evaluate proposed solutions were "impact" of proposed solution and "effort required." In order to ensure that everyone was consistent in their understanding of these two criteria, everyone was informed of the definition of each of these criteria.

Possible solutions were evaluated by assessing each by the magnitude of effort and impact.

▲ High-effort solutions were those that would require input or participation by other business units or might require significant redesign or changes to existing processes.

▲ Medium-impact solutions were those which would involve changes that would take no more than two months and require about two to four hours of retraining for process users.

▲ Low-effort solutions were those which could be implemented almost immediately and would require less than two hours of retraining.

Similarly, impact was assessed based on the following criteria.

▲ High-impact meant that the business would realize large benefits from the associated solution.

▲ Medium-impact meant that there would be some immediate benefit in the form of standardization or process simplifications or defect reduction.

▲ Low-impact meant no significant or visible changes to customer satisfaction, cycle time, quality, or cost.

Categorization of solutions using these guidelines allowed a semi-quantitative method of comparing different possible solutions.

A detailed pain-point and opportunity template (see Table 14.2) was used to tie the process activities to the pain-points and the solutions. The impact (high, medium, or low) and the level of effort (short term, mid term, long term) were used to prioritize the value of execution of each activity.

6. Improve Phase

6.1 Executive Sponsor and Stakeholder Report Out

Within one week of completing the kaizen workout event, the team developed an executive presentation listing the results and recommended actions from the event. The actual value stream created during the event, with all of the pains and opportunities, was displayed in the conference room for the meeting. The leaders were encouraged to come in early to "walk through the process" to see what it actually took to build a production-like environment.

During the presentation, the pain-points and opportunities were illustrated in an impact/effort matrix (see Figure 14.4). The numbers in the grids refer to the individual item numbers in Table 14.2. For example, item 7 in Table 14.2 required medium effort and it had a medium impact; therefore, it appears in the center quadrant in Figure 14.4.

Table 14.2 Pain-Point and Opportunity Matrix

Item No.	Level 2 process	Level 3 process	User needs/ pain-points	Pain-point classification by frequency and impact	Improvement opportunities	Category	Short term/ long term (ST/LT)	Effort	Impact
1	1.0 Receive request	Receive request and verify CR record	Multiple vehicles of receiving request	Medium	Establish and enforce a policy with all customers and the ER team making use of CR as mandatory	Policies	ST	Low	Medium
2	1.0 Receive request	Receive request and verify CR record	Incomplete requirements in CR	High	Partnering with CCMD to gain more insight to upcoming projects	Requirements	ST	Low	Low
3	1.0 Receive request	Receive request and verify CR record	Incomplete requirements in CR	High	Leverage original development environment request to model QA environment using the same specs	Requirements	LT	High	High
4	1.0 Receive request	Receive request and verify CR record	Incorrect requirements in CR	High	During the upstream process, build environment requirements into BR/SES documents. Use criteria identified in the session, along with the ER environment request form details.	Requirements	LT	High	High
5	1.0 Receive request	Receive request and verify CR record	Customer is unaware of the turnaround time (and lead time required) for different request types.	High	Publish menu of services and typical timelines to completion with optimal time leads.	SLA	ST	Low	Medium
6	1.0 Receive request	Receive request and verify CR record	Misunderstanding between customer and ER during project verification for new/ existing (improvements underway).	Medium	ER should publish a list/ inventory of all devices/equip and the standard combination packages	Resources	ST	Low	Medium
7	1.0 Receive request	Receive request and verify CR record	Knowledge issue a) requestor does not have knowledge on types and availability of devices/environments; b) ER person does not know about the device/env	Low	ER should conduct awareness training on devices env packages and processes for request group (QA function)	Training	ST	Medium	Medium
8	1.0 Receive request	Receive request and verify CR record	Typical of a program level initiative–too many requests for the same type of environment	Low	Link MTP and ER env requirements doc to create a checklist/form for CR (to capture complete information at initiation)	Process	ST	Low	High

Figure 14.4

ER payoff matrix.

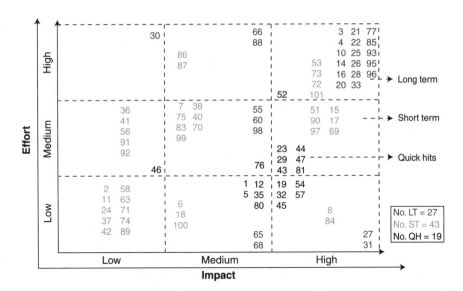

This matrix provided a simplified view of the many solutions that were identified and how quickly some of them could be deployed. But the Lean Six Sigma team and the project sponsor were careful in how the recommendations were presented. Given that there were more than 60 opportunities that could be implemented, the project sponsor challenged the team to identify the top four that could provide the biggest bang for the buck. Along with the top four, those low-hanging fruits that could be implemented within three to four weeks were emphasized.

In addition to the recommended solutions, the Lean Six Sigma team illustrated the relationship between the categories of pain-points and suggestions uncovered in the VOC survey to the categories of solutions to be implemented (see Figure 14.5). The relationship was almost perfectly aligned. In other words, almost all issues identified in the VOC were addressed in the short-term and long-term solutions.

Figure 14.5

VOC response and kaizen event.

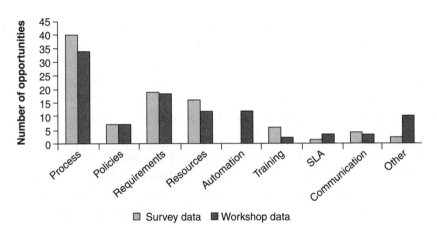

The response from the senior leadership was quite positive. In fact, they verbally supported the recommendations, and as a symbol of their commitment, they ensured provision of dedicated resources to help implement the solutions.

6.2 Action Item Execution

Using the prioritized pain-points and opportunities list, a formal project plan was created accounting for the quick-win and top-four actions.

Subteams were formed to implement the top four solutions, as follows:

1. **Environment maintenance schedule:** Develop and implement a maintenance schedule to ensure that all pre-production environments mirror the production environment.
2. **Standard procedures:** Create checklists (development) for applications so that ER can perform checks to certify nonproduction environments and clients without depending on development to test and certify the environment.
3. **Consistent requirements:** Enhance environment requirements by integrating them into business requirements (BR) documents. Use the ER "Environment Request form" to standardize across retail systems.
4. **Automation:** Create automated certification test scripts for applications to ensure that QA environments can be ready for testing.

The subteams worked over a period of four to twelve weeks (depending on the effort required) to implement the solutions.

For each change, a comprehensive communications plan was implemented, along with training and standard operating procedures.

Senior leaders were leveraged to help deliver the communication and to reinforce that the change could only happen as a team effort.

7. Control Phase

As process changes were implemented, the quality metric trend for high-severity defects was monitored monthly.

These metrics were discussed during regularly scheduled department reviews. Team members within TQA and across departmental boundaries were recognized for positive trends. When negative trends occurred, the problems and root causes were discussed, and plans for corrective action were established and then followed up on during subsequent weekly TQA meetings.

In addition, team members who either participated in the solution or who had demonstrated the new behaviors to sustain the process changes (e.g., consistently provided environmental requirements on time, provided continuous support to development, and performed ongoing maintenance of automated test scripts) were positively reinforced through the department's internal reward system. IT associates can write up an ovation award, praising the specific behaviors that align with the goals and success measures of the project. The recipient not only receives a $20 gift card, but also their name is placed into a pool of other monthly recipients, who are then rated by an ovation committee at the end of the month. The monthly winner receives a trophy, a $250 certificate, and has their picture taken with the CIO and CEO of the company.

8. Benefit Realization

The benefits of this Lean Six Sigma initiative were concerned with quality first and cycle time reduction in the ER process second. The project team began to monitor cycle time trends early in the project after quick-win solutions (low-hanging fruits) were implemented.

As the project progressed, the sponsor emphasized quality as the main imperative—the primary objective of the top four solutions was the elimination of rework. Therefore, the performance metric focused on defect reduction, and thus was continuously tracked beyond the Control phase.

In the early stages of the project, cycle time was reduced by 39 percent (comparison between June 2008, the first month cycle time could be tracked, and the average of August 2008 through January 2009; see Figure 14.6).

During December there was a spike in cycle time. The outliers causing this increase were related to a strategic initiative, which requested multiple store environments at least two months in advance. This caused a spike, because the start time used to calculate cycle time is when the initial request was made. Since the team requesting ten nonproduction environments for this major initiative had put in the request 60 days in advance (a nontypical event), this skewed the average. If those data points were eliminated from the population, the adjusted average cycle time would have been 3.75 days.

After implementing the top four solutions, high-severity ER defects were reduced by 50 percent! The monthly trend stabilized at or below 20 percent of Severity 1 and 2 defects,

Figure 14.6

ER request cycle time.

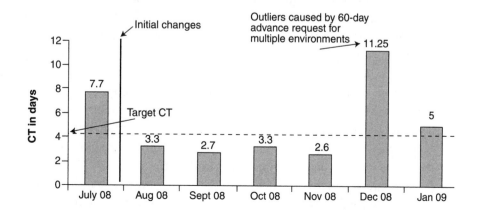

Figure 14.7

ER high-severity defect trends.

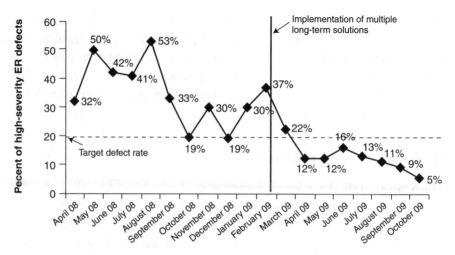

compared to a 41 percent average captured during the initial Measure phase (April 2008–July 2008; see Figure 14.7).

In addition to the cycle time and defect reduction improvements, our client realized a cost savings of approximately US$840,000. This saving was attributed to the reduction of project overruns resulting from nonproduction (e.g., development, QA) environment issues prolonging the project lifecycle. Based on the efficiencies gained in this space, TCS was able to reduce project costs charged to the client by 12 percent.

9. Conclusions and Limitations

Although the project was highly successful overall and demonstrated a cross-functional solution, there were some challenges that at times stifled productivity.

The following observations were key lessons learned:

1. **Cross-functional team participation:** At times it was challenging to get the full and undivided attention of team members both inside and outside of the TQA organization. This was due to a number of competing initiatives that would automatically take precedence, given that everyone's performance was assessed against their own projects. What helped improve participation, regardless of workload, was the midpoint update on the project progress, along with the kick-off for the kaizen workout event. The participants of this kick-off were the senior leaders and their direct reports. By championing the cause, it helped to bolster the involvement of various team members.

2. **Stakeholder resistance:** Even though a tight stakeholder analysis was conducted and the appropriate influencing strategies were executed, there were some internal team members that were not fully supportive or skeptical of the approach. The project manager and project sponsor met with such team members to understand their concerns. Once a personality conflict was detected, the sponsor helped to mediate by making the team

members aware of their personal styles and dysfunctional team behaviors (e.g., stifling the process by not allowing ER team members to provide undivided attention; not effectively collaborating throughout the entire process improvement).

One takeaway was to have a formal team dynamics activity at the onset of a project to ensure that team members are aware of and understand the forming, storming, norming, and performing stages in the development of a new team (Tuckman's stages of group development), but also understand and better appreciate the personalities and work styles of their team members. A DISC assessment is typically a good team-building exercise that can help reduce internal team road blocks. DISC stands for Dominant, Influencing, Steady, and Conscientious, and it defines the behavioral tendencies people elicit in both a steady state and when under pressure. This tool can be extremely helpful in understanding how to work with various behavioral styles.

Acknowledgments

The authors of this case study would like to thank TCS for its permission to allow publication of this case study as a book chapter. The authors would also like to thank the different leaders who, over their careers, had encouraged them to ask the tough questions and to always be curious about why things are performed the way they are and to identify alternative methods to improve performance.

10. References

1. Womack, James and Jones, Daniel T., *Lean Solutions*, Simon & Schuster, October 2005.
2. George, Michael L., *Lean Six Sigma for Service: How to Use Lean Speed and Six Sigma Quality to Improve Services and Transactions*, McGraw-Hill, March 2003.
3. Mikulak, Raymond J. and McDermott, Robin and Beauregard, Michael, *The Basics of FMEA*, CRC Press, April 1996.
4. Cerier, Allison B., "The 4 Dimensional Manager: DISC Strategies for Managing Different People in the Best Ways," *Inscape Publishing*, June 2002.
5. Tuckman, Bruce (1965). "Developmental sequence in small groups." *Psychological Bulletin* 63 (6): 384–99.

CHAPTER 15

Lean Conclusions and Lessons Learned

The preceding five chapters illustrated a different approach to Six Sigma. These Lean Six Sigma projects addressed different types of problems than those highlighted in Part One.

The Lean approach is a bit different from the traditional process improvement focus common to DMAIC projects. Lean projects tend to focus on waste reduction and efficiency improvements. Thus, the use of process maps is key to the simplification and modification of processes that typically result.

Lean projects also tend to involve more players and more people. Kaizen events often involve multiple teams or even entire factories or organizations. Lean projects also strive to achieve faster solutions. Kaizen blitzes can compress the duration of projects from months to days.

The tools used in Lean Six Sigma projects also differ in the following ways:

▲ The use of value stream maps is far more common because of their ability to find nonvalue-added work and waste in established processes.

▲ Lean projects are more apt to identify and incorporate "quick wins" into their solutions. Incorporation of 5-S, poka-yoke (mistake proofing), kanban, muda, or heijunka solutions are relatively common.

▲ Finally, the use of methods to rapidly implement process changes through the use of kaizen events and blitzes reflects a focus on cycle time and speed that is very strong in Lean Six Sigma endeavors.

Many Six Sigma tools, however, are common to both DMAIC and Lean projects. The use of descriptive statistics and comparative methods are essential to the proper analysis of data. Root-cause analysis is important to identify the real causes of process problems, regardless of whether Lean or DMAIC approaches are used. Voice of the customer surveys and Kano analyses are also often used to identify the specific requirements of these target processes. Finally, the implementation of statistical process control charts is common to solutions of all types of Six Sigma projects, especially Lean, because the monitoring of critical process measures is critical to maintaining changes that are made.

Ultimately, the effectiveness of Lean Six Sigma projects is similar to the factors that affect DMAIC projects. You need:

▲ Strong management support
▲ Clearly defined process problems
▲ Champions who can rally support and interest in process improvement
▲ Trained dedicated teams who are skilled in the use of quantitative analysis tools and familiar with the process (kaizen events or blitzes)

The approaches and tools differ only on the surface. Both DMAIC and Lean methodologies focus on finding and fixing problems with processes of all sorts. Both are team oriented and both employ rigorous use of statistical methods in analyzing data and drawing conclusions.

Solutions differ only in style. Ultimately, results are measured in terms of financial benefits to the companies where these projects are initiated and completed.

PART THREE

Design for Six Sigma Projects

PART THREE OF THIS BOOK SHIFTS ITS FOCUS once more to a different type of project. Design for Six Sigma (DFSS) is a more proactive approach, seeking to prevent defects before they occur by introducing a more robust design of products and processes before they are implemented. It is also used to redesign existing products and processes that are significantly underperforming and are in need of complete redesign. DFSS projects (including DMADV) do not strive to fix existing problems by incremental improvements as much as they endeavor to create newer, better processes and products. DFSS projects use different techniques and methods (like Quality Function Deployment or Design of Experiments) to identify and forestall problems with new or redesigned products and processes.

This approach often relies heavily on computer-based simulation or piloting before solutions are implemented. Process and product fixes often seek innovative and breakthrough approaches and solutions, such as those offered by DFSS.

The time and effort associated with such projects is often greater, but so are the paybacks and rewards.

CHAPTER 16

DFSS Primer

Eric C. Maass, Ph.D.
Medtronic
Patricia D. McNair
Raytheon

1. What Is Design for Six Sigma?

Design for Six Sigma (DFSS) is a structured method for developing robust systems (consisting of software, hardware, technology, and/or service) that are aligned with the voice of the customer (VOC), using predictive engineering and anticipating and managing potential issues. Robustness refers to the relative insensitivity of the system and its subunits to variations in production, environment, and usage.

DFSS is applicable to development projects involving a variety of engineering disciplines. Mechanical engineers can apply DFSS in developing simple mechanical devices, like scissors, or a complex mechanical system, like a vehicle or an electromechanical system such as a motor or generator. For example, a team of biomedical engineers, electrical engineers, and mechanical engineers can use DFSS to handle the complex requirements for a prosthetic hand.

There have been several cases where DFSS has been applied to services; published recommendations for applying DFSS to services focus on a subset of the more comprehensive DFSS approach that starts with gathering VOC and developing measurable requirements, but then focuses on developing confidence that the customers will be satisfied with the quality and responsiveness of the service.[1] This approach can be quite relevant for some software projects, including IT projects that focus on providing services.

2. DFSS for Software Development Projects

The DFSS-related success stories in this book relate to software—both IT software and software used in electronic products like computers and communication equipment that involve both hardware and software. DFSS can be applied to software development for an existing hardware platform or in developing the software and hardware for an entirely new product. Iterative development methods such as Agile can align with DFSS, benefitting from the combination of rapid feedback from software testing (perhaps using emulation) with predictive engineering approaches for technical software parameters and quality attributes.

In situations where new product development involves both software and hardware development (Figure 16.1), requirements may flow from the customer, and then the hardware and software teams may address them separately. However, by its very nature, software depends on hardware, and the performance of the hardware tends to depend on the software. The hardware and software interactions and interfaces pose challenges and risks that can best be addressed in a straightforward manner, whether by integrating software and hardware teams, by using emulation effectively, or by empowering special teams to focus on the hardware and software interfaces and interactions.

For a complex electronic system, high-level requirements flow from the customer and are allocated to hardware and software teams, who address these requirements with the appropriate expertise, while supporting the ability of the team to address the interactions and interfaces involved. Teams may focus on a subset of the requirements—the most challenging requirements that involve measurable parameters are referred to as critical parameters. The allocation of critical parameters to subsystems, subfunctions, and components is referred to as critical parameter flow-down (CP flow-down). Critical parameter flow-up involves the prediction of the capabilities for the critical parameters; the process from flow-down through flow-up of optimized critical parameters is referred to as critical parameter management (CPM). Figure 16.2 shows the DFSS flow for a software development-focused project that involves existing, proven hardware.

Figure 16.1

A cellular phone as an example of a system involving both hardware and software.

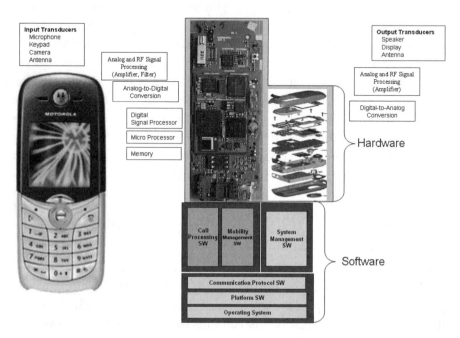

Figure 16.2

DFSS flow for a
software
development
project using
existing
hardware.

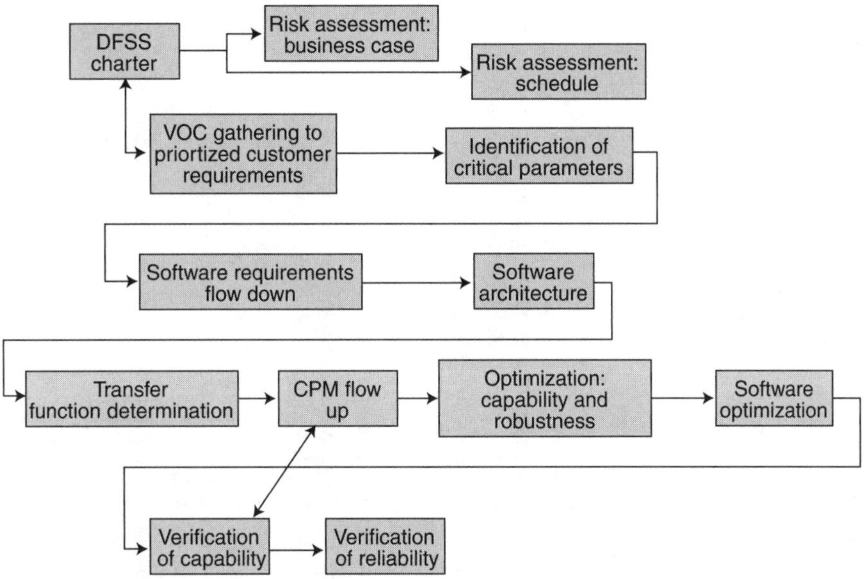

3. DFSS Process Nomenclatures

DMAIC has become recognized as the standard sequence of steps for process or business improvement and problem-solving aspects of Six Sigma. By contrast, no standard process has emerged for DFSS. Alternative DFSS process nomenclatures include Identify, Design, Optimize, Verify (IDOV); Concept, Design, Optimize, Verify (CDOV); and Concept, Design, Optimize, Control (CDOC), which involve the Identify or Concept phase, the Design phase, the Optimize Phase, and the Verify or Control phase. There are also the Define, Measure, Analyze, Design, Verify (DMADV) and Define, Measure, Analyze, Design, Optimize, Verify (DMADOV) nomenclatures that share the Define, Measure, and Analyze phases of DMAIC. DFSS practitioners who feel comfortable with these nomenclatures can use their preferred process with this chapter and associated success stories.

Software practitioners in the industry were dissatisfied with each of these nomenclatures and suggested a nomenclature for DFSS that used terms commonly used during software development. These systems and software engineers suggested the terms Requirements, Architecture, Design, Integrate, Optimize, and Verify; using these terms reduces the need for memorizing a new set of terms associated with "Six Sigma" jargon. This nomenclature provides an easily remembered mnemonic, RADIOV. The alignment of key tools, methods, and associated deliverables to each of these processes is shown in Table 16.1. (Summaries of these tools and methods can be found in a DFSS flowchart that can be downloaded from http://6sigmaexperts.com.)

Let's illustrate each of the phases of DFSS with the example of the RADIOV model.

Table 16.1 Key DFSS Tools and Methods Associated with CDOV, DMAD(O)V, and RADIOV

DFSS step	DFSS processes			Key tools and methods
	CDOV	DMADV	RADIOV	
DFSS charter	Concept	Define	Requirements	DFSS charter, deployment plan
Business case: risk management	Concept	Define	Requirements	Monte Carlo simulation—business case
Schedule: risk management	Concept	Define	Requirements	Monte Carlo simulation—critical chain/TOC-PM
VOC gathering	Concept	Define	Requirements	Concept engineering, KJ analysis, Kano analysis, interviews, surveys, conjoint analysis, customer requirements ranking
System concept generation and selection	Concept	Define	Requirements	Brainstorming, TRIZ, system architecting, axiomatic design, unified modeling language (UML), Pugh concept selection
Identification of critical parameters	Concept	Measure	Requirements	Quality function deployment (QFD), design failure modes and effects analysis (DFMEA), fault tree analysis (FTA)
Critical parameter flow down	Design	Analyze	Architecture	Quality function deployment (QFD), critical parameter management, fault tree analysis (FTA), reliability model
Module or component concept generation and selection	Design	Analyze	Architecture	Brainstorming, TRIZ, system architecting, axiomatic design, universal modeling language (UML), Pugh concept selection
Software architecture			Architecture	Quality attribute analysis, universal modeling language (UML), design heuristics, architecture risk analysis, FMEA, FTA, simulation, emulation, prototyping, architecture tradeoff analysis method (ATAM)
Transfer function determination	Optimize	Design	Design	Existing or derived equation, logistic regression, design emulation, regression analysis, design of experiments (DOE), response surface methodology (RSM)
Critical parameter flow up and software integration	Optimize	Design	Integrate	Monte Carlo simulation, generation of system moments method, software regression, stability and sanity tests
Capability and robustness optimization	Optimize	Design	Optimize	multiple response optimization, robust design, variance reduction, RSM, Monte Carlo simulation with optimization
Software optimization		Optimize	Optimize	DFMEA, FTA, Software mistake proofing, performance profiling, UML, use case model, Rayleigh model, defect discovery rate

| | DFSS processes | | | |
DFSS step	CDOV	DMADV	RADIOV	Key tools and methods
Software verification			Verify	Software testing
Verification of capability	Verify	Verify	Verify	Measurement system analysis (MSA), process capability analysis, McCabe complexity metrics
Verification of reliability	Verify	Verify	Verify	Reliability modeling, accelerated life testing (ALT), WeiBayes, fault injection testing
Verification of supply chain readiness	Verify	Verify	Verify	Design for manufacturability and assembly (DFMA), lead time and on time delivery modeling, product launch plan, FMEA/FTA for product launch

4. Requirements Phase

During this key first step, the requirements and critical competitive differentiators are identified and prioritized, leading to the selection of a superior product concept that is based on VOC. Implicit customer expectations and business expectations in the form of voice of the business (VOB) can also be incorporated as deemed appropriate.

The sequence of steps includes:

1. Gathering, understanding, and prioritizing the VOC and VOB
2. Generating alternative product concepts and selecting among these alternatives
3. Identifying measurable critical parameters requiring intense focus

Key tools and methods for the Requirements phase are summarized in Table 16.2.

Some requirements may be stable, particularly those involving the "must-be" requirements—the basic functions of the product and requirements that handle historical problems with similar products. Some other requirements—particularly those involving optional features—might change over the course of the project. Studies have indicated that software requirements typically change by at least 25 percent during the course of the project.[2,3]

The general goal, then, would be to develop as close to a finalized set of requirements as possible during the Requirements phase at the beginning of the project, and then employ an approach such as Agile development to effectively respond to any subsequent changes. At a minimum, the requirements developed during the Requirements phase should be sufficient to provide guidance for the team when they select the software architecture.

Table 16.3 and Figure 16.3 illustrate some of the deliverables from the Requirements phase of a DFSS project for developing software that customizes the settings for multiple portable communication devices. Table 16.3 is an abridged House of Quality, in which the customer requirements (left column) are translated and linked to three measurable technical requirements for the software: programming time, a security metric, and a user interface complexity metric. The row above these technical requirements indicates the direction of goodness (– for programming time indicates less programming time is preferable, + for the security metric indicates more security is better, and – for the user interface complexity metric indicates a preference for less complexity for the user). The plus and minus signs in the rows above the direction of goodness indicate if the technical requirements tend to improve together (+ or ++) or if there is a tradeoff among them (– or – –). In this House of Quality, programming time and the user interface complexity metrics tend to improve together, whereas improving the programming time could involve a tradeoff with the security metric. Each measurable technical requirement received a score for prioritization, obtained by the sum of the products of the importance scores for the customer requirements (1–10 scale) with the strengths of the relationships between the technical requirements and the customer requirements (0, 1, 3, or 9). In Table 16.3, the programming time receives the highest prioritization score.

Table 16.2 Steps, DFSS Tools, and Methods Associated with the Requirements Phase

DFSS Step	Key Tools and Methods	Purpose
DFSS charter	DFSS charter	Define the business case and expectation for the project
	Deployment plan	Plan the tools and methods to be used, and support needed
Business case: risk management	Monte Carlo simulation—business case	Estimate confidence and prioritize risks to business goals
Schedule: risk management	Monte Carlo simulation—schedule	Estimate confidence and prioritize risks to meet schedule
	Critical Chain/TOC-PM	Handle multiplexing, distractions, Parkinson's law schedule
VOC gathering	Concept engineering	Method to obtain and translate VOC to prioritized requirements
	Interviews	Gather voice of the customer (VOC), rich stories and insights
	KJ analysis	Filter many voices and images from interviews to a vital few
	Kano analysis	Assess impact of customer requirements on purchase decisions
	Conjoint analysis	Prioritize customer requirements and predict impact on sales
	Customer requirements ranking	Prioritize customer requirements
	System level house of quality	Translate prioritized customer to system requirements
System concept generation and selection	Brainstorming	Generate set of concepts that can help fulfill the requirements
	TRIZ	Generate concepts and overcome requirements trade offs
	System architecting	Generate concepts by partitioning the system using heuristics
	Axiomatic design	
	Functional modeling, unified modeling language (UML)	Create architecture-independent model with flows and functions
	Pugh concept selection	Select an optimal concept that helps fulfill the requirements
Identification of critical parameters	System level house of quality	Translate customer requirements to measurable requirements
	Design failure modes and effects analysis (DFMEA)	Access and prioritize technical risks
	Critical parameter selection	Select key system requirements for predictive engineering
	Risk and importance	

371

Table 16.3 First House of Quality (HoQ) for Software to Customize Settings for Individual and Multiple Portable Units

			—>	—
			—>	—	++
Direction of Goodness			–	+	–
Voice of the customer (VOC)	**Importance**	**Programming time**	**Security metric**	**User interface complexity metric**
Ease of initial configuration	10	3		3
Security established to protect data	9		9	1
Minimize use interface complexity	10			9
Improve performance (speed)	3	9		
Improve reliability	7			
Maintain configurations	10	9		
Software needs to always be backward compatible	9	9		
Improve real time performance (speed)	8	9		
Improve help system	7	9		
Improve template management	5	9		1
Easily copy data across units	5	9		
Scoring totals		453	81	134

Figure 16.3 shows the qualitative flow-down for the prioritized technical requirement, programming time, to a set of subordinate measurable parameters (y's) and to factors that are either controllable (x's) or uncontrolled in the actual use-cases (N's).

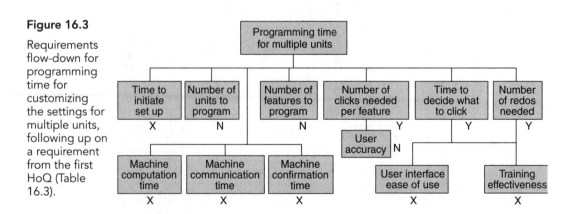

Figure 16.3

Requirements flow-down for programming time for customizing the settings for multiple units, following up on a requirement from the first HoQ (Table 16.3).

5. Architecture Phase

During the Architecture phase, the vital few system-level requirements or critical parameters are refined into measurable requirements for the subsystems. Key stakeholders are engaged, including the test team, to ensure that the requirements can be measured, tested, and evaluated. Concepts and architectures for the subsystems are selected that support the required functions and the features, and customer expectations are derived from the system-level requirements. Some key tools and methods for the Architecture phase are summarized in Table 16.4.

Table 16.4 Steps, DFSS Tools, and Methods Associated with the Architecture Phase

DFSS Step	Key Tools and Methods	Purpose
Critical parameter flow down	Quality function deployment (QFD)	Find measurable requirements to fulfill customer requirements
	Critical parameter flow down	Identify which subordinate parameters (y's), control factors (x's) and noises (n's) affect the critical parameters
	Fault tree analysis (FTA)	Flow down failures: determine how events can cause failures
	Reliability modeling	Use system reliability model to focus on key reliability risks
Module or component concept generation and selection	Brainstorming, TRIZ, system architecting, axiomatic design, Pugh concept selection	Generate concepts that fulfill module or component requirements and select optimal concept
Software architecture selection	Quality attribute analysis	Measure quality aspects: availability, usability, security, performance
	Design heuristics	Develop alternative software architectures using rules of thumb
	Simulation, emulation, prototyping	Evaluate functioning version of candidate architecture
	Architecture tradeoff analysis method	Consider requirement tradeoffs as select software architecture

The goals for the Architecture phase for software involve flowing down the system-level requirements to requirements for the software and developing a high-level, best-fit software architecture (see Figure 16.4). Technical risks inherent in the product architecture are identified using techniques such as design failure modes effects analysis (DFMEA) and fault tree analysis (FTA). Critical parameters from the Requirements phase are aligned with proposed architectures, and quality attributes (nonfunctional requirements) are identified. Concept selection techniques aid in design tradeoff analysis and in selecting the most cost-effective architecture for the product or platform. The selected architecture is evaluated using modeling, simulation, emulation, and prototyping. The architecture is documented, which is crucial for

Figure 16.4

Alignment of a proposed architecture for a wireless base station controller with the critical parameters flowed down from the Requirements phase.

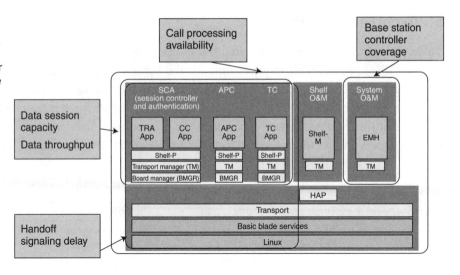

the subsequent detailed design. Architectural patterns, styles, and functional modeling facilitate the employment of standard solutions.

6. Design Phase

The term "Design" in alternative DFSS nomenclatures could generate some confusion. In the CDOV, CDOC, and IDOV processes, the "Design phase" refers to the steps after concept selection and involves the flow-down of critical parameters and functional performance for a beginning design baseline.

In DMADV, DMADOV, and RADIOV processes, the "Design phase" refers to a later stage involving the determination of transfer functions to be used in predictive engineering (see Table 16.5). In DMADV, the Design phase encompasses the entire predictive engineering efforts, while predictive engineering is split among the Design and Optimize phases in DMADOV and RADIOV.

Table 16.5 Steps, DFSS Tools, and Methods Associated with the Design Phase to Determine the Transfer Function

DFSS step	Key tools and methods	Purpose
Transfer function determination	Existing or derived equation	Use equation or mathematical model for predictive engineering
	Logistic regression	Use equation for probability of event from history, experiment
	Simulation	Use a computer simulation program for predictive engineering
	Emulation	Try software functions with substitute hardware and/or software
	Regression analysis	Develop equation for continuous parameter from history
	Design of experiments (DOE)	Analyze planned experiment, identify significant factors, establish equation
	Response surface methodology (RSM)	Analyze planned experiment, develop polynomial for model

R-Sq = 98.73% R-Sq(pred) = 96.82% R-Sq(adj) = 98.16%	**Regression Equation**
Estimated coefficient for callprocessingservicefailure using data in uncoded units	Call_Processing_Failure =
	5.53E-06
Term Coef	+ 1.132 * PAM_Failover_Failure
Constant 5.52569E-06	+ 0.895 * Carrier_Failover_Failure
CageFailover 1.00415	+ 1.004 * Cage_Failover_Failure
PAMFailoverFailure 1.13153	
CarrierFailoverFailure 0.895048	
SSCFailoverFailure -0.0100666	
SAMFailoverFailure -0.0000097	

Figure 16.5 Development of a transfer function for call-processing failures using DOE with emulation on existing hardware.

Figure 16.5 illustrates the development of a transfer function for a software defect—call-processing failure—for the base station described by the software architecture in Figure 16.4.

7. Integrate Phase

The Integrate phase begins the process of combining and integrating the components and ensuring that the interfaces between components meet expectations under the range of use conditions. Software testing, such as software regression, stability, and sanity tests, can be used to verify successful integration and to find defects (see Table 16.6).

Table 16.6 Steps, DFSS Tools, and Methods Associated with the Integrate Phase

DFSS step	Key tools and methods	Purpose
Critical parameter flow up	Monte Carlo simulation Generation of system moments method	Use repeated runs with variations of x's to predict distribution of y Use calculus, statistical variations of x's to predict distribution of y
Software integration	Software regression Stability and sanity tests	Find software defects (after code changes) that impact functionality Check: software functions as expected and doesn't crash

8. Optimize Phase

During the Optimize phase, flow-up is performed to assess the capabilities of the design allowing for variability in use conditions. This flow-up involves predictive engineering, using the transfer function from the Design phase in conjunction with Monte Carlo simulation. Transfer functions for the critical parameters and/or flowed-down measurable technical requirements are determined using appropriate methods such as regression, DOE, and response surface methodology (RSM) for continuous parameters, and logistic regression for discrete parameters. Robust design and stochastic optimization approaches are used to build high confidence that critical parameters will meet expectations and have adequate capability, allowing for variations in use cases (see Table 16.7).

Table 16.7 Steps, DFSS Tools, and Methods Associated with the Optimize Phase

DFSS step	Key tools and methods	Purpose
Capability and robustness optimization	Multiple response optimization	Find set points for x's to co-optimize several y's
	Robust design	Find set points for x's that render y much less sensitive to noises
	Variance reduction	Choose and use a method to reduce variation of the y
	RSM	Analyze experiment, find set points for x's with polynomial model
	Monte Carlo simulation and optimization	Find set points for x's that reduce variation of one or several y's
Software optimization	DFMEA, FTA	Assess, prioritize and manage risks to software functionality
	Software mistake proofing	Anticipate and prevent errors on inputs, interfaces, processes
	Performance profiling	Measure, analyze, benchmark function calls' frequency and duration
	UML, use case model	Model functions by/for the user and other systems, and dependencies
	Rayleigh model	Model and predict failures that will be observed after SW release
	Defect discovery rate	Model and graph number of defects discovered over time

Figure 16.6 illustrates the Monte Carlo results for data session capacity for the wireless base station controller (described in Figures 16.4 and 16.5) before and after optimization, which resulted in an improved Ppk (long-term process performance) from 0.32 to 1.52, and improvement in Cpk (short-term process capability) from 0.32 to 1.54. In this example, optimization was achieved through memory reallocation. Notice how the bell curve shifts to the right. As a result, it no longer intersects the Lower Specification Limit (initially the area to the left of the LSL was all defects) and it moves significantly away from the LSL and to the right after improvements.

9. Verify Phase

At least two sets of expectations should be quantified and assured during the Verify phase for software—namely, that the software will meet performance expectations with high confidence and be reliable.

Verification includes both testing and evaluation to assess the capability and reliability of the product. For software, testing is particularly important in assessing that the software functions as intended, over a range of use conditions and scenarios.

For hardware, the performance capability on an appropriate sample of early production can be used to verify that the optimized parameters perform consistently within acceptable limits, and approaches such as accelerated life testing (ALT) can provide confidence that the hardware aspects will meet reliability expectations. Approaches like fault injection testing enhance confidence in meeting software reliability expectations (see Table 16.8).

Figure 16.6

Monte Carlo
simulation for
data session
capacity for the
wireless base
station
controller
before (top)
and after
(bottom)
optimization.

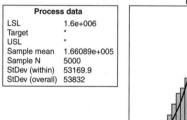

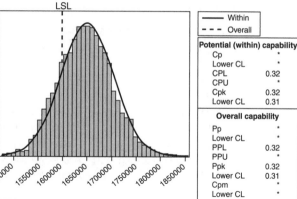

Observed performance		Exp. within performance		Exp. overall performance	
% < LSL	17.74	% < LSL	16.92	% < LSL	17.22
% > USL	*	% > USL	*	% > USL	*
% Total	17.74	% Total	16.92	% Total	17.22

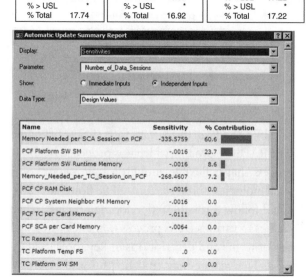

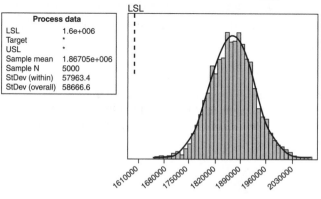

Observed performance		Exp. within performance		Exp. overall performance	
% < LSL	0.00	% < LSL	0.00	% < LSL	0.00
% > USL	*	% > USL	*	% > USL	*
% Total	0.00	% Total	0.00	% Total	0.00

Table 16.8   Steps, DFSS Tools, and Methods Associated with the Verify Phase

DFSS step	Key tools and methods	Purpose
Software verification	Software testing	Provide customers confidence that the software works as intended
Verification of capability	Measurement system analysis (MSA)	Estimate variation introduced by measurement system
	Process capability analysis	Assess confidence y's will meet specifications despite variability
	McCabe complexity metrics	Measure SW complexity, determine test cases needed for coverage
Verification of reliability	Reliability modeling	Predict and assess reliability
	Accelerated life testing (ALT)	Accelerate failure mechanisms to assess reliability in normal use
	WeiBayes	Provide lower confidence limit If reliability testing has no failures
	Fault injection testing	Inject most-likely fault conditions into SW code to detect issues
Verification of supply chain readiness	Design for manufacturability and assembly (DFMA)	Design to simplify and immunize manufacturing and assembly
	Lead time and on time delivery modeling	Apply predictive engineering to a model for the supply chain to provide confidence in lead time and on time delivery
	Product launch plan	Develop supply chain and marketing plans for a successful launch
	FMEA/FTA for product launch	Assess, prioritize, and manage risks to the product launch

10. Summary

DFSS has a wide range of applicability and can be relevant for a variety of engineering disciplines and interdisciplinary efforts. The RADIOV process, which was proposed by software practitioners, provides a comprehensive process that is flexible enough for simple or complex projects for developing software, systems, technologies, and services. A DFSS flowchart can provide guidance for practitioners as they plan and follow through on developing a robust, reliable product. This chapter discussed how DFSS aligns to various DFSS process nomenclatures, including RADIOV.

Tables provided a glossary of terms and descriptions of key tools used through the DFSS software development process. The reader may find helpful and relevant case studies in subsequent chapters in this book, and is encouraged to peruse additional books referenced in this chapter for a deeper dive into the software DFSS development method.

11. References

1. Yang, K. *Design for Six Sigma for Service*, McGraw-Hill, 2005.
2. Jones, C. *Applied Software Measurement*, McGraw-Hill, 1997.
3. Boehm, B. and P. Papaccio. Understanding and Controlling Software Costs, *IEEE Transactions on Software Engineering*, October 1988.

Editors' Note

This chapter is based on material from the book *Applying Design for Six Sigma to Software and Hardware Systems* (Prentice-Hall, 2009) by Eric Maass and Patricia McNair. All figures and tables are reproduced with permission from Prentice-Hall.

CHAPTER 17

How to Radically Streamline Your Business Processes

Bill Cooper

Relevance

This chapter is relevant to organizations that think they have too many or too complex processes and want to streamline them but do not know where to begin and how to proceed. Learn how Motorola overcame this challenge in a concerted 10-month effort to significantly streamline its product and service delivery processes in IT, effecting more than 5,000 practitioners worldwide.

At a Glance

▲ The CIO and IT senior leadership of Motorola directed the IT Process and Quality organization to significantly streamline the product and service delivery processes in IT to make them easier to understand and follow without compromising quality and compliance with regulatory requirements.

▲ Motorola leveraged Design for Six Sigma to systematically redesign current IT processes, focusing on value-added activities.

▲ The project resulted in significant simplification to the processes, reducing time spent on process activities by 15 percent for more than 5,000 IT practitioners and saving the company several million dollars.

Executive Summary

As with many organizations, Motorola's IT organization (hereafter referred to as Motorola IT) is continually evolving. It is always looking for more ways to be efficient and to right-size its processes in order to provide the best value to its business customers. Critical aspects of Motorola IT are the adoption and use of common processes that are used globally to perform key planning, development, and support activities. In 2006 Motorola IT started implementing common processes for product and service delivery and support worldwide based on capability maturity model integration (CMMI) and Information Technology Infrastructure Library (ITIL) frameworks;[1,2] these processes are hereafter collectively referred to as IT Common Process. As the processes became common across all IT functions and outsourcing partners, Motorola continued to look at ways to improve them. As the size and shape of Motorola changed significantly, Motorola IT and its processes needed to change as well.

The CIO and her staff gave clear direction that the IT Common Process needed to be right-sized. It was also clear, based on experience and feedback, that the IT Common Process needed to be significantly streamlined, made easier to follow, needed to provide more flexibility to practitioners, and needed to foster a culture of increased accountability and decision making closer to the practitioner. Industry benchmarking showed that Motorola IT processes were five times larger and more complex than companies of similar size.[3]

In July 2009 a Six Sigma project team was assembled to right-size the IT Common Process. A small team of Six Sigma Master Black Belts, Black Belts, and subject matter experts (SMEs) was assembled and immediately started work on the project. The approach was to not only streamline processes but also to introduce cultural changes necessary to support the streamlined processes. The key principles for process redesign were:

▲ Have a simplified and integrated end-to-end process.
▲ Focus on approvals by key stakeholders with accountability.
▲ Shift focus from reviewing and approving documents to reviewing and approving IT and customer readiness.
▲ Simplify and emphasize regulatory and internal controls.
▲ Focus on what the practitioner needs to do rather than prescribing how the practitioner must always execute the process.
▲ Provide greater flexibility to teams and trust them with greater decision-making authority with regard to how to execute the process for their project or activity. However, ensure key verification controls are in place.

The result was the implementation of a new, fully integrated IT Common Process 10 months after the initial directive given by the CIO. The process is in use by more than 5,000 Motorola IT professionals and outsourcing partners. Motorola now has a process that:

▲ Requires fewer steps
▲ Has fewer required work products
▲ Increases productivity
▲ Improves process compliance

The new process creates a work environment that:

▲ Focuses on *responsiveness and quality* delivered to customers

▲ *Provides flexibility* and empowers IT professionals to execute the process based on risk and complexity of work

▲ *Enables usability* so people understand and follow the process

▲ Does not describe how to handle all exceptional situations

▲ *Defines what to do*, not *how* to do it

▲ Reduces checking and *enhances accountability*

▲ *Is based on trusting people to know and do the right things for the IT customers*

With the new right-sized IT Common Process, Motorola IT is positioned for the dynamic and exciting business opportunities for the next several years.

This chapter provides details as to how the Motorola Six Sigma team executed the project. This will provide a good roadmap and example to managers of any function that need to streamline and reduce process complexity.

1. Introduction

Over the years as Motorola IT grew to support large diverse businesses within Motorola, the IT organizations acted independently and operated with more than 15 different sets of processes. As Motorola started to consolidate functions to support merged and fewer business units, IT had a stronger need to have one set of common processes for its operations. This would enable resources to be more effectively shared across business units and reduce the need for additional resources.

With direct support from Motorola's senior IT leadership, the effort began to move all IT organizations to one IT Common Process. From 2006 to 2007 Motorola IT designed and implemented the IT Common Process, rigorously designed to CMMI and ITIL frameworks. It comprised several processes that were developed by multiple teams with minimal integration. The teams included requirements that originated from specific IT functions around the company, which added complexity to the processes and added many "How To'" activities that the frameworks did not provide. The processes were implemented successfully and enabled IT to adopt one set of common processes.

As the IT Common Process was used, it became clear that it needed to be streamlined to be more integrated, remove nonvalue-added activities, and reduce complexity. Starting in 2008, several Six Sigma projects were conducted to streamline specific portions of the processes based on feedback from practitioners and management. Six Sigma tools used in these projects included value stream mapping, voice of customer, quick wins, poka-yoke, 5-why's, and kaizen, to name a few. These projects were successful in optimizing specific areas of the process, but did not have enough impact on reducing overall process complexity and overhead. The voice of the IT organization grew louder. The processes were far too complex, difficult to follow, and had many redundant and nonvalue-added activities. The organization was saying that many of the requirements of the process did not help with quality, cost, or delivery of products and services.

In June 2009 the leaders of the IT Process and Quality team presented the status on ongoing process streamlining efforts to the CIO and her staff. The CIO was clear that it wasn't enough. She challenged the leaders to significantly reduce processes with regard to steps, activities, documents, and complexity. The CIO empowered the leaders to make the decisions for the organization and to bring in key SMEs that would be an asset to the team. She cautioned not to get too many people involved, as that would add complexity to the process and expand the time to implement the improvements. The CIO also commissioned a small subset of her staff to act as customers and make decisions as quickly as needed by the process team and to approve the final process.

2. Project Background

The Process and Quality leaders immediately started work to significantly streamline the IT Common Process. The first order of business was to develop a charter and a small team to lead the project. The initial project meeting occurred in July 2009 with the two Process and Quality Leaders and two process managers. The team consisted of three Master Black Belts and one Black Belt.

The team immediately recognized that the previous approaches to process streamlining were not enough. A different and more radical approach had to be taken to get the level of change the CIO and IT practitioners were demanding. Improving portions or subsets of the process would not achieve the goals either. The next window of opportunity for implementation of the modified or new process would be early 2010. This gave the team only 10 months at the most to implement the process. It was also recognized that a small empowered work team needed to be established to meet the deadline. The project team consisted of 10 people, with internal and external practitioners and experts included as needed. The team chose to use the Six Sigma DMADV methodology for this project, which is also referred to as DFSS. DMADV is an acronym for Define, Measure, Analyze, Design, and Verify. DMADV was chosen, because this project was more focused on redesign of processes rather than incremental process improvement. The following sections will outline how the team used Six Sigma.

3. Define Phase

The Define phase of a project is focused on defining the business case and opportunity, establishing project goals and scope, drafting a project plan, assembling a project team, and identifying the Six Sigma methodology to be used. The most important deliverable of this phase is the project charter. The following sections describe the project charter and its elements in detail.

3.1 Project Charter

The project charter establishes the clear purpose, goal, scope, and high-level timing of the project. It defines what team members are required to execute the project and the expected benefits of the project. The project charter is typically a one-page document that provides a summary of these elements. It is used throughout the project to keep the team focused on the purpose and expected outcome of the project. The key elements of the charter for this project are described in the following sections.

3.1.1 Business Case Statement

The business case statement describes the current situation that is driving the need for the project. The input for the business case was derived from surveys conducted regarding the IT processes, jump starts with senior leadership, and voice of customer exercises conducted by other Six Sigma projects.

The key points of the business case statement for this project were the following:

▲ The current IT processes, while good, are complex and difficult to follow.
▲ The current processes have activities that do not contribute to the quality, cost, or delivery of IT products and services.
▲ The size and complexity of Motorola is changing, and the IT processes need to change to be right-sized with the corporation.
▲ The business needs are shifting from many large projects to fewer large projects and more small or medium-sized projects. Large projects are defined as projects with a budget (capital and resource) of greater than US$100,000.
▲ Even after tailoring the process for their use, practitioners found the process to be cumbersome, or they failed to correctly tailor the process, thus incurring significant overhead.
▲ The current process was designed with broad participation and a consensus-based approach, which resulted in significant process complexity due to the need to accommodate everyone's input.
▲ IT practitioners and managers were providing feedback that the current processes were too heavy and too complicated to follow.

3.1.2 Opportunity Statement

The opportunity statement defines what the benefits would be if the current situation is improved and the goals of the project are met. It is important to acknowledge that not every project will meet every goal completely. That does not mean the project is not successful. There are many cases where project goals can be partially met while bringing significant benefits to the company.

The expected benefits (opportunities) for this project were:

▲ Reduce several hours a day of time spent by thousands of IT practitioners performing process activities that do not add value to the customers.
▲ Improve quality, cost, and delivery of IT products and services by providing a simpler, more flexible process that focuses on the customer.
▲ Improve process compliance by having a process that is simpler and easier to follow.

3.1.3 Goal Statement

The goal statement should clearly describe what should be accomplished by the project in specific, measurable, attainable, relevant, and timely criteria. Many project teams struggle with establishing these goals without data, which is typically collected in the Measure phase. It is acceptable to have initial goals without the specifics and to update them when the initial Measure phase activities have been completed. Also, some goals may be lacking a clear measurement but meet the other criteria of a goal. If it makes sense for the project and its sponsor, then use it. Do not get stuck on technicalities of goal definitions. Use your common sense and discretion.

The goals for this project were to:

▲ Deploy the process no later than June 1, 2010
▲ Reduce required process steps by at least 50 percent
▲ Reduce post-release defects by 10 percent
▲ Reduce compliance deficiencies by 50 percent
▲ Achieve 70 percent positive reaction to the new process (based on survey)

3.1.4 Project Scope

Drawing boundaries around the project is important to its success. Often management and project leaders will define the scope of the project to be too large. It may seem like an achievable scope at first, but as the team dives into the details, they learn the scope is too large. When defining the scope, it is obvious to clearly state what is in scope. It is also a practice to define the scope in terms of what is out of scope. Stating what the project does not include can drive the project team in a more specific direction and can avoid much wasted time.

The scope of this project was defined to include the streamlining of the IT Common Process from portfolio management to deployment and support. It was also made clear that the project should not include the processes used to operate IT financial functions and IT supplier management.

3.1.5 Project Plan

The project charter listed the high-level milestones and target dates. These included the phases of the DMADV methodology. The high-level project plan depicted in the charter is shown in Table 17.1.

Table 17.1 High-Level Project Plan

Milestone	Target
Define	7/1/2009
Measure	8/10/2009
Analyze	9/20/2009
Design	12/30/2009
Verify	2/10/2010
Deploy	4/19/2010

3.1.6 Project Team

Selecting the right team to execute the project is critical to its success. It is important to have a high-level sponsor to commission the project, a Champion to support the execution and remove roadblocks, and key team members. Be careful not to include too many people. This project had the CIO and a small subset of her staff as sponsors and the two senior leaders of the Process and Quality organization as active Champions. The core project team was limited to 10 people and consisted of the following roles:

▲ Project manager
▲ Process manager
▲ Quality manager
▲ Process architect
▲ Six Sigma Master Black Belts, Black Belts, and Green Belts
▲ Organizational change management and communications expert
▲ Training expert
▲ Regulatory expert
▲ Corporate auditor
▲ Other experts and practitioners as needed

3.2 Detailed Project Plan

When people use the term "project plan," it can mean different things to different people. In the context of this project, a traditional definition was used, as described later. This project utilized three methods to document, communicate, and manage the project activities and milestones.

First, the project charter tracked high-level milestones and target dates.

Second, a detailed project plan was developed and managed throughout the project. The project plan included:

▲ Details about the background and objective of the project
▲ Roles and responsibilities
▲ Controls that would govern the project
▲ References to key plans used in different phases of the project
▲ Financial constraints and objectives

Third, a detailed project work breakdown structure (WBS) and project schedule were created and managed daily throughout the project.

3.3 Organizational Change Management

As part of the project planning activities, the team recognized from previous experiences that organizational impact needed to be managed in a more formal, structured way. It was decided to utilize a formal organizational change management framework.[4] This is a method by which the organization's strategy, processes, culture, and people are realigned from its current state to a different desired future state. An effective change management campaign helps people affected by the change to understand, accept, and commit to the new desired future state. The change management plan is coupled with the communications plan and training plan. These three plans, when integrated correctly, can be a powerful approach to effectively bringing change to the organization. The Motorola change management framework shown in Table 17.2 was used for this project.

The primary tools used in the change management framework were:

▲ Business case for change
▲ Sponsor agreement
▲ Stakeholder management plan
▲ Risk and issue management plan
▲ Communications plan
▲ Project plan (discussed in previous section)

3.3.1 Business Case for Change

The business case for change is the most important document of any change initiative. It describes the current situation and why it needs to change. It is important to describe the current state in enough detail so as to compel the stakeholders to move away from it. This should cause the stakeholder to want the future state. This document also describes the future state and why it needs to be reached. Just as important, the desired future state must be communicated well enough to give people a reason to move to it. A plan for how to close the gap between current state and future state is also identified. Table 17.3 provides an example template of a business case for change document used for this project.

3.3.2 Sponsor Agreement

Coming to agreement with the project sponsor for any project is important. The sponsor's primary role is to support the initiative and help drive the change throughout the organization. When using a Six Sigma approach like DMAIC or DMADV, agreement is usually obtained in the Define phase while developing and approving the project charter.

A tool that can be used to prepare to gain sponsor agreement is the sponsor assessment template. This tool is designed to help the project leader assess the sponsor(s) and prepare for an initial discussion to agree on four key areas:

1. Specific actions needed to engage (i.e., support, guide, or drive) in the project
2. Key communications, timing, and frequency
3. Dealing with and mitigating issues and risks
4. Estimated time for completion of the project

Table 17.2 Motorola Change Management Framework

	Planning change	Designing change	Executing change	Sustaining change
Objective	Establish and gain agreement on the business case for change	Develop a structured, well-governed approach to implementing change	Execute plan with focus and discipline	Evaluate effectiveness, sustain momentum and transfer learning
Activities	Identify and qualify sponsor Establish and summarize the business case for change Current state assessment Future state design Gap-filling strategy Identify high-level project milestones and timeline Identify key stakeholders, engage as appropriate Identify key components of of communications plan Identify potential risks and mitigation strategies	Assess stakeholders; outline management plan Assess organizational readiness Outline roles, responsibilities Determine resource requirements Develop project plan including/incorporating: Identify potentional quick wins Key milestones Prioritized list of actions Risk/issue management plan Communication plan Learning plan	Implement project plan while carefully managing: Stakeholders Risk Communication Governance Course-correct as needed Celebrate "quick wins" Recognize and reward change participants, as appropriate	Evaluate the effectiveness of the initiative Capture learning Conclude the project Develop a sustainability and continuous improvement plan
Governance	Gate review: Gain sponsor/steering committee agreement on initial elements of change plan	Gate review: Gain sponsor/steering committee agreement on final change plan	Gate review: Gain sponsor/steering committee agreement on change results	Gate review: Gain sponsor/steering committee agreement on project close-out and sustainability plan
	← Continuous focus on stakeholder management, communication, and risk/issue management →			
Deliverables	Project governanance Clearly articulated business case for change Initial risk/issue management plan Sponsor agreement Stakeholder identification Initial milestones and timelines	Project plan Stakeholder management plan Communications plan Risk/issue management plan	Completion of deliverables as outlined in project plan	Initiative evaluation Summary of key learning Sustainability plan Project close-out

Table 17.3 Business Case for Change

Current state: Describe the current business situation with a description of the problem and how the change process will address it. Develop this as part of the "burning platform" addressing why the change is important. This should help push people away from the current situation.
Desired future state: Describe the ideal desired future state with the benefits to the organization. This should be a compelling overview that pulls people toward the desired change.

Business factors	ID#	Current state	Future desired state	Gap	Change goals

Customers	ID#	Current state	Future desired state	Gap	Change goals

Competitor/industry	ID#	Current state	Future desired state	Gap	Change goals

Consequences of NOT changing: Provide a summary of what might occur if nothing new is done to change the current situation. This is often a powerful tool for energizing the organization to take action. Take care not to overstate your case, but take the situation to its logical conclusion if nothing new is done to address it.

Opportunities: Consider linkages, activities, initiatives, etc. going on within the company or department that can be leveraged for the current project.

Summary: Briefly summarize the overall business case for change ("elevator speech").

Once agreement has been reached, the project charter can be updated accordingly. Table 17.4 gives an example of a sponsor assessment template used for this project.

3.3.3   Stakeholder Management Plan

A critical element of the organizational change management effort is identifying your stakeholders and preparing for their reaction to the change. A stakeholder is considered to be anyone who has an interest in the success of the project. It may be people causing the change, people receiving the change, or people indirectly affected by the change. Stakeholders need to be identified and assessed for their anticipated reaction (positive, neutral, or negative) to the upcoming change, which would cause them to be barriers, enablers, or neutral to the change.

Table 17.4 Sponsor Assessment Template

Qualification factor	Rating Yes/No/Unknown
Has access to funding for the project	Yes
Can free up people from other responsibilities to devote to the project	Yes
Willing to take a public role in support	Yes
Is accountable and willing to hold others accountable	Yes
Feels dissatisfied with the current situation	Yes
Personally committed to the success of the project	Yes
Follows up on commitments	Yes
Manages consequences from those who facilitate or impede the business agenda	Unknown
Uses data to track progress during implementation	Yes
Understands what resources will be required to make the change	Unknown
The change initiative takes a high priority on this person's agenda	Yes
Will play an active role on change teams	Unknown
Is visible in promoting the change	Yes
Will support the governance/gate review process	Yes
Overall assessment score	29

Scoring criteria

>= Sponsor is engaged
21–26 Sponsor engagement is uncertain
<21 Sponsor may not be engaged

Key engagement actions

Given the desired outcomes and objectives, what specific actions are needed from the sponsor to support the change?

What key communications will be needed from the sponsor?

How will status, issues, and risks be communicated to the sponsor?

What is the estimated duration of the change process?

Based on the assessment and importance of each person or group of people, actions are identified to leverage the positive individuals or groups to influence and convert the neutral or negative individuals and groups into change supporters. Table 17.5 is an example of a stakeholder management plan template used for this project.

3.3.4 Risk and Issue Management Plan

Risk management is focused on identifying and managing potential problems that threaten the achievement of project goals, and issue management is focused on managing risks that have

Table 17.5 Stakeholder Management Plan Template

Identify	Assess				Strategize	
Stakeholder	Importance	Expected reaction		Their needs	Our needs	Strategy	Actions
Who are the individuals or groups involved in this change?	How important is stakeholder to the success of this initiative? High/Med/Low	Will the reaction be positive, negative, or neutral?	Will stakeholder be a barrier or an enabler to change? Provide a brief description of the expected reaction	What needs do they have regarding this change? What do they stand to lose or gain?	What needs do we have for them? What role do we need them to play?	Which strategy is required to manage this stakeholder: convert, leverage, protect, or develop?	What specific actions need to be taken in order to effectively manage stakeholder?

matured into problems (or issues). The project team developed a comprehensive risk and issue management plan to manage project risks and issues. The key steps to managing risks were:

1. Identify and document the risk (or issue).
2. Prioritize the risk (or issue).
3. Create a mitigation action for each risk (or issue).
4. Drive the risks and issues to completion by linking the identified actions to the project plan and schedule.

3.3.5 Communications Plan

The communications plan is a critical document for any project, and becomes even more important when using the organizational change management approach. This plan defines what will be communicated, when, to whom, and how. It is aligned with all phases and activities of the project that require communications outside the team. It is also aligned with the stakeholder management plan and training plan.

Communications had several organizational areas of focus. The primary areas of focus were:

▲ Global communications to 5,000+ IT professionals, including outsourcing partners and contractors
▲ Specific communications to the CIO staff, providing updates and ways they could help support the upcoming change
▲ Guidance to department managers to help communications within their teams
▲ Detailed communications to the IT practitioners providing information on how they would be affected, when, and what they needed to do to prepare

A key element of the communications plan was an overall campaign strategy. The campaign strategy brought a theme to the communications with recognizable and interesting graphics. The communications experts on the project team used the theme "Imagine…". This theme challenged people to think or imagine about the possibilities and benefits of a streamlined process that gave the practitioner more flexibility, empowerment, and a more efficient process with fewer mandated meetings and approvals. As the number of communications increased using this theme, it became recognizable by the IT community and prompted them to read messages and pay more attention to the communications. The overall campaign strategy and theme proved to be a beneficial technique in the change management plan.

4. Measure Phase

The Measure phase is that point in the project where you determine how you will measure the key attributes related to the current problem, the goal, and expected benefits as defined in the charter. The results of the Measure phase may prompt you to adjust, clarify, or even redefine the goals initially established. It is perfectly acceptable and encouraged to make these changes to ensure the team is driving the solutions in the right direction. Your plan for measurement does not need to be complex and long. It should be focused on the problem and goals providing measurement of current state and future state. This will give you the ability to see what difference the project has made sometime after you implement the solutions.

The measurement plan was executed and the data gathered from internal and external resources (see Table 17.6).

Table 17.6 Measurement Results

Measurement	Result	
Number of process steps in current process	996	
Number of documents describing the current process	536	
Customer satisfaction with current process	65% dissatisfied	
Potential savings of reducing process overhead and complexity	US $X* million (95% confidence)	
Process complexity (similar companies vs. Motorola)	**Benchmark**	**Motorola**
Number of procedures	5-21	66
Number of templates	11-50	99
Number of mandatory documents	3-6	50

Note: Confidential data

At this point in a Six Sigma project, the team, Champions, and sponsors need to verify the problem statement, goal, opportunity, and scope. This is the point to decide whether to proceed as planned, proceed with changes to one or more elements of the charter, or cancel the project. By no means is modification of the charter or cancellation a failure of the project. It is avoiding the waste of valuable resources and money if real benefits are not achievable or the cost versus benefit is not justifiable. Based on the results of the data in the previous table, the team validated that the IT Common Process was indeed overly complex and there were potential benefits to streamlining.

5. Analyze Phase

The Analyze phase is the opportunity to step back and take a big-picture view of the data that was collected in relation to the original charter. This is when you ask the question "What does the data tell us?" Based on analysis of the data, several elements are evaluated. The elements for a DMADV type of project include:

▲ Key design factors influencing project objectives
▲ Impact of key design factors
▲ Sources of variability
▲ Defining requirements
▲ Identifying design alternatives and selecting one

In this project, the team analyzed both quantitative and qualitative data collected in the Define and Measure phases. The team was able to draw several conclusions based on the data analyzed. The IT Common Process was more complex and contained more overhead activities than processes in similar companies. Practitioners (customers) of the processes were dissatisfied with the processes. The current processes were designed for a larger company than Motorola was at the time, and the current IT project portfolio of small and medium-size projects required less rigor. A process that is too complex increases time to deliver solutions to customers, contributes to process noncompliances, wastes valuable resources, and causes morale problems within the organization.

The team also determined that the sources of variability in the process stemmed from six sources:

1. There were many paths that could be taken by projects based on a complex set of criteria, which resulted in a large amount of required activities, documents, and approvals.
2. The lack of integration between process areas caused work products to be reworked and performed differently.
3. Duplication and contradiction between process frameworks of CMMI and ITIL required practitioners to perform the same work more than once and increased rework.
4. The large number of process documents and detailed prescription of the activities led to misunderstanding of the process.
5. The process was too prescriptive and was telling the practitioners how to do their jobs.
6. The process asset library (PAL) was not conducive to helping people learn and apply the process to do their daily job.

The next step for the team was to draft requirements for the design. Keep in mind that requirements are defined, refined, and finalized. It is important to balance the refining of the requirements and eventually freeze them so that a design can be completed. The requirements covered the areas of process, training, PAL, user interface, and tools.

The last deliverable of the Analyze phase for the team was to brainstorm design alternatives and select one. Three major approaches were considered:

▲ Streamline key subsets of the existing process
▲ Streamline complete process areas
▲ Comprehensive redesign based on a different paradigm

The team investigated using development lifecycles like Agile and iterative. Since these models are valid software development lifecycles, Motorola's view was to have a complete end-to-end process that would allow the use of these and other methods within the framework of the total IT process. A project team would be given flexibility to use lifecycles like Agile, iterative, waterfall, DMAIC, DMADV, or other methods as determined necessary but within the framework for a full end-to-end process.

The team decided to take an "ideal design" approach using DMADV. This is a clean slate approach to process design that encourages visionary thinking about the best system or process. This approach also legitimizes "letting go" of the legacy process. It starts with a vision for the process and designs the process backwards, establishing the outputs, activities, and inputs required from the previous step to satisfy the next step. The vision for the process was to ensure quality IT products and services were delivered to customers on time and within budget. Even though this fresh approach was used, the team did not want to throw away some key process improvements and mature practices established over the last few years. The team decided to use a "shopping mart" approach wherein current process assets would be pulled off the shelf to get process elements that were needed, and these were then modified as appropriate. There was also a need to ensure that internal and external regulatory policies were met with the process.

One last but important factor was discovered in the Analyze phase that became an integral part of the design approach—trust. We heard loud and clear from our IT professionals that the process was so prescriptive and required so many things to be reviewed, approved, inspected, audited, checked, verified, and so on that they felt the IT leadership did not trust them to know what to do and to do the right things to deliver quality products and services. That was an important message for the project team and the IT leadership team to hear and learn. We have to learn to trust our people and build policy and processes around that trust in a responsible way.

6. Design and Verify Phases

The Design phase contains those activities that most teams want to jump to immediately. By using a structured Six Sigma approach like DMADV, the design activities can be focused more on solving the real problems in more effective ways. This is the point in the project where the solution, process, or system is designed and built.

The Verify phase is where the design is tested, analyzed for potential failures, and verified for capability and functionality. The Design phase and Verify phase can be and usually are iterative cycles of design, verification, changing design, reverification, and so on. This project decided to combine the Design and Verify phases to allow for an iterative approach, which helped to accelerate development activities and align more directly with the customer's needs.

6.1 Design Principles

Before the team started detailed design work, it established design principles to follow to ensure that the new process design would meet the objectives set forth in the project charter and the requirements previously defined. The design principles established were:

▲ Will have one common, simple, and flexible process.
▲ Need high-level depiction of the process that fits on one page and is easy to read and use.
▲ Will have a continuous flow type of process from start to end. Remove the concept of process areas like change management, incident management, and project management.

- ▲ Use CMMI and ITIL frameworks as a guide, as opposed to compliance to the standard. The goal is not to achieve a certain CMMI maturity level, but have a right-sized process for Motorola.
- ▲ Focus the process on IT customers and have them involved throughout the process.
- ▲ Empower the practitioner to utilize their best judgment and skills to do their job.
- ▲ Define *what* to do, not *how* to do it.
- ▲ Process is not a substitute for IT skills.
- ▲ Practitioners are fully accountable.
- ▲ Will not prescribe a recipe for every different scenario.
- ▲ Reduce the number of process variants.
- ▲ Process will be role-based, not organization-based.
- ▲ Give the practitioner flexibility and trust to make decisions that are best for projects and customer situations.
- ▲ Approvals:
  - ▼ Limit approvals to key stakeholders.
  - ▼ Approvals imply understanding and agreement.
  - ▼ Remove process requirement to have meetings.
- ▲ Meet IT general controls as defined by regulatory and internal control standards.
- ▲ Allow the process to become institutionalized by using a good process rather than trying to make it perfect and having it undergo constant change.
- ▲ Reduce process metrics from 50+ to a few key process indicators directly linked to the CIO scorecard.

6.2 Design Iterations

Based on these design principles, and using the requirements gathered in the Analyze phase, the team started the process design with its first iteration. The following sections will discuss the results of each design and verification iteration as the team designed the new IT Common Process.

6.2.1 Iteration #1: High-Level Design

The first design iteration focused on developing a high-level process using the "ideal design" approach discussed earlier. The process was to be organized by phases, with each phase containing the output(s) to be delivered to the next phase, the input(s) needed to produce the output(s), and the activities required to produce the output(s). The process design was started from the latest point (delivery of an IT product or service to the customer) and then incrementally developed backward, phase by phase. All *required activities* for each process phase that were necessary to deliver a high-quality output to the next phase were identified. A required activity meant that good IT practices, internal policies, or regulatory requirements demanded it. It was also determined that the process would be scalable to programs that consisted of multiple projects, projects of various sizes and complexity, and changes. Changes are modifications to the applications and systems that are usually smaller impact and lower risk than modifications managed as a project.

This is a novel and less frequently used process design approach, which is also referred to as COPIS (customer, output, process, input, and supplier). COPIS puts the customer first and all process design work begins with the customer and deliverables to them, and then activities are designed backward in the value chain. The advantage of COPIS over the traditional SIPOC (supplier, input, process, output, customer) approach to process design is that it facilitates a much more focused and lean process design, with just enough process necessary to produce a high-quality deliverable at each step along the way; this is akin to the "pull concept" in kanban.

An important and sometimes controversial topic with any process is approvals. Based on analysis of approvals of the previous process design, there were too many approvals required throughout the process, which diluted accountability. Approvals were identified in the new process design to focus on key stakeholders as projects and changes progressed. It was decided to reduce approvals to those people who were accountable for a deliverable. In the previous process, many approvals were added as a mechanism to notify people. There are other valid forms of notification that do not require formal approval.

A key component to any global process supporting any business function is the use of templates and checklists. For the project team and IT practitioners, this was a popular and sensitive topic that included much debate. The previous process had more than 150 templates and checklists, and more than 50 of them were mandated by the process to be used, and the resulting work products had to be formally peer-reviewed and approved. The project team decided to change the paradigm. The focus should be on providing the practitioners with templates and checklists that would help them do their job more effectively and efficiently. This shift in paradigm led to the classification of templates and checklists to be identified as required, recommended, or reference. Required templates would be those that were required by corporate policies and standards. Only three templates were identified as being required. Recommended templates and checklists would be those that the process team had determined to be good practice and were aligned to the process with embedded help text related to key process activities. Reference templates and checklists would be those that were identified as good practices within and outside of Motorola. Except for the three required templates, practitioners were to be given the flexibility to select templates from outside of Motorola; modify *all* templates as needed (including recommended and reference templates); and to use additional templates that were not included in the set of required, recommended, or reference templates but which the practitioners had found to be useful from their past experience. Except for the four required approvals in the process, it was also decided to allow the project teams flexibility to decide how much formality and rigor would be required in the review and approval, if any, of work products produced using the templates and checklists.

The team decided to use two commercial off-the-shelf applications already in place to manage the process. One application would be used to manage the IT portfolio and projects, the other would be used for change management, incident management, problem management, and configuration management database activities. Minor changes to these applications would be required to ensure they aligned with the new process. The application changes would also bring simplification to the end users.

It was also a requirement of the first design iteration to display the end-to-end high-level process on one page for easy comprehension. An early draft of the Process On a Page (affectionately known to the project team as POP) is shown in Table 17.7.

Verification of the first design iteration was focused on testing, peer review, practitioner review, and customer review. Testing was performed against applicable regulatory laws and industry standards. Testing showed the process did fulfill the laws and standards, but the team knew that Motorola internal controls were even more stringent. Those would be considered in a future iteration. Peer reviews within the team were conducted as the elements of each process phase were defined. Key practitioners were selected to conduct an early review of the process. While the new process had several significant paradigm shifts that required the early reviewer to think "out of the box" in order to challenge the existing paradigm and embrace the new process, their feedback was positive. The process design was then reviewed with the customer representatives (a subset of IT senior management). The team received positive feedback. The customer representatives did question if the process was too simple, too lean, and too flexible. At this point in the design cycle, it was expected that too much change was made and that upcoming iterations would add back the most important activities leading to a right-sized process.

6.2.2 Iteration #2: Table-Top Simulation

The second design iteration focused on developing the next level of detail for the process and making changes based on the verification activities in the first iteration. The project team worked to ensure that the process was focused on partnership with IT's customers, accountability, trust, and delivery and support of IT products and services. The process was modified to add two key concepts:

- ▲ First, IT would be more service focused, accountable for all changes made to the production environment. This entailed ensuring that IT had completed all required activities necessary for a flawless delivery and was ready to deliver the change.
- ▲ Second, customer approval was strengthened by clarifying customer accountability for their readiness to receive and implement the change. This entailed verification with the customer regarding acceptance of testing performed, training and communications received from IT, and completion of necessary customer readiness tasks by the customer prior to accepting delivery. This meant the customer would be accountable to have the business processes, users, and related documentation ready for the change.

The project team also focused on drafting a process manual that would describe the details of the process. The process manual was designed to be 100 pages or less and would replace the 66 procedure documents. Initial training material was also developed to instruct practitioners

Table 17.7 First Draft of Motorola's New IT Common Process

	Portfolio management Gate M15-M11	Project planning M10-M7	Requirements M9-M7	Design and develop M6-M5	Test M4-M3	Deploy M2	Support M1-M0
Outputs	Prioritized projects	Project plan	Approved requirements Submitted SOWs	Developed CIs Design Submitted RFCs Submitted iCSRs	Test results approved Tested CIs	Changes put into production and ready for use Updated CMDB	Service restored User support Permanent corrective action in place
Inputs	Project ideas Architecture standards	Prioritized projects	Project plan Voice of customer Standards (e.g., info protection requirements, SoX, regulatory requirements, etc.)	Requirements Architectural standards	Developed CIs to test Approved requirement Designs	Tested CIs RFC ready for approval Test results approved Communication plan executed Training plan executed	Customer's issues and requests Production CI
Activities	Determine if separate PPM required Create the PPM proposal Architecture assessment Define business case (cost, resource, benefit, etc.) Prioritize projects	Develop plan Select mgmt tool Resource requirements Budget WBS and schedule Project monitoring and control (risk/ issues) Communication Training Development lifecycle (waterfall, spiral, etc.) CM plan and project artifacts list Go live criteria	Get requirements from sources Document requirements Create SOWs Bundle changes for one-shot implementation (if applicable) Define maintenance windows (if any), freeze windows (if any)	Document design Coding Update CM system Create iCSRs Create RFCs (CI list, backout plan)	Plan tests (strategy/types of tests, test cases) Execute tests Capture results (pass/fail)	Confirm customer readiness for the change Execute, validate, record results of the changes Update CMDB	Service desk/ incident mgmt Identify and record Investigate and diagnose Resolve and recover Close Problem mgmt Identify, record, and classify Verify work around Investigate, diagnose, and resolve Identify known error and record Resolve and close
Approvals	Plan leads approve PPM proposal	ARM/BA, build lead, test lead, SDM approve project plan	Customer approves requirements		Customer and test lead approve test results	SDM approves RFC	

on the application of the process. Design of the new IT PAL was started in this design iteration. The PAL is the location where all process assets (policies, procedures, work instructions, templates, job aids, checklists, and training material) are stored and controlled. This is where IT practitioners go to get information on the process and how to use it correctly. The designers had to architect the PAL so that it would be easy to use and focused on helping the practitioner learn what to do at different phases of a project or change.

Verification of the second iteration was focused on peer reviews, customer review, design failure mode and effects analysis (DFMEA), and table-top simulation. The project leaders made the decision at this point to invite the Motorola Internal Audit group to join the project to help with verification and all other remaining project activities. This is the internal organization that conducts audits of business processes and systems throughout the year. They have broad experience and history of auditing IT systems and processes that was valuable to the project team.

Formal peer reviews and inspections of process assets were conducted within the project team, which helped to keep the team focused and productive. Once assets were complete, the process was reviewed with the customer representatives.

The project team initiated a DFMEA to objectively identify risks due to potential failures of the new process design, quantify each risk, prioritize the risks, and define appropriate actions required to reduce or eliminate the risks. A more detailed discussion of FMEA can be found in Chapter 3.

A major aspect of verification for this iteration was the execution of a table-top simulation. This type of simulation brought together 56 practitioners from all roles in IT and exercised the process by stepping through everyday scenarios. The project team developed 21 different use-case scenarios for the table-top participants to execute. Execution meant that a small group of people related to a use-case scenario would walk through verbally and on white boards how they would use the new process to execute everyday transactions. Based on simulated execution of transactions, the practitioners would provide feedback to the project team on what worked and what did not. After a week of running simulated use-case scenarios and transactions through all aspects of the process, the practitioners received 448 inputs for ideas and suggestions regarding the process. Many of these inputs were similar or duplicate, many directly conflicted with other ideas, and many were good ideas. Less than half of the 448 inputs resulted in change to the process. The new process was well on its way to being adopted by the practitioners and the reinstatement of certain activities back in the process that were deemed valuable. The table-top simulation was a positive exercise that was immensely helpful to right-size the IT Common Process.

6.2.3   Iteration #3: Process Pilot

The third design iteration focused on adding and strengthening critical controls and making the process adaptable for all types of IT project activities. Motorola has very well-defined standards of internal controls, security controls, and regulatory controls that affect IT projects and everyday operations. The project team reviewed the latest version of the IT Common

400 I Chapter Seventeen

Process with internal control experts and performed a gap analysis. This analysis resulted in some of the process activities being strengthened and some controls being added. The process team made several changes and additions to the set of templates and checklists. Training material was modified to align with the process changes and to incorporate changes based on the feedback from the table-top exercise.

The concept of classifying different types of work performed by IT based on scope was also added to the process. The team added the concept of differentiating IT activities by programs, projects, and changes. This would allow the practitioner to quickly focus on the parts of the process that were applicable based on the activity being performed. The process was designed to be used for all types of IT activities, from the largest program to the simplest of changes. In addition, certain other activities were added back as SMEs and the project team determined they would add value.

Verification of the third iteration was focused on peer reviews, customer review, CIO review, and piloting of the process. As in previous iterations, the project team conducted internal reviews and inspections of the process assets. The process was reviewed with the customer representatives in detail. In addition, the project was reviewed with the CIO and her staff, and the process was presented in detail as well. The project team received the go-ahead to conduct a pilot with a large group of practitioners.

The project team selected 125 IT practitioners and members of management to participate in the pilot. The training material previously developed was delivered to the pilot participants. The participants used the process to execute transactions that represented all types of activities performed by IT. The pilot was conducted over a two-month period resulting in constructive feedback to help right-size the process.

A survey was taken of the pilot participants to measure their satisfaction with the new process. The results were compared to those of the survey from the prior year to assess satisfaction with the old process. The results of the survey were positive and showed 95 percent of the respondents supported the implementation of the new process. Most of the feedback demonstrated improvement in overall satisfaction with the process and the level of detail of process documentation—overall satisfaction with the process improved from 17 percent before the project to 53 percent, and dissatisfaction with the process declined from 65 percent to only 7 percent (in both cases, the difference between satisfied and dissatisfied represents neutral population). All dissatisfied respondents were people in the process-auditing function, who felt the flexibility in the new process would make it more difficult to audit. The position the project team took was that it is better to have a process that is more effective to execute each day by thousands of practitioners than to have one that is easy to audit a few times a year by a few auditors.

During the execution of the pilot, the project leaders shared the details of the process design with external process consultants. The feedback received was positive and confirmed that the design was sound and solved a problem that many organizations have today—too much process. The external consultants suggested minor changes in the communications strategy and suggested a tighter alignment of the new process with the CIO scorecard measurements.

The results of the verification of the third iteration confirmed the process design was being accepted and was being right-sized to fit the organization.

6.2.4   Iteration #4: Ready for Implementation

The fourth and last iteration of the process design was focused on finalizing process assets, completing the design of the PAL, and completing development of training materials. Based on feedback from the pilot, process assets were added and modified to ensure that the practitioners had clear instruction on what they needed to do to execute the process. The illustration in Figure 17.1 shows the final version of the POP. As you can see, the process is not represented in flowchart form but in a format that is more applicable to a practitioner performing certain process phases, activities, and need-to-know required outputs and approvals.

Much time was spent on completing the layout and design of the POP and the PAL. So as to make it easy and fast for the practitioner to locate and use process assets based on their work, the team decided to implement the PAL as a set of integrated webpages, with the POP providing hypertext links to online documentation of the process and supporting process assets for each activity. This eliminated the need for hundreds of discrete process and procedural documents required by the previous process.

Training materials were finalized using a concept new to Motorola IT. That concept was the use of text-to-speech software to create the audio (voice-over) for computer-based training (CBT) modules rather than human voice. The primary reason for using computer-generated voice-over was to reduce the time to create CBTs and modify them in the future. Motorola found itself in the situation, based on the previous process, whereby CBTs were not kept up to date because it required human voice-over by a limited group of people with that capability. In the past, recording human voice from a script would take up to 15 hours to complete for a one-hour training class. With text-to-speech technology, that time was cut to no more than four hours to generate the computerized voice-over and less than two hours for future modifications. Granted, everyone will not like the computerized voice-over, but the project team was able to improve the quality of the voice-over and deliver a better user experience by applying lessons learned from earlier recordings to subsequent ones. These included building a reusable library of certain words whose enunciation was adjusted to more closely match a human speaker, breaking the monotony of a digital voice by mixing up male and female voices for different slides, and using humor. The team found that the computerized voice-over was well received in countries where English is a second language, because the voice is slower and is spoken at a steady pace that helps with comprehension. Overall, the benefits of using text-to-speech software outweighed the concerns.

Verification for the fourth and final iteration consisted of peer reviews, customer representative approval, user acceptance testing of tool changes, and a CIO go/no-go decision. The project team once again conducted peer reviews and inspections of the process assets and the PAL. The customer representatives were asked to provide formal written approval of the process, which they granted. User acceptance testing was conducted on the two applications

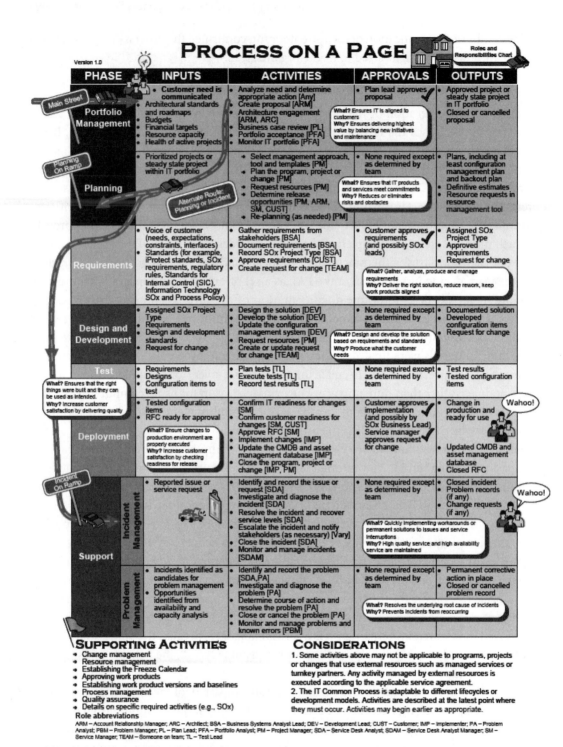

Figure 17.1 Final version of Motorola's IT common process.

402

that were modified to align with the new process. The final process design and project status were reviewed with the CIO and her staff. They were asked to make the go/no-go decision based on the information presented to them by the project leaders. The CIO and staff gave the approval to implement. All the key stakeholders and sponsors agreed that the process was now right-sized to fit the organization going forward. The next section of this chapter will review the activities related to the deployment of the new IT Common Process.

7. Deployment Phase

Usually in the DMADV model, the implementation of the design is performed in the Verify phase. However, since this project had such a large implementation, a separate phase was created—the Deployment phase. The Deployment phase focused on the execution of the communications plan, training plan, and actual go-live of the new process.

7.1 Planning

The project team defined success criteria that were used to monitor the overall progress of the deployment; these are different from the overall project success criteria described in the "Goal Statement" section. The deployment success criteria were:

▲ Seventy-five percent of IT practitioners have completed training before going live
▲ PAL deployed and fully functional
▲ One hundred percent of assets deployed to the PAL
▲ All changes to the two affected applications have been completed and verified
▲ All deployment communications have been completed

To ensure the deployment success criteria was met on time, a deployment readiness plan and checklist was developed. This helped the team focus on executing the deployment step by step.

The project team developed a comprehensive deployment communication plan to ensure the right people were communicated the right information at the right time. The outline of the plan is shown in Table 17.8.

To assist the practitioners in the IT community to adopt and transition to the new process, a transition plan and checklist was created. This checklist would help each practitioner decide what needed to be done to transition existing projects to the new process. This checklist provided detailed instructions on what needed to be done to start using the new process, what did not have to be done any longer, and how to handle open records in the common applications that supported the process.

In case of an issue with deployment or a last-minute change deciding not to implement the new process, a backout plan was developed. The backout plan provided details as to how backout changes were made to the IT PAL, backout process assets and application changes,

Table 17.8 Outline of Deployment Communication Plan

Phase approach (steps)	Time
1. Announcement: Create interest	Day 1 (go/no-go date)
2. Preparation: Educate, gain buy-in	Day 1 plus 1 to 1.5 weeks
3. Manager/SME development: We're on our way	Step 2 plus 1 week (if Step 2 is on a Monday, Step 3 could be on a Thursday) Must perform Step 3 at least 1 week before training starts
4. Practitioner deployment: Reinforce benefits	Step 3 plus 1 week (or closer)
5. Go live! Celebrate!	1 week after Step 4
6. Sustaining	Repeat of Step 3

and communications to the IT community in case of a last-minute change in plans. While backout plans are usually not executed, IT projects should have one in case of a situation where the project needs to be stopped during deployment. When a project team is in the middle of deployment and something goes wrong, it is difficult to think through the options and actions. The backout plan provides the different options based on scenarios and the actions to take. At that point, the plan becomes a checklist to stop the deployment in a controlled manner if needed and restore the original service.

7.2 Training

A critical element of the deployment phase was the execution of the training plan. The training plan was designed to ensure at least 75 percent of the 5,500 IT practitioners worldwide received training prior to going live.

As the project team considered different training options to reach the widespread practitioners, it acted upon a common requirement expressed in the voice of customer exercise. The practitioner community wanted training to be delivered with instructors and wanted face-to-face classes as much as possible. The team needed to decide how to balance the need for instructor-led classes and the use of CBT using text-to-speech software. The project team also had to consider two phases of training:

1. Deployment training.
2. On-going training for new employees and contractors who joined the organization, or for changes in the roles of existing ones. The team devised a strategy to meet both time horizons of training.

Two levels of training were planned:

1. **Overview training:** Provided overview of the new process and the cultural changes required to make it successful. This training was an hour long and was computer-based using presentation slides with computer voiceover utilizing the text-to-speech technology.

2. **Role-based practitioner training:** This training was labeled *track* training. The team defined three tracks based on the process design. Other types of training were also identified for additional roles that were performed within the organization.

Now that the project team had a strategy and detailed training material, it had to decide how to conduct training for deployment around the world and to provide ongoing training in the future. The solution, with great support from the CIO and her staff, was twofold:

▲ First, instructor-led training was to be delivered worldwide. This would consist of face-to-face training in the larger locations and remote instructor-led training using remote presentation tools for smaller locations and telecommuters. The project team established a list of trainers to conduct the instructor-led classes and provided detailed training to prepare them. The trainers conducted 135 instructor-led classes in two months, training 85 percent of the IT practitioners prior to going live.

▲ Second, for ongoing training, it was decided that the instructor-led classes would be converted into computerized voiceover training using text-to-speech technology. This would allow the training material to be more easily updated to match any future process changes. The CBT material was released a month after going live and made available in the IT PAL for anyone to use as needed.

7.3 Go Live

Ten months after the CIO challenged the team to streamline the IT Common Process, the project team implemented the new IT Common Process. Implementation consisted of deploying the new PAL, the related process assets, and changes to the two supporting applications. Implementation occurred on the scheduled date, all components were verified, and practitioners started using the new process.

To support the practitioners, the project team joined key meetings within each organization to answer any questions. It is important that such an outlet be created and left in place while an organization adopts and adapts to a new or significantly modified process. For this project, this support mechanism was planned to be in place for four to six months.

7.4 Next Steps

The project team turned over ongoing support and management for the new IT Common Process to the IT Process and Quality organization. The Process and Quality organization continues to support the IT functions to ensure the process is properly institutionalized. This is done by helping practitioners and first-line managers become knowledgeable of the process and take accountability to be the first line of support to coach their staff on the process, answer any process-related questions, and ensure their staff applies the process correctly to the different types of work they perform. The Process and Quality team provides the organization with coaching on the process in one-on-one sessions with managers and small groups of managers. The leadership of IT is committed to stopping frequent changes to the process, which make it difficult to institutionalize the process. The team agreed that the process would henceforth change no more than once per year. However, minor process and user interface changes may continue to be made to improve usability, readability, and to add process help features as needed. However, the minor changes would not change the process functionality and requirements. Compliance to process execution, besides being monitored by the direct supervisors, would also be monitored using audits. Six Sigma tools and techniques would be used to drive preventative actions to correct noncompliance issues.

8. Benefit Realization

The benefits realized in this project are expressed in reduction of complexity and efficiency improvements resulting from fewer activities and deliverables required from the process. The complexity of the process was reduced significantly by removing redundant and nonvalue-added steps and activities, and eliminating low-value-added templates and process documents. The required process activities and assets have been significantly reduced. Survey results show 90 percent satisfaction rating with the new process. The efficiency improvement was a 15 percent reduction in time spent by more than 5,000 IT practitioners having to produce the right amount of project work products rather than all work products for every project.

Performance to the original goals established in the charter is shown in Table 17.9.

Table 17.9 Performance to Project Goals

Goal statement	Target	Actual
Deploy the process on time	6/1/2010	4/19/2010
Reduce required process steps	50%	80%
Reduce post release defects	50%	70%
Reduce compliance deficiencies	50%	TBD*
Achieve positive reaction to new process (survey)	70%	90%

*Note: At the time this chapter went to print, this data was not yet available but preliminary data for a closely-related metric indicated first-pass rate in compliance audits had improved from a mean of 75% before this project to 90% after this project.

9. Conclusions and Limitations

Although this project had many elements that made it successful, there were notable keys to success. First, the strong and consistent support and participation of senior leadership was a driving force to complete the project successfully. This is essential to any such project where processes are being redesigned. Second, the approach described in this chapter provided a systematic and step-by-step way of approaching such significant process redesign challenges when often companies don't know where to begin or how to proceed. This project demonstrates that companies can design "just enough" processes by taking a customer-centric view using a COPIS approach with the customer as the starting point. Third, an iterative design and verification model involving practitioners using the DMADV approach allowed for incremental fine-tuning and optimization of process design by exercising it with real-world scenarios. Fourth, having a small team empowered to execute the design iterations and make decisions quickly by the two project Champions. Finally, effective organization change management planning and execution was essential to prepare the organization for significant change. This helped to maximize the odds of the organization adopting the change and minimizing the transition time from the old way of working to the new way of working.

As with any approach, there are potential limitations. The limitations identified for this project are in the areas of cost and savings calculations, as well as survey results.

The Monte Carlo simulation used to predict the potential cost savings of the project was limited by the estimates of the project team and internal experts (Table 17.6). The key factors in the estimation were the number of IT practitioners executing processes, the estimated percent overhead spent by each person on process execution, and the estimated reduction in overhead. While such estimates are always stochastic and there can never be one 100 percent accurate estimate, the team did calculate a 95 percent confidence factor by using historical data and experience of experts in the subject area.

Survey results at the time of this writing were based on practitioners that had completed training classes and elected to take the related survey. A survey is planned to be sent to all IT practitioners to improve the accuracy of the satisfaction results.

Acknowledgments

The author of this case study is thankful to Motorola for its permission to allow publication of this case study as a book chapter. The author would also like to thank all the project team members of the IT streamlining project.

10. References

1. Capability Maturity Model Integration (CMMI), www.sei.cmu.edu/cmmi. Software Engineering Institute, Carnegie Mellon University, Pittsburgh, PA.
3. Information Technology Infrastructure Library (ITIL), www.itil-officialsite.com. OGC, UK.
3. Eskerod, P., and Riis, E., Project Management Models as Value Creators, *Project Management Journal*, March 2009.
4. *The Motorola Change Management Framework*, Motorola University, March 2009.

CHAPTER 18

How Motorola Reduced the Effort Required for Software Code Reviews

Fernanda de Carli Azevedo Oshiro
Luiz Antonio Bernardes

Relevance

This chapter is relevant to organizations that want to reduce the effort required to perform software code reviews (or inspections), without compromising the effectiveness of the reviews.

At a Glance

▲ Motorola's Mobile Devices business unit leveraged Six Sigma to redesign the code review process in order to dramatically reduce effort and cycle time within R&D operations.

▲ The project team managed to reduce effort by 60 percent.

▲ The project resulted in annual cost avoidance by reduction or elimination of nonvalue-added tasks of approximately US$500,000 (in addition to intangible cost avoidance in cost of poor-quality items, like warranties).

Executive Summary

In 2008, as part of a regional initiative in the Mobile Devices business unit to reduce time to market, the R&D Latin America (LATAM) department mobilized improvement resources and key projects were launched to streamline core software development processes. In R&D, Six Sigma project proposals were identified from two main sources: top-down and bottom-up.

Top-down projects are identified during a yearly effort led by an MBB in a three-step selection process:

1. Business scorecards goals are traced down to R&D operation and its performance indicators.
2. Performance baselines are calculated for those key R&D indicators aligned with scorecard goals for gap estimation.
3. MBB and functional managers identify and charter key "close the gap" initiatives from which senior management selects a subset for staffing and execution.

The objective is to ensure that the right improvement projects are defined and staffed in order to maximize R&D contribution to scorecard goals attainment.

Bottom-up projects are typically low-hanging fruit opportunities proposed by any contributor. These projects usually fit in the Six Sigma project portfolio as supporting initiatives of lower priority, and can be launched anytime.

Although this project was identified as a bottom-up idea, MBB and senior management foresaw a real benefit to several other software development teams in Mobile Devices. The objective was to reduce effort on software code reviews by adopting new approaches without compromising defect detection effectiveness.

The Green Belt (GB) assigned to lead the project investigated alternative approaches before setting up the project scope and quantitative goals. With the challenge to reduce effort in a short time, the GB considered the adoption of lightweight reviews in closer alignment with Agile methodologies. Two widespread Agile-like lightweight alternatives investigated were pair programming and tool-assisted reviews. Considering the pros and cons, as well as the culture in the organization, tool-assisted reviews were considered a better fit because they would have faced less resistance at all levels in the organization.

The reasons for deciding on a lightweight review approach before project launch were twofold:

1. Considering that this was a Six Sigma project pertaining to software development process, six months was a short time. Our past experience with other projects in software development process had shown that projects with broad scope took, on average, 12 months to complete due to longer cycles in data collection and analysis of causes and solutions.
2. Define, Measure, Analyze, Design, and Verify (DMADV) had proven to be an effective data-driven framework for fast introduction of process innovations in a short time (for example, a new method or a new tool) as long as the change was relatively well defined from the outset.

Therefore, there was a strong argument in favor of the lightweight review approach prior to the launch of the project.

The scope of the project was thus defined. The team should optimally identify, pilot, and deploy tool-assisted reviews in lieu of Fagan-like inspections in a six-month timeframe relying on DMADV framework.

After project kick-off, the team mapped the old process to identify nonvalue-added steps as well as effort reduction opportunities resulting from the adoption of lightweight flows relying on code review tools. The team thus created a lightweight flow to be piloted by some software development teams.

Code review tools from several vendors were evaluated to facilitate a criteria-based selection. Key criteria were:

▲ Ease of use
▲ Capability to adapt to proposed lightweight flow
▲ Integration with current software development suite (e.g., version control system)
▲ Return on investment (ROI) estimates (i.e., translating effort and cycle time reduction to dollar savings)
▲ Ability of the vendor and tool to continuously accommodate process requirements from Motorola

Once the proposed flow and the supporting tool were defined, some software development teams piloted the tool in order to obtain qualitative and quantitative feedback. The pilot provided valuable data that was analyzed to identify and manipulate performance drivers and fine-tune the process and tool throughout the DMADV steps. Statistical evidence enabled informed decision making toward continuous optimization of the new code review process. Eventually, "optimal" process design was achieved and its efficiency and effectiveness were quantitatively verified.

The Six Sigma team managed to complete the project within six months, as planned. The new lightweight tool-assisted code review process resulted in effort reduction that yielded savings of approximately US$500,000 in the first year of adoption.

1. Introduction

In 2008 Motorola launched several Six Sigma projects to simplify processes in order to reduce time to market.

The R&D LATAM department generated top-down project ideas aligned with regional and global goals, as well as bottom-up project ideas. Code review effort reduction was one of the Six Sigma team charters approved for execution due to the significant benefit it could provide to several software development teams.

The ultimate goal was to reduce overall effort with software code reviews by 50 percent with no degradation in defect detection effectiveness. The effort reduction target was estimated based on adoption of a tool-assisted lightweight approach to code reviews. Potential savings were calculated in dollar terms by estimating 50 percent reduction in the effort allocated for software code reviews.

2. Project Background

This project was launched and most project tasks were performed at the Motorola R&D center in Brazil. There was an expectation from project sponsors that the final process redesign would be replicated at other Motorola R&D sites within Mobile Devices. In order to accomplish that, the project team was challenged to gather compelling statistical evidence proving both the effectiveness and efficiency of the redesigned process.

3. Define Phase

The Define phase of the project is when the following question is answered: what is important? By answering this question, the project team will try to find the reason why the project should be executed and what benefits it will bring to the organization.

In order to achieve the objectives of this phase, Six Sigma methodology recommends a set of steps that should be taken:

1. Identify and validate the business opportunity.
2. Identify critical customer requirements.
3. Define and map processes.
4. Form a project team.

For this project, the Define phase comprised the following activities:

1. Definition of Six Sigma project team
2. Creation of project charter
3. Process mapping and analysis

3.1 Six Sigma Project Charter

The first definition that is made in the beginning of a Six Sigma project is the team charter. The team charter summarizes the essentials of the project by stating how it is expected to bring improvements to the company, what are the estimated gains, the project scope, and high-level schedule.

The team charter of this project comprised several elements, which are discussed in the following sections.

3.1.1 Business Case Statement

Before this Six Sigma project, the code review process used by Mobile Devices business unit relied on Fagan's inspection method.[1] It was supported by an internal tool for recording code review data. This approach required significant redundant and nonvalue-added effort. If other, more efficient code review methods that were equally effective could be implemented and supported with best-in-class code review tools, then code review effort could be reduced by up to 50 percent.

3.1.2 Opportunity Statement

Improving the code review process by adopting alternative and more efficient approaches and leading code review tools would enable effort reduction while maintaining fault detection rate with potential savings of US$370,000 per year.

The project cost with resource allocation and purchase of tool licenses was estimated at US$18,600.

3.1.3 Goal Statement

Reduce code review effort by 50 percent of Brazil Mobile Devices Software development teams by November 2008.

3.1.4 Project Scope

All software developed by engineers working on Brazil Mobile Devices Software development teams.

3.1.5 Project Plan

The high-level project plan included estimated dates for completion of all project phases (see Table 18.1).

Table 18.1 Project Plan

Phase	Due Date
Define	Jul. 1 to Aug. 31, 2008
Measure	Aug. 31, 2008
Analyze	Sept. 30, 2008
Design	Oct. 30, 2008
Verify	Nov. 1 to Dec. 31, 2008

3.1.6 Project Team

The core project team was composed of:

▲ One team leader (Green Belt candidate)
▲ Two team members
▲ One Black Belt consultant who was to mentor the Green Belt candidate
▲ One Master Black Belt

The governance team for the project was composed of:

▲ Two project sponsors
▲ Three project Champions
▲ One representative from the finance organization

3.2 As-Is Process Analysis

Next, within the Define phase, the as-is code review process was mapped and analyzed (see Figure 18.1). The project team identified the activities that should be performed in each process phase, and they highlighted the activities they considered nonvalue-added, redundant, or inefficient.

After mapping the process, the phases were analyzed to identify the time spent in executing each of them. A Pareto chart was created with the results of this analysis to show the effort distribution by process phase (see Figure 18.2).

Figure 18.1

Process map.

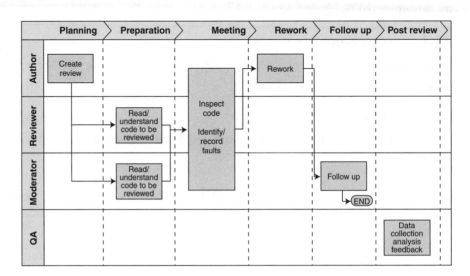

Figure 18.2

Pareto of code review effort by phase.

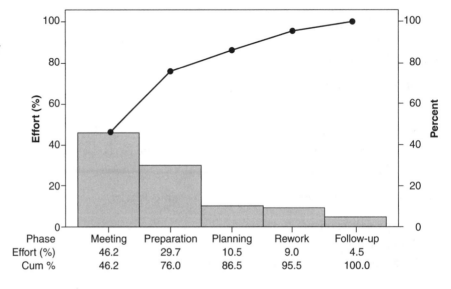

Phase	Meeting	Preparation	Planning	Rework	Follow-up
Effort (%)	46.2	29.7	10.5	9.0	4.5
Cum %	46.2	76.0	86.5	95.5	100.0

Per the Pareto, the Meeting phase was consuming 46.2 percent of the total effort spent in executing the code review process.

Through observation of code review process execution and interviews with the process performers, the project team identified that only a few defects were found during the Meeting phase. Most of the defects were found by reviewers while reviewing and analyzing the code during the Preparation phase.

Based on the Pareto map and conclusions about the Meeting phase, the project team suggested the adoption of a lightweight code review process by merging the Preparation and Meeting phases into one single phase called Review phase (Figure 18.3).

(Editor's Note: It may seem out of place to suggest solutions this early in the project, but that is exactly the opportunity that results from process mapping. Possible solutions, however, should not be implemented until the problems are identified and root causes are validated. Without the subsequent analysis and verification, a project team would only have been guessing and presuming the solution.)

During this new Review phase, each code review participant would review the code offline on their own (asynchronous review).

The project team conducted a benchmarking study on a set of alternative tools in order to choose an appropriate tool to support the suggested lightweight process. For each candidate tool, the following items were analyzed:

▲ Cross-platform support
▲ Ease to configure
▲ Online communication support
▲ System integration support
▲ Notifications to reviewers via e-mail
▲ Automatic data collection
▲ Reports generation support
▲ Support to customization of review fields
▲ Cost per license

Figure 18.3

Lightweight code review process suggestion.

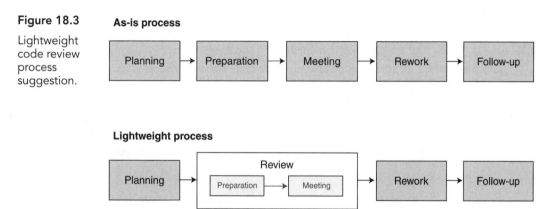

Based on benchmarking results, the Code Collaborator tool from Smart Bear Software, Inc., was chosen to support the lightweight code review process.

Code Collaborator supports the code review process by:

▲ Providing a "Code diff view" that highlights removed and changed lines of code (LOC) with different colors.
▲ Allowing logging of faults and comments tied directly to the code. Defects, comments, and version control files stay linked together and are immediately available to the author and other reviewers.
▲ Maintaining conversation threads on each fault or comment registered.
▲ Supporting automatic data collection (LOC, effort, and faults).
▲ Providing reports containing code review data.

Once the tool was chosen, the suggested lightweight high-level process could be refined and tuned.

4. Measure Phase

The general objective of the Measure phase is to provide quantitative insight into the current status of what is being improved. Usually in this phase, the project team determines what should be measured, how it should be measured, and how data should be collected.

To determine what should be measured, it is necessary to analyze the process to identify its indicators. Once the indicators are identified, the operational definition of each indicator should be developed. The operational definition is a precise description of the criteria used for the measures and the methodology to collect the data. It guarantees consistency while measuring and collecting data. The operational definition is the input to the measurement plan. All data collection performed in the project should adhere to the measurement plan.

The Measure phase of this project started with the identification of performance indicators to be monitored for the code review process and the establishment of a measurement plan. After establishing the measurement plan, the project team ran a pilot with the suggested new high-level lightweight process, which was an output from the Define phase. The objective of the pilot was to gather data to support code review process refinement and tuning. Finally, pilot data was collected and analyzed according to the measurement plan.

The performance indicators identified for the code review process were:

▲ **Total Defect Density (TDD):** Total number of defects found in each review, divided by the number of LOC reviewed in that review.
▲ **Inspection Effort Rate (IER):** Number of LOC reviewed in a review, divided by the total effort spent (in hours, by all review participants) to perform the review.*

---

* To ease understanding of the measures, IER is transformed to "Effort to inspect 1,000 lines" in some charts.

TDD and IER data from the previous code review process was collected from the old code review tool database. With this data, a performance baseline was established for comparison with pilot data for the new code review process.

A measurement plan was created to establish how data would be collected, when, by whom, and in what circumstances (see Table 18.2).

The measurement plan was then executed for pilot data collection, and performance baselines were established for TDD and IER for the lightweight code review process.

Pilot data was compared to the old process baseline, and this analysis showed that the new process, together with the new code review support tool, led to 66 percent reduction in code review effort (see Figure 18.4).

However, it was noted that the effectiveness of the code review process was lower than the old process baseline (see Figure 18.5), since the total defect density (TDD) was lower than the baseline.

The challenge of the project was to maintain the efficiency improvements with the new process while maintaining the same defect detection effectiveness.

Figure 18.4

Effort to inspect 1,000 LOC: Old process effort vs. new process effort.

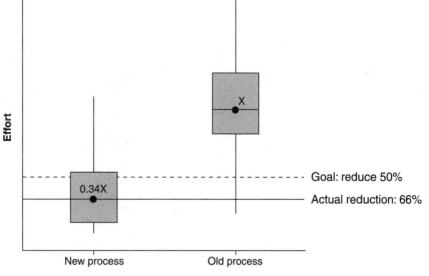

Table 18.2 Measurement Plan

Performance Measure	Operational Definition	Who Will Collect the Data?	When Will Data Be Collected?	How Will Data Be Collected?
Inspection ID	ID of the inspection in Code Collaborator database	SQE and SQE student	Once every 15 days	Code Collaborator Data Report template should be used. Instruction available in Code_Collaborator_Metrics_Procedure.doc
Primary team	Name of the team that developed the code being inspected	SQE and SQE student	Once every 15 days	Code Collaborator Data Report template should be used. Instructions available in Code Collaborator Metrics Procedure.doc
Inspection size SZ)	Number of LOC reviewed in an inspection	SQE and SQE student	Once every 15 days	Code Collaborator Data Report template should be used. Instructions available in Code_Collaborator_Metrics_Procedure.doc
Total effort (TE)	For each inspection, the sum of the hours spent by all participants to perform the inspection	SQE and SQE student	Once every 15 days	Code Collaborator Data Report template should be used. Instructions available in Code_Collaborator_Metrics_Procedure.doc
Inspection effort rate (LOC/h)	IER = SZ/TE	SQE and SQE student	Once every 15 days	Code Collaborator Data Report template should be used. Instructions available in Code_Collaborator_Metrics_Procedure.doc
Total defect density (defects/KLOC)	TDD = (TDC/SZ)*1000 TDC: Total number of defects found in an inspection	SQE and SQE student	Once every 15 days	Code Collaborator Data Report template should be used. Instructions available in Code_Collaborator_Metrics_Procedure.doc

Figure 18.5

Old process
TDD vs. new
process TDD.

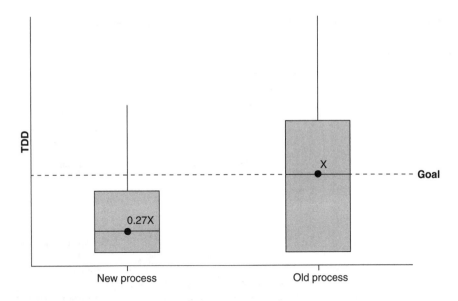

5. Analyze Phase

During the Analyze phase, the project team examined the process in detail in order to understand the sources of variations, identify their root causes, and discover opportunities for improvement.

A brainstorming exercise was performed to identify the main factors that affect code review effectiveness. A cause-and-effect diagram was used to consolidate brainstorming results (see Figure 18.6).

Correlation analyses were performed to verify the relationship between the factors identified in the brainstorming and code review effectiveness.

Considering all the factors, two of them in particular could be quantitatively analyzed and they were identified as having a potential relationship to TDD. These two factors were:

▲ Size of code to be reviewed in a review
▲ Time spent to perform this review

The Pearson correlation coefficients in the correlation analyses showed that there was moderate negative correlation between IER and TDD, and weaker negative correlation between code size and TDD (see Figures 18.7 and 18.8).

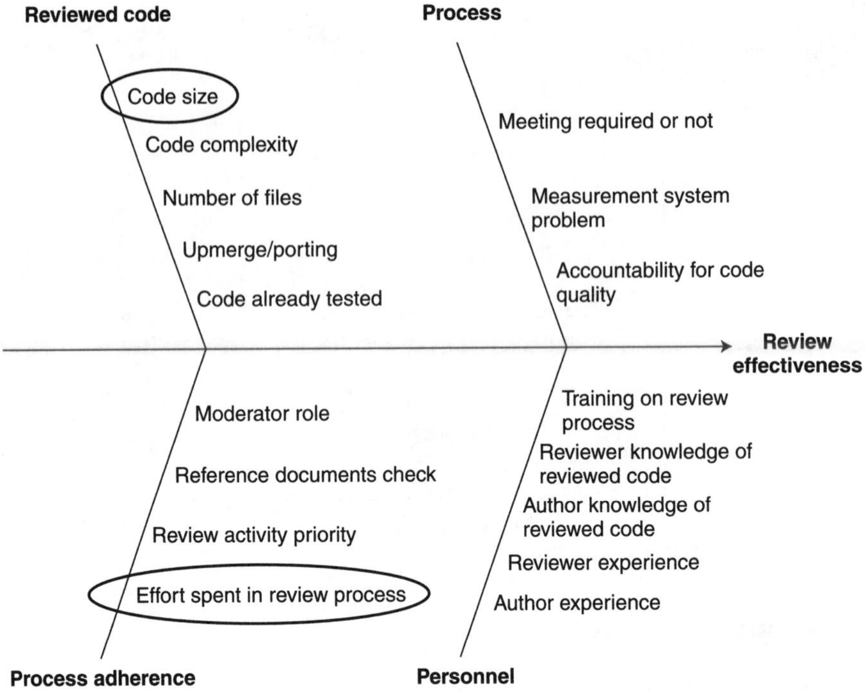

Figure 18.6

Cause-and-effect diagram.

Reviewed code

Code size

Code complexity

Number of files

Upmerge/porting

Code already tested

Process

Meeting required or not

Measurement system problem

Accountability for code quality

Review effectiveness

Moderator role

Reference documents check

Review activity priority

Effort spent in review process

Process adherence

Training on review process

Reviewer knowledge of reviewed code

Author knowledge of reviewed code

Reviewer experience

Author experience

Personnel

Figure 18.7

Correlation analysis: TDD vs. IER.

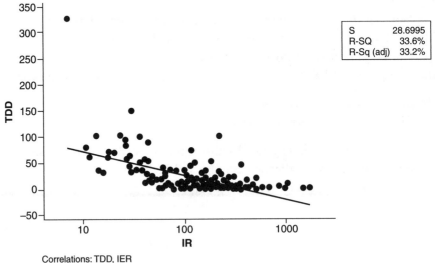

S	28.6995
R-SQ	33.6%
R-Sq (adj)	33.2%

Correlations: TDD, IER
Pearson correlation of TDD and LOG IER = 0.580
P-value = 0.000

Figure 18.8

Correlation analysis: TDD vs. size.

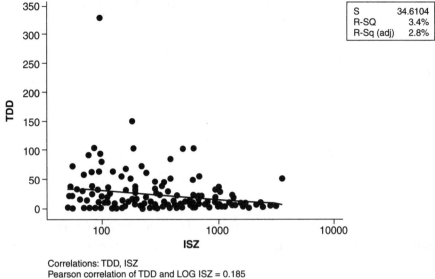

Correlations: TDD, ISZ
Pearson correlation of TDD and LOG ISZ = 0.185
P-value = 0.016

Based on this analysis, it was decided that effort rather than size should be considered for refining and tuning the new lightweight code review process.

This analysis suggested that in order to support definition of recommended ranges of effort to achieve and sustain targeted TDD, further examination of this relationship between effort and TDD was warranted.

To further investigate the opportunities, a model was created to analyze tradeoffs between effort and TDD. The recommended ranges of effort needed to be determined to optimize the process so that it would balance effectiveness and efficiency.

6. Design Phase

During the Design phase, the project team optimized the new process by using modeling and simulation.

For this project, in the Design phase, a spreadsheet-based model was created to represent the relationship between effort and TDD.

This model considered several input factors, in addition to effort, and was characterized by a range of inspection rates with a defined variability.

The input values could then be varied using Monte Carlo simulation to show how the outcome (or dependent variable TDD) would be affected (see Table 18.3). The objective was to estimate "optimal" ranges of the independent variables that were likely to yield predictable good performance of TDD. Those ranges would become references as operating conditions for software development teams to use for designing, planning, executing, and controlling subsequent code reviews.

Table 18.3 Spreadsheet-Based Model

Variable	Description	Type	Probability Distributions
IER	Inspection effort rate	Input	96
Sources of variance	Variance from various uncontrolled sources (e.g., code complexity, code knowledge of reviewers)	Input	1
Zero-fault probability	Probability of occurrence of zero-fault reviews regardless of other factors	Input	1
TDD	Total defect density	Output	23

Regression analysis was performed to determine the coefficients for use in the model. It should be noted that this analysis also confirmed that size was irrelevant to the output variable, TDD.

One of the main advantages of using a model instead of running real-life experiments is that by using simulation, it is possible to evaluate a large number of scenarios and operating conditions of the process in very little time and with limited usage of resources.

The team modeled the process across a wide range of likely scenarios and IERs. Figure 18.9 illustrates two different scenarios.

The left histogram shows potential TDD results that might have occurred if the IER were high (i.e., between 200 and 600 LOC per hour) after a simulation run of 10,000 code reviews.

The right histogram shows likely TDD results for 10,000 simulated reviews where a lower inspection rate would occur (i.e., <200 LOC per hour).

Figure 18.9

Monte Carlo simulation.

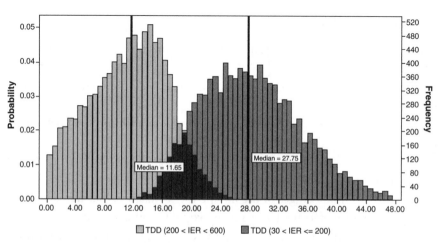

Little overlap between distributions proves by simulation that by shifting IER to the left, there is high probability that TDD will increase.

After running a number of different scenarios, the team concluded that keeping the IER less than or equal to 200 LOC per hour would yield good TDD in most reviews (i.e., more than 85 percent probability of exceeding the current TDD mean).

With this new design goal (quantified in terms of IER) code reviews would be established to manage inspection rates within these new boundaries. New procedures for code reviews could thus be planned and tracked using these new guidelines.

The final version of the new lightweight process map is represented by the flow in Figure 18.11. For comparison reasons, the old process map is presented in Figure 18.10.

New procedures, guidelines, and instructions were prepared. For example, a one-page code review process was documented and published on the intranet.

Instructions on how to install and use Code Collaborator, a new tool to track metrics associated with this new inspection process, were also published and made available to the teams.

Figure 18.10

Old process map.

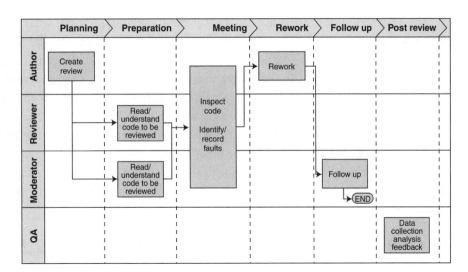

Figure 18.11

New process map.

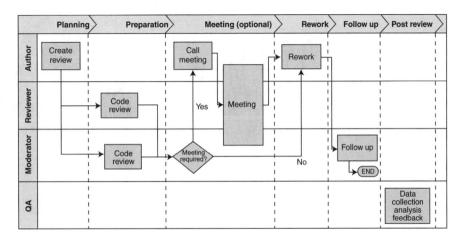

7. Verify Phase

In the final phase, Verify, the objective is to validate and verify that the redesigned process accomplishes its objectives and that it can be deployed successfully in the organization.

For this project, during the Verify phase, the project team piloted the new process across a number of real inspections and measured the results. To document the effectiveness of the new process, the team created control charts.

Figure 18.12 shows a comparison of TDD from code reviews before and after the new process.

As shown by the control chart (see Figure 18.12) and the box plot chart (see Figure 18.13), after process modeling and establishment of IER limits, TDD remained statistically same as the baseline.

The project team used the Mann-Whitney test to statistically compare the results for TDD with the new process and the old process.

The Mann-Whitney test is used to answer the following question: what is the chance of randomly sampling two populations that have the same median and the median of these samples being different? The team chose this test because the distributions of IER and TDD do not follow a normal distribution; thus, a median test should be used instead of a mean test.

If the result of the test shows a low p-value, we can reject the thought that the difference between the medians of the samples is a coincidence, and assume that the medians of the population are really different. A low p-value would indicate that the observed differences are statistically significant and would not be likely to occur by chance alone.

Correspondingly, a large p-value would indicate that the medians are not statistically different. (This is not the same as concluding they are the same. It is only that there is not enough information to conclude they differ.)

Figure 18.12

TDD control chart.

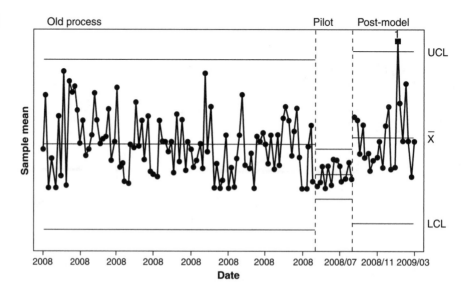

Figure 18.13

TDD box plot: Old process vs. new process.

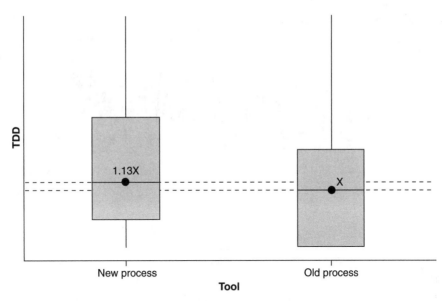

Therefore, the null and alternate hypothesis are as follows:

Ho: TDD_{Old} Process $= TDD_{New}$ Process
Ha: TDD_{Old} Process $\neq TDD_{New}$ Process

As shown in Figure 18.14, the test results for TDD showed a large p-value (0.2955). This means that, with available data, there is no reason to conclude that TDD with the new process is different from TDD with the old process. In this case, we assume that the TDD before and after process improvement remains the same.

Figures 18.15 and 18.16 show that the project achieved a significant reduction (about 60 percent) in the effort spent to perform code reviews. This improvement was sustained even after implementation of the IER limit of $<= 200$.

Again, the Mann-Whitney test was used to statistically compare the results for the effort spent with the new process and the effort spent with the old process (see Figure 18.17).

Therefore, the null and alternate hypothesis are as follows:

Ho: IER_{Old} Process $<= IER_{New}$ Process
Ha: IER_{Old} Process $> IER_{New}$ Process

Figure 18.14   Mann-Whitney Test and CI: TDD, old process; TDD, new process

Comparative TDD.

```
                        N      Median
TDD - old process      288     14.380
TDD - new process      166     11.839

Point estimate for ETA1-ETA2 is -0.000
95.0 percent CI for ETA1-ETA2 is (-3.496, -0.000)
W = 64111.0
Test of ETA1 = ETA2 vs ETA1 not = ETA2 is significant at 0.2955
```

Figure 18.15

Effort control chart.

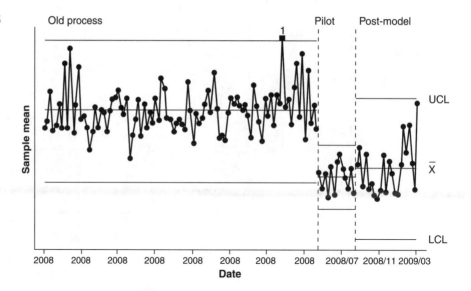

Figure 18.16

Effort box plot: Old process vs. new process.

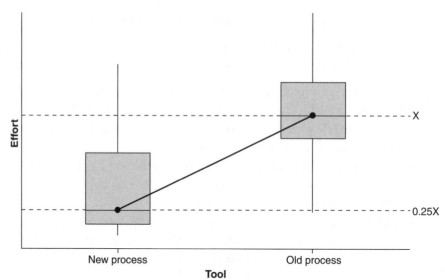

The test results for IER showed a small p-value (0.0000). This means that we can reject the null hypothesis that the medians are the same and conclude that the effort spent with the new process is lower than the effort spent with the old process.

Control charts and statistical tests thus showed that the new process significantly reduced effort without significantly increasing inspection effectiveness.

As a final control measure, the project team prepared a control plan to ensure that Six Sigma project results would last beyond project completion. The control plan included the following activities:

Figure 18.17   Mann-Whitney Test and CI: IER, old process; IER, new process

Comparative
test of effort.

```
                            N        Median
TDD - old process          288       180.90
TDD - new process          166       122.74


Point estimate for ETA1-ETA2 is 48.01
95.0 percent CI for ETA1-ETA2 is (31.28, 64.04)
W = 71662.0
Test of ETA1 = ETA2 vs ETA1 not = ETA2 is significant at 0.0000
```

▲ Performance baseline execution once a year, for TDD and IER, to identify the need for changes in IER reference limit recommended by this project
▲ Monitoring of performance indicators IER and TDD
▲ Review of deviations by software development teams from defined targets for IER and TDD, with actions taken as necessary with those teams

8. Benefit Realization

The cost avoidance associated with this project was calculated in terms of effort savings resulting from achieving a reduction in time required to perform a code review.

By taking into account the number of software engineers who had their day-by-day work affected by this project, and the productivity of these engineers, the amount of code generated in one year by the team was estimated.

In addition, taking into consideration the engineering costs per hour and the time (in hours) necessary to review this estimated amount of code using the old code review process, the costs for software code review would be near US$700,000.

After project implementation, which brought a considerable reduction of 60 percent in code review process, the cost to review all the code estimated to be produced in one year is approximately US$200,000.

In summary, this project led to a cost avoidance of US$500,000. Since the project costs with resources allocation and purchase of tool licenses were calculated as approximately US$50,000, the net savings resulting from code review effort reduction are approximately US$500,000.

The complete breakdown of financial projections was reviewed with the finance representative for approval.

9. Conclusions and Limitations

This Six Sigma project proved statistically that effort reduction can be achieved by adopting a lightweight tool-assisted code review process in lieu of Fagan-like inspections without compromising the effectiveness of the process.

The success of this project was dependent on a number of factors:

▲ Maturity of software development teams on the practice of code reviews

▲ A well-defined and stable process prior to improvement efforts

▲ The software development team's experience with good measurement practices, enabling reliable data for analyzes

▲ Same software development platform and similar complexity and application domain of source code considered in the scope of the project before and throughout improvement cycle

▲ Same or nearly the same software development team members conducting code reviews and generating data before and throughout improvement cycle

The lack of some of these factors would have made it more difficult or perhaps impossible for the improvement team to make truly data-driven decisions while optimizing the process (which are both a *sine-qua-non* condition and an advantage of the Six Sigma approach to problem solving).

Acknowledgments

The author of this case study is thankful to Motorola for its permission to allow publication of this case study as a book chapter.

10. References

1. M. E. Fagan. Design and Code Inspections to Reduce Errors in Program Development, *IBM Systems Journal*, 15(3):216–245, 1976.

CHAPTER 19

Predictive Engineering to Improve Software Testing

Patricia D. McNair*
Eric Maass, Ph.D.†

 MOTOROLA

Relevance

This chapter is relevant to companies that want to improve the cycle time, productivity, and effectiveness of software testing for product development. The chapter describes a highly successful project at Motorola where DFSS principles were used to improve the software testing process in an organization that used Agile software development and test-driven development methodologies.

At a Glance

▲ Motorola's Mobile Devices organization was challenged by its Executive Council to improve software engineering efficiency by:
  ▼ Reducing cycle time for build, integration, and test by 30 percent
  ▼ Reducing resource requirements for build, integration, and test by at least 25 percent
  ▼ Improving the software build environment to support these goals
▲ The project was successfully able to reduce cycle time for build, integration, and testing close to 30 percent and reduce resource requirements for build, integration, and testing by 25 percent.
▲ The resource savings due to this project resulted in annual savings of US$8.2 million for Motorola.

---

* Patricia McNair was the Master Black Belt for this project. At that time she was the Director of Software DFSS Corporate Quality at Motorola.
† At the time of this project Dr. Eric Maass worked at Motorola and he was the Director of Hardware DFSS.

Executive Summary

The pipeline of new product development is the lifeblood of many businesses. Software testing is vital to verifying and validating that the software meets or exceeds customer expectations.

In developing software and hardware for complex electronic products, software development is often the "long pole" in development and thus receives the most attention. Software testing, on the other hand, is often the bottleneck or constraint. However, it is possible to optimize testing activities even in an organization that uses advanced practices such as Agile software development aligned with test-driven development.

This project improved efficiency in the software development process for Motorola's cellular phones. By improving the efficiency of software engineering, Motorola's Mobile Devices business freed resources that could be used for development of the new Android platform and improve gross margins.

During the Define phase, aggressive improvement goals were established and stakeholders were interviewed to map out current testing process steps. The scope of the project affected the end-to-end software testing process and involved the Mobile Development software development, system test, field test, and customer validation teams.

The Measure phase included collecting data to evaluate the testing cycle and the number of test cases executed, estimating the software development cycle time and resources available.

During the Analyze phase, opportunities for possible improvements to the existing testing process were identified when it was confirmed that there was extensive duplication of effort across separate test teams. However, it was not clear whether combining the teams would meet the cycle time or resource effort reduction goals for the project. Therefore, extensive data collection activities were conducted and a robust process model created that enabled *what-if* analysis to evaluate the potential impact of a variety of different new test strategies and methods.

During the Design and Verify phases, the stochastic models were used to brainstorm test process improvements. The models clearly showed the magnitude of improvements that could result from combining parallel integration testing activities. Moreover, they ended up being quite accurate in their predictions about cycle time reduction and the reduction in effort.

The new process improvements were integrated with other "quick win" opportunities associated with changes to tools and methods. The changes were implemented and the separate test teams were combined, resulting in a 25 percent reduction in resources and nearly 30 percent reduction in test preparation and execution cycle time, which equated to more than $8 million in savings in the first year alone!

The use of detailed process modeling identified and validated substantive process changes while minimizing risk to the organizations involved. The detailed process models gave stakeholders a high level of confidence in the recommendations made by the Six Sigma process improvement team and afforded a faster transition to the new process and a faster realization of the resultant benefits.

This project is a classic example of DFSS methods being applied to existing processes to achieve breakthrough-level improvements.

1. Introduction

In September 2007, Motorola's Mobile Device business chartered and challenged a team to improve software engineering efficiency in the software build, integration, and testing cycle by reducing cycle time and resource demand. The business goals were:

▲ Reduce cycle time for build, integration, and testing by 30 percent.
▲ Reduce resource requirements for build, integration, and testing by at least 25 percent.
▲ Improve the software build environment to support these goals.

Traditionally, the degree to which a program is exercised by a test suite can be described with coverage criteria:

▲ **Function coverage**
  ▼ Has each function in the program been executed?
▲ **Statement coverage**
  ▼ Has each line of the source code been covered? A set of test cases is generated to ensure that every source language statement in the program is executed at least once. In many cases, 100 percent statement coverage is infeasible or prohibitive due to time or cost considerations, and it may be simply impossible to test all execution sequences and branching conditions in the program.
▲ **Branch or condition coverage**
  ▼ Has each branch or condition (such as true and false decisions) been executed and covered? This is similar to statement coverage, except that the number of equivalent branches or decisions is counted rather than the number of statements.
▲ **Path coverage**
  ▼ Has every possible route through a given part of the code been executed?
▲ **Entry and exit coverage**
  ▼ Has every possible call and return of the function been executed?

This project considered the following five key principles of testing that are useful when developing test cases:

1. **Define the expected output or result.** More often than not, the tester approaches a test case without a set of predefined and expected results. If the expected result is unknown, erroneous output can be overlooked. This risk can be avoided by carefully predefining all expected results for each test case.
2. **Don't test your own program.** Programming is a constructive activity. It is difficult to suddenly reverse constructive thinking and begin the destructive process of testing, and it is possible, if not downright likely, that the programmer will be blind to some of his or her own mistakes.
3. **Completely inspect the results of each test case.** As obvious as it sounds, this simple principle is often overlooked. There are many test cases in which an after-the-fact review

of earlier test results shows that errors were present but overlooked because the results were not inspected (or were not inspected with serious intent).

4. **Include test cases for invalid or unexpected conditions.** Programs already in production often experience errors when used in some new way. This stems from the natural tendency to concentrate on valid and expected input conditions during a testing cycle.

5. **Test the program to verify it does what it is supposed to do and not what it is not supposed to do.** A thorough examination of data structures, entry and exit criteria, and other output can often find errors.

The purpose of software testing includes but is not limited to:

▲ Reducing defects before moving to the next lifecycle phase
▲ Validation
▲ Reliability testing

Although software testing is often considered complex, test plans and test cases can be methodically developed to reduce complexity and increase confidence in the functionality and performance of the software. Software testing can be a dynamic assessment, in which sample input is run through software programs, and the actual outcome is compared with the expected outcome.

Software testing typically includes:

▲ White-box or unit testing for particular functions or code modules. This requires knowledge of the internal logic of the code.
▲ Black-box or functional testing for requirements. This does not require knowledge of the internal logic of the code.
▲ Performance, stress, or load testing of an application under heavy loads, such as testing a website under a range of loads to determine at what point the system's response time degrades or fails.
▲ Regression testing after fixes or modifications to the software or its environment to ensure the fixes and modifications don't generate new defects.
▲ Security testing to determine how well the system protects against unauthorized internal or external access, or willful damage.

End-to-end testing validates that the system meets overall requirements specification in a situation that emulates real-world use, such as interacting with a database, using network communications, or interacting with other hardware, applications, or systems in a complete application environment.

End-to-end testing comprises front-end and back-end test phases.

Front-end phase includes:

▲ Unit testing
▲ Feature testing

▲ Build or integration

Back-end phase includes:

▲ Interoperability testing (IOT), which checks functionality, appropriate communication, and compatibility with other systems with which it interfaces
▲ System testing (black-box testing based on overall requirements specifications for all parts of a system)
▲ Field testing
▲ Customer validation testing

2. Define Phase

The Define phase of the project entailed:

▲ Formation of the project team (also referred to as the Software Engineering Effectiveness team)
▲ Development of the project charter
▲ Definition of the problem statement
▲ Creation of a detailed project plan (Table 19.1)
▲ Review of the as-is model and process
▲ Gathering the voice of the customer (VOC) and voice of the business (VOB) data

2.1 Project Team

After this project was formally authorized by the project sponsors, the project Champion requested qualified Six Sigma resources to participate in the project.

Team members included:

▲ Master Black Belt to lead the team
▲ Black Belt
▲ Green Belt
▲ Configuration manager
▲ Software process director
▲ Three other subject matter experts

The chief quality officer at Motorola stepped forward as the Champion for this project.

2.2 Six Sigma Project Charter

The project charter comprised the elements discussed in the following sections.

2.2.1 Business Case

By improving the effectiveness of software engineering, the organization is more likely to deliver quality products on time to customers while minimizing the resources required to do so, and thus, improve customer satisfaction and gross margins simultaneously.

2.2.2 Opportunity Statement

Inefficiencies in overall software development process are affecting the organization's ability to deliver new products on time, and unfavorably affecting engineering expenses relative to gross margin and profitability. These affect the organization's competitiveness and financial viability.

2.2.3 Goal Statement

Develop short-term tactical recommendations and long-term systemic recommendations to:

▲ Reduce cycle time for build, integrate, and test by 30 percent.
▲ Reduce resource requirements for build, integration, and test by 25 percent.
▲ Improve software development environment (process) to support these goals.

2.2.4 Project Scope

The scope of the project was limited to the testing activities spanning the end-to-end software development process:

▲ 3 GSM platform development teams
▲ MDS (Mobile Device Software)
▲ System test
▲ Field test
▲ Customer validation

2.2.5 Project Plan

This project was chartered with a firm requirement for completion in six weeks (see Table 19.1).

Table 19.1 Software Engineering Effectiveness Project Plan

Task/phase	Start date	End date	Actual end
Interview key resources	24-Aug-07	12-Sep-07	12-Sep-07
Identify opportunities for improvement	12-Sep-07	18-Sep-07	18-Sep-07
Map as-is process	18-Sep-07	28-Sep-07	26-Sep-07
Map should-be process	28-Sep-07	4-Oct-07	26-Sep-07
Collect detailed data	24-Sep-07	1-Oct-07	9-Oct-07
Analyze data and develop solution	27 Sep-07	3-Oct-07	10-Oct-07
Submit recommendations		4-Oct-07	11-Oct-07

2.2.6   Project Schematic

A project "schematic" is a method to map organizational goals to key drivers and organizational or process components upon which these goals depend. That is, there is a relationship between the desired outcomes or goals (y's) and lesser goals, performance "drivers," and other factors (x's)

$$Y = y_1 + y_2 + y_3 + \dots$$

where each Y is a function of different factors (x)

$$y = f(x_1, x_2, x_3, \dots)$$

The "Big Y" is the primary organizational goal (typically financial in nature).

"Little Y's" are subordinate or lesser included goals.

"Vital X's" are those factors or components that are the key drivers that affect or determine the outcomes for those goals.

In this case, the "Big Y" was to improve software efficiency.

"Small Y's" included:

▲   Reduce software build, integration, and test cycle time and resources.
▲   Reduce software build environment cycle time and resources.
▲   Improve the software build environment to support these goals.
▲   Optional: Identify overall software engineering process improvements.

The purpose of a Six Sigma project is to identify the vital X's and to determine what can be done to change these to produce desired outcomes and goals.

2.2.7   As-Is Process Map

The first step to addressing the problem was to gather details on the current test process activities. To gather the stakeholder requirements (VOC and VOB), the Master Black Belt and team interviewed team leads, managers, directors, and engineers from product and test (including system test, field test, and customer validation).

An as-is, or current, process model was developed from these interviews. It clearly showed that different development teams each created and tested their own pieces of the final software products. This resulted in multiple teams in multiple locations performing nearly identical testing.

These initial interviews revealed far more duplication across these different software development teams than had been previously suspected. However, the degree to which overall testing efforts might be duplicated and the opportunities to eliminate this would need to be quantified.

That effort would involve substantive data collection and analysis activities.

3. Measure Phase

The Master Black Belt led the project team in developing a measurement plan to ensure that the data collected was valid. Team members collected, verified, and validated data regarding resource allocations and number of test cases.

Data about the testing process included a wide range of variables and factors (x_1, x_2, and so on), including but not limited to:

▲ Components developed
▲ Platform
▲ Language
▲ Team size
▲ Code size
▲ Cycle time
▲ Effort
▲ Number of test cases
▲ Number of test cycles
▲ Yield data
▲ Defect data and defect rates

With this data it was possible to quantify the activities in the test process with remarkable accuracy. It was possible to characterize the magnitude and the variation of test activities at every step of the process from code development to product release. This data was used to develop a stochastic model for analysis purposes. However, before the model could be used with any confidence, it needed to be validated.

The historical cycle time data for each step in the test process was fitted to distributions (Figure 19.1). Cycle times for some steps in the software test process were well represented with exponential distributions, others with normal and gamma distributions.

With the as-is process fully characterized, random values were generated and input into the model to calculate possible outcomes of cycle time and effort across thousands of simulated product development cycles. This technique for generating many hypothetical outcomes is called Monte Carlo analysis.

The distribution of forecasted total cycle time based on Monte Carlo simulation overlapped the observed distribution of total cycle time (Figure 19.2). This validated the accuracy of the initial model. The Monte Carlo simulation was then set up for what-if analysis of cycle times.

Resource requirements were also measured from historical data, and a Monte Carlo model was developed to enable what-if analysis of resource requirements. This allowed the team to input different ranges of input values to simulate effects of changes to the end-to-end test process. Thus, the potential effects of combining teams, adding or removing steps to the test process, and changing the number of components could be evaluated quantitatively to

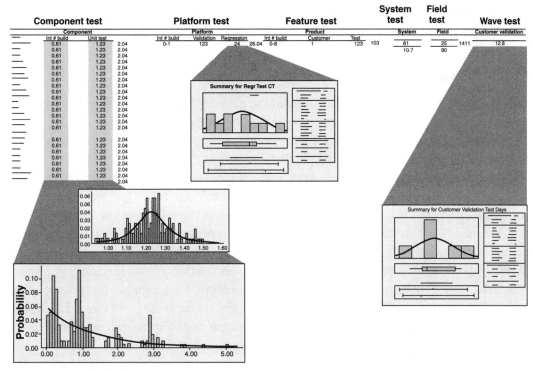

Figure 19.1 Distributions fit to historical data for cycle times for each step in the software test process.

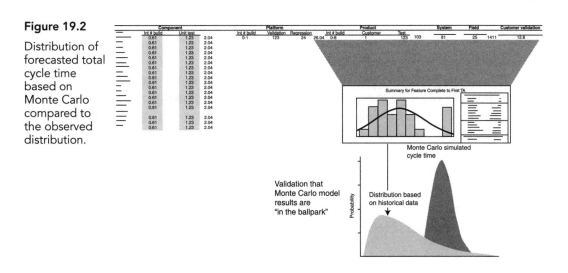

Figure 19.2

Distribution of forecasted total cycle time based on Monte Carlo compared to the observed distribution.

determine the best possible combination of changes to improve the end-to-end test process and achieve overall project goals.

4. Analyze Phase

Based on the Monte Carlo simulation, one of the primary focus areas for improvement was component testing and platform testing. Note that the early stages of the test process involved component tests for sets of features and functions across multiple component groups. Component testing was conducted in parallel for two sets of components (MDS and engine components). Therefore, component testing would not be complete until the last set of features and functions for both of the components were tested. Therefore, in the Monte Carlo simulation, the total cycle time for component testing was dependent on the component test that had the longest cycle time (for MDS or engine components), even if it was conducted in parallel with the testing for the other component. For example, if three component tests for MDS components took one hour each (total component test time = three hours), and two of the three component tests for engine components took one hour each but one test took four hours, then the total duration of component testing for engine components would be at least four hours, and quite likely six hours if the component tests for engine components were conducted sequentially.

One important process variation under initial consideration was a proposed alternative approach in which one cycle of component testing would handle both components (MDS and engine components), thus saving cycle time by having the longest cycle time component test execute in parallel with the component unit tests.

The Monte Carlo model was modified to generate an alternate outcome for this new test process.

5. Design Phase

Figure 19.3 shows the projected improvement in test execution cycle time that would be obtained by changing from the as-is process of separate component, platform, feature, and system testing (software only) into one software validation phase. With the as-is process, cycle time for each type of test was controlled by the longest cycle time among several tests in parallel.

By having one software validation phase, one could observe not only a projected reduction in effort but a significant reduction in cycle time as well!

The model predicted that it would be possible to reduce the total test execution cycle time by 39 days. To validate these results, however, it would be necessary to actually test these results in the field.

After this project was completed and recommendations were formally submitted, this change was implemented on a pilot basis and the actual reduction in total test time was observed

Figure 19.3

Comparison of predicted cycle time between the as-is test process and the proposed combination of different test phases into one software validation phase.

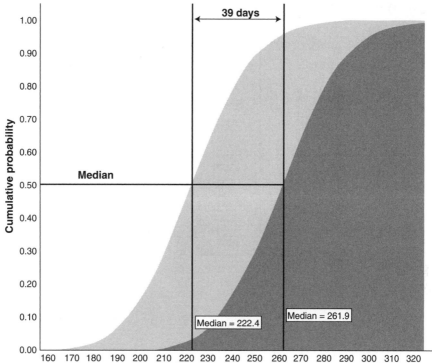

Assumes: All components handled by the same test team and some reduction in test cases

to be 38 days—thus attesting to the accuracy and efficacy of highly detailed process models. Furthermore, by combining different test phases and eliminating duplicate test cases, total number of test cases could be reduced and this resulted in reduction in test case preparation time which was almost as much as the reduction in test execution cycle time (39 days).

Additional simulations were conducted to evaluate the effect of these process changes on overall effort and resource requirements. Figure 19.4 illustrates the forecasted improvement in test resource requirements associated with the combination of all software test phases into one software validation phase—reduction in resource demand from 182.2 to 120.36 people, a savings of 62 people. The new process would reduce the separate test phases into a single build and software validation phase, thus eliminating substantial duplication of effort and testing (Figure 19.5). In addition, having one common integration and build team for multiple products and platforms instead of separate ones would result in reduction in resource requirement from 300 to 242 people, a savings of 58 people.

The new process predicted a *reduction in overall cycle time* by 30 percent (test execution cycle time reduction of 39 days, as shown in Figure 19.3, plus an equal amount of cycle time reduction in test case preparation as described earlier); it also promised to *reduce resource requirements* from 482 (300 plus 182.20 people) to 362 (242 plus 120.36 people) (a 25 percent reduction, meeting the project goal).

Figure 19.4

Comparison
of predicted
resource
requirements
between the
as-is process
and the
proposed
combination
of component
and platform
testing into
one integration
test cycle.

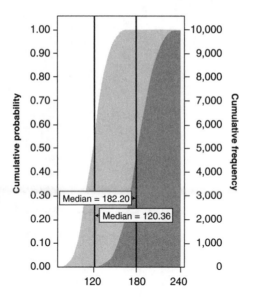

As-is number of platforms +
product with current cases

One integration test team with
duplication removed

Total resources for one integration
test team to release 9 products
simultaneously

Staff required: 121

Delta: 62

Figure 19.5

As-is process
and new
combined
proposed
process.

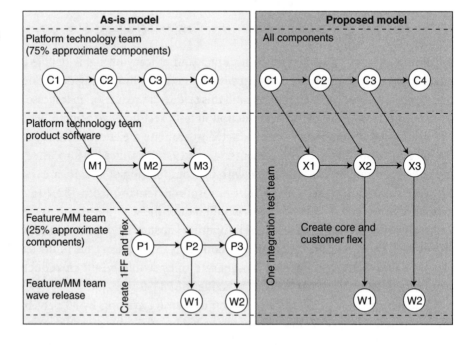

As a result of the varied simulations, interviews with stakeholders (at all levels) and research on external industry best practices, some additional changes were also proposed to accompany these overall process changes. These included:

▲ The implementation of new test methods (orthogonal array-based testing)
▲ Automated regression testing
▲ Model-driven feature interaction test suite to reduce regression cycle time

6. Verify Phase

The improved process was institutionalized with new specifications and procedures. One common integration and test team would build and integrate all components for 3GSM rather than separate teams.

Additional improvements included the following:

▲ The build and integration team would physically reside closer to the component team.
▲ The integration team would handle creation of the environment.
▲ One overall integration test plan would be created.
▲ One single binary (executable file) for testing would be created and daily software builds would be performed to deliver fixes for defects found during testing.

These improvements allowed the test teams to:

▲ Eliminate duplication of regression test cases between integration, system, and field testing
▲ Build and integrate all components together
▲ Reduce the number of test cases that needed to be documented and executed
▲ Reduce the number of test cycles
▲ Reduce the number of test resources

However, to implement this across all existing test teams would be a monumental task. A separate implementation plan had to be developed involving all other organizations and regions.

Subsequent implementation activities included:

▲ Piloting of the new process
▲ Development of new testing procedures and standards
▲ Acquisition and implementation of new automated tools
▲ Development of training materials
▲ Reorganization to combine (and eliminate) teams
▲ Development of consolidated reporting and metrics

7. Conclusion

The project quantified and validated the effects of major changes to the organization's overall test development process. Simulation and analysis allowed the organization to consider different improvement scenarios in a short time, without adverse effect on software development teams, with less risk, and with greater confidence.

Ultimately, the implementation effort associated with rolling out the new process greatly exceeded the project effort itself, but the final payback on this endeavor was quite substantive, as noted in the "Benefit Realization" section that follows.

8. Benefit Realization

This project provided a substantial improvement in software development cycle time, product launch timeliness, and resourcing. Although the benefits associated with the 38-day development time reduction were not quantified in financial terms, the resourcing benefits were quantified with the support of the accounting and finance organization, as follows:

▲ Resource optimization in integration and build: 58 people
▲ Resource optimization in software validation phase: 62 people
▲ Total resource optimization: 120 people (distributed worldwide)
▲ Cost per resource = $5,700/staff month
▲ Savings per year = 12 months × 120 resources × $5,700 per staff month = US$8.2 million per year

Acknowledgments

The authors of this case study are thankful to Motorola for its permission to allow publication of this case study. The authors would also like to thank the team members, stakeholders, and supporters: Kathy Winter, Rey More, Mike Potosky, Daniel Green, Kathy Feld, and Suresh Kumar.

9. References

1. Haapanen, Pentti and Atte Helmunen, *Failure Mode and Effects Analysis of Software-Based Automation Systems*, STUK-YTO-TR, August 2002.
2. Maass, Eric and McNair, Patricia, *Applying Design for Six Sigma to Software and Hardware Systems*, Prentice-Hall, 2009.

CHAPTER 20

Improving Product Performance Using Software DFSS

Patricia D. McNair*
Edilson Albertini da Silva
Alex Garcia Gonçalves

Relevance

This chapter is relevant to organizations that want to optimize product performance and improve product reliability, testability, and usability.

At a Glance

▲ Before the deployment of a new version of push-to-talk over cellular applications, Motorola leveraged Software Design for Six Sigma (SDFSS) to proactively identify key application features that best reflected the voice of the customer and to identify design improvements that could positively affect those application features.

▲ This project's Big Y was to improve startup time on push-to-talk (PTT) P2K platforms by 50 percent. The quantitative improvement due to this project was a 87 percent improvement in startup time.

▲ This project resulted in cost avoidance of approximately US$1.8 million on each similar software development project in the future.

---

* Patricia McNair was the Master Black Belt for this project. At that time she was the Director of Software DFSS Corporate Quality at Motorola.

Executive Summary

Push-to-talk (PTT) over cellular (POC) is intended to provide rapid communications for business and consumer customers of mobile networks, allowing voice communications between single recipients (one-to-one) or between groups of recipients as in a group session (one-to-many).

The POC client applications on the mobile phone communicated with different servers using data services (e.g., GPRS, 3G) to initiate calls (PTT Server), store contact lists (XDM Server), and get presence status from contacts in the contact list (Presence Server). The Latin America (LATAM) country customer requested that the PTT team port the Platform 2000 (P2K) solution to other platforms that were requested by three major carriers.

Customer feedback from user trials and field tests raised the following concerns about the performance of PTT on P2K:

▲ As PTT was intended for rapid communications, it was important that the application have a short boot time and fast response time. Due to the significant amount of data transfer, contact list and groups download time varied, and a key customer requested Motorola to reduce it. The long startup time gave the impression that the PTT service was slow because the service would not be available until the contact list and groups were fully downloaded.

▲ As all the communications were performed using client-server architecture and a packet data network, it was important to minimize network traffic to improve performance.

A Software Design for Six Sigma (SDFSS) team was assembled to define the charter and implement a solution. The project team had a challenging goal to reduce to the startup (sign-in) time from 97 to under 20 seconds within three months.

The team used the Requirements, Architecture, Design, Integration (RADI) model and began gathering the voice of the customer in the form of "must have" features. These were then ranked in order of importance and translated into technical requirements. Using Six Sigma tools, sign-in time was identified as the critical parameter to improve for performance.

The next step was to investigate the product's architecture to understand factors that influenced this critical parameter. Sign-in time was further decomposed into and mathematically related to its functional components.

Next, the project team conducted a Design of Experiments (DOE) to determine which of these components were significant factors in contributing to total sign-in time. This enabled the software engineers to focus on only those factors that needed to be adjusted (optimized) in order to achieve the sign-in time reduction goal.

The project team then mapped out the current PTT design in detail to understand what bearing the current design had on the couple of factors that were most influencing sign-in time. Design modifications were brainstormed, selected, and piloted. Additional simulations using statistical tools such as Monte Carlo confirmed that the new implementation reduced the

PTT sign-in time by more than 80 percent! The estimated savings in future development projects as a result of this project was US$1.8 million, and this was verified by the finance department.

The customer was impressed with the SDFSS approach used to formally gather requirements and identify and resolve the issues. The improved design has been leveraged in future platforms, such as Android.

1. Introduction

From 2007 through 2008, Motorola's Corporate Quality group, along with business leaders and Motorola University, developed a new SDFSS program and curriculum, including a four-phase RADI model training course. The objective of the SDFSS program was to improve software development and prevent the occurrence of defects in earlier phases of the development lifecycle (see Figure 20.1).

Common flaws usually found in each software development phase are:

▲ Requirements:
  ▼ Missing and incomplete requirements and key performance indicators
  ▼ Missing feature interaction and other key scenarios
▲ Architecture/Design:
  ▼ Risks and failure modes not properly identified and managed
  ▼ Feature architecture not well defined and managed
  ▼ Lack of resilient and robust design

Figure 20.1

Software work flow phases improved by SDFSS.

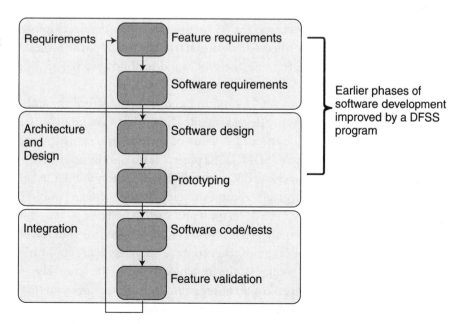

▲ Integration:
  ▼ Too many test cases during verification and validation phases
  ▼ Missing test coverage for some key scenarios

1.1 How Does SDFSS Help?

SDFSS tools can greatly reduce the risk of defects by rapidly deploying a roadmap, simple tools, and behaviors designed to dramatically reduce requirements defects and the cycle time to complete the requirements phase of a typical software project. The main focus was to reduce the introduction or injection of defects rather than the containment and repair of defects. The SDFSS RADI model is a four-phase model that includes basic statistics, software metrics, data presentation, modeling and simulation, requirements, architecture, design, and integration. This enables software designers to:

▲ Design the product with quality in requirements, architecture, design, and feature integration phases
▲ Improve on-time delivery and reduce defects
▲ Effectively reduce cycle time and improve performance of feature interactions
▲ Implement metrics-based decision making for technology development

Deliverables for the RADI phases were used as the baseline requirements for SDFSS Belt certification.

2. Project Background

Once the opportunity to leverage SFDSS methodology for the development of a new POC version was identified, the project Champion started to assemble the project team. In order to address the need for SDFSS training, the program director of the SDFSS program was invited to train the project team.

At the beginning, the project Champion faced resistance from the software engineers in using Six Sigma during software development. The project Champion was challenged by the project team to demonstrate the added value of introducing statistical tools on top of the software development lifecycle (SDLC). The project team questioned if it was worth the effort, as they might be required to create documentation in addition to their usual software project deliverables. They were uncertain about how they or the software would benefit from this additional effort spent on using SDFSS tools and methodologies.

The project Champion managed this resistance by suggesting that the project team run a pilot focusing on the redesign of the contact management application. The contact management application manages the downloading of the contacts into the contacts application. In order to address concerns regarding additional workload that may be incurred because of the implementation of SDFSS methodology, the project Champion and the Master

Black Belt incorporated the redesign in the class exercises. Therefore, as the team learned SDFSS concepts and tools during the training, it applied the knowledge gained during classroom exercises for requirements, design, and development using the real-world example of the contact management application.

3. Define Phase

During this phase, the main focus was on understanding the problem definition; gathering the requirements; and developing the business case, goals, opportunity statement, scope, and benefits.

3.1 Six Sigma Project Charter

3.1.1 Business Case Statement

The business case of this project was as follows:

▲ A major carrier has requested the porting of the PTT P2K solution to new P2K handsets using new chipsets.

▲ The PTT solution on new P2K chipset phones shall have the same functionality as legacy P2K and also improved performance, minimizing user response time and network traffic.

▲ The use of SDFSS shall minimize the performance risks associated with this new PTT development.

▲ The solution adopted must be platform-independent so that it can also be applied on other platforms (for example, Windows Mobile and Android).

3.1.2 Opportunity Statement

Following were the key anticipated benefits:

▲ The application of SDFSS tools will enable the PTT team to identify the critical parameters related to customers' performance needs.

▲ Once the critical performance parameters are identified, the SDFSS tools can be used to make sure they are tracked in the PTT development lifecycle, resulting in a product that meets the customer's performance expectations.

3.1.3 Goal Statement

The project goal was to mitigate performance risks associated with PTT application by reducing the sign-in time to less than 20 seconds (from a baseline of 97 seconds).

3.1.4 Project Scope

The project scope was to include prototyping the product improvements on two specific phones, but the final improvements were to be implemented on all Motorola phones.

3.1.5 Project Plan

The key project milestones were:

Deliverable	Due Date
Requirements and architecture phases	
Requirements and critical parameter identification and transfer function	Sept. 2008
First design of experiment (plan, execute, and analyze)	Sept.–Oct. 2008
Design and integration phases	
Design analysis and prototype	Oct. 2008
Second design of experiment	Oct.–Nov. 2008
Design verification	Nov. 2008
Lessons learned and project closure	Nov. 2008

The Green Belt candidates created a project plan listing project deliverables that would be developed during the project using the RADI model. See Tables 20.1 and 20.2 for an example of the deliverables of the Requirements and Architecture phases.

Table 20.1 Deliverables and Tools in the Requirements Phase

Phase	Owner	Green Belt deliverables	Tools
Requirements	Edilson Silva	VOC, requirements development and analysis	Brainstorm, HOQ, UML use case
Requirements	Edilson Silva	Perform and document critical parameter flow down and analysis	Critical parameter tree, cognition cockpit
Requirements	Edilson Silva	Establish initial transfer function	Critical parameter tree, cognition cockpit

Table 20.2 Deliverables and Tools in the Architecture Phase

Phase	Owner	Green Belt deliverables	Tools
Architecture	Edilson Silva	Evaluate architecture	Design of Experiments (DOE)
Architecture	Edilson Silva	Measure non-functional critical parameters (e.g., performance)	

3.1.6 Project Team

The project team consisted of:

▲ Sponsor
▲ Champion

▲ SDFSS Master Black Belt
▲ Software Black Belt
▲ Two PTT software developers (Green Belt candidates)

4. Requirements and Architecture Phases of RADI

4.1 Gathering Voice of Customer

If the team hoped to have satisfied customers, they had to seek input from Motorola's customers before designing the product. The voice of the customer (VOC) facilitates an understanding of customer needs and quantifying them through tools such as KJ analysis and quality function deployment. KJ analysis is a formal technique developed by Jiro Kawakita for developing requirements based on free format interviews. Basically, it is a semantic based method of developing affinity diagrams for classifying comments and phrases and grouping them hierarchally.

4.1.1 VOC Purpose and Objectives

Before gathering the VOC for a new product, some initial effort may be required to review relevant inputs, including business requirements, portfolio requirements, engineering inputs, lessons learned, and information that can be derived from customer returns and customer feedback for products that may be similar to the proposed new product. One key deliverable from this effort in reviewing and understanding the business requirements can be a set of voice of the business (VOB) criteria, which can be used later in the quality function deployment (QFD) method and as criteria for the concept selection process. The VOB was often rather predictable in terms of including expectations, such as that the project should be completed on schedule, the development and marketing costs should stay within budget, and there should be no unpleasant surprises—including major recalls after the product was launched. The latter requirement might flow down into reliability, availability, and quality requirements that might be important to customers, but might not be specifically mentioned when the VOC was gathered.

4.1.2 The VOC Gathering Team

Once the relevant inputs had been reviewed, the team was empowered to gather the VOC. This provided a fantastic opportunity for marketing and engineering to work together in the beginning to define the new product. In many organizations, there seems to be some level of tension between these two organizations. While the new product is being developed, that tension may devolve into doubts about the product definition, questions as to whether certain properties or features of the product are worthwhile, and even finger-pointing when problems arise or when managers question whether the product will be viable and successful in the marketplace.[1] This team was able to get through this phase with less tension than expected.

The project team collected VOC feedback from people working directly with the POC customer. Brainstorming sessions were conducted with the following team members:

▲ Software designers and system engineers working in PTT development
▲ Account manager responsible for PTT in LATAM
▲ Deployment team member responsible for coordinating PTT installation at customer
▲ Test team members—field and feature

As a result of brainstorming sessions, a list of possible "must have" features and improvement areas were defined and then translated into VOC, and the brainstorming team also voted to rank items based on their importance to the customer (see Table 20.3).

Table 20.3 Voice of the Customer Importance

Voice of Customer	Importance
Minimize network traffic	8
Minimize PTT application start-up time	10
Minimize PTT call setup time	9
Maximize audio quality (e.g., minimum delays and package loss) during PTT calls	6
Minimize PTT user interface (UI) response time	5
Minimize PTT application recovery time after out-of-coverage (OOC) scenarios	7
Minimize PTT application failures (e.g., no response from server) during startup	10

4.2 Transforming Customer Requirements into Technical Requirements

After the VOCs were completed, the team performed a KJ analysis for customer requirements. The KJ method is for:

▲ Discovering and clarifying customer needs
▲ Organizing qualitative VOC data
▲ Focusing attention on the critical new, unique, or difficult requirements:
  ▼ **New:** A requirement that your customer has never asked you to fulfill before and is completely new to you.
  ▼ **Unique:** A requirement that is distinctive or highly desired; may also be a requirement that is being fulfilled in the market, but not by you; perhaps by a competitor or by a different product/service.
  ▼ **Difficult:** A requirement that is difficult for your company to fulfill, but not necessarily new or unique.

Requirements gathered using KJ analysis were analyzed for relative importance using two methods. The first was a subjective assessment to identify whether any requirements were NUDs, that is requirements that are New, Unique, or Different. Next, the requirements were analyzed and classified using Kano analysis where individual requirements were classified as "Satisfiers," "Must-Do's," or "Delighters." These steps helped clarify which requirements were

Figure 20.2

Twelve steps to complete QFD.

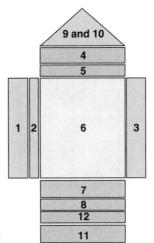

12 Steps to a QFD

1. What is wanted?
2. How important are these things?
3. How well are we, or others, satisfying the requirements now?
4. How is it done?
5. What is the direction of improvement for each requirement?
6. What are the relationships?
7. What is the weighted relationship score?
8. What is the rank order of requirements?
9. What conflicts in design might we anticipate?
10. What complementary effects in the design might we expect?
11. What competitive or internal benchmark/metric shall we compare to as we go through development?
12. What are the largest target ranges of measurement for each requirement?

the most critical and the most important requirements were used to start QFD analysis (as shown in Figure 20.2).

As shown in Figure 20.3, based on its weighted scores in the House of Quality (HoQ) matrix, the system requirement of *sign-in time* was identified as the critical parameter to improve for performance.

4.3 Critical Parameter Management Decomposition

Critical parameter management (CPM) is a methodology for managing, analyzing, and reporting technical product performance, which helps development teams understand and quantify how well their design will be able to meet critical parameter specification limits. The goal here was to determine the key relationship (transfer functions) that correlates a product's characteristics with the critical parameter identified in the previous section. For architectural references and capability flow-up and flow-down of critical parameters, refer to the CPM tree in Figure 20.4.

Through its dynamic relationships between requirements and parameters, CPM provides real-time performance feedback about requirement or parameter changes, as well as the sensitivities of individual parameters on the overall system.

As stated previously, sign-in time, which includes PTT registration and downloading of contacts, was the critical parameter chosen from the QFD. The next step was to investigate the product's architecture to understand the factors that influenced this parameter. The tool used for this task was the use-case diagram, allowing the team to decompose sign-in time into its functional components (see Figure 20.5).

Rating links legend
H (9) = High effect; M (3) = Medium effect; L (1) = Low effect;
0 = No effect

Root links legend
Equal effect: =; No effect: 0 or blank string; Relative effect: +, ++, −, − −

Direction of goodness legend
Increases customer satisfaction: +; decreases customer satisfaction: −
On target for customer satisfaction: 0 or blank string

System requirements

Direction of goodness								
Voice of customer, VOCs (VOCs and its importance gathered from cross-functional team brainstorming)	Importance	System requirement 1	Sign-in time	System requirement 2	System requirement 3	System requirement 4	System requirement 5	System requirement 6
VOC's requirement 1	8		H	M	H	L	M	H
VOC's requirement 2	10	L	H				M	H
VOC's requirement 3	9			M				
VOC's requirement 4	6			L		H		
VOC's requirement 5	5	H	L	L	M		L	L
VOC's requirement 6	7		H				M	L
VOC's requirement 7	10		H			M	M	
Scoring totals		55	320	62	87	92	110	174
Relative scores		6.1%	35.6%	6.9%	9.7%	10.2%	12.2%	19.3%
Normalized scores		2	10	2	3	3	3	5
Units		Seconds	Seconds	Seconds	Seconds	Numbers	Seconds	Percentage

Figure 20.3 QFD identifying sign-in time as key component to redesign for maximum optimization.

Figure 20.4

Critical
parameter
tree—Flow-up
and flow-down.

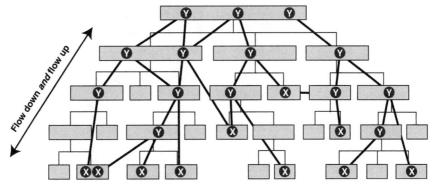

This is the overlap of architecture and critical parameter management

CPM is a *mathematically* interconnected hierarchy of requirements and
parameters allowing flow down *and* flow up of design sensitivities

Figure 20.5

Sign-in time
use-case
decomposition.

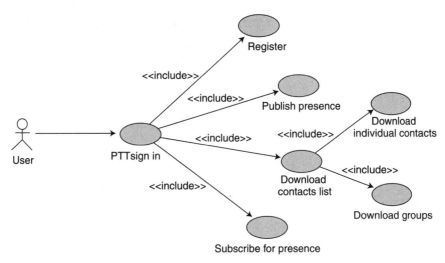

4.4 Critical Parameter Analysis and Requirements Decomposition

In this phase the team created a critical parameter (CP) tree that mathematically
interconnected PTT's system requirement of sign-in time with the functional components
identified in Figure 20.5. The output shows the VOC's relationship to the system requirement,
functional architecture, transfer functions, and all decomposed subsystems and components,
along with capability flow-up (see Figures 20.6 and 20.7). Using the CP tree diagram, the team
decomposed each of the top-level Y's into subordinate Y's and X's that contributed to the
performance of the top-level Y. The key question was:

What subordinate parameters contribute to the performance of the top-level parameter?

As described in the next section, the project team conducted DOEs to determine which of
the following factors contributed to the total sign-in time:

Figure 20.6

CP tree flow-down diagram— Level 1.

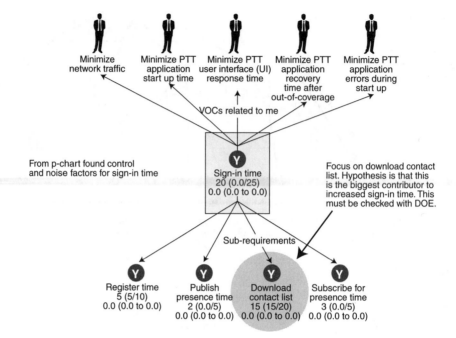

Figure 20.7

CP tree diagram— Level 2 showing transfer function.

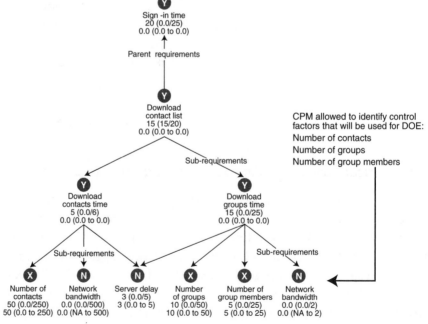

- ▲ Register time (from 5 to 10 seconds)
- ▲ Publish presence time (from 0 to 5 seconds)
- ▲ Download contact list time (from 15 to 20 seconds)
- ▲ Subscribe for presence time (from 0 to 5 seconds)

4.5 Design of Experiments

DOE allowed the project team to:

- ▲ Determine the factors that would have the most impact on satisfying the sign-in time requirement
- ▲ Determine the relationship between the factors
- ▲ Identify the transfer function between the factors and this requirement

The vital few factors used for the experiment and DOE settings are shown in Table 20.4.

Table 20.4 DOE Factors and Settings to Evaluate Startup Time

Factor	Low setting	High setting
Number of contacts	25	250
Number of groups	1	50
Number of group members	1	25

These vital few factors were identified based on historical data and the architect's knowledge. A full factorial design was chosen with eight runs.

4.5.1 DOE Measurements

Using a benchmark that follows the POC contacts management protocol, the following DOE measurements were taken (see Table 20.5):

- ▲ Full factorial design created in Minitab for three factors
- ▲ Factors: Three-base design: 3, 8
- ▲ Runs: 16 replicates: 2
- ▲ Blocks: None center pts (total): 3

4.5.2 DOE Analysis of Determinant Factor and Model Validation

After feeding the model with the measured values from the existing system presented in the previous section, the team was able to create the chart shown in Figure 20.8.

The team reviewed the results of the estimated effects and coefficients for PTT sign-in time, and based on the p-value, which was less than 0.5, the team was able to conclude that a correlation existed between the critical parameter of sign-in time and the parameters of "number of groups" and "number of contacts."

Table 20.5   Design of Experiment Factor Details

Std. Order	Run Order	Center Pt.	Blocks	Contacts	Groups	Cont. per Group	PTT Sign-in Time (sec.)	Config.	Measurement
1	1	1	1	25	1	1	14	1	1
2	2	1	1	250	1	1	23	5	1
3	3	1	1	25	50	1	143	3	1
4	4	1	1	250	50	1	171	7	1
5	5	1	1	25	1	25	13	2	1
6	6	1	1	250	1	25	29	6	1
7	7	1	1	25	50	25	117	4	1
8	8	1	1	250	50	25	167	8	1
1	1	1	1	25	1	1	13	1	2
2	2	1	1	250	1	1	28	5	2
3	3	1	1	25	50	1	163	3	2
4	4	1	1	250	50	1	178	7	2
5	5	1	1	25	1	25	12	2	2
6	6	1	1	250	1	25	31	6	2
7	7	1	1	25	50	25	147	4	2
8	8	1	1	250	50	25	161	8	2

Figure 20.8

Box plot of PTT sign-in time.

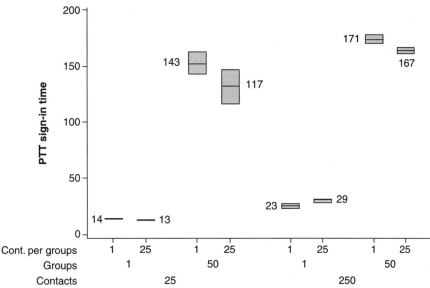

4.5.3 Transfer Function Determination

Having identified the most significant factors, the team could see that the significant effects are further away from the line in a normal probability plot (these effects are shown in bubbles in Figure 20.9). A new factorial was done isolating only the relevant items, that is, number of

Figure 20.9

Normal probability plot of groups and contacts.

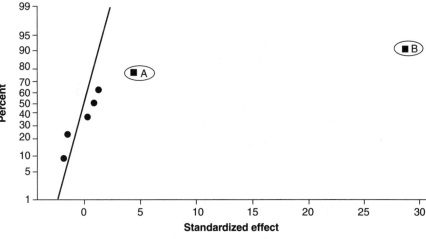

groups and number of contacts that affected PTT sign-in time, and then used it to calculate transfer function coefficients.

Figure 20.9 shows that Factor A has some impact; however, Factor B (number of groups) has a greater impact.

Results from the new factorial:

▲ Term Coef constant 4.92914

▲ Contacts 0.0922222

▲ Groups 2.76531

▲ S = 10.4679 PRESS = 2157.82

▲ R-Sq = 98.14% R-Sq(adj) = 97.85%

Therefore, the transfer function for this model was:

$$\text{PTT sign-in time} = 4.92914 + 0.0922222(\#\text{Contacts}) + 2.76531(\#\text{Groups})$$

4.5.4 Results Analysis and Conclusions

After modeling the transfer function, the team was able to run simulations of the current architecture to determine if the software would meet the customer's requirement of a sign-in time less than 20 seconds. This was done with Monte Carlo simulation using the transfer function developed in the DOE phase (see Figure 20.10). The figure shows that if the current

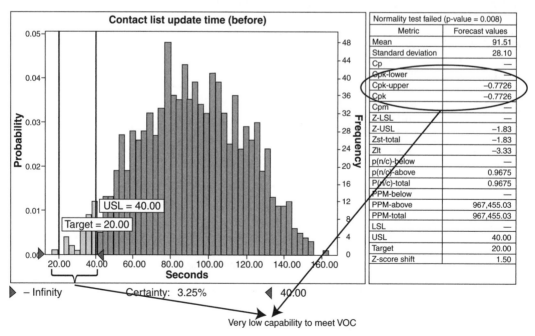

Figure 20.10 Monte Carlo simulation using transfer functions.

architecture was used, there is only a 3.25 percent probability that the customer requirement of less than 20 seconds would be met. This is clearly unacceptable.

From the DOE analysis, the team statistically showed that the control factors that affected PTT sign-in time were number of contacts and number of groups, and the number of groups had the highest impact on PTT performance at startup.

The CPM modeling and transfer function determination based on quantitative data helped software engineers focus their efforts on the significant factor instead of expending time and resources on insignificant factors.

5. Design and Integration Phases of RADI

The previous phase showed that the number of contacts and, especially, groups has a huge impact on the PTT sign-in time, indicating that the Design phase must focus on implementing a solution that significantly reduces the impact that the number of contacts and groups has on the sign-in time.

Note that although this project followed the RADI model, use of the DMADV model would have been appropriate as well. In fact, this point would signify the end of the Analyze phase in the DMADV model.

5.1 Current Design

This activity entailed mapping out the current PTT design in detail in order to fully understand the implications of the current design on sign-in time and the network traffic involved in contact list and group download.

5.2 Choosing the Best Solutions for the Problem

Four possible solutions were considered to resolve the problem of the sign-in time being too long. In order to select the best of the possible solutions presented, a prioritization matrix was used. This mechanism assigned scores for how well an alternative (possible solution) met the weighted evaluation criteria. In this case, the criteria were performance, network traffic, data availability, and cost implementation. This results in a total score for each possible solution, and the one with the highest score was selected.

5.3 Implementing the Selected Solution

In this activity, the existing design was changed in order to apply the selected solution and improve the contacts/groups download process. Because the improvements are proprietary to Motorola, they are not detailed here.

5.4 Verifying the Results

After the conclusion of the Design phase described earlier, a prototype was implemented following the new design, and the sign-in process was tested again. The results obtained are shown in Table 20.6.

Table 20.6 Results of the Sign-in Process from DOE

Std order	Contacts	Groups	Contacts per group	% PTT sign-in time reduction
1	25	1	1	42.86
2	250	1	1	43.84
3	25	50	1	93.01
4	250	50	1	88.89
5	25	1	25	38.46
6	250	1	25	58.62
7	25	50	25	90.6
8	250	50	25	89.22
1	25	1	1	7.69
2	250	1	1	60.71
3	25	50	1	93.87
4	250	50	1	89.89
5	25	1	25	33.33
6	250	1	25	61.29
7	25	50	25	91.16
8	250	50	25	87.58

Comparing the new sign-in times to the times before the optimization and after applying the changes, the process of downloading contacts and groups is much faster than before (that is, there is considerable performance improvement).

With the new results, the following transfer function was obtained:

$$\text{PTT Sign-in Time} = 8.63662 + 0.0129025(\#\text{Contacts}) + 0.0300454(\#\text{Groups}) + 0.000430839(\#\text{Contacts}*\#\text{Groups})$$

Using this transfer function, the project team could simulate several different scenarios with the Monte Carlo simulation tool (see Figure 20.11).

This chart shows that the obtained results are better than the customer requirement that sign-in time take less than 20 seconds.

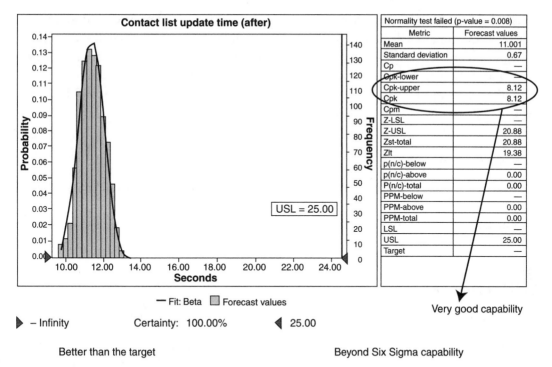

The chart shows "Contact list update time (after)" with the following table:

Normality test failed (p-value = 0.008)	
Metric	Forecast values
Mean	11.001
Standard deviation	0.67
Cp	—
Cpk-lower	
Cpk-upper	8.12
Cpk	8.12
Cpm	—
Z-LSL	—
Z-USL	20.88
Zst-total	20.88
Zlt	19.38
p(n/c)-below	—
p(n/c)-above	0.00
P(n/c)-total	0.00
PPM-below	—
PPM-above	0.00
PPM-total	0.00
LSL	—
USL	25.00
Target	—

USL = 25.00

— Fit: Beta   ▢ Forecast values

Very good capability

▶ – Infinity   Certainty: 100.00%   ◀ 25.00

Better than the target   Beyond Six Sigma capability

Figure 20.11   Monte Carlo simulation analysis.

6. Benefit Realization

As stated earlier, the return on investment (ROI) associated with this project is calculated in terms of future savings in software development. Since the methodology helped to identify and focus on the relevant areas to meet customer requirements, it was estimated that this project saved approximately US$1.8 million in future development projects. The success of the project convinced other teams that SDFSS works and would provide the quality and performance our customers desire.

7. Conclusions and Limitations

The use of the SDFSS methodology significantly reduces the subjectivity of the decision-making process during the software development lifecycle by providing means to combine measurable data with developers' expertise. SDFSS tools and concepts increased the efficiency of the complex task of translating customer requirements into a product that effectively met customer requirements.

However, a notable limitation of this methodology is the prescribed sequence of steps and statistical tools, which a typical software designer is typically not familiar with and thus may be intimidated by. Indeed, this was a challenge that this project team faced. The solution lies in establishing an active guidance and support mechanism for the software designers by assigning a knowledgeable SDFSS coach to mentor and guide them through the process. By taking a real-world example of a problem that is preferably from the software designer's daily work, and by helping them apply the SDFSS methodology, the coach can help demonstrate the benefits of applying SDFSS. This, in turn, provides software designers with the confidence and belief to independently apply the methodology in future projects.

Acknowledgments

The authors of this case study are thankful to Motorola for its permission to allow publication of this case study as a book chapter. The project Green Belts would also like to thank Patricia McNair, Luiz Bernardes, José Barletta, and Eder Alves for their support, insights, and knowledge of the subject.

8. References

1. McNair, P.; Maass, E. *Applying Design for Six Sigma to Software and Hardware Systems*, Prentice Hall, 2009.
2. Dvir, D. Transferring Projects to Their Final Users: The Effect of Planning and Preparations for Commissioning on Project Success. *International Journal of Project Management*, v.23, pp. 257–265, 2005.
3. Goldratt, E. *The Haystack Syndrome: Sifting Information Out of the Data Ocean*, Great Barrington-MA: North River Press, 1991.

CHAPTER 21

High-Speed Product Development at Xerox

Robert Hildebrand

Relevance

This chapter is relevant to organizations that want to reduce product delivery cycle times by accelerating integration and test processes.

The holistic approach to cycle time reduction described in this chapter applies to the development of products requiring high levels of integration.

At a Glance

▲ To enable the rapid release of the next-generation iGen digital production press, Xerox leveraged Design for Lean Six Sigma (DfLSS) to drive image quality improvements and time to market.

▲ The project team reduced cycle time from 12 to 14 weeks to 3 to 4 weeks, while increasing flexibility and responsiveness to shifting requirements.

▲ The project resulted in a savings of $12.5 million, an ROI of 49 percent, excluding follow-on uses of the models and processes created.

Executive Summary

In 2006, a Xerox engineering team was tasked with developing and delivering a new digital production press that would redefine the industry's standards of color printing.

The new press, dubbed the Xerox iGen4, needed to be delivered quickly. With complex technical deliverables throughout the product, it was clear that legacy processes would not achieve the desired results. A new method of product delivery was needed. Design for Lean Six Sigma was gaining acceptance and momentum. The question was if we could leverage DfLSS to drive real performance improvement in the product and deliver on time.

When a product is brought to market, the development team often builds and tests several iterations. These iterations allow for learning and can be the difference between success and failure. With iGen4, the team estimated that a one-month iteration cycle was required to magnify the learning cycles and hit our launch targets. The problem? This would mean nearly a 70 percent reduction from the historical status quo.

A team of DfLSS Black Belt practitioners was assembled to "shoehorn" Xerographic development into the one-month iteration cycle. The task was daunting. The extended team involved more than 50 engineers, designers, and technicians and at least 30 singular models of critical parameters. The goal was to manage the product optimization and testing process in the most efficient and effective way, while considering cost to Xerox and product performance.

Performance metrics were broken down into two categories: image quality and productivity.

Cost metrics also were broken down into two categories: unit material cost (equipment) and post-sale cost (parts and labor per thousand prints).

The team used the Identify, Design, Optimize, Validate (IDOV) DfLSS method to manage its critical parameters in a Xerox-derived process called CPM[1] (critical parameter management). CPM is an engineering process to flow requirements down from the system to the individual piece parts and then roll capability back up to the system level, providing transparency and alignment in engineering efforts. The details of CPM and the evolution of the modeling work are discussed later in the chapter.

Applying CPM brought two tangible benefits:

1. It refined system performance by addressing customer and business requirements with improved precision and accuracy. It also improved agility by enabling faster concept to implementation cycle times. This improvement was made by reusing data relationships (transfer functions) and streamlining testing of configurations.
2. The resulting transparency into how performance was affected by inputs improved the predictability of the launch schedule and performance.

DfLSS modeling allowed the hundreds of relationships between design factors to be boiled down to just a few system outputs. These outputs were directly tied to customer-critical requirements. As a result, the team was able to align resources and stay focused on the end customer. This alignment was a critical factor in the ultimate success of the project. As the maturity of the model grew, nonvalue-added steps in the engineering process were eliminated. Once a transfer function, the functional relationship between inputs and outputs ($y = f(x)$), was in the model, only the system-level outputs required continuous validation runs. *This reduced complexity in test and validation and reduced test cycle time from 12 to 14 weeks to 3 to 4 weeks per configuration.*

Savings were measured by comparing like numbers of iterations and adjusting cost based solely on the length of each iteration. The result was $12.5 million in savings, a 49 percent return on investment. In addition, the Xerox iGen4, introduced in 2008, is considered the most productive and highest-quality cut-sheet digital press in the printing industry.

In addition to this accomplishment, future benefits of the modeling effort are anticipated because as designs are reused, their transfer functions can be injected easily into other system models. The data and support that come from a proven process and system model can keep an organization's momentum going and capture extended returns on the initial investment.

CPM under the IDOV structure is a powerful method that changes both mechanical processes and behavior, resulting in better alignment with customer needs. The successes outlined in this case study are directly applicable to companies that want to deliver complex products with best-in-class performance using tight resources. The methodology is not specific to Xerox or printing systems. It can be applied to any product that suffers from long development cycles or heavy use of empirical testing to validate performance.

1. Introduction

Xerox's innovative line of digital production presses has redefined the standards of the printing industry. Commercial printers, photo finishers, book printers, direct-mail houses, and digital service providers depend on digital production presses to produce high-end collateral, direct marketing, and photo specialty products—the type of print jobs that typically generate the most profits for print providers.

As Xerox prepared to develop a new-generation product for the portfolio, the Xerox iGen4 press, the team recognized the need to enhance color control and image quality to meet and exceed customer expectations. The project goal was to distill configuration test cycles by nearly 70 percent to enable time to market. No change could affect the customer experience in a negative way. The iGen4 press needed to deliver superior image quality and vibrant colors—building on the huge success of its predecessor, the iGen3 press.

2. Project Background

2.1 Defining the Problem

Consider a bee hive. A hive represents a collective of workers aligned toward a common goal. The hive is efficient and functional. The jobs are simple and straightforward. Workers have similar skills and are interchangeable.

Now consider the development of a complex product. Skills are varied, and workers have deep but narrow expertise. The rapid rate of change prevents system-wide understanding of cause and effect. Testing is extensive, yet rework cycles are prevalent.

As a manager, I faced this challenge. How can we function like a hive? How can we integrate deep knowledge with broad understanding, and do it efficiently? The iGen4 press needed to be delivered quickly, and our legacy processes were not going to get us there.

Within the production color-printing segment, image quality, productivity, and cost are predominant considerations. Each area has its own complexities, so our legacy processes tended to take a "batch" approach to analysis and improvement recommendations. These "batches" would be bundles of fixes and feature content aimed at addressing the goals of the product. A typical cycle, shown in Figure 21.1, included the following basic components:

1. Engineering DOEs (Design of Experiments)
2. Analysis
3. Software requirement development
4. Software implementation
5. Software testing
6. Xerographic testing
7. System integration testing

Typically, the Xerographic test cycle time dominated the critical path, taking up to *80 percent of the usable machine time from a given release.* Given constraints on iteration delivery for software, it was not uncommon for content to slip from one release to another because of a gap in subsystem or integration test data. For example, a design change may pass the subsystem stage but create an adverse effect on another part of the system. This effect would be caught downstream in the system test and create a rework cycle. The time required to complete the rework cycle pushed the content to the next system release.

At the engineer level, we began with just two goals:

1. Improve engineer awareness of each other's work.
2. Make the DOE runs as effective as possible.

At the management level, we expected to realize:

Figure 21.1

A typical iteration cycle.

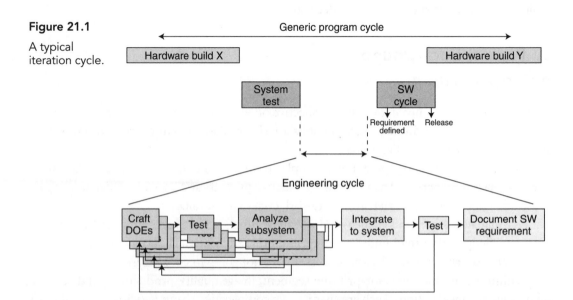

1. Shorter cycle times
2. Fewer rework cycles
3. Better-quality predictive models

In short, we needed to create an environment that found system interactions before we implemented the design change.

2.2   The Proposed Solution

Xerox had been undergoing a transformation in recent years, a refresher, if you will, of statistical tools and methods. This effort, driven by our leadership team, was rolled out as Design for Lean Six Sigma. Within the production color space, and specifically within Xerographics, the program covered nearly 80 percent of the engineering population, with more than 30 percent of them Black Belt–certified. The iGen4 press project was a ripe opportunity to put a holistic method in place.

2.3   The Project Team

The team consisted of subsystem and integration engineers, building transfer functions for their areas. There were also technicians and data clerks running tests and processing print data. Finally, a core process team integrated the transfer functions and created the system-level model. The optimization was run by one of two Black Belts. Requirements and target performance came from the project lead (author), and input ranges came from the subject matter experts for each subsystem. In total, 50 people were involved with the iGen4 press from the Xerographic team, and about half were directly tied into one of the transfer functions in the model.

The development team used DfLSS tools and methods to reduce the length of the testing process.

Voice of the customer surveys were implemented to understand what key audiences expected out of Xerox's next-generation digital production press. The team also analyzed information taken from the database where the iGen3 press customer service calls were logged. This helped them understand the usage trends from active customers and determine where to focus efforts. These steps ultimately played a role in getting the product to market on time, hitting customer quality metrics, and ensuring that the new product improved customer productivity.

Models and simulations were critical to understanding product performance across the entire system. Instead of trying to optimize each subprocess independently, a significant effort was made to examine the ways the subprocesses work together and use that information to optimize the system as a whole. The "holistic" approach enabled the team to develop a product that offered the best possible customer experience.

Transfer functions were the building blocks of the entire process. Simple $y = f(x)$ relationships were born from engineering experiments. More complex representations of the system emerged as these building blocks were fitted together.

T-tests and f-tests were used to understand how configuration changes would affect image quality. Air Academy Monte Carlo simulation tools were used to automate the simulation process, investigating each possible input variation to generate the ideal output in a greatly reduced cycle time.

2.4 Initiating a Culture Shift

The hand-off of technical set points, or the specific values of design inputs, and requirements between teams was the initial focus. The Black Belts on the team began collaborating and quickly realized that churn in the version being optimized by design engineers was outpacing the results of empirical tests, creating confusion and rework cycles. The team decided to build a consolidated model, or single source of record, for the technical set points of the product. By taking this approach, the engineers shifted their focus from what they individually owned to the system as a whole. They recognized that simply optimizing their area of ownership would not be sufficient.

Identify, Design, Optimize, Validate (IDOV) is the most common nomenclature for DfLSS projects. In this case, because such a large percentage of the engineering team was certified in DfLSS, a specific method or use of IDOV was developed, called CPM.[1] CPM encompassed the mathematical tools needed as well as guidance concerning behavior and personal interactions between engineers. An overview of the CPM process steps is shown in Figure 21.2.

Each phase consisted of an exit review, and throughout the process the team constantly iterated and optimized the design to mature the product. The result was a product that followed IDOV as a delivery model but also used IDOV for each phase of the delivery. The

Figure 21.2

Critical parameter management process steps.

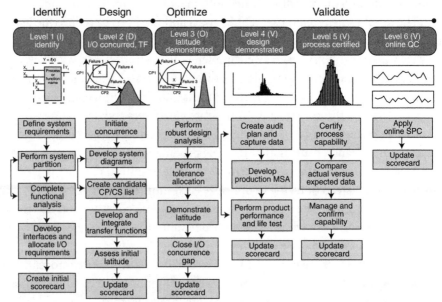

rest of the chapter articulates the steps taken for Xerox iGen4, and highlights some of the transfer function work that was undertaken.

3. Identify Phase

The Identify phase is much like the Define phase in LSS/DMAIC. With the goals of reducing cycle time and the creation of a single source of record, the team needed to start with an understanding of how the system would be put together. With respect to IDOV/CPM, there were two primary objectives:

1. Clearly lay out the target performance in terms that were measurable by our customers.
2. Partition the system to define clear ownership boundaries within the product development team.

These objectives were used to focus attention, not to define optimization levels. Once ownership was established and the identification of the hand-offs was completed, everyone understood who was delivering what and to whom. This map may seem trivial, but it is a critically important factor in becoming lean in product development. The map allows customer requirements to flow to the system, subsystem/module, and piece parts with transparency. Remember, the testing and transfer functions created downstream in the process would need to be integrated. Predefining the interface and inputs and outputs made that process far less cumbersome.

Diving deeper into the Identify phase, we now discuss some specifics. Note that the output of these steps is considered evergreen. Learning does take place in the later phases, and updates to the requirements of a product are common.

3.1 Defining System Requirements

System requirements start with the customer and are translated into engineering specifications through House of Quality (HOQ) or other tool sets. There is a strong tie here to process mapping used in DMAIC projects. The difference here is that the process steps are mathematical as opposed to time-based. In our case, we used HOQ coupled with a solution matrix to maximize customer utility, given the skills and resource constraints of the team.

3.2 Perform System Partition and Complete Functional Analysis

The system partition can be completed in one of two ways: by module or by functional area. In this case, we were focused on a single functional area, image quality, so the appropriate view of the partition was by module. Figure 21.3 articulates the Xerographic modules involved. Note that electrical, software, and other modules also existed but they have been omitted here.

Figure 21.3

System partition for Xerographics.

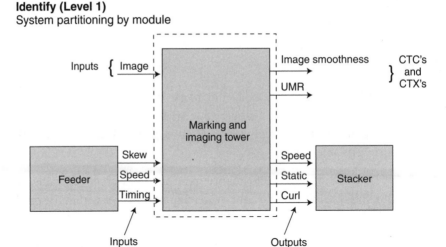

Identify (Level 1)
System partitioning by module

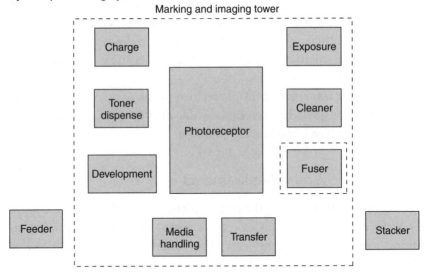

Identify (Level 1)
System partitioning by module

Dashed boxes denote interfaces
Enable interface management and define clear ownership

Image quality was broken down using a FAST diagram, or Functional Analysis System Technique. This diagram, shown in Figure 21.4, described the "How" and "Why" descriptors of the intended system. Using "How" and "Why" statements helps engineers understand how their components fit into the system. These qualitative statements will become the bases of the Input-Process-Output (IPO) diagrams in the next step.

Figure 21.4

Image quality
FAST diagram.

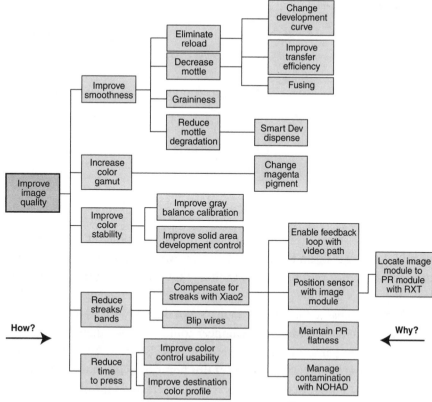

3.3 Develop Interfaces and Allocate I/O Requirements

The next step in the Identify phase is to develop interfaces and allocate requirements. To facilitate this, we refine the FAST diagram so that specific requirements can be tied to the appropriate owner. An IPO diagram was the tool of choice and became the official map of the system.

At the most basic level, an IPO represents one or more transfer functions. Figure 21.5 shows a generic portrayal of an IPO. Note that each transfer function is represented in $y = f(x)$ form in the IPO.

Within a subsystem, the inputs and outputs (I/O) are used to tie piece parts to the subsystem outputs. At the subsystem level, the I/O are focused on hand-offs across teams. At the system level, the I/O are used to tie engineering metrics to customer-based requirements.

The consistency of the IPO generated a number of advantages.

1. Allowed engineers to work in parallel and to link their work later on.
2. Kept the scope well defined.
3. Early DOEs were simple and straightforward, improving their usefulness and reducing rework cycles.

Figure 21.5

Generic portrayal of an IPO diagram.

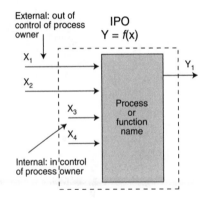

Definition of terms

- Inputs (x's): source of variation that affect the outcome of the process or function
  - Definition of internal or external x's depends on where you draw your context boundary
  - Design factors are inputs that can be defined and set
  - Noise factors are inputs that cannot be defined and set (the process must be robust against)
    - Piece to piece variation
    - Degradation over time
    - Customer usage and abuse
    - Duty cycle
    - Internal environment
    - External environment
- Output (Y): the measureable outcome of the process or function (CTCs)

Figure 21.6 shows an early iteration of the Xerographic IPO. The shaded blocks represent modules (or teams), and the arrows are known hand-offs. The hand-offs also are called intermediate y's, or outputs of one module that input to another module. Historically, these intermediate y's were the scrutiny of optimization and the engineering effort. Design inputs would be set to ensure the intermediate y's met their specifications. Engineering performance was measured based on what each engineer could control—their intermediate y performance. Under CPM, these intermediate y's were the links between simulation models. Note that not all hand-offs were completely defined at this point. This is not uncommon. Until the team got a more complete picture of how the system interacted and what parameters drove system performance, the hand-offs would not be completely determined.

3.4   Create Initial Scorecard

The final step in the Identify phase was to create the initial scorecard. The initial scorecard was nothing more than a list of specifications from the HOQ activity. In addition to the specification itself, we added information around competitor performance, unit of measure, and measurement systems analysis (MSA) maturity. Because the contrast between current and desired performance was not known at this point, the prioritization of the metrics was based on marketing input and engineering risk assessment.

At the completion of the Identify phase, the following artifacts were recorded:

1. Defined set of customer requirements
2. System partition with clear ownership boundaries defined
3. Initial IPO diagram
4. Initial scorecard

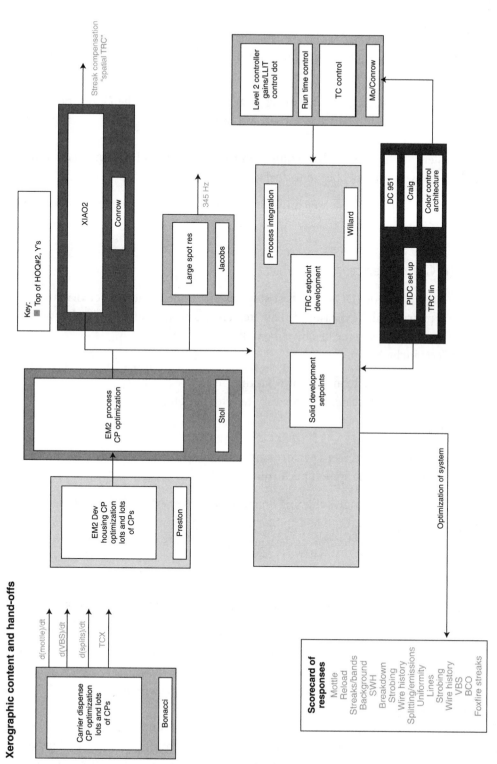

Figure 21.6 Simple view of the Xerographic IPO.

473

4. Design Phase

The Design phase focused squarely on establishing initial performance. Within the Design phase, the following specific outputs were realized:

1. The IPO diagrams were used to construct DOEs. The output of the DOEs included the mathematical relationship between inputs and outputs, or the transfer functions where $Y = f(x)$.
2. These transfer functions were then connected to each other and the first version of the integrated simulation model began to surface.

Similarly to the Identify phase, we chunk the progress toward these outputs into smaller process steps.

4.1 Initiate Concurrence

Within the Design phase of IDOV, the first step toward understanding performance is to concur on what metrics will be measured. This step creates both positive and negative effects on a manager's effort to create the bee hive mentality.

Positive effects include:

1. Everyone has a clear understanding of what they need to deliver.
2. A record of decisions is created for the team to refer back to.

Negative effects include:

1. Reinforcement that an engineer's performance could be measured based on his/her lower-level output, instead of the system performance.

Business theory often promotes aligning financial incentives with metrics that most closely tie to individual performance. Care had to be taken to explicitly break this thought process. Instead, engineers were rewarded for understanding their subsystem's effect on the system performance.

4.2 Develop System Diagrams and Candidate Critical Parameter List

With concurrence understood and the initial IPO diagrams in hand, a more complete system view can be created. System diagrams are the linkage of IPOs, where the output of one module becomes the input to another. Using the concurred metrics, these modules were pieced together until each of the critical hand-offs was labeled with a *measurable output*. Figure 21.7 shows a generic view of how IPOs can be pieced together.

These outputs were used initially to correlate the lower-level transfer functions in the model. Later in the process, these intermediate measures were treated as secondary to the overall system outputs that drove customer satisfaction. The act of constructing this system-

Figure 21.7

Generic
connection of
IPO diagrams.

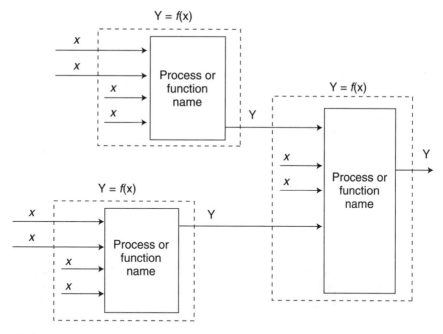

The outputs of one IPO become the inputs to another IPO
creating a multiple IPO view of the system

level IPO in itself helps improve awareness and reinforces the idea that some outputs are not the end state of the engineer's responsibility, that is, debunking the belief that an optimized subsystem is "good enough." Figure 21.8 portrays a more realistic, and complex, view of a system-level IPO diagram.

Critical parameters (CP) are defined as outputs (one or more intermediate y) that have statistically significant impacts on the system-level outputs. Critical specifications (CS) then become the inputs, at the very lowest level, that statistically drive the CP. The system-level IPO helped distinguish and navigate the CSs/CPs of the system.

4.3 Develop and Integrate Transfer Functions

Finally, we reach the point of adding quantitative information to our process. Transfer functions represent the quantitative relationship between inputs and outputs. In this step, we overlay transfer functions onto our IPO diagrams.

A transfer function is a mathematical equation. The format of a transfer function is $y = f(x)$, much the same as the IPOs we discussed earlier. The difference is in the math. An IPO statement might suggest force as a function of mass and acceleration, while the transfer function would include the fact that force is equal to the product of mass and acceleration. Transfer functions, therefore, hold *the level of relevance* of inputs in the expected outcome of the output.

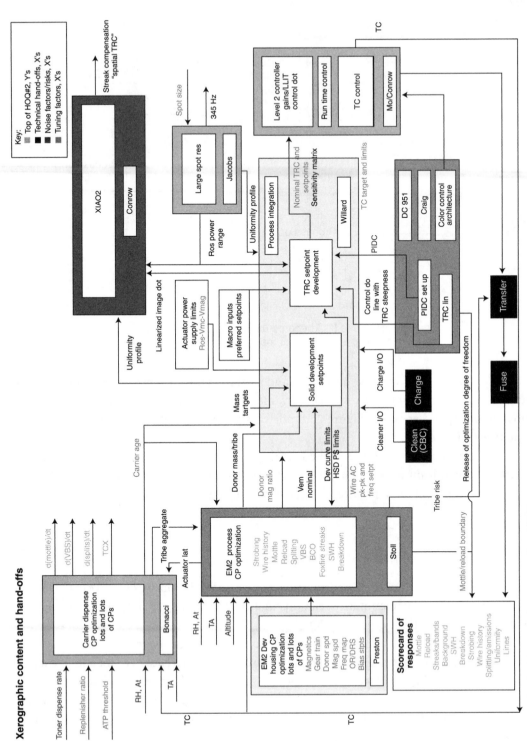

Figure 21.8 Xerographic system-level IPO.

Just as the IPO diagrams were built individually and then merged, so were the transfer functions that represent them. From the system-level IPO, integration engineers were able to develop DOEs and test plans that spanned across modules, identifying interactions between module set points. Transfer functions were typically derived through DOE testing, but also could be established empirically or by first principles, for example, laws of physics.

Each transfer function was proposed, generated, and validated at a local level and then added to what we called "The Big Model." The Big Model spanned across subsystems and allowed variation of CS to drive system-level outputs. The results of these tests moved CS and CP from being considered as candidates to being substantiated. This set the stage for the Optimization phase. An example of a transfer function is shown in Figure 21.9. The developer housing temperature control (DHTC) is an intermediate output of the developer subsystem that affects image quality. The figure shows the regression with coefficients on each input variable as well as the statistical relevance that variable has on the output. The plots show output performance over a Monte Carlo simulation with respect to input variation and proposed specification limits.

From the system IPO, DHTC latitude is an input to the process integration module, and flows through to the system output of image quality. The Big Model absorbed the DHTC latitude information, and the connection between cooling air set points (the CS) and image quality was made.

In total, the Big Model encompassed 30 transfer functions made up of 14 system-level outputs and 16 integrated intermediate transfer functions, 21 input distributions, and 18 noise distributions. The model was complex, to say the least, but elegant in the fact that each piece of the puzzle came from a simple IPO mold. This allowed any participating engineer to navigate the multiple levels, or hierarchy, of relationships between inputs, intermediate y's, and outputs.

The model also allowed weighting of the system-level outputs in the optimization, which then were used to evolve the design. Specifically, when an input or intermediate y participates in multiple system-level output distributions, the relative weights of each output allow the simulation to maximize system-level utility.

The key here is that all system outputs must be viewed together as a portfolio. This approach has two distinct advantages.

1. First, correlation of the model to empirical testing could be completed using system-level outputs. These typically are much easier to measure since they are customer based. The intermediate y's of the system could require added sensors and equipment.
2. Second, there is only one confirmation test, preventing inputs from receiving multiple set-point versions. For instance, a color stability test could show that the best developer housing set-point temperature is 85 degrees Fahrenheit, while the image smoothness test could generate an optimal set point of 92 degrees Fahrenheit. Testing both color stability and smoothness together means that for a single given developer housing temperature, both smoothness and color stability are measured.

Test matrix

Cell	Cooling air temp F	Cooling air temp cfm	Housing target temp F
1	50	10	75
2	50	10	95
3	50	20	75
4	50	20	95
5	60	10	75
6	60	10	95
7	60	20	75
8	60	20	95
9	55	15	85
10	55	15	85
11	60	15	85
12	55	15	85
13	55	10	85
14	55	20	85
15	55	15	75
16	55	15	95

Y-hat model

Factor	Name	Avg PWM Coeff	P (2 Tail)	Tol	Active	Max dev - therm Coeff	P (2 Tail)	Tol	Active	Max dev - tooup Coeff	P (2 Tail)	Tol	Active
Const		142.67	0.0000			-0.07813	0.7158			-4.257	0.0000		
A	Cooling air temp	-22.100	0.0004	1	X	0.72500	0.0795	1	X	1.430	0.0003	1	X
B	Cooling air flow	29.400	0.0000	1	X	-0.77500	0.0139	1	X	-1.40	0.0002	1	X
C	Housing target temp	91.500	0.0000	1	X	-1.005	0.0005	1	X	-3.790	0.0000	1	X
AB										0.68750	0.0414	1	
ABC		-21.667	0.0110	1	X	0.84375	0.0150	1	X	1.677	0.0029	1	X
BB													
R2		0.9795				0.7799				0.9667			
Adj R2		0.9720				0.6999				0.9501			
Std Error		13.7416				0.8394				0.8313			
F		131.0841				0.7443				58.0642			
Sig F		0.0000				0.0013				0.0000			
F		20.195				NA				4.1544			
Sig F		0.5023				NA				0.3645			

Source	Avg PWM SS	df	MS	Max dev - therm SS	df	MS	Max dev - tooup SS	df	MS
Regression	99010.6	4	24752.7	27.5	4	6.9	200.6	5	40.1
Error	2077.1	11	188.8	7.8	11	0.7	6.9	10	0.7
Error	98.0	1	98.0	0.0	1	0.0	0.2	1	0.2
Error	1979.1	10	197.9	7.8	10	0.8	6.7	9	0.7
Total	101087.8	15		35.2	15		207.5	15	

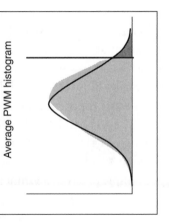

Max dev - therm — Average PWM histogram

At nominal blower speed (10–20 cfm) and the 95F hsg temp target, PWM is railed for a portion of the population.

Max dev - tooup — Average PWM histogram

Lower blower rpm(8–15 cfm) and the 95F hsg temp target, there is more latitude for the controls.

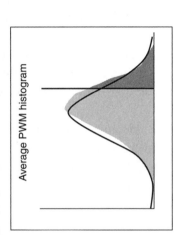

Average PWM
Y-hat contour plot housing target temp versus cooling air flow
Constraints: cooling air temp = 55

Legend:
- 225-250
- 200-225
- 175-200
- 150-175
- 125-150
- 100-125
- 50-100
- 0-50

Cooling air flow: 20, 10
Housing target temp: 75, 95

Temp target has biggest impact on heater PWM. Cooling air is still significant and may help buy some latitude.

Source: B. Phelps, K. Stoll

Figure 21.9 DHTC transfer function.

In the Optimization phase, these outputs were balanced with each other based on their importance, and the best housing temperature was chosen for the portfolio of system outputs.

Another consideration in the Big Model is the control system. Control states can be modeled using the same input/output relationship discussed earlier. This allows the control algorithm to be embedded as a layer of hierarchy in the transfer functions. The initial machine state becomes the input, the post-control machine state becomes the output, and the closed loop performance is now part of the model.

4.4 Assess Initial Latitude and Update Scorecard

The final step in the Design phase is to assess the current latitude. Initial latitude is about identifying gaps in performance against the system-level outputs. Intermediate y's, again, are useful insights, but the target of the activity is to understand current system-level performance against the customer requirements. Initial performance is recorded by testing the full system against variation in CP that the customer base is in control of, often called noise factors from an engineer's point of view. The output data is added to the scorecard developed during the Identify phase (see the section "Create Initial Scorecard"), as shown in Table 21.1.

At the completion of the Design phase, the following artifacts were recorded:

1. Concurred set of requirements at each level (system, module, piece part)
2. System-level IPO with candidate CP identified
3. Initial transfer functions and resulting gaps in performance against customer requirements
4. Updated scorecard

5. Optimize Phase

The Optimize phase is where the true value exists in hierarchical transfer functions. In this phase, the team can *optimize the subsystem-level configuration* and design set points in light of the noise factors and *to achieve desired system-level outputs*. If we look at how this process would have happened before CPM, we'd see the following:

A sequential series of steps, including:

1. Subsystem DOEs for a narrow understanding of design factor impacts
2. Subsystem analysis of data
3. Subsystem optimized set points, with feedback loops based on intermediate y's
4. Subsystem confirmation runs
5. System-level confirmation testing using subsystem set points

Prior to CPM, this sequence would occur for each change. Given a test fleet of 8 to 10 machines and 50 engineers, the configuration change rate was extremely high, and testing the interactivity of each change was not being controlled.

Table 21.1   Scorecard with Initial Latitude Assessment

Delphinus 3.2TTM – Xero. status

- **Scorecard of CTC like elements from Xerographics**

2006 Benchmarks

CTC #	Performance	HOQ#2 Target	Continuous Variable M	Y	C	K	UOM	Type of Data Surrogate	Des. Intent	Gauge R&R
	Smoothness									
2	Mottle	MYC = 17 +/-3, K = 22 +/-3	▲		▲	▲	NMF @ 0.4 OD	X	X	
1	Graininess	no worse than base		Post 3.2 evaluation			VNHF @ 0.3 OD			
	Color Stability									
8	ΔE mc 2 mc	5 (before calibration)		Post 3.2 evaluation			ΔE 2000	X		
9	ΔE run 2 run	1.2					ΔE 2000	X		
7	ΔE w/i run	1.1		▲	▲	▲	ΔE 2000	X	X	
6	ΔE w/i page	<1.5 time zero pk-pk dE cie, <3.0 for all colors up to 200kp life			▲		ΔE CIE		X	
10	Reload	<0.3	▲	GAP		▲	L*	X	X	
	Streaks									
4	VBSv	<1.5 to 2.0 (200K life, no maint.)		▲			L*	X	X	
4	eVBSv	<.5 to .75 (200K life, no maint.)					L*		X	
4	fsVBSv	<1.5 MYCK at 200kp					L*	X	X	
10	harmonic Strobing	0.3		Post 3.2 evaluation					X	
10	Wire History	1					ΔE CIE		X	
10	Super Wire History	1					ΔE CIE		X	
10	Foxfire Streaks	same as base	▲						X	
	Bands									
5	by freq	no visible banding		Post 3.2 evaluation			L*	X	X	fsVBSh
10	fundamental Strobing	<0.15				▲	sRMS	X	X	
5	345 Hz	<0.24				▲	L* FFT	X	X	
	Line Width Growth									
10	2, 8, 24 pixel vertical	40+/-20; 20+/-20; tbd		▲			μm	X	X	in process
10	2, 8, 24 pixel horizontal	40+/-20; 20+/-20; tbd					μm	X	X	in process
	Other									
10	OI registration	mean shift 65 to 40	▲				μm	X	X	in process
10	Spitting	25/9Mp					Spits/Mprints		X	

As mentioned in the Design phase, when the interactivity of CS is not captured in individual tests, it requires complete rework cycles of the entire configuration test and adds waste to the process. Further, if the range on inputs is not equally explored in the Design phase, the transfer functions used are not necessarily valid. Recall the example of the developer housing temperature control. If color stability was tested with a developer temperature range of 80 to 95 degrees Fahrenheit and smoothness tested a range of 90 to 100 degrees Fahrenheit, set points outside 90 to 95 degrees could not be validated without going back and repeating the subsystem tests to confirm the transfer functions were valid over the broader range.

5.1 Perform Robust Design Analysis

During this phase, the configuration optimization was completed. The Monte Carlo simulation used the integrated transfer functions and determined the optimal set of design set points. Remember that the goal of the project was to reduce cycle time by 70 percent. Relying on simulation, in the Big Model, each iteration was now done in hours, and confirmation runs took three to five days. Historically, these steps took weeks to complete. Adding the reduction in rework cycles based on a single source of record for the product design set points, the goal of 70 percent reduction in cycle time was becoming achievable.

The team varied output weights to understand the shifts in set points and ran several use-cases to simulate specific machine performance. Intermediate y's were still measured in these tests but the optimization of the intermediate y was not a priority because the optimum product configuration (choice of set points) was based on overall performance of the system.

Figure 21.10 shows how the system partition and IPO diagrams from earlier phases can play a role in the optimization process. There are two distinct advantages.

1. In the first figure, the partition shows which teams own the outputs. As discussed earlier, defining ownership drives accountability. Optimization means change, and to implement changes in set points or configurations, clear ownership of the required change is a key enabler for speed.
2. In the second figure, the IPO diagram shows how outputs from one team become inputs to another. It also shows inputs that come from external sources. This pictorial view can help articulate the relationship between terms in the model, especially when noise and design factors are highlighted, as shown in the figure.

Visually, the optimization occurred in cycles. The design configuration or requirements were updated, the simulation in the Big Model was completed, and the set points were determined. Those set points were then analyzed, and new interactions or gaps in understanding were uncovered, leading to another iteration of the process. This process is shown in Figure 21.11. Each cycle improved the usefulness, and usually the size, of the Big Model. Because of the rapid learning and significantly reduced testing costs, several cycles could occur within a given hardware and software release.

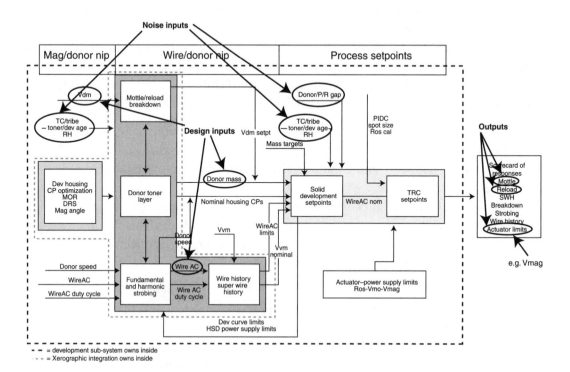

Figure 21.10 Partition and IPOs as seen during Optimization phase.

Figure 21.11

Optimization using the Big Model occurred in cycles.

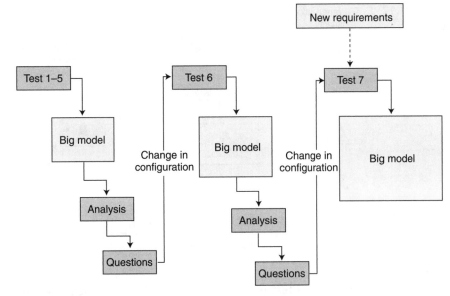

5.2   Perform Tolerance Allocation

Once a system is performing adequately, two questions become critical.

1. How much cost have I incurred to achieve this performance?
2. Where can I reduce cost and minimize the impact to performance?

The first question can be answered through normal methods, but the second requires an elegant solution. Tolerance allocation is a type of sensitivity analysis. The elegance of this process lies, once again, in the hierarchy of the Big Model. With the full interactivity within the system, the lowest-level inputs that drive cost are varied and the movement of the system-level outputs is measured. A cost/performance tradeoff could then be established and decisions could be made regarding the launch design. A visual example of this effect is shown in Figure 21.12.

In reality, given more than 30 system-level outputs, massive intercorrelation, and constant pressure to improve image quality, the idea of excess latitude could be elusive. Often, optimization is about achieving the best configuration given the financial and time constraints of the project.

5.3   Demonstrate Latitude

So far we've created large Excel models and pretty pictures of performance. Demonstrating latitude is when analytical model meets tangible results. Senior managers will want to see proof that your product can do what your model predicts. In this context, demonstrating latitude means showing system-level performance.

As we have seen earlier, not all inputs are within our control. Therefore, the demonstration of system performance must include data points that span the expected demographic of the

Figure 21.12

Tolerance allocation.

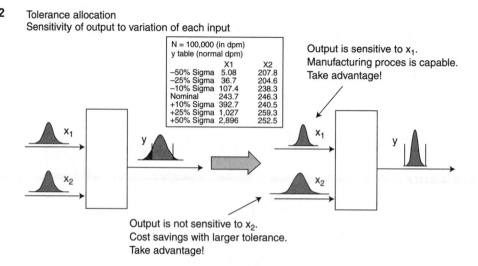

Tolerance allocation

Sensitivity of output to variation of each input

N = 100,000 (in dpm) y table (normal dpm)		
	X1	X2
−50% Sigma	5.08	207.8
−25% Sigma	36.7	204.6
−10% Sigma	107.4	238.3
Nominal	243.7	246.3
+10% Sigma	392.7	240.5
+25% Sigma	1,027	259.3
+50% Sigma	2,896	252.5

Output is sensitive to x_1.
Manufacturing proces is capable.
Take advantage!

Output is not sensitive to x_2.
Cost savings with larger tolerance.
Take advantage!

customer population. This test is completed over a fleet of machines, where system-level outputs as well as their supporting CP are measured. These CPs are then put back into the analytical simulation, and correlation between empirical and analytical results is calculated.

Gaps in expected performance are then prioritized and decisions are based on significance and cost to fix. Figure 21.13 shows an example of the Big Model output distributions. As previously stated, not all outputs will be within specification. However, given the variation in each input and output coupled with the interaction of intermediate y's and weights of the outputs, the best portfolio option can be calculated. If a particular output is not achieving desired performance, rerunning the model with a higher weight on that output can be exercised painlessly, providing flexibility when you can't afford to execute another test cycle.

5.4   Close I/O Concurrence Gaps and Update Scorecard

As was done for each of the prior phases, we close the Optimization phase with a record-keeping step.

Closing I/O concurrence is a documentation step with two distinct components.

1. First, it relieves the subsystem owners of their anxiety. Recall the discussion on individual performance measures and financial incentives. This update in I/O ensures that the targets subsystem engineers are accountable for match those used in the system optimization, reinforcing alignment and driving cooperation.

2. Second, it sets the stage for manufacturing process steps that will take hold going forward.

Subsystem targets are set based on intermediate y performance. These intermediate y's are the measured outputs of the demonstrated latitude tests just completed. This ensures that each

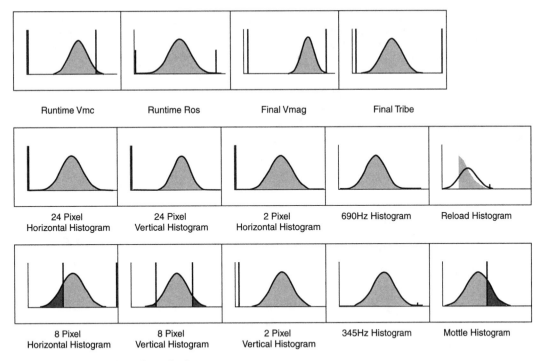

Figure 21.13 Analytical results for system outputs.

subsystem has a target and measure of how they affect the system. This will help debug and control quality as the product is being manufactured.

Finally, as is the case with each phase, we document our findings in the scorecard. The evolution of the scorecard is staggering. The content and maturity reflect the depth of understanding that has been gained, and the tie to customer requirements allows for easy translation to marketing and sales groups who must promote the product.

At the completion of the Optimize phase, the following artifacts were recorded:

1. Optimized set of design inputs and the resulting system-level performance
2. Understanding of how variation in the inputs affects output performance
3. Updated piece part and module specification targets
4. Updated scorecard

6. Validate Phase

The Validate phase often is overlooked, but could be considered the most important of any of the IDOV activities. In this phase, all prior sins in models, correlation, and empirical study come to light.

Key attributes of the Validate phase include statistical sampling and representative demographic testing. Statistical sampling is done at various levels of the design, from the lowest

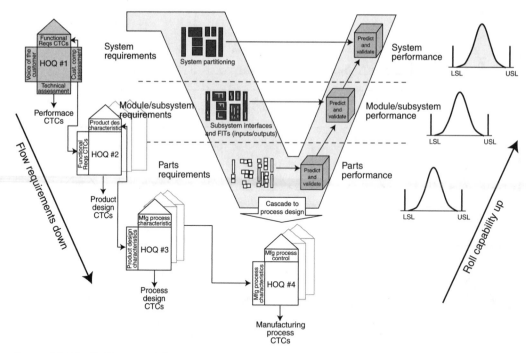

Figure 21.14   Rolling capability up.

piece part all the way up to the system. This progression of "rolling capability up" is shown in Figure 21.14. As data is acquired, the actual distributions replace the assumed variability in the transfer function hierarchy. This step uncovers errors in assumptions as well as in modeling accuracy.

When the customer base is heterogeneous (meaning the way customers utilize the machine varies greatly), the validation testing must take into account the specific customer types as use-cases. Internal testing targets the customer types directly, and then is coupled with external beta tests. Results are correlated and gaps are fed back to the design teams, where they are prioritized and worked as required.

During validation is when the awesome power of the Big Model is restated. As gaps in performance are uncovered, the model begins to serve as a map of the system. An output can be traced back through the hierarchy of interactions and module performance metrics to debug the source of the problem. This "guide book" *explicitly reduces debug time and accelerates fix response and flexibility.*

7. Benefit Realization

The benefit of CPM and hierarchical transfer functions takes many forms, all of which lead to financial gain. Three subsets are worth discussing in brief detail.

Requirement migration is something that all companies face to some extent. Market requirements shift as economic conditions, competition, and technology evolve. A simulation tool that mimics product performance allows an immediate understanding of how the current product would perform under new demographics, without capital expense in new equipment or testing. The model also can narrow the scope of design change if current product performance needs to be augmented to reach new markets.

The second benefit is in the opportunity cost of engineering time, in particular, when specific skill sets are in short supply. Faster iterations in testing enable engineers to either add more iterations on the current product or work multiple projects in parallel. This "do more with existing resource" strategy leads to efficiency gains that allow more projects to get completed.

Finally, time-to-market benefits can be achieved when the test and integration cycle is the critical path to program launch. The team can achieve target performance in fewer iterations and accelerate the product schedule. Faster time to market means increased market share and revenue.

In this project, complexity in test and validation test cycle time was cut from *12 to 14 weeks to 3 to 4 weeks per configuration*. Dollar savings were then calculated, holding product performance constant. That meant that we held the number of design cycle iterations constant and compared the cost of long and short design cycles. The result was $12.5 million in savings, a 49 percent return on investment. In addition, the Xerox iGen4, introduced in 2008, is considered the most productive and highest-quality cut-sheet digital press in the printing industry.

8. Future Uses, Lessons Learned

The CPM process is applicable across many other systems and products. Specific implementation of the process will depend on the culture of the firm and complexity of the product in development. The overarching process is explicitly applicable.

In the case of the iGen4 press, CPM was predominantly used to optimize image quality. In the future, CPM could be expanded to include productivity, cost, or other customer and business metrics of interest. Beyond the iGen4 press, the team now has experience in hierarchical transfer functions, and has seen the benefit of connecting and modeling the entire system. Future products developed by this engineering team will continue to use the methods discussed here and further improve its effectiveness.

Key lessons of the project relate to:

1. **Modeling and simulation tools.** Modeling and simulation tools have the power to significantly reduce testing cycle time by producing accurate results and validations in minutes. Modeling and simulation tools are powerful when used correctly. While all models have limitations, using the models within the bounds of those limits can save time and money.

2. **Critical mass in skills.** Critical mass of skills refers to sufficient knowledge about Six Sigma and its statistical tools, especially DOE, transfer functions, and robust design. If engineers and managers are not educated in Six Sigma tools, they can stall or even derail the effort. Control over the methods and objectives is vital. The team should have as high a Green Belt certification rate as possible. Black Belts should be a part of every module and stationed throughout the integration team—especially where the transfer functions are linked and the Big Model is created.

3. **Having the right people in the right jobs.** Having the right people in the right jobs requires both technical skills and leadership skills.[2] Technically, the owners of the system-level model need to be respected and have influence within the engineering ranks. Key management positions must be filled with leaders who are willing to shield the team from negativity and naysayers. The leaders must be willing to celebrate successes and continually reinforce the vision to help maintain momentum. The leadership also must be committed to communication and visible participation.

4. **Focusing process overhead so it pays back the highest return.** Process rigor can be costly. Applying that rigor in areas that do not require it is a great way to defeat your efforts. CPM and hierarchical transfer functions add value to complex systems, where interactions between modules are high and engineers are already struggling to keep the inputs, outputs, and design constraints straight.

These areas will see value in utilizing DfLSS, IDOV, and CPM. Engineers and managers responsible for simple modules will view these tools as excess overhead, and will not return on the investment required. It is completely acceptable to focus hierarchical transfer functions narrowly where they have the most benefit, at least initially, and grow the model and pervasiveness over time. In fact, utilizing the learning cycles of how your organization adapts to and internalizes CPM helps ensure your return continues to be positive.

9. Conclusions and Limitations

As with any approach, there are potential limitations. Modeling and simulation is not a cure-all, nor can these models perfectly mimic the systems they represent. They can be extremely useful, however, in breaking down complex systems into manageable chunks. The resources required to build and integrate hierarchical transfer functions, or a "Big Model," are significant, and often exceed what many managers expect. As we have shown here, the payoff is lean, flexible, and rapid deployment of design change during development and traceability during production and service after the sale.

It is important to note that while these transfer functions are powerful, they are only applicable over the range of input variation they were constructed upon. Extrapolation of transfer functions to design space beyond the tested range is the most common pitfall. This can have disastrous effects, causing extensive rework. Managers must be vigilant in understanding the content and applicability of the simulations to their engineering teams.

Finally, the CPM approach in support of hierarchical transfer functions is a behavior, a definition and alignment of goals across a large group of engineers and managers. CPM, and pervasive use of DfLSS, is hard to implement effectively, and not for the casual user. The return on investment is substantial, but when order-of-magnitude improvements in efficiency are required, DfLSS and CPM can be an appropriate answer.

Acknowledgments

The author of this case study is thankful to Xerox for its permission to allow publication of this case study as a book chapter. The author would also like to thank Elizabeth Cronin, Brad Willard, Mike Martin, and the iGen4 Xerographic team for their guidance, brilliance in modeling, leadership, and cutting-edge engineering, respectively.

10. References

1. Hildebrand, E. G., "Critical Parameter Management Process", Property of Xerox Corporation; Xerox Corporation, Webster, New York 14580; Xero Corporation. All rights reserved; 2006–2008.
2. Hildebrand, R., Manager Need Know-How, iSixSigma Magazine Open Mike article; *iSixSigma Magazine*, September/October 2008.

CHAPTER 22

How Seagate Technology Reduced Downtime and Improved Availability to 99.99 Percent

Peter J. Clarke

Relevance

This chapter is relevant to IT organizations that support a 24/7 manufacturing facility or business operation and who want to minimize disruption to business due to application downtime.

At a Glance

▲ Throughout the considerable growth of the Seagate Technology Springtown manufacturing facility, certain components of the IT infrastructure were not capable of coping with the changes. This led to service availability issues for customers (manufacturing and process engineering), which in a 24/7 plant compromised both opportunity and revenue

▲ The Springtown IT organization took the important step of initiating an IT infrastructure control strategy to provide a roadmap for the implementation of a number of data capture, monitoring, and analysis systems to help maintain service levels and predict future performance.

▲ The strategy resulted in improved service availability (almost 100 percent!) of the IT services, providing the opportunity for the organization to achieve its customer commitments and to maximize revenue.

Editor's Note: This project is an example of how close the different Six Sigma methodologies truly are. This project started out as a DMAIC project with the intent to improve a problem: low availability. However, as the nature of the problem was revealed and as solutions were identified, it transitioned to a set of tasks that might have fit the DMADV methodology better.

This happens sometimes as discovery progresses throughout a project. In this case, the project remained in its initial DMAIC format, though many argue that it is more suited to DMADV because of the tools and solutions that ultimately resulted. When this type of situation occurs, it is the call of the project leader whether to continue in its initial model or to switch to a better-fitting lifecycle model.

To reconcile this project into the DMADV model, note that the DMA (Design, Measure, and Analyze) sections are virtually identical in both DMAIC and DMADV models. However, the Improve phase in the DMAIC would map to the Design phase of the DMADV as new solutions are identified. Similarly, the Control phase of the DMAIC would change to the Verify phase of the DMADV as the new metrics system was implemented and tested.

Again, this project could be showcased under either DMAIC or DMADV sections, but it has been placed here in the DFSS/DMADV section because the changes were not so much incremental improvements as much as they were a complete overhaul of the metrics and monitoring system.

Executive Summary

IT departments throughout Seagate Technology worldwide use disciplined methodologies, data collection, and root-cause analysis to improve and control processes. This chapter provides a case study from the IT organization in Springtown (Northern Ireland), which in 2005 developed a strategy to quantify and better understand the impact that the IT infrastructure had on the satisfaction of its customer (Seagate Technology manufacturing and process engineering).

The purpose of the strategy was to implement efficient IT support processes along with supporting tools and data to ensure that as new business products were introduced and supporting tools and processes changed, the IT infrastructure did not become a bottleneck in the Seagate Technology manufacturing process.

The IT infrastructure is an integral part of the support of the manufacturing processes at Seagate Technology. Certain IT infrastructure failures, such as those related to network, storage, databases, servers, or applications, have the potential to bring the whole manufacturing facility down. Historically, there have been incidents where IT infrastructure failures had stopped the factory and it had taken a disproportionate amount of time to pinpoint the root cause prior to resolution. This Seagate Technology factory runs 24/7, and every minute of downtime represents lost opportunity in disk drive sales and lost revenue.

By implementing the required IT processes, measures, and controls, this strategy aimed to eliminate IT downtime, and where that was not possible, to minimize the time to repair and get the factory running as soon as possible.

A further salient aim of the strategy was to ensure that the metrics that were to be implemented would help pinpoint root cause(s) and to prevent any recurrence of an outage. Using the Information Technology Infrastructure Library (ITIL) methodology, it is clear that in the event of a factory outage, the first priority is the incident management process to get the factory up and running again. Equally important is the problem management process to identify the root cause of an incident to ensure that it is never repeated. To paraphrase Oscar Wilde, *to take the factory down once may be regarded as a misfortune, to do it twice, for the same root cause, looks like carelessness.*

The scope of the strategy was far more than would normally be tackled in one Black Belt project, in that it spanned the entire IT infrastructure. However, the preferred approach was to develop a standard process that could then be rolled out for all of the infrastructure components. It was during this rollout that the project team (initially Black Belt, Brown Belt, and IT managers) expanded substantially as specific areas were "subcontracted" to the subject matter experts with guidance from the project team.

From the outset, there was a clear need to fully understand the high-level relationships between all services provided by IT and all the underlying infrastructure components (servers, storage, applications, network, and databases) to analyze all the strands of the strategy. The approach was to first map out the relationships between the factory-critical services and the IT infrastructure components. These relationship diagrams helped to clarify how each of the services was supported. This paved the way to drill down to each of the components for metrics capture and analysis before rolling back up to service performance.

As the service performance metrics were already in place, the focus was on each of the IT infrastructure components and their critical health indicators. The key process input variables (KPIVs) for each component were prioritized and correlated with service metrics and business drivers. The KPIV metrics and their associated correlations formed the basis for the capacity planning required to support the changing business environment.

The key outcome of the project was an improved understanding of the relationship between the changing business environment, its impact on the IT infrastructure, and the associated change in service availability/performance for the business. This information is updated on a regular basis to ensure that the IT infrastructure is continuously enhanced and is capable of supporting the increasing business requirements without affecting service performance/availability.

1. Introduction

The IT Infrastructure Library (ITIL) underpins this project. This best-practice framework for the provision of quality IT services provides a systematic approach to service management. ITIL defines service management as "a set of specialized organizational capabilities for providing value to customers in the form of services."

The IT department at Seagate Technology in Springtown provides factory-critical services to its Seagate Technology business partners, each of which has a direct impact on the ability to manufacture and ship good-quality recording heads. If any of these services is unavailable, production will either stop completely or be critically affected at certain stages.

Service value to a customer is driven by two factors. First, a service must perform what it was designed to do, that is, be "fit for purpose." Second, it must also be available when needed and perform as expected, in other words, be "fit for use."

To ensure that a service is "fit for use" and to validate customer satisfaction, monthly reviews take place with the business partners to review the service level agreement (SLA). The SLA contains details of the service, its components, and performance and availability targets agreed upon with the business partners. Historically, during SLA reviews there were far too many occasions where the service was failing to meet the agreed availability targets. Achieving the SLA targets is dependent on the capability of the underlying infrastructure. The key challenges are to ascertain which of the underlying infrastructure components affect the services and to understand how each impact manifests itself.

The Seagate Technology plant in Springtown is spread across two buildings within which there are 12 factory-critical services, which are underpinned by a complex network and storage infrastructure: 20 Oracle databases, 95 servers (Unix, Sun, and Windows), and more than 100 applications. Understanding the complexities and interactions of the IT architecture and their business impact in a changing business environment is not a short-term assignment. Therefore, an IT Infrastructure Control Strategy (IICS) was implemented to map out and fully capture, monitor, and model the metrics over a five-year period.

2. Project Background

It is not a pleasant exercise to chair quarterly presentations with the customers to explain why IT is the root cause of manufacturing downtime. It is certainly not one that any quality IT organization should be doing on a continuous basis. Hence, this project was initiated by the IT director at the Springtown facility. He appointed an IT Master Black Belt to lead this initiative, with the support of the local IT organization and business organizations.

A DMAIC project was initiated to address the problem, explore root causes, and implement appropriate solutions.

3. Define Phase

Due to the size and complexity of the project, there was a risk that the scope could become too large and the required outcome would never be achieved. The Define phase of the project helped to focus the team on the key area of the process for data capture, analysis, and predictive modeling prior to a rollout across all the individual components.

3.1   Six Sigma Project Charter

The IT organization in Seagate Technology is a service department supporting the manufacturing process in a 24/7 facility. The department has a target of 99.5 percent availability for each of the factory-critical services (although 100 percent is expected). On a quarterly basis, IT's performance against the SLA is reviewed with business partners, and IT is required to explain why any service did not achieve 100 percent availability.

3.1.1 Business Case Statement

As the manufacturing process grew in volume and complexity, the IT infrastructure struggled to maintain 100 percent availability. This problem was expected to get worse with the predicted increases in the manufacturing schedule and the increased length of the manufacturing process with the new products (from 450 steps to 800 steps). In order to support the changing business environment, the complexity of the IT infrastructure increased, and this, in turn, compounded the challenge of achieving 100 percent availability. Every minute of factory downtime not only disrupts the manufacturing process but it also results in additional recovery time in relation to any tool interruptions. Wafers that were in the manufacturing process prior to the disruption must also be restaged.

3.1.2 Opportunity Statement

The monitoring and reporting of IT service performance to the business organization highlighted an opportunity to improve the availability of the services to the business. Availability figures of 99.5 percent represent 11 hours of downtime to the business every fiscal quarter. The Service Availability Report in Table 22.1 shows that the target of 99.5 percent availability was not achieved for a number of services.

3.1.3 Goal Statement

The project goal was to "increase service availability to the Seagate Technology business partners to 99.99 percent by improving the monitoring and capability of the IT infrastructure by August 1, 2009."

The project due date was reviewed and agreed to with the project team and the project sponsors, given that extensive preparation work was necessary prior to the implementation of any monitoring systems and infrastructure improvements.

3.1.4 Project Scope

The scope of the project was to:

1. Develop and embed a process within the IT organization to monitor and control the service and component performance indicators.
2. Correlate the IT performance data with the associated business drivers to predict potential breakpoints in IT service performance before they affected the business.
3. Implement permanent corrective actions to sustain these improvements in availability.

Table 22.1 Service Availability

Service	Harm Class	FY07 Q3 Overall Availability	FY07 Q3 No. Incidents	FY07 Q4 Overall Availability	FY07 Q4 No. Incidents	FY08 Q1 Scheduled Downtime	FY08 Q1 Unscheduled Downtime	FY08 Q1 Overall Availability	FY08 Q1 No. Incidents	FY08 Q2 Scheduled Downtime	FY08 Q2 Unscheduled Downtime	FY08 Q2 Overall Availability	FY08 Q2 No. Incidents
FIS													
WIP reporting	BC	99.87%	1	100.00%	0	0	270	99.78%	1	80	540	99.53%	1
Recipe/spec management	BC	99.87%	1	100.00%	0	0	270	99.79%	1	80	0	99.94%	0
Tool/operator interface	BC	**99.13%**	**5**	100.00%	0	20	270	99.78%	1	80	0	99.94%	0
WIP tracking	BC	99.87%	1	100.00%	0	0	580	99.56%	**3**	80	0	99.94%	0
Dispatching	BC	99.87%	1	100.00%	0	0	270	99.79%	1	80	100	99.86%	1
Process control	BC	99.87%	1	100.00%	0	0	270	99.79%	1	80	0	99.94%	0
Equipment reporting	BC	99.87%	1	99.98%	1	0	315	99.76%	2	80	20	99.92%	1
Equipment tracking	BC	99.87%	1	100.00%	0	0	450	99.66%	2	80	0	99.94%	0
Process reporting	BC	99.87%	1	99.98%	0	0	270	99.79%	1	80	0	99.94%	0
Data transfer	MC	100.00%	0	1000.00%	0	0	270	99.79%	1	0	0	100.00%	0
E-messaging													
Mail	BC	99.87%	1	99.97%	1	0	15	99.99%	1	0	0	100.00%	0
File and print													
File and print	BC	99.86%	3	99.98%	2	0	0	100.00%	0	0	0	100.00%	0
LAN													
LAN	MC	99.87%	1	100.00%	0	0	0	100.00%	0	0	0	100.00%	0
WAN													
WAN	MC	100.00%	0	100.00%	0	0	0	100.00%	0	0	0	100.00%	0
Telecom													
Telcom	DI	100.00%	0	100.00%	0	0	0	100.00%	0	0	0	100.00%	0

Note that 12 factory-critical services are routinely monitored and that they are tracked by percent downtime and minutes of downtime per quarter.

3.1.5 Project Plan

The project plan was set up as part of a long-term strategy to fully automate the control, predictability, and optimization of the IT infrastructure supporting the manufacturing processes at Seagate Technology in Springtown.

The high-level project plan in the charter included milestone dates for the end of each of the phases in the DMAIC lifecycle of the project: Define, Measure, Analyze, Improve, and Control.

This plan was further detailed to help the SMEs focus on their specific key metrics.

3.1.6 Project Team

This project encompassed every area of IT and every area of the Seagate Technology business. The project Champion was the IT director at the Springtown facility. The project team also included an MBB and subject matter experts (SMEs) in a diverse number of areas who contributed throughout the project.

The SMEs from IT comprised:

- ▲ Unix engineers
- ▲ Network engineers
- ▲ Windows engineers
- ▲ Database administrators
- ▲ Software engineers
- ▲ Storage engineers

The SMEs from business comprised:

- ▲ Manufacturing engineers
- ▲ Manufacturing managers
- ▲ Equipment engineers
- ▲ Process engineers

3.2 Detailed Project Plan

The detailed project drilled down into each of the phases of the project and listed start and end dates along with who was responsible for each of the underlying tasks.

4. Measure Phase

The starting point for this strategy was to use a top-down approach by performing configuration management and aligning the components (assets) to the services. Next, the project team researched best practices from industry for service performance metrics and worked with the component SMEs to identify appropriate service performance metrics. A risk assessment on these metrics was performed to identify those that were most likely to affect service performance.

The initial steps focused on mapping out the relationships between all of the IT services and their underlying components (servers, network, applications, databases, storage). Figure 22.1 identifies all of the IT infrastructure components managed by the Springtown IT organization that support the 12 factory-critical services.

For each of the 12 factory-critical services and each of the component categories (servers, network, applications, databases, and storage) the project team created a relationship diagram, shown in Figure 22.2, to identify, at a high level, potential areas of interaction.

The relationship diagram highlighted a number of focus areas that needed to be considered when brainstorming for potential inputs. Brainstorming sessions were held by the SMEs to identify the KPIVs that could potentially be monitored for their service or infrastructure component. This process, in most cases, provided a large number of KPIVs, which was a good starting point, and in order to narrow down the KPIVs for detailed analysis, the project team and the SMEs performed an FMEA risk assessment. Table 22.2 shows the top portion of an FMEA performed for one of the factory-critical databases. The KPIVs were listed as the "Task/Function" and the potential failure mode for each KPIV was listed. The severity score

Figure 22.1

Service-component relationship.

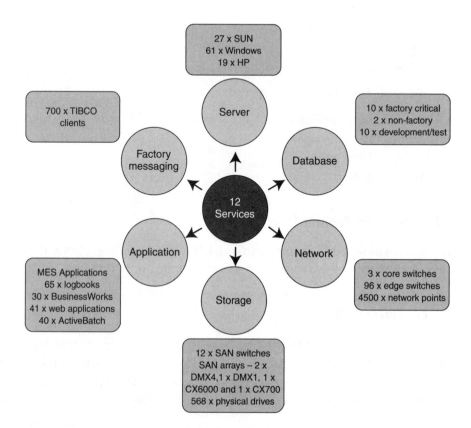

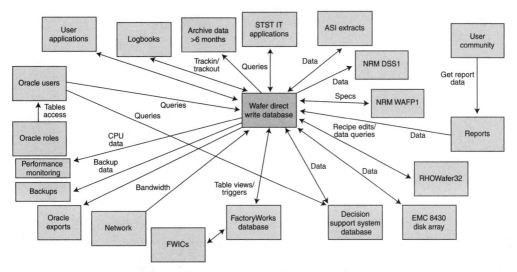

Figure 22.2 Relationship diagram for Seagate Technology parametric database (wafer direct write).

selected was based upon the impact to the business, the likelihood of occurrence based on historical data, and the controls based on existing monitoring systems or IT processes.

The KPIVs having the highest risk priority number (RPN) in the FMEA were prioritized, and plans were put in place to capture these KPIVs. One major factor that the project team had to consider when capturing this large amount of data was to plan for the Analyze phase. To aid the analysis, a consistent sample rate for each of the metrics was defined, and a standardized set of database tables was designed to provide a centralized store of data.

The standard sample rate was set at one hour to give sufficient granularity to highlight trends and be able to correlate against the business metrics (factory moves, product type, product complexity, number of equipment sets, amount of wafers in the line, and yield percentage). For troubleshooting purposes, monitoring tools captured the performance in more granular fashion, but to iron out the short-term spikes and be able to correlate trends, a sample period of one hour seemed to be optimum.

In a large number of cases, the key inputs (KPIVs) for components that were deemed to pose the greatest risk to service disruption (based on the FMEA) were either not being monitored or the data was inaccessible and could not be used for analysis. A considerable effort was invested in capturing the KPIV data and storing it in the centralized database. This was an important milestone, and allowed us to set limits on the KPIVs and provide alerts when limits were breached. The tracking of KPIVs facilitated the early identification of changes in IT performance, but this still didn't provide any predictive modeling to facilitate capacity planning.

Basically what is being done here was identifying selected measurements and data elements, defining them, defining operation procedures for collection and storage, and defining operational procedures for analyzing them. This is basically a measurement plan, and it is a key step of the Measurement stage.

Table 22.2 FMEA for Seagate Technology Parametric Database (Wafer Direct Write)

Process or product name:	Wafer direct write database	Prepared by: Ann Marie MacDonald	Page ___ of ___
Responsible:	Ann Marie MacDonald	FMEA date (Orig) ___20-Oct-2008___ (Rev) ___ A___	

Task/function	Potential failure mode	Potential failure effects	SEV	Potential causes	OCC	Current controls	DET	RPN
Find blocks in memory 99% of the time (buffer cache hit ratio)	Slow return of data/slow application, got to go to disk	More reads from disk, slower application, the database has a bigger overhead, database crash	8	SQL out of control	8	Change management	3	192
Count the amount of sessions which are actively using CPU (active sessions using CPU)	Application slow, database slow	More the number of sessions using CPU, more likely the transactions will have to contend for CPU	8	Existing applications being used by more users	6	OEM session count	4	192
Active sessions waiting on information	Slow read and write of data from disk	Slow database, slow application	8	Poorly written SQL, e.g., full table scan	6	Change management	3	144
Find blocks in memory 99% of the time (buffer cache hit ratio)	Slow return of data/slow application, got to go to disk	More reads and writes from disk, slower application, the database has a bigger overhead	8	More steps	5	Factory plan	3	120
Active sessions waiting on information	Slow read and write of data from disk	Slow database, slow application	8	Queue for frequently accessed blocks	7	OpenView, stats pack	2	112
Count of the amount of sessions which are actively using CPU (active sessions using CPU)	Application slow, database slow	The longer the transaction holds onto the CPU the slower the transactions will be	8	SQL running less efficiently so holding onto CPU for longer	6	SQL respond time, session_long_ops_table	2	96
Active sessions waiting on information	Slow read and write of data from disk	Slow database, slow application	8	Heavily used parts off disk	6	OpenView, stats pack	2	96
Find blocks in memory 99% of the time (buffer cache hit ratio)	Slow return of data/slow application, got to go to disk	More reads and writes from disk, slower application, the database has a bigger overhead	8	Increase in WIP	3	Factory plans	3	72
Find blocks in memory 99% of the time (buffer cache hit ratio)	Slow return of data/slow application, got to go to disk	More reads and writes from disk, slower application, the database has a bigger overhead	8	Slow to read and write data from the SAN	4	OpenView	2	64
Find blocks in memory 99% of the time (buffer cache hit ratio)	Slow return of data/slow application, got to go to disk	More reads and writes from disk, slower application, the database has a bigger overhead	8	Buffer full—workload bigger than database can manage	2	Change management	3	48

To help ascertain the business drivers that had driven changes in IT performance, it was necessary to work closely with the business partners to gather their business drivers (schedule, product mix, product complexity, equipment sets, wafers in the line, and yield).

Essentially, this helps define the relationship/correlation between KPIVs, service performance data, and higher-level business drivers. Collecting both the business and IT metrics provided the baseline data to allow the project to enter the Analyze phase.

The Measure phase ended with a formal milestone review that involved a presentation of phase results to the project governance team.

5. Analyze Phase

The approach described previously offered only a reactive paradigm. A more proactive approach was necessary, one that would connect the IT and business metrics to the customer satisfaction factors. How could a predictive model be defined that could be used for capacity planning? The consistent one-hour sample rate for the services and components metrics meant that correlations were "easier" to generate.

When analyzing IT performance over a long period, correlations can be made between the customer satisfaction factors, the IT system performance, and the changes in the business environment (e.g., increased activity). Using regression analysis, these correlations were built up to facilitate understanding of historical degradations or improvements and provide the formula to predict future performance. The correlations needed to explain the effect that business drivers had on the IT infrastructure and how changes in the performance of the IT infrastructure affected the customer experience.

The project team focused on one of the key business drivers: the number of process and metrology steps (known as moves) executed in the factory each day. The analysis in Figure 22.3 shows an example of a strong correlation (80 percent) between moves in the factory and the CPU of one of the factory databases (wafer direct write). This analysis was repeated for moves versus each of the other IT infrastructure components to provide a list of all components affected by a change in factory moves.

Now that we had identified the components (IT infrastructure) affected by the number of moves, a further analysis was performed to assess how the IT infrastructure affected the factory services (customer experience). Figure 22.4 shows a correlation identified between one of the components (wafer direct write database CPU) and the customer experience (performance of the process control service). Note that the wafer direct write database correlated to a number of services, but for clarity we will just consider this example.

The correlation in Figure 22.4 indicates that the customer experience is affected by the IT infrastructure. The correlation gives a 62 percent relationship, which indicates that other factors are influencing the service but the database CPU is a major factor. We can relate this to the previous correlation that the IT infrastructure is affected by factory moves. This suggests the hypothesis that the factory moves affected the wafer direct write database performance,

Figure 22.3

Correlation between factory's daily gross moves and database CPU.

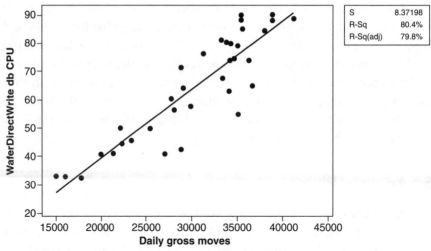

Fitted line plot: WaferDirectWrite db CPU = −9.031 + 0.002422 Daily Gross Moves

Figure 22.4

Correlation between process control service (customer experience) and database CPU.

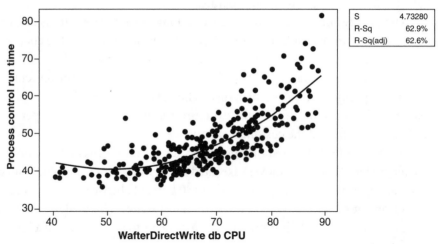

Fitted line plot: Process Control Run Time = 81.56 − 1.641 WaferDirectWrite db CPU + 0.01637 WaferDirectWrite db CPU **2

which in turn affected the process control service. To validate this, an analysis was run to correlate moves directly with the process control service.

Figure 22.5 demonstrates how the fitted line plot was used in Minitab to correlate the number of daily moves in the factory with the performance (customer experience) of the process control service.

In the fitted line plot there was an 86 percent correlation between the process control service and the factory moves. This correlation also provided an equation to use as a predictive model for capacity planning:

Process control = 2059 − 0.01752 Daily Gross Moves + 0.000001 Daily Gross Moves ** 2

Figure 22.5

Correlation between process control service and factory daily gross moves.

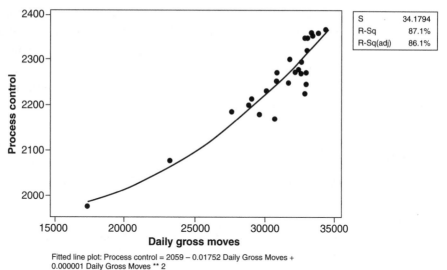

Fitted line plot: Process control = 2059 − 0.01752 Daily Gross Moves + 0.000001 Daily Gross Moves ** 2

In addition to correlating the data, some extra factors in the form of "event data" needed to be considered when developing the models. Since the business and IT environments are continually evolving, there is a need to be aware of "event data": certain events will occur in the environment that cause a large fluctuation in the performance of a service. This could be a patch rollout, a new software release, or a database backup. All of these events also needed to be captured in the change management process in order to understand their impact and, where necessary, exclude them from the predictive model correlations.

The predictive model from Figure 22.3 is one example of a business driver (daily gross moves) with an impact on a customer service metric (process control performance). These and other analyses were performed for each of the customer service metrics against each of the business drivers and IT component KPIVs. In a large number of cases, there was no correlation, but where one or more inputs affected the customer service metric, that relationship (predictive model) was recorded. This set of predictive models has been the foundation behind improved service availability at the site and the ongoing capital planning.

One of the main business drivers for the Springtown factory is the production schedule. This defines the plan for the factory in terms of output, product mix, and process steps.

Using the predictive models in conjunction with the production schedule meant it was possible to predict the performance of the IT systems and their subsequent effect on customer satisfaction. Since the business drivers (in this case, the production schedule) were going to change, the factor that IT could influence was the IT infrastructure.

Such predictive modeling helps to justify investment in future capital, or, where capital investment is not available, makes all stakeholders aware of the potential risk to the business and gives them the opportunity, where possible, to mitigate the risk. The mitigation might be through a workaround, relieving some of the pressure from the infrastructure at risk, or the business partners may make an informed decision to run the process "at risk" and deal with incidents as they occur.

The long-range plan (LRP) of the production schedule for the business provided the figures to fit in to the predictive models, which facilitated the prediction of when the capability of a system would be exceeded. This in turn fed the organization's capital plan, with the analysis providing the justification for the new capital. It also provided the ability to illustrate the impact on the factory if no improvements were implemented and predict the date at which a certain piece of infrastructure would fail.

6. Improve Phase

Over the years, as the site has evolved, it has experienced a number of major factory schedule increases. In summer 2005 the Springtown site ramped up its manufacturing output by 30 percent, combined with a more complex product (process steps increased by 40 percent). This meant an increase in factory activity by roughly 50 percent. These changing business parameters were fed in to the predictive models to highlight infrastructure that would not be able to cope, either through IT performance or customer experience.

Over US\$1.2 million was invested in new storage, database, and server infrastructure in advance of the manufacturing output increase. This ensured a continuous high service level to the business throughout this ramp. In 2006 the schedule increased once again, this time by another 30 percent, and the predictive models were once again used to predict the impact on the IT infrastructure and the required upgrades, or risks to the business if the infrastructure was not upgraded. This process has been repeated on an ongoing basis to support the business for revised schedules, new products, and new functionality.

As KPIVs are monitored through performance analysis, they indicate that components are starting to struggle. Crucially, the correlations of the KPIVs with the customer performance metrics provide a model that tells us when a component is likely to fail. These two analysis strands may be combined to generate a list of IT infrastructure that is at risk of constraining the manufacturing process. The list of potential constraints was then prioritized using the following criteria:

▲ Severity
▲ Likelihood of occurrence
▲ Urgency (a combination of when an item is likely to fail and how long it would take to mitigate the risk)

The Springtown IT organization continues to add IT capability into the Springtown wafer factory to meet our business partners' ongoing requirements. IT complexity (and the demand on the service infrastructure) increases as a function of time, irrespective of business complexity and activity increases. Therefore, excess IT infrastructure has effectively been consumed by new functionality added to the IT infrastructure over time. All of this increased business complexity and activity (current and planned) has caused instability and outages in

Figure 22.6

Customer performance metric showing service improvement after an upgrade.

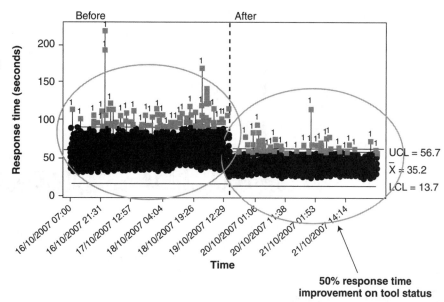

some of the critical IT services. There is a sizeable effort required to calculate the current capacity of the IT services, taking into consideration the business complexity and scheduled increases in manufacturing activity. A high-level impact analysis has been completed, and based on services that were displaying performance degradations, immediate constraints have been identified and relieved. As complexity increases further, comprehensive phases will be required going forward to ensure that future constraints are identified before they occur.

After infrastructure upgrades, any improved performance provided by the new infrastructure is communicated back to the business using the metrics previously implemented (shown in Figure 22.6). This closes the loop back to the business and cements their confidence in the IT department's proactive infrastructure upgrades.

7. Control Phase

The service performance metrics and the infrastructure KPIVs are now monitored using a statistical process control (SPC) system, shown in Figure 22.7, which alerts key personnel when one of the SPC rules is broken. In addition to the standard SPC rules, there are limits set by the customer as part of an SLA. If the service performance exceeds 80 percent of the SLA target (in most cases), an alert is generated to allow time for an IT response before the customer is affected.

As per any control plan, when an "out of control" event triggers an alert, there are associated action plans to resolve each scenario. A similar set of process steps is defined for each of the KPIVs monitored that generate alerts.

Figure 22.7

Service
performance
metric.

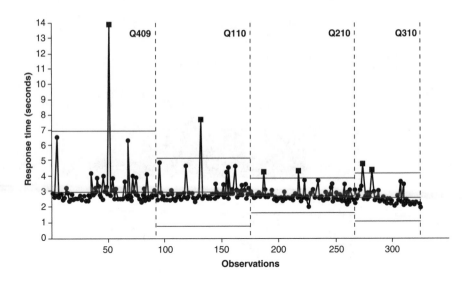

8. Benefit Realization

The improvement to the IT services now means that the IT organization in Springtown is able to offer "near 100 percent" availability (Figure 22.8). Any impact to availability is reported as lost "moves" in the factory. Within a 24/7 factory, the return on investment (ROI) for the proactive infrastructure improvements is achieving the schedule of shipments that are needed downstream, which in turn will provide Seagate Technology with increased revenue from disk drive sales.

9. Conclusions and Limitations

Over time, with the changing business and IT parameters, the predictive models have had to be adapted or rebuilt. With a consistent IT infrastructure across Seagate Technology, and with similar products being manufactured, one would expect that the predictive models would not change drastically. However, that is rarely the case in the technology sector. Currently, the site is in transition from products containing 38,000 recording heads and a 650-step manufacturing process to products with 76,000 recording heads and a 900-step manufacturing process. These products are new to the site and are being hosted, to a large extent, on new infrastructure. This means that there is no local historical data on which to build our model. However, our sister plant in Minnesota has launched these products, and their experience is being captured through the system and performance data to assist Springtown in building the performance models. Since Springtown is a higher-volume manufacturing plant, these models need to be extrapolated to match the levels of manufacturing output expected in Springtown.

Predictive modeling within the IT sector supporting a state-of-the-art semiconductor factory is a continuous and complex process. The main challenges are presented through

Service Availability Metrics

○ Monthly for a Quarter ◉ Quarterly for a year

Site | ☐ All | ☐ AMK ☐ Korat ☐ NRM ☐ Penang ☑ Springtown ☐ Suzhou ☐ Teparuk ☐ Wuxi
Operation | ☑ All | ☐ Drive ☐ HGSA ☐ HIS ☐ Slider ☐ Wafer

Quarter: FY10
Type | ☑ All | ☐ BC ☐ BI ☐ DI ☐ MC
Downtime | ☑ All | ☐ Schedule ☐ Unschedule   [GO]

Service Availability Metrics

No.	Service	HARM Class	FY09				FY10-Q01				FY10-Q02				FY10-Q03				FY10-Q04				FY10			
			Unsch DT	Sch DT	Overall Avail	No. of Incident	Unsch DT	Sch DT	Overall Avail	No. of Incident	Unsch DT	Sch DT	Overall Avail	No. of Incident	Unsch DT	Sch DT	Overall Avail	No. of Incident	Unsch DT	Sch DT	Overall Avail	No. of Incident	Unsch DT	Sch DT	Overall Avail	No. of Incident
Wafer																										
1	Data Transfer	BC	0	0	100%	0	0	0	100%	0	0	0	100%	0	0	0	100%	0	0	0	100%	0	0	0	100%	0
2	Dispatching (150mm)	BC	250	0	99.95%	1	0	0	100%	0	0	0	100%	0	0	0	100%	0	0	0	100%	0	0	0	100%	0
3	Dispatching (200mm)	BC	250	0	99.95%	1	0	0	100%	0	0	0	100%	0	0	0	100%	0	0	0	100%	0	0	0	100%	0
4	Durables Management	BC	250	0	99.95%	1	0	0	100%	0	0	0	100%	0	0	0	100%	0	0	0	100%	0	0	0	100%	0
5	Equipment Reporting	BC	250	0	99.95%	1	0	0	100%	0	0	0	100%	0	0	0	100%	0	0	0	100%	0	0	0	100%	0
6	Equipment Tracking	BC	0	0	100%	0	0	0	100%	0	0	0	100%	0	0	0	100%	0	0	0	100%	0	0	0	100%	0
7	File and Print	BC	0	0	100%	0	0	0	100%	0	0	0	100%	0	0	0	100%	0	0	0	100%	0	0	0	100%	0
8	Process Control	BC	250	0	99.95%	1	0	0	100%	0	0	0	100%	0	0	0	100%	0	0	0	100%	0	0	0	100%	0
9	Process Reporting	BC	250	0	99.95%	1	0	0	100%	0	0	0	100%	0	0	0	100%	0	0	0	100%	0	0	0	100%	0
10	Recipe/Spec Management	BC	0	0	100%	0	0	0	100%	0	0	0	100%	0	0	0	100%	0	0	0	100%	0	0	0	100%	0
11	Tool/Operator Interface	BC	803	0	99.85%	5	552	0	99.58%	3	0	0	100%	0	0	0	100%	0	0	0	100%	0	552	0	99.89%	3
12	WIP Reporting	BC	0	0	100%	0	0	0	100%	0	0	0	100%	0	0	0	100%	0	0	0	100%	0	0	0	100%	0
13	WIP Tracking (150mm)	BC	0	0	100%	0	0	0	100%	0	0	0	100%	0	0	0	100%	0	0	0	100%	0	0	0	100%	0
14	WIP Tracking (200mm)	BC	250	0	99.95%	1	0	0	100%	0	0	0	100%	0	0	0	100%	0	0	0	100%	0	0	0	100%	0

Legend: ☐ 99.5% AND ABOVE/ LESS THAN 9 ☐ 99.5% TO 99.89%/ 9 TO 16 ■ BELOW 99.5%/ 17 AND ABOVE

Figure 22.8 Service availability reporting.

increased production schedules and the introduction of new products in the manufacturing plan. Such challenges are driven by increases in customer demand, and new models need to be developed as a reflection of corporate growth. The IT infrastructure control strategy must be ready to meet these challenges.

The potential limitations typically associated with an IT infrastructure control strategy are that with such a complex environment, a lot of KPIVs are outside the control of the IT organization. It requires a good relationship between the IT organization and its business partners to identify the KPIVs, and provides a further challenge with regard to how to capture the identified KPIVs. In addition, with a large number of users, it is often difficult to capture all the KPIVs in the first pass—an iterative process is required to maintain the integrity of the strategy.

Acknowledgments

The author of this case study is thankful to Seagate Technology for its permission to allow publication of this case study as a book chapter.

10. References

1. Seagate Technology: www.seagate.com
2. IT Information Library: www.itil.org.uk

CHAPTER 23

DFSS Conclusions
and Lessons Learned

The six stories highlighting Design for Six Sigma and DMADV projects are quite different from the earlier chapters. The methodology is more proactive and utilizes different types of tools and techniques.

While still focused on process improvement, DFSS and DMADV projects do not seek to correct or fix existing processes and products as much as they endeavor to create new processes and products that are robust and that will prevent defects before they occur.

Since they are intended to develop new ways of doing things, they depend on innovation, creativity, and breakthrough ways of doing things and in taking a "big picture" point of view. They often revisit customer requirements and critical to quality factors more than Lean or DMAIC projects. They often introduce Kano analysis or K-J methods to further analyze requirements as well.

To head off or forestall variation or defects, the analysis of factors that affect these processes and products are also studied in greater depth. Tools like quality function deployment (QFD), principle component analysis, regression analysis, and curve fitting help to reveal the relative importance and interaction of different factors.

Similarly, since the processes and products being developed are new, the methods of verifying or testing them are also different. Where data is available from the analysis of factors, it is common to build statistical process models so that the behavior of systems can be studied in more detail. Monte Carlo simulation is frequently used to further explore and evaluate possible solutions before they are implemented. Where possible, piloting or Design of Experiments techniques are employed to collect new data on the proposed solutions.

DFSS and DMADV approaches tend to have greater depth and typically use different statistical methods that are more forward-looking and predictive.

Finally, as noted in our last success story, DMAIC projects can transform into DMADV projects as teams address existing problems and brainstorm new "breakthrough" solutions.

As illustrated in Figure 23.1, DFSS and DMADV approaches are just one end of the spectrum of ways to undertake process and products improvements. Combined with Lean and DMAIC methodologies, Six Sigma remains one of the predominant and successful strategies for business and engineering organizations.

Figure 23.1

Suitability of
Six Sigma
methodologies
based on
risk and
complexity.

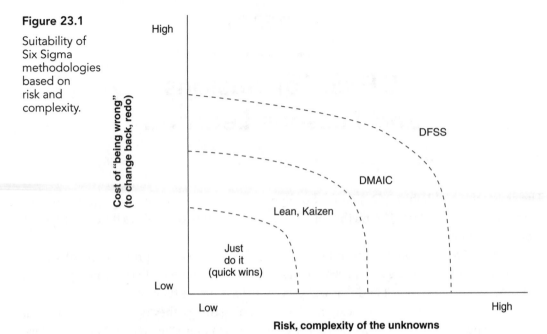

Together these approaches complement and overlap each other. They offer strategies and techniques that provide insights and solutions while reducing risk and uncertainty in complex environments. Six Sigma methods have become standards in quality and process disciplines.

Ultimately, the reason they have endured over the decades since they were first developed is that they work, and they work well.

PART FOUR

Six Sigma Programs

Part Four, the final section of this book, digresses and looks at projects from a "big picture" perspective. This section offers testimonials and examples of entire programs that span years and scores of Six Sigma projects. Some of the stories here relate to how Six Sigma was introduced into companies. Others provide accounts of the difficulty in sustaining and maintaining Six Sigma programs over time, or relate how programs were restarted or revitalized within companies.

Just as Lean and Six Sigma approaches are often recursive in nature (as problems are found and fixed and new problems arise or appear to be addressed in turn), so, too, are programs iterative in their struggle to sustain success over time.

The final chapter of the book pulls together lessons across all the approaches discussed in the earlier parts of this collection and offers recommendations and lessons for creating and sustaining successful organizational programs that will endure over the years.

CHAPTER 24

Cisco Successfully Reinvents Its Six Sigma Program

Jason Morwick

CISCO.

Evan H. Offstein

Frostburg State University

Relevance

This chapter is relevant to organizations that either have been unsuccessful in effectively deploying Six Sigma or have a Six Sigma program that needs to be revitalized. After a false start in 2004, see how Cisco learned from its prior experience and used an innovative approach to successfully deploy a robust Six Sigma program that continues to grow to this day.

At a Glance

▲ Cisco launched a Six Sigma program in 2004, but after encountering deployment challenges, it soon had to return to the drawing board to rethink its deployment approach.

▲ Cisco performed an organizational diagnosis to assess its readiness for Six Sigma deployment, performed a needs analysis, decentralized responsibility and accountability for the Six Sigma initiative, developed a vast network of Green Belts, and embraced a Six Sigma culture.

▲ The Six Sigma program and number of completed projects at Cisco continue to grow exponentially, with 17 completed projects in the first year (with savings per project of US$200,000) and the number of completed projects nearly doubling in each of the subsequent years.

1. Introduction

San Jose–based Cisco Systems, Inc., may be better known for its innovative products than its focus on operational excellence. With products and services that span market segments ranging from consumer to commercial to service providers, the network gear manufacturer has seen tremendous growth in its 26-year history, evolving from a Silicon Valley startup to a global company of almost 70,000 employees. However, the company's rate of growth sometimes outpaced existing systems and processes. In addition, the struggling economy highlighted the need to focus on internal operations.

"It is more critical than ever in the economic downturn and inevitable upturn that we renew our focus on improving processes," said Chief Operating Officer Randy Pond. "We must remove inefficiencies and eliminate waste to ensure we remain a market leader."

Despite the need for Lean Six Sigma, the company faced several key challenges, such as gaining leadership support in a culture that approached initiatives in an open, collaborative structure versus top-down mandates from corporate headquarters. The company culture would need to change too. Managers and employees would need to become more data-centered and process-oriented for the program to really work. In addition, many organizations within the company were already leveraging governance models like the software development lifecycle (SDLC) and were confused as to how to integrate process improvement methodologies into their governance structure.

"We actually started this journey in 2004," explained Rick Walus, leader of the company's Lean Six Sigma Center of Excellence. "We sent numerous people to training, both at the Green Belt and Black Belt level. Unfortunately, few were able to complete their projects and get certified. We had to go back to the drawing board and understand what happened.

"In many pockets around the company, there was no formal structure, nor did the right environment exist for Six Sigma projects to be successful or sustainable," added Pond. "Although the people we selected for training were all talented, intelligent, and motivated, without the right infrastructure in place, they would never have the opportunity to apply their newly acquired skills," said Walus.

2. Creating a Six Sigma Infrastructure

Unlike other companies that have successfully deployed Six Sigma, Cisco had no intention of forcing the methodology onto its employees. There would be no Jack Welch-like edicts from corporate headquarters about expected certification levels, number of Black Belts, or even the number of projects. Instead, Six Sigma would take root in each organization as organizational leaders expressed the desire and commitment to roll out the program. The deployment strategy had to match the decision-making norms that were defined as collaborative and consensus building.

To ensure that this strategy would be more successful than the company's initial efforts, a small Center of Excellence (COE) was established at the corporate level. The COE was staffed with Master Black Belts, each with several years of experience deploying Six Sigma prior to joining Cisco. The role of the COE was to prepare and help the individual organizations with their Six Sigma deployment.

First, the COE team members initially performed as auditors, rating the organization's readiness to adopt Six Sigma based on 12 key criteria (see next section). The outcome of the readiness assessment was reported back to the organizational leadership and translated into a deployment strategy. By working with assigned members in the organization to shepherd the fledging program, the strategy was broken down into more granular details to create a detailed roadmap covering the next several fiscal quarters. This operational plan, signed off by the organization's leadership, created executive-level buy-in and also created ownership by management since the plan was specifically tailored to the needs of the organization.

2.1 Key Learning 1: The Importance of a Needs Analysis

When we reflect on the success of the Six Sigma initiative here at Cisco, we continually return to this first step—a step that many organizations either ignore or view as perfunctory. In short, leaders and Six Sigma practitioners conducted a needs analysis. Both the academic and applied literature are rife with data and stories where organizations, in general, and HR departments, in particular, fail to conduct a thorough needs analysis. In so doing, any training or process improvement campaign usually suffers as the program is applied to organizations that do not need the initiative or, more likely, are not ready for its implementation.

It is important to note that the diagnostic included a diverse mix of criteria spanning the strategic to the tactical, to the hard and technical skills such as results tracking, to the softer skills such as leadership commitment and the importance of having a communication plan. The needs analysis evaluated factors such as:

1. Leadership commitment
2. Resource bandwidth and availability
3. Training strategy and plan
4. Project accountability
5. Process monitoring and metrics
6. Core processes identified, documented, and owned
7. Integration with project management office
8. Voice of the customer collected and available
9. Proactive benchmarking
10. Communication plan
11. Results tracking
12. Sustained results

Herein lay two particularly valuable takeaways:

▲ First, when embarking on a Six Sigma program, begin with a needs analysis. Is the team or department ready for this new process improvement initiative? Ask whether the people are truly ready to embrace this departure from the status quo. Move beyond the requests to simply send people off to Green Belt or Black Belt training. Successful Six Sigma implementations involve significant commitment. Ensure leaders are ready for the commitment required, such as supporting process improvement projects through budgeting or resource allocation. Moreover, carefully consider whether the product or service is apropos for Six Sigma. Although Jack Welch was noted for implying Six Sigma could apply anywhere at any time, some teams, organizations, or processes will more readily adopt Six Sigma initiatives than others. The task, then, is to find which one of those departments is fertile ground for such an initiative.

▲ Second, while Six Sigma and the DMAIC methodology are inherently technical and quantitative, it is the softer skills, or the qualitative, that need to be consulted and emphasized for the technical, systematic Six Sigma program to work.

In any regard, after the operational plan was approved, leaders would communicate and cascade the information with their support. Training would begin at the executive level first, followed by Champion training. Champions, the keystone of the organization's program, would learn how to select and prioritize appropriate Six Sigma projects and Green Belts. The focus was on project selection in order to emphasize the organizational benefit over the individual certification. With projects selected, Green Belt training would be scheduled.

Again, Cisco departed from other Six Sigma implementations in placing the burden of project leadership on Green Belts versus Black Belts. Rather than staffing the organization with many full-time, dedicated Black Belts, the majority of the projects would be led by Green Belts. Experienced Black Belts could still be found throughout the company; however, the ratio of Black Belts to regular employees was far lower than other companies. Training was targeted at the Green Belt level and the training content was highly detailed in order to properly prepare the Green Belt to be primarily self-sufficient. Black Belts would mentor but not actively involve themselves in Green Belt projects. Consequently, the resource commitment would be high. Green Belts were expected to spend up to 25 percent of their time leading projects. The result, company leadership hoped, would create a broad employee base with process improvement ingrained in the organizational DNA in lieu of a small population of subject matter experts.

2.2   Key Learning 2: Exposing the Missing Variable of Leadership

On the surface, it may seem that this is more of a staffing issue—the decision for the Green Belts to do the "heavy lifting" as opposed to Black Belts. In retrospect, though, there are several important leadership takeaways that deserve notice and attention. Most notably, there is a macro leadership belief that it is acceptable and preferable to decentralize knowledge, responsibilities, and decision making. While many organizations may attempt to centralize power and responsibility in regard to the design and delivery of their Six Sigma programs, Cisco again departed from the norm. The technical expertise and knowledge would be disseminated broadly rather than residing in the power of the few. And on first blush, this may sound like a staffing decision; it is far more a leadership decision. Simply put, Cisco believed in its human capital enough to decentralize responsibility and accountability of the Six Sigma initiative. The leadership requisite here is to have faith and trust in your people. Without these two endogenous variables of trust and faith, there is only centralization and compartmentalization of Six Sigma people, expertise, and ideas.

Second, a major leadership lesson revealed here is the need for mentoring. Considering that project success hinged on the capability of the selected Green Belts, Green Belts were paired with experienced Black Belt or Master Black Belt mentors. These mentors helped increase the Green Belts' technical mastery of Six Sigma tools and by sharing experiences, they helped overcome nontechnical challenges associated with Six Sigma projects, such as resistance to change or changing behaviors. To adequately support Green Belts with mentors, the company identified internal personnel with Six Sigma skill sets or staffed the organization with experienced Black Belts and Master Black Belts by hiring externally.

Of course, considerable patience had to be exercised for the mentoring to work. The two variables—mentoring and patience—are inseparable. It is difficult to have one without the other. In this case at Cisco, the mentors were the experts—both technically and organizationally. In grooming the Green Belts, the Master Black Belts and Black Belts had to maintain some level of patience and tolerance as the Green Belts would make mistakes in the execution of the Six Sigma program. However, and in the very best Black Belt to Green Belt relationships, a mistake was viewed as a learning opportunity or a teachable moment. Drawing both on anecdotal along with applied and scholarly works, it is mentioned that mentoring often fails because organizations do not want to invest the time or energy into developing a cadre of competent succession leaders. The paradox is equally stunning. Six Sigma, at its very core, is about reducing mistakes or errors. However, training and tolerance for mistakes by leadership is necessary for the successful transfer of knowledge and expertise from the Black Belt mentor to the Green Belt protégé.

3. Transitioning to a Six Sigma Culture

Overcoming organizational inertia and changing culture is often a slow, difficult task. The Cisco approach to transitioning from an innovation-centric culture to one that valued both innovation and operational excellence was to follow a manageable and methodical adoption strategy. Rather than deploying Six Sigma en masse, the corporate COE team concentrated on making the program a success in each organization it engaged before rolling it out in a new organization. Even within organizations, a pilot was often conducted, focusing on three to eight projects, before engaging more of the organization. The company learned that momentum, created through project success, was the biggest influence on the speed of adoption.

To ensure project success, the active involvement of the Champions was vital. Ultimately, the project Champions were accountable for the project success or failure. As one of their key roles, project Champions proactively reviewed projects and communicated results to senior management. These Champions also served as the first line of defense in terms of removing roadblocks, assigning resources, or obtaining executive-level support. Furthermore, the Champions became the evangelists within the organization to help increase adoption.

To aid the Champions and upper management, many tools were deployed to maintain clear visibility to project progress and results. Web 2.0 tools, such as wikis and collaborative workspaces, were leveraged for project teams to update project performance in real time. Communication throughout the organization was done on a frequent and consistent basis via e-newsletters, e-mail updates, video-on-demand (VoD), websites, and all-employee meetings. Communications ensured all members of the organization were aware of the project progress and also reinforced the value proposition to leadership. The focus on hard dollar savings and tangible benefits helped constantly remind leaders of what was in it for them.

Finally, project success was celebrated. Green Belts and their project teams were publicly recognized for their efforts. Again, Web 2.0 tools and other communication vehicles were leveraged to broadcast benefits achieved. Although much of the recognition was localized within specific organizations, selected projects were nominated for recognition at a company level. The recognition events, whether local or enterprise level, helped to create continued momentum for the Six Sigma program.

"Letting the Green Belts and project teams receive recognition for their hard work and effort not only leaves them with a positive feeling, it shows the rest of the organization that process improvement efforts are valued more than reactionary, firefighting efforts," said Joy Lin, a Sydney, Australia–based Black Belt. "Over time, people will realize that it is better to be proactive and solve for the root causes of problems than just address the symptoms. And that creates the momentum we are looking for."

3.1   Key Learning 3: Building and Fostering a Culture that Wins

As mentioned previously, Cisco's initial attempts at Six Sigma initiatives were not always fruitful. This is common to many organizations and is the driving force behind this collection of Six Sigma chapters. However, one can see from the detail of events that Cisco's luck began to change as their culture began to change and adjust. Three "S" variables were instrumental in moving the culture toward supporting the Six Sigma regimen.

The first critical "S" variable is that of *selection*. The selection of Green Belts became a critical explanatory variable that correlated directly with program success. Here, too, the paradox surfaces. As with many organizations that deploy Six Sigma, many Six Sigma enthusiasts tend to enjoy working with data as opposed to people or things. Although this attention to detail mindset and data-crunching appetite is good for the tactical implementation of any Six Sigma program, it can work against the strategic success of the program. Put differently, the Green Belts selected to lead Six Sigma projects were good with "people" first and "data" second. Successful Green Belts were those that possessed superior interpersonal, project management, and change management skills. They were skilled communicators. They could talk to individuals as well as motivate teams. They were influencers who could negotiate the conditions for Six Sigma success, particularly important in a matrix environment where project leaders had no formal authority over others. Almost forgotten, but extremely important, many of these Green Belts showed courage. They were willing to hold the line and, if need be, demand resources and incubation time for Six Sigma projects to succeed. But, make no mistake, these Green Belts were not selected haphazardly. Rather, their selection involved careful study, due diligence, and forethought.

The next critical "S" variable was that of the expert and nuanced use of *symbols* to carry the day and advance the Six Sigma culture forward. Evidence of this can be found in the communication patterns where successes were celebrated. Not only were successes celebrated, the news of victory was spread far and wide through a variety of sources to include wikis, interorganizational announcements, and, of course, face-to-face dialogue. The symbols of success, mainly the symbols embedded in positive communication, were given nutrients and the resources to flourish and spread. In short, success gained traction. Cases of failure were never dismissed out of hand. Shortcomings or unsuccessful projects were consistently examined for lessons learned or for pitfalls to avoid in the future, but were not given the same "air time" as successes. Said differently, people began to see, observe, and hear the sounds and symbols of success. Success became a norm. Not the same with failure.

Finally, the last critical "S" variable is that of promoting *subcultures*. This notion of subcultures is often lost on many different audiences—applied and scholarly alike. What happened at Cisco, however, was prescient. It is true folly to believe that a culture as diverse and expansive with upwards of 60,000 employees could enjoy a single, monolithic culture. Cisco leadership seemed to intuitively grasp this truth. Rather than trying to force a single culture upon an entire organization, they realized the value of allowing organic growth to occur. Rather than discourage subcultures to exist, they promoted this phenomenon. What transpired was uneven growth—some units and departments embraced Six Sigma before others. Although some may consider this a drawback, it is more reflective of reality, and the cultures that did take shape and form were truer to the Six Sigma cause. Allowing subcultures to form in regard to Six Sigma requires considerable tolerance of ambiguity and a general acceptance of variance— again, one of the primary things that Six Sigma is supposed to rid the organization of—variance!

4. Integrating Six Sigma with Project Governance

Launching Six Sigma in Cisco presented another key challenge: how to integrate a process improvement methodology with established governance structures such as product lifecycle (PLC) or the software development lifecycle (SDLC). Governance in this case refers to the consistent management, policies, processes, and decision-rights for an area of responsibility.

Simply, it is a specific process, like PLC or SDLC, which provides decision points at critical junctures in a process flow. Although this governance structure was not required for every project within the company, it was a typical standard for IT projects. Furthermore, even though the company managed many IT projects or projects with IT components through slightly different governance structures or processes, each governance process used the same taxonomy. For example, each governance structure usually consisted of the same milestones or toll-gates with similar high-level objectives:

▲ **Business Commit:** Align initiative with strategic objectives of the business. Ensure business case and justification for the project is clear. Funding and high-level timeline complete.

▲ **Concept Commit:** Outline initial (draft) requirements. Ensure initial assessments and project scoping are complete.

▲ **Execute Commit:** Finalize project plan ("point of no return"), scope, resources needed, and schedule. Systems analysis complete.

▲ **Design Review:** System design complete; specifications approved.

▲ **Readiness Review:** Solution built and tested. Ready for go-live.

Introducing a new methodology with foreign terminology ran the risk of confusing employees or, worse, inadvertently creating bureaucracy and negatively affecting projects. For process improvement projects that required IT investment, thought was given to how to overlay the DMAIC methodology of Six Sigma with existing governance models. By integrating the methodologies and ensuring that one model did not replace the other, the company was able to add clarity, increase adoption, and better support projects.

During the implementation of Six Sigma, the initial emphasis was on the DMAIC methodology, and the DMAIC phases and toll-gates were compared to standard governance milestones. Practitioners found many similarities between accepted company practices and the requirements of each DMAIC phase. The DMAIC methodology complemented, rather than conflicted, with established protocols. In fact, many found that the DMAIC methodology strengthened existing governance.

For example, the Pre-assessment phase of PLC or SDLC usually culminated in a Business Commit toll-gate (Figure 24.1). During this milestone, key stakeholders or business leaders

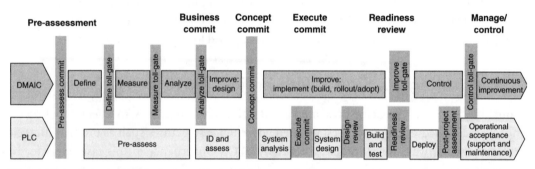

Figure 24.1 Integrating Six Sigma with the PLC.

would evaluate the merit of the project and decide if the business case or value proposition warranted initial investment of resources to pursue further. Typical requirements for this phase review included high-level project plan, financial analysis, value proposition, and initial resource plan, to name just a few. By aligning the Define, Measure, and Analyze phases to this Pre-assessment phase, decision-makers could feel more confident in their judgments of the initiative because more upfront rigor and due diligence had been applied. At the Business Commit milestone, the problem had been clearly articulated (from the Define phase), the process baseline performance was known (from the Measure phase), and the root causes to performance issues were understood (from the Analyze phase). Subsequent phases in the PLC or SDLC lifecycle could be quicker due to the front-loading of effort and business leaders better informed for key project decisions.

5. Results That Matter

Most Six Sigma leaders are well aware of the intangible benefits of Six Sigma, such as creating a common language across the business, creating a culture that values change, and empowering employees to change how they conduct their work. However, Six Sigma leaders at Cisco knew they had to emphasize and focus on the quantifiable benefits, given the struggles of the initial implementation. They were committed to ensuring leaders felt the positive financial impacts as a primary benefit while gradually realizing the softer, intangible advantages as time went on.

In the first year of the revitalized Six Sigma program, the company piloted its new approach in just two organizations: one from manufacturing and one from supply chain. The intent was to monitor the rollout closely and ensure the newly trained Green Belts and their projects had the highest possible chance for success. It also helped that many of the leaders within these organizations were already familiar with Lean, Six Sigma, or other related tools and concepts. Within these two organizations, the COE team concentrated their effort on training 19 Green Belts with an average hard dollar savings per project of just over $200,000. Out of the 19 Green Belts, 17 were certified by meeting internal certification requirements that included the official closure of their projects.

During the second year of the reimplementation, an additional four organizations were included, training an additional 42 Green Belts. Again, the average hard dollar savings per project was a little more than $200,000. Currently in its third year, the company has added another 8 organizations with more than 70 additional Green Belts and another 12 organizations that have shown interest in deploying Six Sigma in their business area. The hard dollar savings per project is expected to match the projects executed in the first two years; however, additional benefits from projects, such as improving the customer experience or cost avoidance, are also being realized.

The following list of recommendations summarizes some of the more important lessons that were learned in the development of this program.

▲ Do

  ▼ Ensure buy-in and support from senior leaders.

  ▼ Generate "quick wins" to gain momentum.

  ▼ Select projects that matter and are aligned to business goals.

  ▼ Select projects before selecting Green Belts.

  ▼ Ensure visibility to project and program progress.

  ▼ Train all levels of the organization, not just Green and Black Belts.

  ▼ Publicly recognize successful efforts.

  ▼ Assign experienced mentors to Green Belts.

  ▼ Ensure selected Green Belts can spend at least 25% of their time on the project.

  ▼ Engage finance representatives to validate project benefits.

▲ Don't

  ▼ Focus on individual certification levels or number of belts.

  ▼ Focus on numbers of people trained (more is not always better).

  ▼ Choose only projects where data is readily available.

  ▼ Select Green Belts simply because they want to go to training.

  ▼ Train Green Belts prior to training leaders and champions.

  ▼ Expect a change in culture to come from the bottom up.

  ▼ Believe that every project or problem should be solved using Six Sigma.

  ▼ Try to implement Six Sigma everywhere in the business at once.

  ▼ Ignore leadership and project management skills when selecting Green Belts.

In conclusion, there are several meaningful themes that emerge based on an analysis of the Six Sigma adoption within Cisco. One of the most critical is that leadership will always matter. Six Sigma is a revolutionary tool for so many firms. However, Six Sigma initiatives often fall well short of expectations. A key variable that can help explain either success or failure is that of leadership.

Finally, Six Sigma is seemingly a bundle of contradictions. And it is the best leaders that are able to either reconcile these contradictions or at least allow them to exist peacefully. The execution of Six Sigma is a formal, data-driven, methodical, quantitative exercise that exterminates or reduces variance from an organization. However, the strategic adoption of Six Sigma within an organization appears to be an informal, social, people-driven, and qualitative exercise. When these two forces are reconciled, we see Six Sigma success, as we did here within Cisco.

CHAPTER 25

Six Sigma Practice at Thomson Reuters

Jian Chieh Chew

Relevance

This chapter is relevant to senior managers who are interested in starting up a Six Sigma program in their own organization and Master Black Belts (MBBs) who wish to learn about program growth strategies within a transactional or service environment.

> ## At a Glance
>
> ▲ Thomson Reuters Content Organization implemented a Six Sigma program that has grown steadily since 2005. Project initiations grew from 6 in 2005 to 176 in 2009.
> ▲ Learn how Thomson Reuters followed a disciplined approach to initiate, execute, and close meaningful projects that directly contributed to operational improvements.
> ▲ Learn how the Six Sigma program at Thomson Reuters gathered significant momentum, with the average number of projects closed per month leaping from three in mid-2008 to 15 by late 2009.

1. Introduction

In this chapter, we share how we have managed and grown the Six Sigma program in Thomson Reuters Content Organization. This chapter is organized around the three phases of a Six Sigma program, namely:

1. Initiate projects
2. Execute projects
3. Close projects

We will describe some of the challenges we faced at Thomson Reuters with the program and practices we employed to address these three challenges.

2. Phase 1: Initiate Projects

There are two main challenges in this phase:

1. Initiating the right projects
2. Initiating enough projects

2.1 The Challenge of Initiating the Right Projects

The types of projects that a Six Sigma program chooses to pursue are crucial to the health of the program. Only projects that are of high value to management will lead to program growth, so it is important to consider the following scenarios:

1. A project of high value that is initiated and closed successfully increases management and employee buy-in for the Six Sigma program.
2. A project of high value that is initiated but fails to close results in diminishing (or complete loss) of management and employee buy-in and increased skepticism of the Six Sigma program.
3. A project of low value that is initiated and closed successfully results in no change to management or employee buy-in for the Six Sigma program.
4. A project of low value that is initiated and fails to close also reduces management and employee buy-in for the Six Sigma program.

Hence, only the first scenario will lead to program growth.

Ideally, projects should come from management and by definition be a strategic concern. In fact, it is common for "how to" books on Six Sigma to say that project selection should not be delegated by senior management.[1] In practice, however, we seldom see this unless Six Sigma has achieved the status of an embedded business-as-usual (BAU) organization program, as is the case in GE, Motorola, and Honeywell.

For most organizations initiating a new Six Sigma program, trainees tend to be given "training projects" that are usually of low value with limited impact to the larger organization. Because of their low value, the success or failure of these projects is not really felt. By doing this, the organization has destined its Six Sigma program for failure from the start because even if the projects achieve their goals, the program is not seen to deliver value to the organization.

In Thomson Reuters, we assume that management does not care about Six Sigma, per se, but cares about what Six Sigma delivers, that is, if the results favorably affect the strategic goals of the organization. Therefore, projects that are seen to deliver real value for the business and customers make the program healthier, while those that are completed successfully but provide no real value can actually harm the Six Sigma program.

Based on this understanding, we implemented a couple of mechanisms to prevent low-value projects from entering our portfolio of Six Sigma projects:

▲ **Project initiation forum:** This monthly forum is facilitated by a Master Black Belt (MBB). During these meetings new projects are presented to the management team for approval for project initiation. Strategic alignment is ensured as this forum acts as a filter against low-value projects.

▲ **Linkage of projects to strategic objectives:** All new projects are formally linked to at least one of the strategic objectives of the organization.

It stands to reason that if management's interest in Six Sigma does not grow, the number of projects initiated will not grow. Therefore, growth in project initiation is a good gauge of management's interest in Six Sigma. As you can see from the time series plot in Figure 25.1, we have grown project initiations over the years and this tells us that we have been successful in nurturing our Six Sigma program.

2.2   The Challenge of Initiating Enough Projects

A limited use of Six Sigma would be to apply it as an option for problem solving. This kind of implementation involves sending people for Six Sigma training and assigning them Six Sigma–type problems to solve. There is no desire in such implementations to spread the use of Six Sigma to the rest of the organization. In such deployments, the practice of Six Sigma tends to fade over time as there is no chance for a Six Sigma culture to develop. For a Six Sigma culture to develop, a critical mass of people has to participate in and buy into Six Sigma. That's why every serious Six Sigma deployment must also initiate enough projects to embed a Six Sigma culture.

Figure 25.1

Projects initiated by month.

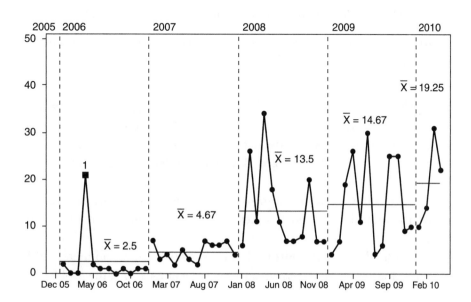

Six Sigma is not culturally neutral. To sustain and grow healthily, it requires a culture that assumes at least the following are true:

▲ Organizations are not perfect and will never be perfect and there is always room for improvement.

▲ Data-driven, fact-based decision making is better than gut-based or experience-based decision making.

▲ An organization comprises processes, and work is done through processes.

▲ Whether a process is good or not depends on whether it is able to produce an output that consistently meets or exceeds customer requirements.

▲ It is possible to measure the health of any process using metrics, and this is an important activity.

▲ Metrics and measurements can contain errors and when such errors are detected, it is important to correct them.

▲ It is okay to expose process problems. It is not okay to hide them.

▲ Everything cannot be improved at once. Lasting gains will be achieved by pursuing improvement on a targeted project-by-project basis and by building this norm into the fabric of the organization.[1]

▲ Leadership should understand the fundamentals of any change initiative (such as Six Sigma) and actively champion and support it.

▲ It is important to grow and develop a pool of change agents at every level of the organization, and it is important to give them time to devote energy to improving processes.

▲ It is much better to build in quality when designing and implementing processes than to fix them after they have been implemented.

▲ Processes should be both stable and capable. One or the other alone is not sufficient.

In practice, most organizations start their Six Sigma journey with a mixture of agreement and disagreement with these cultural assumptions. Some common Six Sigma countercultures are:

▲ **The hero culture:** People who clean up problems are rewarded and people who implement systems and processes that prevent problems from occurring are never noticed.

▲ **The silo culture:** People refuse to think in terms of processes because that would entail silo crossing, which is discouraged. Such cultures have a strong "mind your own business" mentality.

▲ **The punishment culture:** Someone always has to take the punishment for failures. There is little incentive to expose process problems, as senior management only likes to hear good news.

▲ **The father knows best culture:** The boss is always right and there is little incentive to do Six Sigma because the boss already knows what is causing the problem.

Embedding the Six Sigma culture within the existing organization's culture can be an extremely challenging task for any MBB team.

Our fundamental working assumptions about building Six Sigma culture are:

1. If you do Six Sigma projects, you have to adopt the assumptions of Six Sigma.
2. The more projects you do, the more you will adopt the assumptions of Six Sigma as your own assumptions of how to improve processes and solve problems.
3. If enough people do Six Sigma projects and critical mass is achieved, a Six Sigma culture will start to grow.

As we want a culture of continuous improvement in Thomson Reuters, increasing the number of projects initiated and the number of people with Six Sigma project experience is a major program management priority. As shown in Figure 25.1, we grew the number of projects initiated from an average of 2.5 per month in 2006 to an average of 19.6 per month in 2010.

We think that if an organization wants to embed a Six Sigma culture, it has to initiate many strategically aligned projects. We used the following strategies to achieve this:

1. We moved from a Black Belt-centric to a Green Belt-centric program. In some organizations, there is an assumption that only Black Belts can lead complex, cross-functional or cross-site projects. Most Six Sigma "how to" books say that Green Belts can only be trusted with smaller-sized projects within a department instead of working with a cross-functional focus.[1] The reason is Black Belts receive more training and are better equipped with the tools and knowledge needed for cross-functional or cross-site projects.

   In Thomson Reuters, we also had the same assumption. Initially, Black Belts were assigned to more strategic, cross-functional and cross-site projects and there were no Green Belts. But with a limited pool of Black Belts we could not grow the program exponentially and needed a new strategy.

   In 2007, we trained our first Green Belts, beginning our shift toward a Green Belt-centric program. In 2008, we completely changed the way we managed projects and the roles of Black and Green Belts. For Black Belts to become force multipliers, we assigned all Green Belts projects to Black Belts. The Black Belt role went from 80 percent project leader and 20 percent project coach to 80 percent project coach and 20 percent project leader.

   We also stopped making distinctions between Green Belt and Black Belt projects based on scope and whether they are cross-functional or cross-site. In fact, being a highly globalized organization, it is virtually impossible to scope a meaningful project that affects just one site or one function. Therefore, Green Belt projects are led by Green Belts and Black Belt projects are led by Black Belts, and we have no evidence that there is any difference in complexity (Table 25.1).

   The Black Belt became the coach to the Green Belts and the leader behind the leader. We made the Black Belt play the MBB role and the Green Belts play the Black Belt role. The MBBs, in turn, played the role of coach to the Black Belts. They provided help when the

Table 25.1 Transition from Black Belt-Centric to Green Belt-Centric Program

Start Year	BB-Led Projects	GB-Led Projects	Total
2005	6	0	6
2006	30	0	30
2007	25	31	56
2008	5	157	162
2009	25	151	176

Black Belt did not know which roadmap to use, how to proceed, and what tool to use or how to use it. In this manner we used our limited resources to a much greater effect and consequently the program grew by almost 190 percent from 2007 to 2008.

2. We gave stretch targets to every Six Sigma project initiated. We did not specify if these were to be Black or Green Belt projects. Most Black Belts opted to create Green Belt projects, as it was far easier to execute them with the assistance of Green Belts than to attempt them alone.

3. We aligned all the Black Belts with business units. All Black Belts were aligned with the three content verticals at the end of 2008. Before that, all the Black Belts reported to the MBBs and ultimately to the global head of process excellence. One of the challenges of the former structure was the perception that Six Sigma projects belonged to the Six Sigma organization and, therefore, project initiation targets as well. The paradox of this is that it is impossible for the Six Sigma organization to initiate any projects by itself—only the content verticals can do this.

   By aligning the Black Belts, we also aligned the project initiation targets. After alignment, the Black Belts reported to the vertical's service and quality (S&Q) leader who owned the Six Sigma targets for the vertical and led its Six Sigma program.

   The MBB team remained a centralized function and took on the program-level leadership, program governance, and provision of MBB services such as Six Sigma training, coaching, development of Black Belts, toll-gating projects, and advisory support. Every key production site (Bangalore, Manila, Beijing, and India) has a site MBB and each develops the site's Six Sigma program across verticals by working with the Black and Green Belts, project sponsors, and onsite production teams.

4. We started tracking project initiations monthly. This made it transparent which parts of the program were lagging in project initiations. Before Q4 2008 this information was not widely available. Initially, the metrics were supplied and tracked only by the MBB team, but by Q2 2009, almost every vertical was publishing their own metrics and had self-governance around them.

5. We initiated governance reviews. We created vertical engagement meetings where the MBB team would meet with the vertical Six Sigma teams to talk about project initiation ideas and review program metrics. We used these meetings to reinforce the importance of initiating projects.

6. We published Six Sigma success stories. In 2009, we employed push strategies to ramp up project initiations as we believed that there were lots of Six Sigma project opportunities within the verticals that required pushing to the surface. We knew that this was not the most sustainable way to achieve this and looked for pull strategies to increase the project numbers. One of the most successful initiatives in this respect was "Project Closure Communications," a short e-mail written with almost no Six Sigma speak, with the following generic structure:

▲ Who led the project
▲ What was the problem; what was the state of the process before and after improvement
▲ What were the root causes of poor performance
▲ What was done to fix the problem

This e-mail was sent to managers in the three content verticals and made the successful closure of projects rewarding, the contributions of the Six Sigma program more concrete, and, from the perspective of project initiations, educated a vast pool of potential project sponsors about what Six Sigma projects are and could achieve.

3. Phase 2: Execute Projects

For an organization starting Six Sigma, even with the best of intentions, the core theory of how to execute projects is:

1. Identify and select the most urgent and important issues in the organization and make these projects.
2. Select the best and the brightest resources and assign them to projects.
3. Send them for Six Sigma Black Belt training.
4. Form Six Sigma teams that are staffed by people from the process improvement organization and led by the newly qualified Black Belts.
5. Provide management attention and support to remove organizational obstacles.
6. Expect the team to complete the projects.

The normal experience is that most projects do not complete on time and/or achieve their goals, and many are suspended or cancelled. This can happen for a number of reasons.

1. It is not easy to convert the most urgent and important issues of an organization into projects. The challenges are:

▲ How to convert the voice of the business or customer into a project
▲ How to scope the projects
▲ How to identify exactly what needs improving
▲ How to decide whether to have one or multiple projects

These are not skills that one immediately has after Six Sigma training. Many Six Sigma projects cannot be easily completed simply because the way the problem is framed does not lend itself to easy completion. It is also common for new programs to initiate "boil the ocean" projects—a wrong belief that a large project can remove a large problem in a short time. Although it sounds and is logically incoherent, it is surprising how common this belief is.

2. The first Six Sigma training anyone receives is DMAIC (Define, Measure, Analyze, Improve, and Control) or on process improvement methodologies. However, in the real world, the problems that organizations assign to Black Belts are not all DMAIC problems and may in fact require other methodologies. At Thomson Reuters, in the initial years, there was a mindset that said only DMAIC projects were real Six Sigma projects and, consequently, every project was forced into a DMAIC mold, whether it was correct or not.

3. Even if DMAIC is the correct methodology to use (which in around 80 percent of cases is true), it does not mean that someone who has just completed 20 days of DMAIC training is equipped to successfully guide a project team in using the DMAIC methodology. The most common mental state of a new Black Belt is one of extreme information overload. He or she has just learned a full year's worth of undergraduate statistics plus another year's worth of quality engineering plus other topics like change management and project management in just 20 days.

Unless we are talking about someone exceptional, generally, newly trained Black Belts are not equipped to lead project teams. When forced to, this Black Belt will guide the team through a series of tools completion and when we look at their project storyboards we see the form of Six Sigma but not the essence. Often, tools are used in the right sequence and the right phases but the Black Belt and team may not know why such a tool was used and what value the tool brought to the project.

3.1 Good Master Black Belts Are Critical for Program Success

What is missing from this picture? What is missing is a mentor—someone who can guide and teach. In our experience, newly trained Black Belts need to be guided through at least five projects from start to completion before they can even begin to attempt to play the full role of a project leader without support. Who are these guides, coaches, and teachers? It is someone:

▲ Who has actual hands-on project experience and who has led many Six Sigma projects to successful completion

▲ With Six Sigma coaching experience who has coached many Six Sigma projects to successful completion

▲ Who knows the various process excellence-related methodologies such as DMAIC, DFSS, Lean, and theory of constraints

▲ Who can explain complicated concepts simply and correctly

▲ Who has experience of Six Sigma across many types of organizational settings

▲ Who is a change leader and who works well with people

Such a person is called a Master Black Belt, or MBB, and it is almost impossible to build a Six Sigma program without them. Our core theory is good MBBs create good Black Belts, good

Black Belts create good Green Belts, and good Green Belts result in well-executed projects, all of which equals program success. Therefore, staffing a Six Sigma program with good MBBs is critical to success, and we have MBBs with this profile in all our key production sites.

3.2 Adapting Six Sigma for Your Business

In the consulting industry, there is a famous saying: "When all you have is a hammer, every problem looks like a nail." As explained earlier, in the early years of our deployment, our Black Belt community was only trained in DMAIC methodology and thus every project was a DMAIC project. This impeded execution because our Black Belts lacked the variety of approaches needed to deal with the operating environments they faced.

To begin with, we found that even the DMAIC approach we were using had to be modified. The DMAIC roadmap our Black Belts were taught was really the manufacturing-style DMAIC made famous by Motorola and GE. Although the logic can be transplanted into a transactional environment, it had to be amended innovatively to make it more appropriate for our operating environment.

When we wanted to introduce Design for Six Sigma (DFSS) into our program, we also encountered the same problem. The predominant DFSS approach grew from a product design environment whereas we operated in a transactional process design environment. As a result, we also found we had to be innovative and amend the DFSS methodology to create the right approach for our operating environment. Over time, we developed and deployed our own unique style of DMAIC and DFSS for our business.

We also found that our customers demanded process-related interventions that do not neatly fit into DMAIC or DFSS. For example:

▲ What do we do when we do not have any process management and control system for a particular process? What kind of project is that?
▲ What do we do when we want to audit a measurement system for a particular process? Is measurement system analysis alone enough?
▲ What do we do when we want to do an audit of whether or not a process is over- or under-resourced?

Within our Thomson Reuters practice, we have a concept called bespoke project. This is a project that is executed using a roadmap that has been designed for that particular project and that is employed when the sponsor's requirement(s) do not fall neatly in either DMAIC or DFSS. For such projects, we design bespoke roadmaps that deploy a logical sequence of activities that meet the necessary requirements. Here are some examples:

▲ Process management and control systems implementation using Identify, Develop/Design, Execute and Assess (IDEA), a type of specialized DFSS
▲ Process metric system audit using Customer Alignment Check, Accuracy Check, Sampling Methodology Check, and Execution Check (CASE)

▲ Process full-time effort rationalization using Standardize, Troubleshoot, Estimate, Plan, and Sustain (STEPS)

Table 25.2 shows an example of the activities associated with a "bespoke" roadmap following the CASE model noted above.

Table 25.2 CASE as an Example of a Bespoke Roadmap

Phase	Key Question	Key Tasks
Customer alignment check	Do the metrics currently employed measure the customer's experience?	Identify and map the process under review. Obtain the metrics and specifications currently employed for the process under review. Identify the customers. Obtain the voice of the customer (VOC). Translate the VOC into critical to quality (CTQ) (metrics and specifications). Compare the CTQ derived from the VOC with the metrics and specifications currently employed. Produce deviation report and recommendations.
Accuracy check	Is the measurement system accurately measuring what it's supposed to?	For every metric, check if there is an operational definition. For every metric, check if the operational definition matches the deployed method of data collection. For every metric, check if the operational definition is in line with the VOC. Check if there are Gage repeatability and reproducibility issues. Check if the measurements are accurate compared to the standard or master. Produce deviation report and recommendations.
Sampling check	Is the sampling done correctly?	Find out the current sampling logic and check if it is theoretically sound. Check if the sample sizes collected are sufficient to detect the changes we want to detect.
Execution check	Are the metrics being used for decision making?	Is there a control plan for this process? Check how management uses the metrics for process review. Are they being used at all? Are the correct control charts being used? Is poor performance relative to specifications being pipelined for improvement projects?

We do not believe that Six Sigma must be just about DMAIC or DFSS projects, nor do we feel the need to force everything into a DMAIC or DFSS framework. In short, we found that we had to innovatively amend Six Sigma to make it work better with the demands of our operating environment to make execution of projects better.

4. Phase 3: Close Projects

The philosophy of our Six Sigma program is this: Whenever we successfully close a strategically aligned project, we create value for the business. When the business receives this value, it will become motivated to invest more in Six Sigma. As a result, more strategically aligned projects will be initiated.

Central to this philosophy is the understanding that a trained Six Sigma practitioner alone is useless unless he or she executes and closes projects that are of value to the business. Unlike other Six Sigma deployments, we place little emphasis on reporting how many people we train, because trained personnel alone do not deliver value—they just have the potential to do so.

4.1 Certification Policies

We deliberately moved the program toward a strong emphasis on project closures. We made project closures the key definition of success at program and individual levels. Our certification and progression policies all centered strongly on execution and project closure.

- ▲ **Green Belt certification:** Attended Green Belt training and successfully led two Six Sigma projects to closure.
- ▲ **Black Belt certification:** Attended Black Belt training, passed the Black Belt exams, and successfully led at least two Six Sigma projects and coached at least five to closure.
- ▲ **MBB certification:** Attended MBB training, successfully led at least five Six Sigma projects, and coached at least 30 Six Sigma projects to closure and passed the interview panel.

Figure 25.2 shows the sequence of accomplishments required for an individual to progress through certification from Green Belt to Black Belt and on to Master Black Belt. Certification criteria include completion of project and training, to mentoring and coaching for higher levels of Six Sigma Certification.

Anyone who wants to attend Green Belt training also has to come with an approved project; otherwise, training will not be granted. Black Belt training has so far only been granted to full-time Six Sigma practitioners, who are expected to deliver a targeted number of closed projects each year.

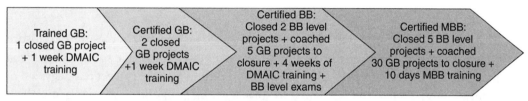

Figure 25.2 Journey from Green Belt to Master Black Belt.

In order for Six Sigma to develop a reputation for professionalism, generating closed projects alone is not enough. We had to ensure that every project we closed delivered value. In the case of DMAIC projects, we had to ensure that the process had improved performance and was sustainable. In the case of DFSS projects, we had to ensure that the new process was able to meet its design goals in a sustainable manner. We had to ensure that when we recommended a project for closure it had delivered value to the business.

4.2 Project Closure Forum Process

Before mid-2008, Thomson Reuters did not have a standard process for project closures and practices differed from site to site. Often, it was not clear whose decision it was to close a project, and we had no control mechanism to ensure the authenticity of closures. Therefore, we implemented a program-wide closure process that consisted of:

1. **Technical closure:** We created a forum called the technical project closure forum (PCF), where every project seeking closure had to be presented. The forum is normally chaired by the global head of process excellence and attended by at least two other MBBs. During the forum, the Black Belt or project leader presents the project. The MBBs review the project for technical correctness and, in particular, check if there are technically sound reasons to believe that goals of the project have been achieved.

   For instance, for a DMAIC project, we would check if the root causes of poor performance have been uncovered, if the improvement actions relate to the root causes, and, crucially, whether or not the process has improved and is sustainable. At the end a vote is taken among the MBBs in attendance and the decision to recommend closure is taken. Initially, approximately 50 percent of projects presented in the technical closure forum were granted technical closure at the first try, but this has improved to 90 percent.

   The PCF also served as a developmental platform and standard bearer. Black Belts learned from the PCF the standards of closure that were required by the MBB team. Over time, the quality of projects improved significantly.

2. **Business closure:** This is the final closure forum where the project is presented to the sponsoring stakeholder. The sponsoring stakeholder evaluates whether or not a project has successfully delivered on the business goals set in the project charter. This meeting happens after the PCF.

This practice also ensures that we have a common global standard of closure and closures are transparent. This practice also creates program credibility because the sponsors know that when the MBB team says a project can be closed, it is certain that the project has delivered on its promises. Over time, using Six Sigma to solve a problem has become an organizational byword for taking a problem seriously.

4.3 Performance Tracking

To drive project closures, we implemented monthly performance reporting for the Six Sigma program. These reports are published to all management in the Content Organization, making the performance of the program transparent:

▲ Projects closed vs. targets
▲ Projects initiated vs. targets
▲ Hard savings vs. targets
▲ Soft savings vs. targets

5. Results

We have described so far our approach to the management of Six Sigma and the deployment of that approach. Our approach can be summarized as follows:

5.1 Initiate Projects

1. Develop mechanisms to ensure Six Sigma projects are strategically aligned to the strategic business objectives.
2. Develop mechanisms to grow a culture of Six Sigma within the organization. We believe that if more people in the organization do Six Sigma, then this culture would grow.

5.2 Execute Projects

1. Make everyone in the organization potentially a Six Sigma project leader. Therefore, make the Green Belt the project leader and the Black Belt predominantly the project coach.
2. Create a performance culture by setting goals and targets for the program, the verticals, and the Black Belts.
3. Don't force everything to fit in DMAIC or DFSS. Innovatively amend Six Sigma and create customized methodologies that fit the needs of your business.
4. Hire good Six Sigma MBBs. Without them, you can't possibly start to do anything else.
5. Communicate the success stories.

5.3 Close Projects

1. Make performance transparent by publishing the results of the program periodically.
2. Emphasize execution—make project closure the most important metric.
3. Use certification policies to drive project closures.
4. Enforce a common and high standard of closure. Ensure that all closed projects benefit the business.

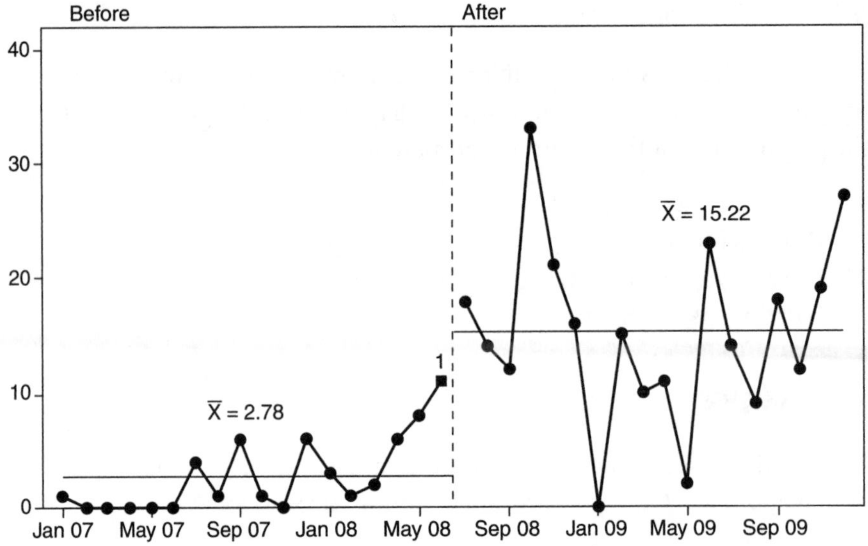

Figure 25.3

Projects closed
by month.

Before

After

$\overline{X} = 15.22$

$\overline{X} = 2.78$

40

30

20

10

0

Jan 07   May 07   Sep 07   Jan 08   May 08   Sep 08   Jan 09   May 09   Sep 09

Most of these changes were implemented after mid-2008. As shown in Figure 25.3, the performance of the program shifted from an average of 2.8 projects closed per month to 15.2 projects closed per month.

Our Black Belts also improved their performance. In 2007, each Black Belt, on average, was only capable of closing 0.6 projects. In 2008 this increased to 5.58 projects. By 2009 this number had increased to 8.9 projects.

6. Conclusion

We hope that through our story the reader has gained an insight into the complexities of leading a Six Sigma deployment, which is so much more than just training staff and pairing them with projects.

We hope also to have demonstrated that Six Sigma deployments require leadership and management and it is the job of leaders to build enabling mechanisms for the program to grow. Some of the things that helped our program grow (like the PCF) might also help your program, while others may not be relevant at all. Leading Six Sigma is not a cut-and-paste process where one size fits all. Your program is or will be unique and will require both common and unique strategies to grow.

7. References

1. Thomas Bertels (ed.), *Rath & Strong's Six Sigma Leadership Handbook*, John Wiley, 2003.

CHAPTER 26

How Convergys Injected Six Sigma into the Company DNA

Manisha Kapur

Relevance

This chapter is relevant to companies who want to implement a Six Sigma program from scratch. The chapter describes a comprehensive, step-by-step approach for successful Six Sigma deployment at a company.

At a Glance

▲ Convergys launched its Six Sigma program in 2007, and Six Sigma steadily took firm root in the company.

▲ Learn how Convergys did it the right way, with detailed insights into its Six Sigma training program; governance model; approach to selecting, initiating, and tracking Six Sigma projects; and tips on leadership and employee engagement.

▲ In the three years since its adoption, Six Sigma has become ingrained in the corporate DNA and the company has realized a 4 percent return on revenue and more than 10 times the return on investment in its Six Sigma program.

---

1. The Journey Begins

In August 2007, Convergys adopted Six Sigma as its key quality methodology. This marked the start of a journey of continuous improvement for Convergys that has been both rewarding and full of learning.

The primary business of Convergys, a global leader in relationship management, is to service inbound contacts from our clients' customers. Convergys India is part of the thriving information technology (IT) and IT-enabled services (ITES) sector that contributes US$50 billion to India's economy and provides employment to more than 700,000 people. The ITES model is to service back-office work on behalf of clients in an effective and efficient manner.

Convergys's objective in adopting Six Sigma, a world-class quality methodology, was to fulfill the vision of making quality a competitive advantage for Convergys. The expectation of the Six Sigma program was that Convergys, over time, would reap significant performance and financial benefits from adopting Six Sigma. Thus, Convergys decided to start on a journey of continuous improvement.

The journey started with the appointment of a new CI (continuous improvement) leader with in-depth knowledge of implementing Six Sigma at a company that nurtured quality professionals and where Six Sigma thrived.

Before setting out to implement Six Sigma, the CI leader focused on establishing the foundation of the CI deliverables. Implementing a robust and a proven quality methodology does not ensure success in itself. What is needed is a good implementation plan, prepared with the business needs in mind. All too often, companies implement Six Sigma unsuccessfully because they don't align the implementation to business needs and set objectives that are not business imperatives. Companies often fault quality implementations for *complicating simplicity*, whereas the role of any quality implementation is to *simplify complexity*.

To begin with, the CI leader focused on understanding the expectations of the leadership and "feeling" the culture of Convergys India before developing an implementation plan. This included speaking with each member of Convergys's leadership team to understand individual expectations and perspectives, as well as linking with people across the various levels in the organization to understand the company's culture. The time spent in connecting culturally was critical to building a clear view of what would work for the company.

The guiding principle for the CI implementation plan at Convergys India was to *add value to the business*. The CI leader's observation, along with the discussions with the leadership team, clearly brought out the importance of linking continuous improvement to the core operations of the organization. Convergys's core operations objective is to deliver high-quality work at reasonable cost to our clients. Convergys achieves this objective by providing quality interactions between the clients' customers who call and the agents who service them. Thus, adding value to the business meant adding value to our agents, to our clients, and to the customers of our clients. This, in turn, would add value to Convergys India's profit and loss (P&L).

With this guiding principle in place, Convergys defined four elements for driving value through CI:

1. Convergys leadership established and endorsed the overall approach to get *close to the business and be a part of the operating environment of the organization*. This meant all CI initiatives were to have a direct link to the core operations of the organization. We rejected projects that did not have a business case linked with business benefits.
2. We emphasized that Six Sigma should be a part of Convergys India's culture through a formal mission statement: *Make operating excellence our mindset and Six Sigma our DNA*. The premise behind this mission statement was that any improvement methodology could

only sustain results if it became the culture of the organization. The core CI team would only be able to do a handful of improvement projects, and the cycle of organizational improvements would be slow. The expectation was to make quality improvements pervasive with sustainable results. In an organization like Convergys, where work is human intensive, it was even more critical to focus on building the Six Sigma culture. Therefore, operations and business subject matter experts led most improvement initiatives, not CI team members. Instead, the CI team members worked as mentors and Six Sigma guides.

3. The CI team was primed and tasked to work as *partners and not as mere consultants* to the business. Leadership made CI team members responsible for project success and closure rates. It encouraged the team to participate in all team meetings, provide Six Sigma guidance, and help project owners. A critical element of project success is the selection of an appropriate project objective and project owner. CI team members provided support and advice in the project initiation stage to create buy-in with project participants and improve successful project closure rates. Weekly reporting of project status was also the responsibility of CI team members, thus providing unbiased project updates.

4. Convergys measured CI engagement success through *business benefits quantified in US dollars* and signed off by the chief financial officer (CFO) of the company, and not the number of projects certified, as is often the case in many companies. We clearly linked project financial benefits to P&L benefits. As a criterion for certification, Convergys expected each project owner, including the CI-assigned resource, to compute the financial benefits of the project. The team then worked with the finance manager to link it to the appropriate P&L line item and reviewed the results with the Financial Planning and Analysis (FP&A) lead before declaring a project complete.

We initiated the journey for a successful Six Sigma implementation by identifying the customers and listing the four elements for driving value. This is, by far, the most important stage in the Six Sigma journey of an organization. While there is always the temptation to skip this and get started with the details of the Six Sigma implementation plan, many Six Sigma implementations fail due to lack of management or employee support, which results from lack of buy-in or *unmet expectations*. It is best to overcommunicate at this stage in the journey and, if required, revisit this stage to make sure that you are embarking on a road that travels in the same direction that your customer wants to travel.

2. On the Road to Continuous Improvement

After defining and establishing the vision and mission for the new Six Sigma journey, the next step was to put together an execution plan to implement Six Sigma successfully throughout the organization. The Six Sigma implementation plan detailed the training plan, governance model, communication plan, and Six Sigma programs that the CI team planned to undertake. A Six Sigma implementation plan should not be developed once and followed forever. A good implementation plan is dynamic and has some foundational elements and others that evolve over time in response to organizational needs. Let's look at each of these in detail in the sections that follow.

2.1 Six Sigma Trainings

As we set out to implement Six Sigma at Convergys India, we put together a training plan that linked the Six Sigma trainings offered to all the levels of the organization with a clear view that continuous improvement is the job of all employees and not just the CI team (see Figure 26.1). We designed the graded training program in such a way that entry-level employee training created awareness, while training aimed at supervisory and senior management levels built their skills to deliver improvements in the area of work they managed. In this way, Convergys empowered the organization with Six Sigma skills and knowledge and provided support to drive improvement projects wherever the opportunity existed.

While we conducted the first Six Sigma Black Belt (BB) and Green Belt (GB) trainings utilizing external support, since the middle of 2008 we have developed an in-house curriculum for training up to GB level in a phased manner. This has helped us customize our training to our needs with examples, games, and case studies that relate directly to our work. Given that we employ young professionals with an average age of 23 years, our training is contemporary in nature with a focus on learning through fun. While our White Belt (WB) training is a web-based, self-paced learning module with about 10 percent theory interwoven into a storyline,

Figure 26.1

Six Sigma training structure of Convergys India.

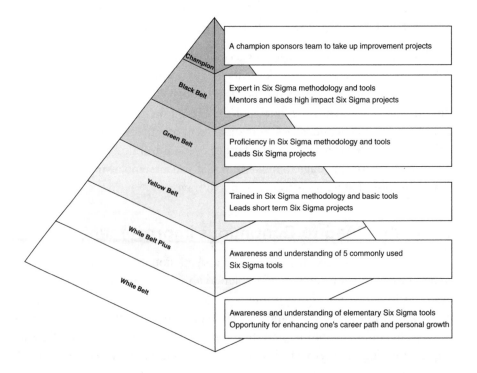

our Yellow Belt (YB) and GB trainings are instructor-led classroom trainings with tools and concepts supported by games and case studies.

The objective of our Six Sigma training is not about teaching the organization but to impart learning on an individual level. Thus, we reached out to the intended audience to understand their training expectations and choice of learning methods before preparing the various modules. Our training penetration has far exceeded our expectations. As of July 2010, more than 70 percent of our agents, which translates to about 4,500 employees, took WB training in the first year of its launch. Our WB module has also garnered external appreciation. The Quality Council of India, the apex quality body of the government of India, uses our WB module on its official website to encourage people to learn Six Sigma.

In beginning of 2010, the CI team built bridge training exercises between WB and YB called WB Plus (WB+) because agents who had undergone WB training were keen to learn more and engage in improvement projects. Close to 1,000 employees have taken the WB+ training in the first three months since its launch.

A new trend that has emerged in the Six Sigma trainings space is to make *all* trainings (aside, perhaps, from the BB level) self-paced on the Web. This is perhaps not the best way to engage people, however. Web-based training should be used carefully as an introductory or foundation training because it doesn't provide an opportunity for interaction and more enriching learning experience due to lack of a live instructor. Six Sigma is not only about learning tools and concepts; it is more about building a mindset to continuously look for improvements.

2.2   Governance Model

To support the culture of continuous improvement, Six Sigma projects were initiated and supported through a *formal Six Sigma governance model*, which was well understood and recognized in the organization. The objective of the governance model was to provide a structure for project selection and execution and *balance formal reviews with time to close projects*. In a company handling customer contacts, our clients measure us by our success or failure to meet critical metric targets every month. A three-month period in this environment is considered long-term. Therefore, Six Sigma project duration also tends to be shorter than in other industries. With this in mind, the governance model we established ensured appropriate checks and balances while also providing flexibility to allow for speed. Our governance model is a simple, four-stage process of project selection, initiation, progress tracking, and formal closure, as shown in Figure 26.2.

2.2.1   Project Selection

Project selection at Convergys India is not a one-time exercise. We initiate projects when any critical metric is underperforming for approximately three months, or if an opportunity is identified in discussion with clients in quarterly or monthly business reviews (QBR/MBR).

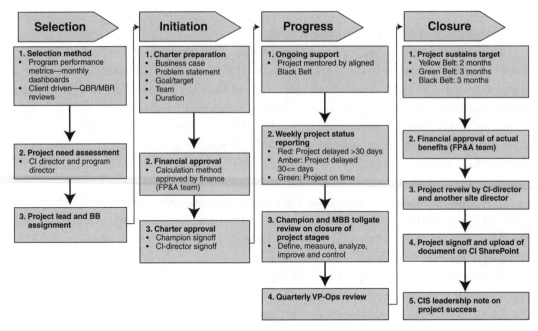

Figure 26.2 CI governance model of Convergys India.

Ours is a client-driven business where projects are primarily selected based on how strongly (both positively and negatively) the problem and opportunity statement influence our client relationship. Projects that influence Convergys's internal goals are second in priority. For example, a project to meet a client customer satisfaction (CSAT) requirement carries priority over meeting an efficiency target for the same program, even though the CSAT attainment may not be linked to any direct financial gains, whereas an efficiency improvement is most likely to result in lower costs of servicing the client. Projects affecting clients are jointly prioritized with them in QBRs. Some of the projects undertaken are clearly designed to exceed a client statement of work goals with a view to positively influencing the client's position in its own market by reducing their cost of outsourcing (e.g., taking more calls for the fixed agents paid by the client) or plugging revenue leakages (e.g., reducing billing waivers given to customers).

One must be careful in starting too many initiatives across the organization. There is a tendency to adopt a carpet-bombing approach when the overall objective is to drive the Six Sigma culture and show improvement in a very short time. Project selection should keep in mind the engagement level and availability of the sponsoring leader and project team.

The question of how many projects are too many is best answered by the leadership of an organization. The size of the CI or quality team and the scope of opportunities are the critical inputs that the quality leadership needs to make a call. With 9,000 employees, we have had about 50 projects at all levels running at any point of time, with no more than 5 to 7 projects assigned per BB. These are good numbers to address improvement initiative needs sufficiently for client programs. We have been able to show success in the first year of Six Sigma implementation across critical programs against a set target timeframe of 18 to 24 months.

2.2.2 Project Initiation

A formal project charter heralds the initiation of a Six Sigma project. The project Champion and CI leader sign the charter. The project Champion is the program director entrusted to provide support and clear any hurdles during the project execution. The Champion is present in all project reviews, including the final project readout for certification.

We also encourage cluster projects, including YB with GB and BB, where a lower-level objective is undertaken as a quick-hit YB linked with a program level objective being driven as a GB or BB project. For example, we successfully ran an initiative to turn a program profitable by initiating it as a BB project and linking it to a GB project that involved reducing cost in the biggest line of business for the program. We also linked the BB project to a couple of YB initiatives to reduce the average handle time of two lines of business for the program (see Figure 26.3). This worked very well for us, as we were able to turn around the program profitability in three months after initiating the projects. Each of these four projects had their own project rigor and closure, but since we linked them, the same BB managed and delivered the projects to a common objective.

Usually, projects are identified at a business need level, which has many contributing factors; for example, in a project to reduce facilities costs, there are many contributing factors and functions, such as office consumables, infrastructure, electricity, housekeeping, and so on. It makes sense to break this large objective into initiatives, which can then be driven by separate cost owners

Figure 26.3

Six Sigma project linking for a program.

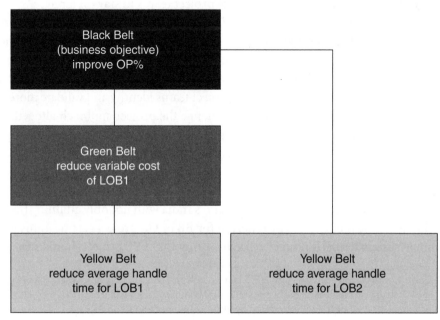

and linked to each other through an overall program to reduce facilities costs. This allows multiple teams to work in tandem on separate smaller initiatives, which add up to a larger objective.

It is important to have an overall program lead. A BB owns the complete initiative and leads one of the many subprojects, while GBs and YBs drive the subprojects. The BB is the mentor for all the subprojects and provides overall guidance to the team. The sponsor of the program is also the sponsor for all the underlying subprojects; however, Champions for the subprojects can be different. For example, the facility head may be the facilities cost reduction champion, while the procurement head may be the champion for office consumables. The BB ensures that the goals of all the subprojects add up to ensure the attainment of the overall program-level goal. The sponsor and leadership review progress of the program and subprojects together in order to get a complete picture of program status, interdependencies, and risks.

In our experience, linking initiatives in a tree structure is a good way to drive business benefits and allow speedy closure of a large objective by breaking it into clear and self-standing smaller initiatives. In addition, it motivates people as they drive improvement projects with clearly defined targets and earn certification on closure. It also helps employees build on their project management skills.

2.2.3 Progress Tracking

Sending weekly updates on Six Sigma initiatives to all leaders has worked as a motivator for project owners—this includes rolled-up counts of red, amber, and green status of BB, GB, and projects (overall, as well as by site). Red denotes the project is delayed more than 30 days, amber denotes the project is delayed up to 30 days, and green denotes the project is on schedule. High-level summary for each project is also provided in a separate spreadsheet. This has worked as a good soft tool to encourage project teams to stay on target.

Every quarter, the vice president (VP) of operations reviews the top few projects based on business impact. This provides the push for teams to make sure they have achieved critical mass in their projects before coming up for review, and also allows the VP of operations to get a good view of the organization's engagement with Six Sigma. The perspective provided by the VP of operations has helped some project teams identify and validate more causes for long-standing problems and, in some cases, change the project approach altogether. Engaging the highest level of operational leadership in project reviews has helped link Six Sigma closely with operations and ensure that projects deliver business benefits.

2.2.4 Project Closure

We link project success criteria at Convergys India with not only attaining the project goal but also sustaining the results. The criterion for BB or GB project certifications is that the project should sustain results for three* *consecutive* months. A YB project should sustain results for two

---

* Three months is considered a complete business cycle in ITES, after which changes in core processes may occur, such as change in credit card fees by the credit card company, new product features introduced by telecom service providers, new support rates introduced by technical clients, and so on.

consecutive months before it is successfully closed. A shorter benefit realization period for YBs was allowed initially to encourage employees to take up improvement opportunities and get certified. Now that we have a good uptake of Six Sigma in the organization, we plan to increase the benefit realization period for YBs to be three months as well.

In an ITES company, where the client determines operational priorities, traditionally, we have witnessed a change in focus every three to six months. With this as a business reality, we have set targets for sustaining results such that teams put processes in place that deliver month after month while ensuring that changes in business realities do not adversely affect Six Sigma certification.

Formal project closure requires the project team to present its project to the CI leader along with another site director or VP of operations, depending on the level of the project. Inviting another impartial leader to review Six Sigma projects ensures objectivity in reviewing projects for certification and also enables the sharing of best practices across the organization.

2.3 Leadership Engagement

An *organization's culture is driven from the top*, whereas the level of acceptance by the lower levels of the organization determines the success of execution of a strategy. With this in mind, CI strategy has leaned heavily on leadership support to demonstrate commitment in action. More than 60 percent of Convergys India leadership has taken at least one form of Six Sigma training in the last three years, and they are engaged in the Six Sigma initiatives governance at various stages. The managing director (MD) and CEO review half-yearly and yearly CI US dollar benefit numbers and share these with Convergys's global leadership. The MD and CEO have also participated in CEO forums on business excellence and shared the Convergys CI success story with others.

2.4 Communications

Often, Six Sigma is considered too complicated, time consuming, and therefore best left to the "nerds," who understand statistics and might have no idea of the business needs. To break this myth, it is important to simplify Six Sigma and make it a part of everyone's work life. At Convergys India, we have adopted the approach of making Six Sigma fun. Keeping this in mind, in 2008, the CI team created a cartoon character as a brand ambassador and named him Kaizen. Kaizen is the established face of CI across Convergys India. He comes with the regularly delivered joke based on Six Sigma and related to the latest hot topic (see Figure 26.4). Kaizen launches quizzes and contests and announces Six Sigma contest winners. Kaizen has made Six Sigma a part of water cooler conversation, and this is truly driving awareness and acceptance of Six Sigma throughout the organization.

Figure 26.4 A biweekly cartoon from "Kaizen Says" series.

2.5 Employee Engagement

Kaizen has provided the CI team the much-needed link with the large employee base of Convergys India. Kaizen reaches out to 1,500 employees by sharing the latest CI updates when employees log on to the corporate network. Every two weeks, a new "Kaizen Says" cartoon is sent out to 2,500 employees on the Convergys network domain to share a new Six Sigma concept in a fun manner. Kaizen is also a coveted feature of our quarterly employee touch magazine *Linked*, where he brings a new Lean Six Sigma quiz and announces Lean Six Sigma programs.

Connecting with the core employee base is a critical element of making Six Sigma successful in an organization. Six Sigma teams in the West have traditionally worked as business excellence groups who mainly interact with top and middle management. This can alienate them from the employee base, and therefore, a large number of improvements, which can be done quickly by the subject matter experts, may be missed. On the other hand, the Japanese quality teams primarily focus on the shop floor (or *gemba*) and thus, they sometimes miss out on the need for directional improvements. In our view, it is important to link with all levels of the organization to ensure incremental as well as exponential improvements in the organization. With our Six Sigma governance model and the large-scale employee touch programs, we have attempted to stay connected with all the levels of the organization at Convergys India.

2.6 Benefit Realization

There are no surprises that year-on-year critical business operating metric improvements have also resulted in more and more financial gain certified by the CFO of the company. While 1 to 3 percent return on revenue and eight to ten times return on quality implementation cost is the standard in the market, *Convergys India has recorded more than 4 percent* return on revenue and *more than 10 times return on CI costs* in the three years since we undertook the CI journey. In line with Convergys's strategic thinking, more than one-third of the benefits are client facing, where clients' total cost of ownership (TCO) with Convergys has been affected positively, and this has helped Convergys India in building a partnership relationship with its key clients.

The proudest moment was when Convergys India won the D.L. Shah National Award on Economics of Quality in Service Sector in 2009 awarded by the QCI (Quality Council of India) in recognition of its Six Sigma initiative to drive improvements in issue resolution for a telecommunications client. We realized more than US$1 million in collective benefits for the client, Convergys, and our agents. It is a great validation of the direction and pace of progress in Convergys's quest to be the best in operational excellence in the ITES sector in India.

3. Gathering Momentum

Often, great starts don't lead to long-term successes. Aversion to accepting the need for change is the cause for entropy to set in and leads to a loss of momentum. Just as any probability curve has its point of inflection, so does any process or practice. It is important for CI leaders to stay linked with the business and introduce new programs at regular intervals to address the evolving business needs.

3.1 Performance Dashboards

As the CI practice has matured and become well established at Convergys India, we have enlarged the scope of Six Sigma from focusing on improvement projects to supporting *business as usual* with dashboards on metrics that matter. This evolution was made possible as critical client-driven metrics stabilized, thanks to strong performance aided by the Six Sigma project. This stabilization allowed Convergys India to focus on consistently delivering high performance. The objective of these dashboards is to enable management to correct negative trends before they result in issues requiring concerted improvement initiatives. This has resulted in more consistent and predictable performance across the organization.

The dashboards we have designed to support operations are not typical control charts. In a people-driven business like ours, performance variation is natural. Variation occurs in day-to-day delivery as well as between agents. Thus, to ensure consistently high performance, it is essential for an ITES company to manage variation in delivery. Standard deviation influences control limits in charts. Therefore, control charts assume inherent stability in the process under review. In a people-driven delivery, standard deviation is not consistent over time and across teams. Thus, typical control charts do not fit the bill for managing variation across time and team members; instead, box plots are used for weekly reporting on critical metrics such as customer satisfaction.

In the ITES industry, often effectiveness and efficiency metrics are averages, but the problem with averages is that they are inadequate to examine the impact of individuals on averages. For example, if a team of 10 people collectively met their customer satisfaction target but achieved this through the extraordinary efforts of two individuals, we have a risk in repeating this performance, which is not as good as it appears on the surface. In an industry with significant attrition, it is prudent to attain performance targets through more agents than a few exceeding it substantially. To counter this risk, we introduced the concept that at least 50 percent of the team members should meet the target and we should try to have more people meeting the target to ensure it is met and exceeded consistently. This has brought more focus on both team and individual performance, since everyone's performance is being reviewed and not just the team average. Basically, what you measure tends to improve.

The CI team has designed and implemented a system called Agent Performance Dashboard (APD), which provides online and real-time dashboards of critical metrics for operations (see Figure 26.5). This facilitates review of performance trends over time and across teams, and provides managers the tools to monitor and manage performance on a daily basis.

Figure 26.5

An example of Agent Performance Dashboard.

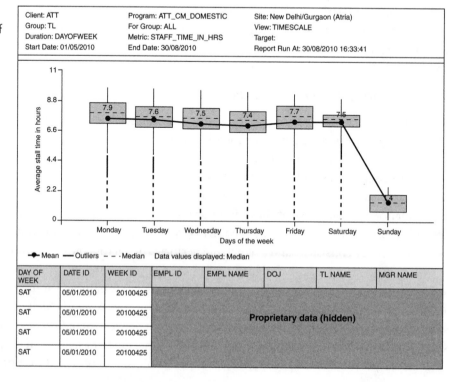

Focus on variation in performance has helped the middle management of Convergys India to align more with clients rather than managing performance issues. This has enabled Convergys India to build better relationships with key clients and work on client-focused improvements.

3.2 Client-Focused Improvements

While Six Sigma and CI practice focused on operational performance improvement earlier, beginning in 2010, the primary focus of Six Sigma practice at Convergys India has been on *client-focused improvements*. This evolution has come about as we have built stronger relationships with our clients and has provided us a "seat at the table" when our clients discuss improvements. We collect ideas on how we can improve client processes regularly from subject matter experts and present them to client leadership teams. We also leverage Six Sigma tools and techniques to understand the vast performance and customer data. This information enables us to provide valuable feedback in quarterly business reviews and follow up with joint action and client task force teams. The expectation is that the joint action teams will not only deliver beneficial results to clients, but also provide vital insights about our client business and take the relationship to the next level.

Our close association with our clients' business improvements has been beneficial in expanding our relationship with some of our clients. Figure 26.6 shows the relationship map of one of Convergys India's top five clients by revenue. In 2007, through a Six Sigma BB project, we improved performance for the client's program in one of our sites and were able to meet and exceed all performance metrics thereafter. In 2008, we worked on reducing the client's TCO with Convergys through various initiatives, some of which were aided by Six Sigma. In 2009, this resulted in Convergys winning more work from the client, both in India and the Philippines.

Figure 26.6

Convergys India's relationship chart with one of its top 5 clients by revenue.

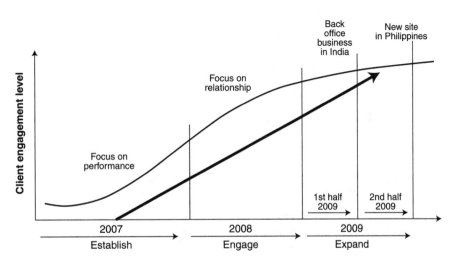

3.3 Business Intelligence

There is a growing trend for new clients to want *analytics as a part of the delivery* of work from Convergys India. The cherry on the cake is the fact that they are willing to pay for it, since they see value in understanding their customers' spending patterns and product and service recommendations in order to improve their market share. As more and more post-sales work is outsourced, clients are also outsourcing interactions with their customers. Thus, it is natural that more and more clients are looking for their back-office partners to leverage their experience with their customers along with customer spending patterns to provide critical customer insights. Figure 26.7 is an example of analytics used on customer demographic parameters (including customer age) to identify what affects customer satisfaction with client products and services.

3.4 Lean

Starting in 2010, Convergys India has embraced *Lean* in addition to Six Sigma and is now proceeding to leverage this program to reduce waste in its upstream processes to positively affect Convergys's bottom line. The plan is also to leverage Lean in engaging and empowering employees across sites to identify and reduce waste in their work area. The objective is to provide a new impetus to the people who work for Convergys India in 2010.

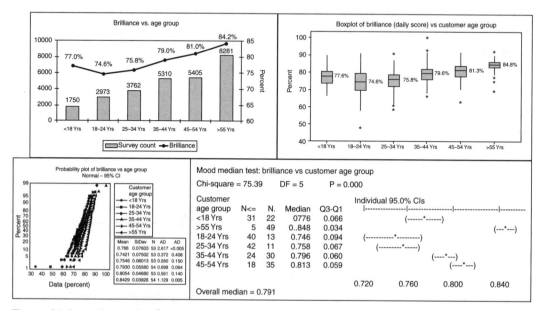

Figure 26.7 An example of analytics used for sharing business intelligence for one of Convergys India's top 5 clients by revenue.

3.5 CI Team Constitution

As the CI practice has matured in Convergys, the focus on CI team constitution has not been lost. Earlier, only experts were a part of the team; however, now one seat has been kept for a deserving employee in the business to take a rotational job assignment with the CI team for 18 to 24 months to work as an expert and go back to the business with better skills. The idea behind this move is to make Six Sigma the culture of the organization. This movement is planned to be bidirectional. In order to build better business knowledge, we also encourage CI professionals to take up assignments with core operations. In the beginning of 2010, one resource from CI moved to operations to lead a site-level client engagement and another joined the team from operations. Also, every 6 to 12 months, the CI team members are realigned to different sites and programs. In this way, the CI team retains its objectivity and gets an opportunity to spend time with different functions and sites to build a better overall understanding of the business. Three years of this deliberate churn has reaped good benefits, and the CI team has now become the team to go to for troubleshooting when extraordinary support is required.

3.6 Six Sigma and Organization Changes

To strengthen the culture of continuous improvement and make Six Sigma part of the organization, we have linked movement within the organization, whether lateral or vertical, with demonstrating Six Sigma knowledge. While we expect agents to complete WB training assessment successfully, supervisors also are expected to take the YB training and pass an assessment. Moving forward, managers and senior managers will also be required to take a data assessment test, which determines their data interpretation and data analysis skills. Thus, there is awareness and acceptance that Six Sigma and data analysis is an important skill set in Convergys India. In contrast, other organizations have pushed Six Sigma through the completion of projects. If not done correctly, the risk of taking up initiatives as Six Sigma projects is that it may encourage all improvement efforts, even trivial ones, to be executed as Six Sigma projects. At Convergys India, we have leveraged the learning of other companies in this regard, and our approach has been largely accepted and supported at all levels of the organization.

3.7 CI Practice Leveraged with Prospective Clients

We showcase the CI practice to all prospective clients, and the feedback has been encouraging. We hear more prospective clients considering our Six Sigma practice and successful case studies in making their outsourcing decisions. Our Six Sigma framework and case studies are also a part of all Convergys India's request for proposal (RFP) responses.

3.8 Relationship Building with Quality Fraternity

Convergys's CI team has built strong relationships with the quality fraternity in India and actively engages in forums, discussions, and action groups in the interest of quality movement in India. We regularly invite respected CEOs and quality leaders to share their views with us, which helps shape the quality deliverables for our organization. Likewise, we are also invited to address other companies' leadership and quality professionals to talk about how we have implemented Six Sigma at Convergys India. This best-practice sharing is helping Convergys and other recognized organizations come closer to and evolve better-quality practices.

4. Conclusion

While Convergys India's CI practice is now truly evolved, we are aware that we need to continuously keep ourselves engaged with our business and the environment it works in so that we are the true change leaders for our organization. We recognize that the road to success never ends. It's a journey well started, well traveled, and it has been fun. The plan is to continue to learn and evolve. As Arthur Ashe said, *"Success is a journey not a destination."*

CHAPTER 27

Bumps in the Road

As we wrap up our review of Six Sigma practices, methods, and programs, we have offered some important conclusions.

▲ Six Sigma methodologies, in their various forms, work; and they work very well. They are quantitative methods with repeatable and well-documented effectiveness.

▲ We heard testimonials and have been witness to examples of how these methods can be applied and have cited the financial and technical successes of Six Sigma projects.

▲ We have seen firsthand accounts from companies about the profound impact that Six Sigma has had on operational performance with documented financial results. Such programs, when successfully established within organizations and inspired by good leadership, can revitalize organizations, keep them well tuned, and give them a competitive edge in the markets and businesses where they operate.

However, there are other lessons that remain to be learned.

Key to success depends on more than possessing the right motivations and goals, developing the right skills, and performing all of the unique Six Sigma activities within a well-managed project thoroughly and completely. It also requires that you avoid the common pitfalls and mistakes and learn from failures as well.

Sometimes to get to your destination, even with the best driver and car, you need to avoid the potholes and "bumps" in the road.

Six Sigma is not a panacea. It is not a cure-all or a magic bullet.

Six Sigma projects can fail, and even the most dedicated practitioners and the most motivated organizations can fail. There are limitations to Six Sigma and what it can do. There are some common pitfalls that we should note, if only to help us in our own endeavors so that we can avoid them.

In previous chapters we discussed critical success factors and focused on how project success can be achieved by doing things right, using structured processes, and quantitative techniques. We now look at the "other side of the coin" to examine how to avoid doing things wrong and repeating mistakes that can lead to failures of Six Sigma projects and programs.

Just as we perform failure mode effects analysis for processes and products, we need to examine things that can go wrong with Six Sigma projects and programs.

The following is some advice based on common failure modes that need to be recognized and avoided.

1. **Start slow.** Just because you might have been inspired by success stories, here or elsewhere, or you are excited about bringing these new strategies and solutions home, don't rush too fast.

   There is a natural resistance to change that occurs whenever new endeavors are suggested or proposed. You can expect that this will happen when you recommend Six Sigma at your company. Therefore, start slow, influence key stakeholders—one at a time—and gradually build allies. After planting the seed, training is the next logical step—not for the masses, but for senior management. They are the ones whose hearts and minds you need to win first. Teach them about Six Sigma concepts. Familiarize them with new terminology and jargon. Talk about problems that currently exist or that have occurred in the recent past and with success stories, such as those in this book. Show them how other companies have successfully overcome these problems with Six Sigma. While you may embrace the idea, there is a lot of organizational and psychological momentum that you have to overcome before you can make progress in a new direction.

   You will inevitably hear cries that "We're not ready for this!" or "Why do we have to do things *this* way? What's wrong with the way things are working now?"

   You have to address these concerns, one at a time. Patience is essential. Like a good teacher, you will need to coach, mentor, and guide with an abundance of patience. If you try to just plow ahead without some level of acceptance, if you try to implement a strategy called success-by-mandate, you will create more resistance than you started with. You may well scare people away or disenfranchise those you need to have on board.

   One common mistake is to try to start too quickly and force people to become Black Belts. This generally doesn't work. You need to find and empower people who have solid change agent and influencing skills, excellent technical skills (often with a love for math), and a passion for problem solving. These people are often hard to find and such skills take time to develop. You may start with a small team that will grow over time as your successes are recognized. You cannot, after all, boil the ocean. Normally, organizations start small and grow over time.

   However, resistance can become entrenched if you narrow your scope too much. Departments and organizations may build cultural or organizational roadblocks to keep your team away. They may say "Six Sigma won't work in HR" or "It won't work in sales or services." If they say it enough, others will eventually believe it. You need to open opportunities wherever possible to solve problems and keep them open, but just as you have to prioritize projects, you may have to prioritize the portions of your business that need Six Sigma the most. Avoid stereotypes. Six Sigma solves problems. Focus on the problems and your opportunities will follow.

2. **Secure management commitment.** One of the primary reasons new Six Sigma programs fail is a fundamental lack of management commitment. This unwillingness to commit can manifest itself in different ways and in differing degrees, but all of them ultimately lead to Six Sigma programs that have limited effectiveness or programs that stall or wither and die.

A common lack of commitment is characterized by supporting only "part-time" Black Belts. In this case, your Champions serve two masters (their regular boss and the Six Sigma program leader) and with a basic inability to focus on Six Sigma activities, they multitask and projects stall and slow down accordingly. We know from personal experience that projects with inadequate resources and lacking in momentum more often than not miss commitments or end in spectacular failure.

If a company will not commit to someone in a full-time role, it is because they do not feel it is important enough, and this priority is recognized by the rest of the company and your would-be Six Sigma practitioners as well. This lack of commitment can also manifest itself in a dearth of certified Black Belts available to work on projects. Too few Black Belts will be unable to address anything except crises, and the proactive potential of Six Sigma in addressing problems while they are small will be lost. Your Six Sigma professionals will be 24-hour firefighters and, as such, they may burn out and further reduce your pool of insufficient resources.

Yet another key indicator of weak management support is a lack of recognition or incentives for Black Belts. The best programs have significant bonuses or financial incentives for individuals dedicated enough to earn these certifications. (The certification process is typically one year for Green Belt candidates and two years for a Black Belt candidate.) Without significant recognition, including financial bonuses or incentives, most candidates drop out of the program before completion. If these people are important and if their unique skills are valued, this needs to be reflected in the organization's policies and actions.

Lack of participation by management is another common shortcoming. If senior managers never participate in project reviews, it will become apparent that these projects are not that important and management support is mere lip service. This pattern of noninvolvement can permeate projects and result in teams that do not show up for meetings or that do not follow up on their work.

The low regard for Six Sigma in an organization can also manifest itself with attitudes like "Six Sigma is only for the quality department." In such environments, Six Sigma can be trapped and constrained to the point that the most important opportunities will be unaddressable.

Another diverting tactic is to put the success or failure on the consultant who was hired to bring in the program. By distancing themselves from risk of failure or disfavor, management can sabotage their Six Sigma program before it even gets started.

Without visible management commitment, the naysayers, skeptics, and cynics in any organization may thwart all efforts to bring in Six Sigma, and thus inherent organizational resistance that opposes any change will defeat even the most ardent champions of change.

3. **Handle data carefully.** Another interesting way that Six Sigma programs fail results from the misuse of data. In some cases, too much time is spent collecting and analyzing data. This can stall and greatly delay projects. Practitioners need to recognize that data is never

complete or totally accurate. Decisions ultimately will be made in the presence of uncertainty, and there will always be some level of risk involved. Trying to gather all possible data and attempting to render decisions with the absolute minimum of uncertainty and risk can lead to a phenomenon commonly known as *analysis paralysis.* Remember, Six Sigma projects are supposed to be fast and efficient. Taking many months to gather and analyze data can belie those goals.

Another strange behavior that sometimes emerges is the other extreme of inappropriate use of data. Sometimes, in an effort to expedite projects, teams jump to conclusions or solutions based on brainstorming activities alone. Under the pretense of employing Six Sigma, this approach reinforces the pre-Six Sigma method of just going with your best guess or hunch. In such cases, solutions may not be appropriate, problems may be perceived rather than real, and improvements may not be quantifiable or measurable. To do it right, you need to maintain the rigor of Six Sigma methods and techniques.

The last problem with data is that it can be manipulated. Sometimes people will claim improvement, not because they reduced defects, but rather because they simply redefined what a defect really is. The result is that it may look like defects have been reduced, but this is just a change in how they are counted and may not reflect real improvement at all. Similarly, financial results can be manipulated and counted differently to make it look like projects are far more successful than they really are. Without financial oversight on financial returns, benefits can be counted multiple times and savings may not be real. In extreme situations, departments with five million dollar budgets may claim ten million in savings, but these improvements exist only on paper and are often not real. Financial experts should sign off and approve any financial benefits that are claimed.

4. **One size does not fit all.** A single methodology is not sufficient. Just as problems come in many shapes and sizes, so do different organizations have different personalities and characteristics.

Six Sigma is a tool, not a solution. Just as no toolbox should contain only one tool, so, too, should no organization limit itself to just Six Sigma methodologies. Six Sigma complements other methods—the 8D method and Kepner-Tregoe problem-solving methods are just two other examples.

Keep in mind that Six Sigma is a proven problem-solving method and process improvement methodology. However, don't limit yourself to it; don't blind yourself to other opportunities. Look for other ways of doing things. Keep an open mind; explore; discover new best practices; learn, grow, and change over time.

Indeed, after Motorola developed Six Sigma in the 1980s, it became popular with other companies and it evolved. Six Sigma has grown a lot from its beginnings at Motorola. As it was accepted, adopted, and adapted by other organizations, it changed, grew, and improved. New methods were developed and existing ones (like Lean) were incorporated.

Six Sigma is a good tool, but it is not the only one. Don't get so focused on one solution or method that you become blind to other paths to success. Adopt other methods that work. Do not become staid and complacent with how you manage projects just because they are Six Sigma projects. Use the tools that work best and adapt your processes as you and your companies grow in maturity.

5. **All projects are not Six Sigma projects.** The Six Sigma method works great for some types of problems, but applying Six Sigma methodology to all projects is unnecessary and inappropriate. Not all projects are suited for Six Sigma.

Applying DMAIC or DFSS to simple projects is overkill. DMAIC should be limited to process improvement projects that are complex in nature. DFSS or DMADV should be limited to projects that need breakthrough designs and creativity.

Only about 10 percent of projects really need Six Sigma techniques. Focus your Green Belt and Black Belt resources on those projects that really need those skills.

Applying Six Sigma to all projects is terribly wasteful. Unfortunately, when you have a hammer, everything looks like a nail. A strong commitment to Six Sigma may be admirable; overusing this method can actually discredit it.

The Six Sigma approach should only be used for projects where statistical methods and formal methods are really needed.

Don't pass off routine projects with known solutions as Six Sigma projects. Six Sigma is best suited for projects with unknown solutions.

6. **Don't oversell your program.** As you build on successful Six Sigma projects, remember to keep your claims and expectations realistic. Today, there are several large Fortune 500 companies where Six Sigma has become a "bad word" because it was oversold and overhyped. After initial success and growing popularity, too many people were trained and too many projects were run as Six Sigma projects when it was obvious that it wasn't always appropriate. Over time, people became disillusioned with the sales pitch and managers backed away from the program to use simpler methods.

Keep your projects focused and in scope. Don't create Six Sigma projects just for the sake of having some. Don't set quotas for your organization (e.g., "We will have 100 Six Sigma projects completed this year"). Don't start projects that are not linked in some way to your organizational goals. Don't require that *all* processes must achieve a certain Six Sigma process capability. Don't create projects that are too large in scope or that have "fluid" project charters. Focus on well-defined problems of manageable scope and address them efficiently.

By overselling Six Sigma and applying it where it is not needed, programs can discredit themselves and drive potential users away.

7. **Train your people and keep them trained.** Another reason that Six Sigma programs often fail is that programs can become bureaucratic over time and become too entrenched. Six Sigma requires specific skills and practitioners need to be familiar with tools and techniques. It takes time to train and develop good Green Belts and Black Belts to manage projects, but keep adding new practitioners over time.

Unfortunately, your best Green Belts and Black Belts may leave your organization over time. Some may be promoted to leadership positions as many companies groom Six Sigma leaders to be business leaders. Some may migrate to other positions where they cannot work on Six Sigma projects. Turnover is a natural state of business; you need a part of your program to focus on training replacements and nurturing new people to take the place of the ones leaving. Indeed, the migration of Six Sigma–trained professionals elsewhere in the organization can only be a good thing. They become evangelists. They help make Six Sigma pervasive.

Six Sigma programs often fail because the wealth of talented practitioners disappears over time and the program has no mechanisms to deal with natural attrition. With no plans established to replace lost resources, essential talent dries up and the program evaporates as well.

On the other hand, Six Sigma programs have also failed because they became so bureaucratic that it became too difficult to reach certification as a Black Belt. It took so much time to complete training and projects that no one ever reached their certification goal. If training and certification is too long or too hard, people give up or drop out before completing their certification and the pool of practitioners will gradually diminish and disappear, leaving the organization with no one who uses these wonderful tools.

Planning for turnover must be a part of a Six Sigma program. In some companies, those who achieved Black Belt status became so locked into these roles that they were not allowed to be promoted or transferred. Newcomers, not wanting to get locked permanently into such roles, declined training that they would have immensely benefited from. In some programs, full-time Green Belts and Black Belts rotate to non-Six Sigma roles on a regular basis to retain their skills and to offer alternative opportunities for career advancement. Such rotations also reduce burnout that can occur. If turnover is not planned, it will occur anyway, but at the cost of a compromised Six Sigma program.

8. **Don't sit on your laurels.** Remember, just because you have achieved success, this doesn't mean that you are done and that you have reached your destination. Aesop's story of the tortoise and the hare is relevant. Don't become too satisfied with success or you can lose it.

Take a lesson from the Lean Six Sigma philosophy. Six Sigma is a recursive process. Look for problems, fix them, then go back and look for more. If you stop, problems will reappear and grow into crises over time. If you have a really good program, you will find yourself chasing smaller and smaller problems. This is a good thing, but just because there are no crises occurring, just because you have put out all the fires around you, doesn't mean that you are finished.

Entropy, both in processes and organizations, is real and Six Sigma is your tool to address the inevitable decay and decline that will occur if you stop trying to seek more and better ways of doing things.

Some of the largest companies who practiced and applied Six Sigma lost focus when the big problems were solved and their programs slowly eroded and disappeared.

An early sign of this can be seen in projects where process monitoring stopped when the project closed. Remember that control plans specify activities to sustain and maintain success … and that is one important goal of Six Sigma. Success needs to be sustainable. Improvements should be based on permanent corrective actions, and monitoring should be continuous and ongoing. Success should be enduring and not just a "flash in the pan" that is gone when the project team adjourns.

9. **Don't deny failures when they occur.** Finally, recognize that Six Sigma is not a panacea. Six Sigma projects are not foolproof. Six Sigma projects can and do fail.

   Anecdotal industry data suggests that 75 percent of all projects (not just Six Sigma projects) fail (50 percent do not reach completion; half of those that do complete do not achieve their goals). Because of the rigor and discipline of Six Sigma, DMAIC, DMADV, DFSS, and Lean Six Sigma projects may have a higher success rate, but they are not infallible.

   You should expect a reasonable percentage of Six Sigma projects to fail. Face it, you won't be able to fix some problems … they are too big, or too complex, or they change too rapidly over time; or the solutions may not be financially, technically, or operationally feasible to implement. Like it or not, some problems are bigger than you are.

   Unfortunately, in some organizations Six Sigma is held in such esteem that failures cannot be admitted. In such organizations, Six Sigma projects cannot be cancelled and, if they don't work, failure cannot be admitted. Where this occurs, the Six Sigma program can slowly discredit itself and fall into disrepute.

   Therefore, when Six Sigma projects fail, don't hide them. Track them and count your percentage of successes, just like you would other projects. Admit your failures; learn from them.

   Use the different phases as checkpoints, just as in regular projects. The end of each phase is a decision point: "Should we proceed or not? Does the problem still make sense? What have we learned? Is it appropriate to continue?" Don't continue a project that is not working just because it is a Six Sigma project.

   Don't turn Six Sigma projects into sacred cows. Don't become Pygmalion and fall in love with your model. Be realistic and remember that Six Sigma is a good method that improves chances of success, but success is neither certain nor guaranteed.

In summary, keep your programs realistic. Don't oversell or overhype the methodology. Avoid too much bureaucracy. Embrace other methods when appropriate. Plan on turnover. Don't become complacent. Keep things simple and use Six Sigma where appropriate, but do not overuse it.

Focus on the problems, not the tool or the solution … and Six Sigma is just a tool. Handle it with care.

APPENDIX A

Chapter Tools Matrix

Part One: DMAIC Chapters

	Chapter 3	Chapter 4	Chapter 5A	Chapter 5B	Chapter 6A	Chapter 6B	Chapter 6C	Chapter 7A	Chapter 7B
Charter	X	X	X	X	X	X	X	X	X
VOC/CTQ		X	X						
Process Map	X	X		X		X	X		X
Measurement Plan	X	X	X	X	X	X	X	X	X
MSA			X	X					
FMEA	X	X					X	X	
Kano Analysis								X	
SIPOC		X	X		X			X	
Cpk			X						X
RCA	X							X	
SWOT		X							
Cause and Effect		X	X			X		X	X
VOC Survey		X						X	
Descriptive Statistics		X					X		X
Test of Normality		X							
Correlation	X	X	X						
Pareto Chart	X	X	X	X	X	X	X		
Comparative Methods		X	X	X	X				
ANOVA		X							
Kruskal-Wallis		X							
Box plots		X		X		X	X		X
Mann-Whitney						X			
Effort-Risk Matrix		X				X			X
SPC		X	X	X		X	X	X	X
T-test			X	X			X	X	X

562

	Chapter 3	Chapter 4	Chapter 5A	Chapter 5B	Chapter 6A	Chapter 6B	Chapter 6C	Chapter 7A	Chapter 7B
Trend Chart			X		X	X		X	
Scatter Plot			X						
Control Plans—Corrective Action Plans	X	X	X	X		X	X		X
Financial Benefits	$568,000+ Millions (intangible)	$500,000	$225,107	$760,000	$600,000	$4 Million	$9 Million	$3.8 Million	$500,000

Part Two: Lean Six Sigma Chapters

	Chapter 10	Chapter 11	Chapter 12	Chapter 13	Chapter 14
Charter	X	X	X	X	X
Measurement Plan	X	X	X	X	X
Process Map	X	X	X	X	X
Value Stream Map	X	X	X	X	X
FMEA	X				X
Cpk	X				
RCA	X			X	
Kaizen	X	X	X	X	X
Kaizen Blitz	X		X	X	
Kano Analysis		X	X		
Kanban				X	
Heijunka				X	
Muda Analysis				X	
Cause and Effect	X				
VOC Survey			X		X
Descriptive Statistics	X	X	X	X	X
Monte Carlo Simulation		X	X		
Time Series Analysis	X				
Correlation	X		X		
Pareto Chart	X		X		
Contingency Charts	X				
Comparative Methods	X		X		
Box Plots			X		
Mann-Whitney					X
SPC		X		X	
T-test		X		X	

	Chapter 10	Chapter 11	Chapter 12	Chapter 13	Chapter 14
Trend Chart		X		X	X
Control Plans— Corrective Action Plans	X		X	X	X
Financial Benefits	$150,000–$1 Million	$72,000	$1.08 Million	Several Million	$840,000

Part Three: DFSS Chapters

	Chapter 17	Chapter 18	Chapter 19	Chapter 20	Chapter 21	Chapter 22
Charter	X	X	X	X	X	X
Measurement Plan	X	X	X	X	X	X
MSA					X	
Process Map	X	X	X	X	X	
Value Stream Map						X
FMEA	X					
Kano Analysis	X					
SIPOC	X					
Cpk				X		
Kaizen Blitz						X
Heijunka						X
Cause and Effect		X		X		
QFD				X		
DOE				X		
VOC Survey	X				X	
Descriptive Statistics		X		X		
Test of Normality				X		
Monte Carlo Simulation	X	X	X	X	X	
Sensitivity Analysis				X	X	
Time Series Analysis					X	
Correlation		X			X	
Pareto Chart		X				
Comparative Methods		X				
Box Plots		X				
Mann-Whitney		X				
Prioritization Matrix				X		

	Chapter 17	Chapter 18	Chapter 19	Chapter 20	Chapter 21	Chapter 22
SPC		X				
T-test		X				X
Trend Chart	X					
Scatter Plot		X				
Control Plans			X		X	X
Communication Plan	X				X	
Training	X					
Pilot	X	X	X			
Financial Benefits	Several Million	$500,000	$8.2 Million	$1.8 Million	$12.5 Million	—

APPENDIX B

Computing Return on Investment

Quite often, Six Sigma project teams struggle with quantifying the financial benefits for their project. In its simplest form, the financial benefit might just be a dollar figure, but a true measure of the financial success of a project is provided by what is called return on investment, or ROI. Simply stated, ROI is a ratio of the benefits realized divided by the cost of executing the project. A project that delivered $50,000 in financial benefits may seem like a good Green Belt project, but if the investment was $60,000 and there is no other strategic advantage to executing the project (other than the $50,000 benefit), then it really might not be such an advantage after all.

In the first chapter we talked about some key program success factors, such as valuing results and recognizing all financial benefits. We saw several success stories in this book where the authors elaborated on how they computed financial benefits, and some even reported ROI.

This appendix will help you understand what ROI is and how to compute it properly.

Goal

The purpose of ROI is to measure the *value* of a funded initiative, project, or endeavor.

The goal is to ensure that activities that consume money, time, or resources (i.e., labor and material) are in fact *value-added* activities.

Question

To arrive at a measure that would quantify value, one may begin by asking the following questions:

- ▲ What is the payback of the activity (project, initiative, or endeavor)?
- ▲ Does the value or payback exceed the effort to perform the activity?
- ▲ Is the improvement sufficient to justify the time, effort, or resources spent in performing the activity?

Metrics Definition

In the simplest terms, *return on investment* is the ratio of benefits over costs.

Expressed as a ratio, values less than 1.0 represent projects or endeavors that do not return as much in benefits as was spent to achieve them.

Ratios of greater than 1.0 represent projects or endeavors whose completion returns a quantitative value that exceeds the cost of executing the project.

While all ROIs greater than 1.0 are technically profitable ones, not all ROIs of this type may be justifiable.

ROI calculations and estimates are often used to evaluate the effectiveness or value of projects during the proposal phase and may be used as a mechanism for prioritizing projects not yet approved (this is because the greater the ROI, the greater the payback or "bang for the buck"). Nevertheless, other factors may exceed the use of ROI in such decision making (e.g., cost of capital, risk, or probability of success.) Refer to the section "Supplemental Information" for more details.

Attributes

Since ROI is a ratio, it must necessarily consist of at least two parts:

▲ The numerator is a value that quantifies the benefits of the project or endeavor.
▲ The denominator is a value that represents the cost or investment associated with the work.

However, each of these measures (returns and investments) may consist of multiple elements.

For instance, *investments* may consist of:

Direct Costs

▲ Effort or labor
▲ Materials
▲ Travel
▲ Software licenses
▲ Hardware (servers)
▲ Administrative costs (mail, shipping, insurance, etc.)

Indirect Costs

▲ Training
▲ Consulting
▲ Meetings and teleconferences
▲ Administrative overhead (additional meetings)

Similarly, *returns* may consist of:

Actual Increased Revenue

▲ Increased sales or revenue
▲ Increased market share
▲ Other collected payments (licenses, fees, and penalties received from suppliers)
▲ Rebates

Savings

▲ Reduced expenses (costs as listed earlier)
▲ Reduced headcount
▲ Reduced licenses, fees, and insurance

Cost Avoidance (Penalties, Losses, Spills)

▲ Reduced maintenance or development costs
▲ Avoided future expenses
▲ Reduced consulting fees
▲ Reduced travel and training
▲ Avoided additional increase in headcount
▲ Reduced cost of poor quality (COPQ), that is, repairs, spills, rework, etc.

Risk Reduction

▲ Quantifiable reduction in expected opportunity loss (use expected payoffs to quantify risk reductions)

Cost Reallocation

▲ Reallocation of work from one task to another (e.g., less time spent on coding, but that time is shifted or transferred to testing activities instead)

Refer to the section "Supplemental Information" for more details.

Equations and Calculations

Thus:

$$ROI = \frac{Returns}{Investments}$$

Data Collection Method

ROI should be estimated before a project (as part of a project's approval process). ROI savings should also be measured during a project and at its completion.

Also, ROI savings may be accrued for a period from six months to two years following a project, although most Six Sigma projects typically have a modest benefit realization period of three to six months just so that the Green or Black Belt candidate can conclude their project and achieve their certification. Requiring them to wait for one or two years after project completion to attain their certification would be discouraging to future candidates.

ROI is reported on a project-by-project basis, though total savings may be summed up across multiple projects in a given portfolio.

If this aggregation occurs, the ROI should not be averaged. The aggregate ROI should be based upon the ratio of total returns over total investments.

$$\text{Aggregate ROI} = \frac{\text{Total Returns}}{\text{Total Investments}}$$

Frequency of Collection

Collection of ROI varies based upon operational or project reporting schedules within an organization. Typically, project ROIs are reported quarterly, if they are available (with year-end roll-ups).

Guidelines for Interpretation

If project ROIs are high (typically greater than 150 percent (i.e., ROI = 1.5)), projects are viewed positively. The higher the ROI, the more favorable the project, in general.

ROIs lower than 115 percent (1.15) may not be a viable use of money or time, since return is low—only 15 percent or less over invested costs.

Projects with negative ROIs are usually not considered viable endeavors and are usually only approved out of necessity (e.g., changes in regulatory requirements or contractual obligations).

Guidelines for Action

ROI estimates should be reexamined periodically over the lifetime of a project. If cost savings do not appear to be on track, cancellation of a given project may be warranted. If project costs are higher than expected, then project cancellation may be the greatest way to save money. The targeted returns may simply not be worth the effort.

ROI should also be tracked after project completion, particularly when payback is not immediate and may be accrued over a year or more.

Again, ROI may not be the sole criteria for project selection.

A project with the best ROI may simply require too many resources and have very high risk, and instead a few small projects with more modest ROI may have a greater chance of success than one or two very large ones.

Other factors that might be considered in addition to ROI for project prioritization include:

▲ Probability of success (ease or complexity of the project)
▲ Total project cost
▲ Project time to completion
▲ Intangible benefits or costs (e.g., morale, brand loyalty, etc.)
▲ Synergy with other projects or initiatives
▲ Technical or managerial ability to support the project (resource availability)

Guidelines for Goal Setting

Typically, goals for projects are established using estimated ROI. However, in the case of projects or endeavors with high uncertainty, it is common to perform three levels of estimates: best case, worst case, and most likely. The use of this model allows not only calculation of a target ROI, but (because of the range of values) the estimation of variance of the project as well.

This is a form of scenario analysis in which the range of possible outcomes is evaluated and estimated. The disparity between these different cases helps define the "variance" of the project and its associated tasks.

a = optimistic value
b = pessimistic value
m = most likely value

Once these have been estimated, the *expected value* for each task can be calculated using the following formula:

$$\text{Expected value (VE)} = (a + 4m + b)/6$$

This is a binomial approximation to a normal distribution where the most likely value is weighed four times more heavily than the best and worst cases.

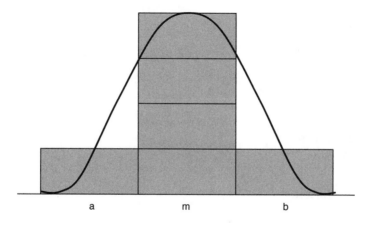

a   m   b

Correspondingly, the variance is given by:

$$\sigma^2 = ((b-a)/6)^2$$

Finally, the standard deviation, σ, is the square root of the variance, or

$$\sigma = (b-a)/6$$

Additional modeling can be performed using tools such as crystal ball. The use of these tools permits Monte Carlo analysis and simulation, which can be used for more detailed estimation, prediction, and analysis.

Roles and Responsibilities

▲ The data collector is typically the financial representative on a project or team.
▲ The data coordinator is typically the project leader or project manager. He is the one who will report ROI to management.
▲ The data owner responsible for ROI is typically the project sponsor or a member of the management team that approved the project.

Supplemental Information

ROIs can be supplemented by several other statistics that also reflect that value of investments. Some companion or alternate measures include:

▲ NPV: Net Present Value
▲ IRR: Internal Rate of Return
▲ Payoff Period (or Payback Period)
▲ Expected Payoff (probabilistic returns)

Some examples of ROI scenarios are provided next for purely illustrative purposes.

Case 1: Increased Revenue

In this scenario, ROI is justified mainly in terms of increased money that will result from the project or endeavor.

Values in the numerators may include:

▲ Increased sales/revenue
▲ Increased market share
▲ Other collected payments (licenses, fees, penalties)
▲ Rebates
▲ Other sources of income

If ROI estimates are performed and there is a high uncertainty of achieving these gains, the expected payoffs can be substituted (i.e., the value of the return times the probability of the return).

In all cases, actual numbers should be used for final ROIs (i.e., at project end or after project closure).

In all cases, the organization's financial representative (who is typically part of a Six Sigma project team) should validate these figures before they are used to report on a project's status or success.

Case 2: Cost Savings

In this scenario, ROI is justified mainly in terms of reduced expenses or savings.

The values that may be used in the numerators of the ROI calculation include:

▲ Reduced expenses (costs as noted earlier)
▲ Reduced headcount
▲ Reduced licenses or fees

Note: *Increased productivity is not a cost savings* unless it reduces expenses (perhaps in the form of reduced headcount or reduced labor costs).

Again, if ROI estimates are performed and there is a high uncertainty of achieving these cost savings, the expected payoffs can be substituted (i.e., the value of the return times the probability of the return).

In all cases, actual savings or reduced expenditures should be used for final ROIs (i.e., at project end or after project closure).

In all cases, organizational financial representatives should validate these figures before they are used to report on a project's status or success.

Case 3: Cost Avoidance

In this scenario, ROI is justified mainly in terms of anticipated costs that can be avoided as a result of the specific project or endeavor.

Values in the numerators may include:

▲ Reduced maintenance or development costs
▲ Avoided future expenses
▲ Avoided additional increases in headcount
▲ Avoided lost sales or revenue (e.g., capacity or regional marketing constraints)

Cost avoidance should only be claimed if it reflects expenses or lost revenues that would have definitely occurred if not for the project or endeavor.

For instance, if a lack of manufacturing capacity would have resulted in actual cancellation of booked orders, then a solution that increases capacity could claim those sales that would have otherwise been lost as cost avoidance.

Only in special circumstances can increased productivity be considered cost avoidance (see notes on productivity later in this appendix).

If there is a probability of costs or expenses occurring, then the guidelines under risk reduction (Case 5) should be used instead, since there is only a chance that the full cost or penalty would have applied.

Case 4: Cost Reallocation

In this scenario, ROI is justified mainly in terms of costs that are transferred from one portion of a process or organization to another. This is a specialized variant of cost avoidance.

1. **One example of this is increasing the productivity** of code generation activities in a software development organization.

   If time or labor savings are achieved in the coding phase of a project, ROI can be affected by the shorter resultant cycle time (as in Case 2).

   However, if the time saved in coding is reallocated (or reinvested) in the testing phase of the same project, then more testing can be conducted without reducing overall project cycle time.

   This type of cost reallocation within (or across) projects may constitute a real savings and immediate reinvestment back into other authorized work. However, this is difficult to quantify.

   For instance, one would have to be able to argue that the additional time spent testing caught more defects than would have otherwise not been caught. Next, one would have to assess the cost of poor quality (COPQ) that would have been associated with fixing those defects if they had not been caught.

   The savings that could be claimed would not be the time saved in coding, but the benefit of performing the additional testing and avoiding the cost of those defects that would have otherwise gone undetected.

   Simply spending less time coding and replacing it with other coding or tasks would not constitute a substantive savings. Cost reallocation is a valid ROI justification *if and only if* the substituted tasks have more intrinsic value than the tasks that were eliminated.

2. **Another example is process simplification** in which tasks or activities are eliminated altogether but where the people affected perform tasks that would otherwise not have been performed.

   Again, the justification must be based upon the case that the new tasks that are performed add value (or savings) that would not have otherwise been realized.

3. **A third example is to reduce COPQ** in which the quality of a process is improved, reducing the amount of time spent repairing, reworking, and analyzing root causes or retesting of defects.

When such quality improvements occur, work is often reallocated to other activities (creation, appraisal, prevention, the development of new functionality, or enhancing services to customers).

Again, such cost reallocations are only justified if the new work is effort that would not have otherwise been performed if the increased productivity had not occurred.

In all cases, actual numbers should be used for final ROIs (i.e., at project end or after project closure).

Case 5: Risk Reduction

The last case of ROI returns, and the most difficult to apply, is risk reduction. In this scenario, ROI is justified in terms of possible costs or expenditures that *might* have occurred.

An example of how to reduce risk is the purchase of hardware to improve system availability. For instance, a system might have a documented and measured system availability of 99 percent, costing the company $100,000 per year.

If a secondary or parallel machine can be purchased and installed in such a manner that it improves system fault tolerance by allowing failover from one machine to the other in the event of a failure, then the risk associated with that downtime can be reduced. If the new availability improves tenfold (to 99.9 percent), then the cost of the associated downtime would be reduced tenfold as well and could be applied as an ROI return.

In another example, if there is a 10 percent chance of a major quality problem on a new product line and processes or products are changed to reduce such occurrences to a mere 5 percent, then the difference between the expected (i.e., probabilistic) costs could be claimed.

For example:

$$\text{ROI risk reduction return} = (\text{Potential cost of quality problem} \times 0.10) -$$
$$(\text{Potential cost of quality problem} \times 0.05)$$

Note: For major quality problems, such savings can be significant. Such ROI strategies are commonly used to justify disaster recovery plans.

In cases where probabilities are low, risk reduction estimates must be used in place of actual savings values, since, by definition, you cannot quantify with absolute confidence the cost of things that did not happen.

Notes on Increased Productivity

Claims of savings related to productivity are problematic.

As noted earlier, there are cases where productivity can be applied to cost reallocation. However, it is rare that it can be used to justify cost avoidance.

Increased productivity *may* be considered cost avoidance *if* increased production results without a corresponding increase in labor or headcount. However, this is relatively rare, since a manufacturing facility, for instance, would have to be at 100 percent capacity for this to be successfully claimed.

If an organization does *not* increase production, and is *not* at 100 percent capacity, then the increased productivity merely increases the spare capacity of the facility, with no corresponding increase in revenue or reduction in labor or expense.

Similarly, if a facility were functioning at less than 100 percent capacity and it suddenly becomes more productive, it can claim increased sales (if any) and savings (if headcount is reduced), but cost avoidance due to increased productivity cannot be claimed. The reason is that the extra work could have occurred without the productivity increase, since there was already available capacity for that additional work.

On the other hand, if production does increase as a result of productivity improvements *and* the facility is at and remains at 100 percent capacity, then the increased revenues can be claimed, as well as cost avoidance that would correspond to the number of headcount that would have been necessary to achieve that increased production (i.e., we made more money without having to hirer X number of people—both increased revenue and cost avoidance apply).

Thus, increased productivity can only be used in ROI if revenues increase, if headcount or labor rates decrease, or in the rare case that output of a 100 percent capacity facility is increased without increase in headcount (therefore claiming both increased revenue and cost avoidance for the people who did not have to be hired).

Most commonly, productivity must be justified in terms of cost reallocation if it can be shown that the new tasks have more intrinsic value than the ones that were eliminated due to the productivity improvements.

Usually, productivity improvements *are* improvements, but are not claimed unless some other quantifiable return is documentable.

Typically, increased productivity should not be used in ROI estimates except as a justification (or explanation) for increased revenue or reduced labor costs.

It may be used as an explanation for cost reallocation, but should not (except in the rare cases mentioned) be used for cost avoidance.

Notes on COPQ Reduction

Cost of poor quality reductions is similarly fraught with difficult justifications. Normally, COPQ is associated with reallocation of work within an organization (Case 4). Sometimes it may be the rationale behind observed savings (i.e., reduced headcount or labor), since fewer repair activities need to be performed, resulting in lower maintenance costs.

However, COPQ savings should never be claimed as both savings (reduced labor) and cost reallocation, since this would be *double* counting the same eliminated labor.

Notes on Double Counting

As a final word of caution, those calculating ROI should be sure not to double-count benefits.

A particular reduction on time to complete a task in a specific process step may be interpreted in several ways.

▲ One might count this as a cost savings.
▲ One could argue that this is cost avoidance.
▲ Alternatively, one might consider this cost reallocation.

Whatever the determination, care needs to be taken to ensure that it is not argued or entered into calculations multiple times.

Indeed, one needs to determine if this is a significant benefit at all. If a task on a process or project is not on the critical path, then reducing that time may not save anything. The project may still take the same amount of time (critical path), the same people may continue to work on the process or project (so salaries and expenses are unchanged).

Benefits need to be real, not hypothetical. If real dollars and budgets are not saved, these details should not be counted as benefits at all.

Glossary*

Accuracy: The characteristic of a measurement that tells how close an observed value is to a true value.

Affinity diagram: A management tool used to organize information (usually gathered during a brainstorming activity).

Alignment: The actions taken to ensure that a process or activity supports the organization's strategy, goals, and objectives.

Analysis of variance (ANOVA): A basic statistical technique for analyzing experimental data. It subdivides the total variation of a data set into meaningful component parts associated with specific sources of variation in order to test a hypothesis on the parameters of the model or to estimate variance components. There are three models: fixed, random, and mixed.

Assignable cause: A name for the source of variation in a process that is not due to chance and therefore can be identified and eliminated. Also called "special cause."

Attribute data: Also called "discrete data"; refers to data measurements that are not quantified on an infinitely divisible numeric scale. Includes items like counts, proportions, ratios, or percentage of characteristics that have measurements like pass or fail, small, medium, or large, go or no go tests.

Autocorrelation: A technique in time series analysis that checks to see if patterns in a time series repeat themselves over time, i.e., tests whether a data series correlates to itself over different time intervals or delays.

Availability: The ability of a product to be in a state to perform its designated function under stated conditions at a given time.

Baseline measurement: The beginning point, based on an evaluation of the output over a period of time, used to determine the process parameters prior to any improvement effort; the basis against which change is measured.

Benchmarking: An improvement process in which a company measures its performance against that of best in class companies, determines how those companies achieved their performance levels and uses the information to improve its own performance. The subjects that can be benchmarked include strategies, operations, processes, and procedures.

Benefit-cost analysis: An examination of the relationship between the monetary cost of implementing an improvement and the monetary value of the benefits achieved by the improvement, both within the same time period.

---

* Note: Some of the definitions are obtained from the following sources with permission:
1. Quality Glossary, *Quality Progress*, July 2002, pp. 43–61.
2. Lean Glossary, *Quality Progress*, June 2005, pp. 41–47.

Best practice: A superior method or innovative practice that contributes to the improved performance of an organization, usually recognized as "best" by other peer organizations.

Big Y, Vital X: A term describing primary business goals (Y's) and the factors that contribute to these outcomes (x's).

Black Belt (BB): Full-time team leader responsible for implementing Six Sigma process improvement projects within the business to drive up customer satisfaction levels and business productivity.

Blemish: An imperfection severe enough to be noticed but that should not cause any real impairment with respect to intended normal or reasonably foreseeable use.

Bottleneck: Any resource whose capacity is equal to or less than the demand placed on it.

Box plots: A simple graphical technique for showing data sets; the "box plot" shows the median, 25 percent and 75 percent quartiles and whiskers with minimum and maximum values. Sometimes outliers and mean are also shown.

Brainstorming: A technique teams use to generate ideas on a particular subject. Each person in the team is asked to think creatively and write down as many ideas as possible. The ideas are not discussed or reviewed until after the brainstorming session.

Breakthrough event: A dramatic change in process during which a team gets past an old barrier or milestone to achieve a significant increase in efficiency, quality, or some other measure.

Burning Platform: A term coined from a true story in which four men were left stranded on the burning platform of a Piper Alpha oil rig on fire in the North Sea in 1967. The men faced the choice of staying where they were and facing certain death or taking the risky step of jumping into the freezing ocean and risking death from hypothermia. The two men who decided to remain behind perished. There are two elements to the story. The two who stayed put died. The unacceptable option is staying the same and hoping things get better. Against the odds the two who jumped into the sea survived. The message is that sometimes radical risky change is essential.

C chart: A type of SPC charts that displays counts of defects.

Cause and effect diagram: A tool for analyzing process dispersion. It is also referred to as the "Ishikawa diagram" (developed by Kaoru Ishikawa), or the "fishbone diagram" (the complete diagram resembles a fish skeleton). The diagram illustrates the main causes and sub-causes leading to an effect (symptom).

Cause effect matrix: A matrix variant of an Ishikawa or fishbone chart; useful when there are a large number of factors that are difficult to draw using fishbone diagrams.

CDOV: Concept, Design, Optimize, Verify—another variant of DFSS popular for use in robust product design.

Champion: A business leader or senior manager who ensures that resources are available for training and projects, and who is involved in project tollgate reviews; also an executive who supports and addresses Six Sigma organizational issues.

Change agent: An individual from within or outside an organization who facilitates change within the organization. May or may not be the initiator of the change effort.

Charter: A written commitment approved by management stating the scope of authority for an improvement project or team.

Classification of defects: The listing of possible defects of a unit, classified according to their seriousness. Note: Commonly used classifications: class A, class B, class C, class D; or critical, major, minor, and incidental; or critical, major, and minor. Definitions of these classifications require careful preparation and tailoring to the product(s) being sampled to enable accurate assignment of a defect to the proper classification. A separate acceptance sampling plan is generally applied to each class of defects.

Common causes: Causes of variation that are inherent in a process over time. They affect every outcome of the process and everyone working in the process (see also "Special causes").

Communication plan: A part of an overall control plan developed in the Control phase of a DMAIC project.

Comparative methods: Different methods of comparing to data sets; including correlation analysis, t-tests, ANOVA, SOV studies, and nonparametric tests.

Conjoint analysis: A statistical technique that requires analysts to make a series of trade-offs and analyze these trade-offs to determine the relative importance of component attributes.

Constraint: Anything that limits a system from achieving higher performance or throughput; also, the bottleneck that most severely limits the organization's ability to achieve higher performance relative to its purpose or goal.

Continuous improvement (CI): Sometimes called continual improvement. The ongoing improvement of products, services, or processes through incremental and breakthrough improvements.

Control chart: A chart with upper and lower control limits on which values of some statistical measure for a series of samples or subgroups are plotted. The chart frequently shows a central line to help detect a trend of plotted values toward either control limit.

Control limits: The natural boundaries of a process within specified confidence levels, expressed as the upper control limit (UCL) and the lower control limit (LCL).

Control plan: A document that describes the required characteristics for the quality of a product or service, including measures and control methods.

Corrective action plan: A part of an overall control plan developed in the Control phase of a DMAIC project.

Correlation: The degree to which two factors are related to one another, statistically represented by Pearson's correlation coefficient. Correlation does not mean causation since two factors may be related but not cause one another (e.g., height and weight).

Cost of poor quality (COPQ): The costs associated with providing poor quality products or services. There are four categories of costs: internal failure costs (costs associated with defects found before the customer receives the product or service), external failure costs (costs associated with defects found after the customer receives the product or service), appraisal costs (costs incurred to determine the degree of conformance to quality requirements), and prevention costs (costs incurred to keep failure and appraisal costs to a minimum).

Cp and Cpk: Capability indices to compare the output of a process to the specification limits. Cp is the ratio of permissible process variability divided by actual process variability. Cpk is a complimentary measure that takes into account the closeness of the mean of the sample to the target.

Critical to quality (CTQ): A method of listing and prioritizing different factors that affect the quality of a product, process, or service. May be determined through use of models or even through VOC surveys and requirements analysis.

Cumulative sum control chart (CUMSUM): A control chart on which the plotted value is the cumulative sum of deviations of successive samples from a target value. The ordinate of each plotted point represents the algebraic sum of the previous coordinate and the most recent deviations from the target.

Cycle time: The time required to complete one cycle of an operation. If cycle time for every operation in a complete process can be reduced to equal takt time, products can be made in single-piece flow (see "Takt time").

Delighter: A feature of a product or service that a customer does not expect to receive but that gives pleasure to the customer when received.

Design for Six Sigma (DFSS): A product and process development methodology related to traditional Six Sigma, focusing on robust design and defect prevention. Rich in the use of quantitative methods, it seeks to develop products and services that provide greater customer satisfaction and increased market share. DFSS is largely a design activity requiring specialized tools including: quality function deployment (QFD), axiomatic design, TRIZ, Design for X, design of experiments (DOE), Taguchi methods, tolerance design, robust design, and response surface methodology (see also "DMADV").

Design FMEA (DFMEA): An FMEA performed on new designs or products to proactively identify failure modes and take necessary preventive action (see also "FMEA").

Design of experiments (DOE): A branch of applied statistics dealing with planning, conducting, analyzing, and interpreting controlled tests to evaluate the factors that control the value of a parameter or group of parameters.

Deviation: In numerical data sets, the difference or distance of an individual observation or data value from the center point (often the mean) of the set distribution.

DMADV: A data driven quality strategy for designing products and processes, it is an integral part of a Six Sigma quality initiative. It consists of five interconnected phases: define, measure, analyze, design, and verify.

Dissatisfiers: The features or functions a customer expects that either are not present or are present but not adequate.

Distribution (statistical): The amount of potential variation in the outputs of a process, typically expressed by its shape, average, or standard deviation.

DMAIC: A data driven quality strategy for improving processes and an integral part of a Six Sigma quality initiative. DMAIC is an acronym for define, measure, analyze, improve and control.

Effort risk matrix: A chart similar to a bubble chart that is used to show the trade-offs between risk and effort when evaluating solutions or alternatives.

Exciter: See "Delighter."

External failure: A nonconformance identified by external customers.

Failure cost: The cost resulting from the occurrence of defects.

Failure mode effects analysis (FMEA): A procedure in which each potential failure mode in every sub-item of an item is analyzed to determine its effect on other sub-items and on the required function of the item.

Fishbone diagram: See "Cause and effect diagram."

Five S's: Five Japanese terms beginning with S (with English translations also beginning with S) used to create a workplace suited for visual control and lean production: Seiri (sort, structure, or sift) means to separate needed tools, parts, and instructions from unneeded materials and remove the latter; Seiton (set in order or systematize) means to neatly arrange and identify parts and tools for ease of use; Seiso (sanitize or shine) means to conduct a cleanup campaign; Seiketsu (standardize) means to conduct seiri, seiton, and seiso at frequent, indeed daily intervals to maintain a workplace in perfect condition; and shitsuke (sustain or self-discipline) means to form the habit of always following the first four S's. Collectively, they define an orderly, well-inspected, clean, and efficient working environment.

Five whys: A technique for discovering the root causes of a problem and showing the relationship of causes by repeatedly asking the question, "Why?"

Frequency distribution (statistical): A table that graphically presents a large volume of data so the central tendency (such as the average or mean) and distribution are clearly displayed.

Gantt chart: A type of bar chart used in process planning and control to display planned work and finished work in relation to time.

Gap analysis: The comparison of a current condition to the desired state.

Gauge repeatability and reproducibility (GR&R): The evaluation of a gauging instrument's accuracy by determining whether the measurements taken with it are repeatable (there is close agreement among a number of consecutive measurements of the output for the same value of the input under the same operating conditions) and reproducible (there is close agreement among repeated measurements of the output for the same value of input made under the same operating conditions over a period of time).

Goal: A broad statement describing a desired future condition or achievement without being specific about how much and when.

Goal-Question-Metric paradigm (GQM): A formal technique developed by Victor Basili to ensure selection of appropriate metrics for a process or project. The acronym for "goal, question, metric," is an approach to software metrics that ensures alignment of metrics to organizational goals.

Green Belt (GB): An employee of an organization who has been trained on the improvement methodology of Six Sigma and will lead a process improvement or quality improvement team as part of his or her full-time job.

Heijunka: A method of leveling production, usually at the final assembly line, that makes just-in-time production possible. It involves averaging both the volume and sequence of different model types on a mixed model production line. Using this method avoids excessive batching of different types of product and volume fluctuations in the same product.

Heijunka box: A heijunka box is a visual scheduling tool used in heijunka, a Japanese concept for achieving a smoother production flow. While heijunka refers to the concept of achieving production smoothing, the heijunka box is the name of a specific tool used in achieving the aims of heijunka. It is generally a wall schedule that is divided into a grid of boxes or a set of pigeon-holes (rectangular receptacles). Each column of boxes representing a specific period of time. Lines are drawn down the schedule/grid to visually break the schedule into columns of individual shifts or days or weeks. Colored cards representing individual jobs (referred to as kanban cards) are placed on the heijunka box to provide a visual representation of the upcoming production runs.

Histogram: A graphic summary of variation in a set of data. The pictorial nature of the histogram lets people see patterns that are difficult to detect in a simple table of numbers.

House of quality (HOQ): A product planning matrix, somewhat resembling a house, that is developed during quality function deployment and shows the relationship of customer requirements to the means of achieving those requirements.

IDOV: A popular methodology mostly used in the manufacturing industry, particularly with DFSS Six Sigma approaches. The acronym stands for Identify, Design, Optimize, and Validate (or Verify). These development phases are similar to the traditional Six Sigma methodology MAIC (Measure, Analyze, Improve, and Control).

In-control process: A process in which the statistical measure being evaluated is in a state of statistical control; in other words, the variations among the observed sampling results can be attributed to a constant system of chance causes (see also "Out-of-control process").

Inputs: The products, services, material and so forth obtained from suppliers and used to produce the outputs delivered to customers.

Internal failure: A product failure that occurs before the product is delivered to external customers.

Ishikawa diagram: See "Cause and effect diagram."

Kaizen: A Japanese term that means gradual unending improvement by doing little things better and setting and achieving increasingly higher standards. Masaaki Imai made the term famous in his book, *Kaizen: The Key to Japan's Competitive Success.*

Kaizen blitz: A method of starting and completing process improvement projects in a very short time, often requiring 100 percent dedication of resources to a project until it is completed.

Kanban: A Japanese term for one of the primary tools of a just-in-time system. It maintains an orderly and efficient flow of materials throughout the entire manufacturing process. It is usually a printed card that contains specific information such as part name, description, and quantity.

Kano analysis: The Kano model is a theory of product development and customer satisfaction developed in the '80s by Professor Noriaki Kano that classifies customer preferences into different categories (three to five different types). This analysis is helpful when analyzing and prioritizing requirements for projects.

Key performance indicator (KPI): A statistical measure of how well an organization is doing. A KPI may measure a company's financial performance or how it is holding up against customer requirements.

Kruskal-Wallis test: The Kruskal-Wallis test is a nonparametric test to compare three or more samples. It tests the null hypothesis that all populations have identical distribution functions against the alternative hypothesis that at least two of the samples differ only with respect to location (median), if at all.

Lead time: The total time a customer must wait to receive a product after placing an order.

Lean: Producing the maximum sellable products or services at the lowest operational cost while optimizing inventory levels.

Line balancing: A process in which work elements are evenly distributed and staffing is balanced to meet takt time (see "Takt time").

Lower control limit (LCL): Control limit for points below the central line in a control chart.

Mann-Whitney test: A non-parametric statistical hypothesis test for comparing the medians of two independent samples of observations.

Master Black Belt (MBB): Six Sigma or quality experts responsible for strategic implementations within the business. The Master Black Belt is qualified to teach other Six Sigma facilitators the methodologies, tools, and applications in all functions and levels of the company and is a resource for utilizing statistical process control within processes.

Mean: A measure of central tendency; the arithmetic average of all measurements in a data set.

Measurement systems analysis (MSA): An analytical method that examines what is measurable and what is not, e.g., Gage R&R studies.

Median: The middle number or center value of a set of data in which all the data are arranged in sequence.

Mistake proofing: A process used to prevent errors from occurring or to immediately point out a defect as it occurs. If defects are not passed down a process, throughput, and quality improve (see also "Poka-yoke").

Monte Carlo Simulation: A computer-based technique that generates pseudo-random data for a model to evaluate outcomes over hundreds or thousands of potential situations (called "trial runs").

Muda: Japanese for waste (see "Waste").

Multivariate charts: A Multivariate chart is a type of SPC control chart for variables data. Multivariate charts are used to detect shifts in the mean or the relationship (covariance) between several related parameters.

Nominal group technique: A technique, similar to brainstorming, used by teams to generate ideas on a particular subject. Team members are asked to silently come up with as many ideas as possible and write them down. Each member is then asked to share one idea, which is recorded. After all the ideas are recorded, they are discussed and prioritized by the group.

Nonparametric tests: Nonparametric tests are often used in place of their parametric counterparts when certain assumptions about the underlying population are questionable. For example, when comparing two independent samples, the Mann-Whitney test does not assume the difference between the samples is normally distributed, whereas its parametric counterpart, the two-sample t-test, does. Nonparametric tests may be, and often are, more powerful in detecting population differences when certain assumptions are not satisfied. All tests involving ranked data (data that can be put in order) are nonparametric.

Non-value added (NVA): Activities or actions taken that add no real value to a product or service, making such activities or actions a form of waste (see "Value added").

Normal distribution (statistical): The charting of a data set in which most of the data points are concentrated around the average (mean), thus forming a bell-shaped curve.

Np-charts: A type of SPC chart that tracks the number of nonconforming units within one subgroup. The number nonconforming units (np), rather than the fraction of nonconforming units (p), is plotted against the control limits (see "P-charts").

Objective: A specific statement of a desired short-term condition or achievement; includes measurable end results to be accomplished by specific teams or individuals within time limits.

Operational metrics definitions: The formal definition of metrics so that they can be collected, stored, used, and interpreted consistently across an organization and over time.

Orthogonal defect classification: A method of classifying defects that generates rich data for analysis.

Out of specification: A term that indicates a unit does not meet a given requirement.

Out-of-control process: A process in which the statistical measure being evaluated is not in a state of statistical control. In other words, the variations among the observed sampling results cannot be attributed to a constant system of chance causes (see also "In-control process").

Outputs: Products, materials, services, or information provided to customers (internal or external), from a process.

Pareto chart: A graphical tool for ranking causes from most significant to least significant. It is based on the Pareto principle, which was first defined by J. M. Juran in 1950. The principle suggests most effects come from relatively few causes; that is, 80 percent of the effects come from 20 percent of the possible causes.

Payoff tables: A type of quantitative risk analysis that combines possible outcomes and probabilities to determine the best course of action.

P-charts: A type of SPC chart that tracks the nonconforming units as a fraction within one subgroup (see "Np-chart").

Pilot: A common practice of testing solutions in a limited venue before widely disseminating to production or across an organization.

Poka-yoke: Japanese term that means mistake-proofing. A poka-yoke device is one that prevents incorrect parts from being made or assembled or easily identifies a flaw or error.

Precision: The aspect of measurement that addresses repeatability or consistency when an identical item is measured several times.

Probability (statistical): A term referring to the likelihood of occurrence of an event, action, or item.

Process: A set of interrelated work activities characterized by a set of specific inputs and value added tasks that make up a procedure for a set of specific outputs.

Process capability index: See "Cp and Cpk."

Process control: The methodology for keeping a process within boundaries; minimizing the variation of a process.

Process FMEA (PFMEA): An analysis method used to identify potential problems with processes where steps can be skipped, performed out of sequence, or performed improperly.

Process map: A type of flowchart depicting the steps in a process, with identification of responsibility for each step and the key measures.

Process owner: The person who coordinates the various functions and work activities at all levels of a process, has the authority or ability to make changes in the process as required and manages the entire process cycle to ensure performance effectiveness.

Productivity: A measurement of output for a given amount of input. Increases in productivity are considered critical to raising living standards.

Prototyping: A common method to test new systems before committing them to production or sending the systems to customers.

Pull system: An alternative to scheduling individual processes, in which the customer process withdraws the items it needs from a supermarket, and the supplying process produces to replenish what was withdrawn.

Push system: The traditional method of manufacturing where material or lots are sent to the next step regardless of whether they are able to process the lots or materials. This type of flow is susceptible to bottlenecks, the formation of large work in process queues, and long cycle times.

Quality function deployment (QFD): A structured method in which customer requirements are translated into appropriate technical requirements for each stage of product development and production. The QFD process is often referred to as listening to the voice of the customer.

Queue time: The time a product spends in a line awaiting the next design, order processing, or fabrication step.

Range (statistical): The measure of dispersion in a data set (the difference between the highest and lowest values).

Range chart (R chart): A control chart in which the subgroup range, R, is used to evaluate the stability of the variability within a process.

Regression analysis: A statistical technique for determining the best mathematical expression describing the functional relationship between one response and one or more independent variables.

Response plan: A part of an overall control plan developed in the Control phase of a DMAIC project.

Rollout/deployment plan: A part of an overall control plan developed in the Control phase of a DMAIC project

Root cause: A factor that caused a nonconformance and should be permanently eliminated through process improvement.

Root cause analysis (RCA): Different ways of determining the originating cause of a problem or defect. Sometimes this is achieved by analyzing data preceding an event; sometimes it is achieved by brainstorming.

Run chart: A chart showing a line connecting numerous data points collected from a process running over a period of time.

Sample standard deviation chart (S chart): A control chart in which the subgroup standard deviation, s, is used to evaluate the stability of the variability within a process.

Satisfier: A term used to describe the quality level received by a customer when a product or service meets expectations.

Scatter diagram (or scatterplot): A graphical technique to analyze the relationship between two variables. Two sets of data are plotted on a graph, with the y-axis being used for the variable to be predicted and the x-axis being used for the variable to make the prediction. The graph will show possible relationships (although two variables might appear to be related, they might not be: those who know most about the variables must make that evaluation).

Sensitivity analysis: An analytical method that changes factors to determine which are the most important to the desired outcome.

Seven tools of quality: Tools that help organizations understand their processes to improve them. The tools are the cause and effect diagram, check sheet, control chart, flowchart, histogram, Pareto chart, and scatter diagram.

SIPOC: A process mapping method that examines each step in a process flow and characterizes it by its Inputs, Outputs, Customer (next step or end user), Supplier (source or predecessor), and the Process (the work performed or done at this step); sometimes done in reverse and called COPIS.

Six Sigma: A methodology that provides organizations tools to improve the capability of their business processes. This increase in performance and decrease in process variation leads to defect reduction and improvement in profits, employee morale, and quality of products or services. Six Sigma quality is a term generally used to indicate a process is well controlled ($\pm 6\ \sigma$ from the centerline in a control chart).

Solution matrix: Also called a decision matrix, it is a matrix used by teams to evaluate possible solutions and selecting the best one.

Source of variation (SOV): Statistical analysis to determine what factors are the greatest contributors to variation.

Special causes: Causes of variation that arise because of special circumstances. They are not an inherent part of a process. Special causes are also referred to as assignable causes (see also "Common causes").

Stakeholder analysis: The process of identifying the individuals or groups that are likely to affect or be affected by a proposed action, and sorting them according to their impact on the action and the impact the action will have on them. This information is used to assess how the interests of those stakeholders should be addressed in a project plan, policy, program, or other action.

Standard deviation (statistical): A computed measure of variability indicating the spread of the data set around the mean.

Statistical process control (SPC): The application of statistical techniques to control a process. The term "statistical quality control" is often used interchangeably with "statistical process control."

SWOT: A risk analysis technique; acronym stands for Strengths, Weaknesses, Opportunities, and Threats.

Takt time: The rate of customer demand, takt time is calculated by dividing production time by the quantity of product the customer requires in that time. Takt, the heartbeat of a lean manufacturing system, is an acronym for a Russian phrase.

Tests of normality: Statistical tests like Levine's test or Bartlett's test that test a data set to determine if they have a normal or Gaussian distribution.

Theory of constraints (TOC): Also called constraints management, TOC is a lean management philosophy that stresses removal of constraints to increase throughput while decreasing inventory and operating expenses.

Time series analysis: Analysis that examines data for patterns over time (see "Autocorrelation").

Tolerance: The maximum and minimum limit values a product may have and still meet customer requirements.

Training plan: A part of the overall control plan developed in the Control phase of a DMAIC project.

Tree diagram: A management tool that depicts the hierarchy of tasks and subtasks needed to complete an objective. The finished diagram bears a resemblance to a tree.

Trend: The graphical representation of a variable's tendency, over time, to increase, decrease, or remain unchanged.

Trend control chart: A control chart in which the deviation of the subgroup average, X-bar, from an expected trend in the process level is used to evaluate the stability of a process.

TRIZ: Russian acronym for the "Theory of Inventive Problem Solving." TRIZ is an international science of creativity that relies on the study of the patterns of problems and solutions; more than three million patents have been analyzed to discover the patterns that predict breakthrough solutions to problems, and these have been codified within TRIZ.

T-test: A statistical technique used to determine whether two different data sets are statistically different or whether the differences could have occurred by chance alone. Used for small sized data sets.

U chart: A type of SPC chart used to track defects or non-conformances per unit within one subgroup.

Unit: An object on which a measurement or observation can be made. Note: Commonly used in the sense of a "unit of product,"the entity of product inspected in order to determine whether it is defective.

Upper control limit (UCL): Control limit for points above the central line in a control chart.

Value added: Activities that transform input into a customer usable output. The customer can be internal or external to the organization.

Value stream: All activities, both value added and non-value added, required to bring a product from raw material state into the hands of the customer, bring a customer requirement from order to delivery, and bring a design from concept to launch.

Value stream mapping: A pencil and paper tool used in two stages: 1. Follow a product's production path from beginning to end and draw a visual representation of every process in the material and information flows. 2. Then draw a future state map of how value should flow. The most important map is the future state map.

Variable data: Also called "continuous data," it is the opposite of attribute data and it refers to data measurements that are quantified on an infinitely divisible numeric scale. It includes items like lengths, diameters, temperatures, electrical measurements, or hours.

Variation: A change in data, characteristic, or function caused by one of four factors: special causes, common causes, tampering, or structural variation.

Vision: An overarching statement of the way an organization wants to be; an ideal state of being at a future point.

Vital few, useful many: A term used by Joseph M. Juran to describe his use of the Pareto principle, which he first defined in 1950. (The principal was used much earlier in economics and inventory control methodologies.) The principle suggests most effects come from relatively few causes; that is, 80 percent of the effects come from 20 percent of the possible causes. The 20 percent of the possible causes are referred to as the "vital few"; the remaining causes are referred to as the "useful many." When Juran first defined this principle, he referred to the remaining causes as the "trivial many," but realizing that no problems are trivial in quality assurance, he changed it to "useful many."

Voice of the customer (VOC): The expressed requirements and expectations of customers relative to products or services, as documented and disseminated to the members of the providing organization.

Waste: Any activity that consumes resources and produces no added value to the product or service a customer receives (see "Muda").

Weighted attribute analysis: A method of evaluating the preference or priority of different solutions based upon weightings that are assigned to the different factors that contribute to the overall system or solution (also called net attribute analysis).

Work in process (WIP): Items between machines or equipment waiting to be processed.

World-class quality: A term used to indicate a standard of excellence: best of the best.

X moving average charts: A type of SPC chart that plots averages across multiple data points instead of individual values.

Xbar-R charts: A type of two-panel SPC chart that shows both average measures of samples and ranges of those samples.

Xbar-S charts: A type of two-panel SPC chart that shows both average measures of samples and the standard deviation of those samples.

Zero defects: A performance standard and methodology developed by Philip B. Crosby that states if people commit themselves to watching details and avoiding errors, they can move closer to the goal of zero defects.

Contributor Biographies

Luiz Antonio Bernardes is a Six Sigma consultant at Motorola Mobility in Brazil. Luiz is a Certified Master Black Belt. He has more than 13 years experience in software development and quality. He has led and coached more than 30 Six Sigma projects and mentored more than 25 Belts. He has a BS in computer engineering from University of Campinas, Brazil, and an MS in operations research from INSA de Lyon, France.

Rich Boucher is a Lean Six Sigma Deployment lead for the IT organization at EMC. He has been directly involved with the Lean Six Sigma initiative since its introduction at EMC in 2002. Rich is a certified Black Belt and certified ASQ quality manager. He has trained 3,200 people, including 400 Green Belts and 200 sponsors, and participated in 150 projects, mostly in IT. Before joining EMC, Rich worked in finance at Data General for 20 years, including 10 years as a controller for operations. Rich graduated from Holy Cross, received an MBA from Boston College, and earned an MS in quality from The National Graduate School.

Jian Chieh (JC) Chew is the Global Head of Process Excellence for the Lean Six Sigma Deployment within Content in Thomson Reuters, Market Division. He is a certified MBB and specializes in Lean Six Sigma within transactional and service organizations. Prior to Thomson Reuters, JC worked primarily as a Six Sigma and change management consultant and has consulted for international organizations like Coca Cola, Ciba Vision, Schneider Electric, B/S/H, Panasonic, and Microsoft as well as several arms of the Singapore government. He has a BA (Hons), 2nd Upper in psychology and English language from the National University of Singapore, a graduate diploma in organizational learning from the Singapore Civil Service College, and a master's degree in human resource management from Rutgers University, New Jersey.

Asheesh Chopra is assistant vice president in the Process Excellence Group, leading the Lean Six Sigma deployment for the retail banking operations at TCS. He has more than 11 years of cross-functional experience in manufacturing as well as the services sector. He worked for Sony, Philips, Bajaj, Genpact, and Citigroup before joining TCS. Asheesh is a certified LSS Master Black Belt and Project Management Professional. Asheesh has a keen interest in innovation and LSS training and has trained more than 250 Green Belts. Asheesh has a bachelor's degree in electrical engineering from C.R. Technical University, Haryana, India, and a master's in business administration from Sathya Sai Unversity, Prashanti Nilayam, India. Asheesh also holds a post-graduate diploma in Indian culture and philosophy from the same university and has worked on the influence of culture on work-related behavior of Indian managers. He is currently pursuing education in the theory of inventive problem solving (TRIZ).

Tim Clancy is a principal consultant in the Lean Six Sigma (LSS) practice at IBM. He specializes in the results-oriented application of LSS in transactional environments such as IT, healthcare, financial, and military sectors. Mr. Clancy is a certified LSS MBB and also received a separate Lean Mastery certification and is the primary developer of the Kaizen-IT methodology in use at IBM. His experience covers Fortune 50 or equivalent nonprofit organizations (NPO) engagements. He is currently on assignment with both the Department of Defense and the U.S. Army, developing and deploying strategic plans and performance management systems of enterprise transformation efforts. Mr. Clancy graduated summa cum laude with degrees in business and history from Reinhardt College.

Peter Clarke is an IT-certified Master Black Belt at Seagate Technology in Northern Ireland. He is a Project Management Professional (PMP) and is ITILv3 certified. He has worked in software development, control system implementation, and process improvement for 23 years, designing and implementing control systems in such fields as manufacturing, resource extraction, and power generation. He is a Chartered Engineer (C.Eng.) through the Engineering Council and a member of the Institute of Measurement and Control. He has a B.Eng. (Hons.) in computer and control systems from Coventry University, UK and a post-graduate diploma in computing for commerce and industry from the Open University, UK.

Bill Cooper is a senior IT Manager at Motorola Solutions. Bill has 25 years of experience that have spanned the areas of banking, manufacturing, logistics, customer service, consulting, engineering, information technology, and process and quality. His Six Sigma qualifications include a Master Black Belt Certification from Motorola with more than 10 years experience leading complex Six Sigma projects. He has a BS in computer science, information technology, and mathematics from Mobile College. He also has an MBA from the University of Mobile, USA.

Edilson Albertini da Silva is a certified Project Management Professional (PMP) at Motorola Mobility. He has 18 years of experience in software development, working at IBM and Nortel Networks R&D as software developer and software architecture leader. He currently manages software projects at Motorola R&D. He has a BS in computer science and anthropology from University of Campinas (Unicamp). He is a graduate of the Executive MBA Program from Fundação Getúlio Vargas (FGV), Brazil, and is currently pursuing post-graduation studies in social networking at Unicamp, Brazil.

Alecsandri de Almeida Souza Dias is a senior software engineer at Motorola Mobility. Alecsandri is a certified software test professional and a certified Six Sigma Green Belt. He has 11 years of experience in IT infrastructure optimization and software testing, working as a consultant and technical leader for major telecommunications and banking companies in Latin America. He has a BS in electronic engineering from Aeronautical Institute of Technology (ITA–Brazil) and a certificate in project management from Management Institute Foundation (FIA/USP–Brazil).

Sanjay (Sonny) Dua is a Master Black Belt for Tata Consultancy Services (TCS). In his role he focuses on delivering Lean Six Sigma solutions to TCS clients throughout North America. He is a seasoned GE trained Black Belt with over 12 years of experience assessing business needs and designing performance improvement solutions. Sanjay's experience spans multiple industries and business functions. Sanjay holds a MBA from the University of Georgia, and a MS in Instructional Systems from the Florida State University.

Alex Garcia Gonçalves is a senior software engineer at Motorola Mobility. Alex has nine years of experience in software development, working for Nortel Networks and Motorola, acting in multisite software development projects. He has a BS in computer science from Federal University of Sao Carlos (UFSCar), a post-graduate diploma in telecommunications from National Institute of Telecommunications (INATEL), and is currently pursuing a master's degree in computer engineering from State University of Campinas (Unicamp), Brazil.

Robert Hildebrand is the service supply chain strategy manager at Xerox. He has 12 years experience at Xerox focused on design and integration, with additional experience in sales and marketing support, customer support, and in developing and teaching Design for Lean Six Sigma critical parameter management classes. Robert holds seven patents and is an editorial advisor for iSixSigma. He has been published twice in iSixSigma and was also awarded iSixSigma's "Best Manufacturing Project" award at the 2009 iSixSigma Summit. Robert was also recently published in *Industry Week* and *Quality Digest*. He has BS and MS degrees in mechanical engineering from Rochester Institute of Technology and is a candidate for an MBA in finance from University of Rochester.

Manisha Kapur is the quality leader and director of continuous improvement for Convergys Corporation in India and EMEA. Manisha has more than 16 years experience in the field of quality and IT. She is a certified Master Black Belt, certified Black Belt, certified PMP, trained Lean trainer, and trained change acceleration process (CAP) trainer with proficiency in project management, Lean, and Six Sigma methodologies. Manisha is among the first 35 certified Black Belts in GE Capital worldwide and is well recognized in the quality community in India. She was invited to present her paper on the evolving role of quality in India's IT-enabled industry to the Asia Network of Quality (ANQ) in Tokyo in 2009. She frequently contributes her views and articles on quality in many magazines. She has a bachelor's degree in commerce from Delhi University and an MBA in finance. She also holds a post-graduate advanced diploma in systems from NIIT.

Siddharth Kawoor is currently pursuing his MBA at Northwestern University's Kellogg School of Management. He is an ASQ-Certified Six Sigma Black Belt. Previously, he worked as an electronics and telecommunication engineer at TCS. He has also worked in telecommunication deployment, software and hardware design, and R&D in electronics, manufacturing, and business process outsourcing environments. He has worked in various capacities as designer, team lead, quality prime, and project lead. Siddharth has a bachelor's degree in electronics and telecommunications engineering from Pune University, India.

Ambuli Nambi Kothandaraman is a delivery manager at TCS for one of its retail clients in North America. He has more than 12 years of experience in various IT projects spanning the retail and automotive industries. With certifications from the QAI (CSQA in 2004) and PMI (PMP in 2005), Ambuli successfully led multiple project/program initiatives before stepping in as the delivery manager (QA) for a large retail client in 2008. In this role, he was responsible for strengthening the quality assurance organization through various process improvement initiatives. Currently, he is managing a major e-commerce transformation initiative for a large retail client. Ambuli has a bachelor's degree in engineering from Anna University, India.

Eric Maass, Ph.D., is a senior program manager and Master Black Belt for DRM/Design for Six Sigma (DFSS) at the Medtronic Tempe Campus (MTC), leading its Design for Six Sigma/DRM strategy and deployment. Dr. Maass previously worked at Motorola in R&D and semiconductor manufacturing, and he was director of operations for a $160 million business and director of design and systems engineering. A co-founder of Six Sigma methods at Motorola, Eric was Lead Master Black Belt for DFSS. He is the co-author of *Applying DFSS to Software and Hardware Systems* (Prentice Hall, 2009), which provides clear, step-by-step guidance on applying DFSS for developing innovative and compelling new products and technologies while managing business, schedule, and technical risks. Eric received his bachelor's degree in biology from the University of Maryland, Baltimore County; his master's degree in biomedical engineering and his doctorate in industrial engineering from Arizona State University.

Patricia McNair is a senior manager of Risk & Opportunity at Raytheon. She is a certified Motorola Six Sigma Master Black Belt, SEI Introduction to CMMI v1.2 instructor, and served as the director of Motorola's Software DFSS and was responsible for the certification of the program's Belts. She recently authored the book, *Applying DFSS for Software and Hardware Systems* (Prentice-Hall, 2010). She has more than 25 years in software and systems engineering and has worked for various companies such as Motorola, GE Healthcare, and IBM Federal Systems. She also has served as an adjunct professor at De Paul University in Chicago, State University of New York-Binghamton, and University of Phoenix. She holds a BA in computer information systems, University of Houston, an MS in computer science, State University of New York-Binghamton, and an MBA from Lake Forest Graduate School of Management.

Jose A. Mechaileh is a certified Green Belt and senior architecture and software test engineer at Motorola Mobility. He has 23 years of experience in systems engineering and software testing in telecommunications and IT companies (Alcatel, Lucent, and Motorola), working with mobile technologies, PSTN switching, IP/ATM networks (carrier Ethernet, IP/MPLS, ATM networks) and next-generation networks (VoIP gateway and Softswitch). Mechaileh worked as a consultant for the Brazilian federal digital TV commission, helping to define alternatives of interactivity for the Brazilian digital terrestrial TV platform. He has extensive experience in testing, characterization, certification, and product development, and in the analysis of ITU-T/IETF/ETSI/OMA/TELCORDIA standards and protocols. He graduated with a bachelor's degree in electronic engineering from Unicamp University, Brazil (1986), and post-graduate degree in mobile technologies from the USF University, Brazil (2002).

Jason Morwick is a Master Black Belt with Cisco Systems, Inc. Morwick has led Six Sigma teams while working as a Black Belt and Master Black Belt at General Electric and CHEP USA, a Brambles Company. Jason has published in numerous journals, including *Quality Progress, Review of Business, Business Review,* and *Strategic HR Review.* Jason is also the co-author of *Making Telework Work: Leading People and Leveraging Technology for High-Impact Results* (Nicholas Brealey Publishing, 2009) and *Gridiron Leadership* (Praeger, 2009). He is a graduate of West Point with an MBA from Regis University.

Evan H. Offstein, Ph.D., is a professor of Business Management at Frostburg State University. Dr. Offstein has published in numerous journals, including *Human Resources Management Review, Human Resources Management Journal,* and *Group and Organization Management.* Dr. Offstein is also the author of *Stand Your Ground: Building Honorable Leaders the West Point Way* (Praeger, 2006) along with *Making Telework Work: Leading People and Leveraging Technology for High-Impact Results* (Nicholas Brealey Publishing, 2009) and *Gridiron Leadership* (Praeger, 2009) with foreword by Art Rooney II, the president and majority owner of the Pittsburgh Steelers. *Gridiron Leadership* has also been endorsed by General Ray Odierno, Commander of Forces in Iraq, and Sergeant Major of the Army Ken Preston. Dr. Offstein's clients include the U.S. Army, Raymond James Financial, Northwestern Mutual Life, the Human Resources Development Commission, the Maryland Transportation Authority, City of Pensacola, City of Hagerstown, Christopher Newport University, Texas A&M, and Century Furniture. He is a certified Senior Professional in Human Resources (SPHR). He is a graduate of West Point with a doctorate in business from Virginia Tech.

Fernanda de Carli Azevedo Oshiro is a software process and quality engineer at Motorola Mobility. Fernanda is a Certified Motorola Green Belt. She has five years of experience in software process and quality, participating in projects planning and tracking and working in process definition and improvement using Six Sigma methodologies. She also supports Motorola's internal audit process for ISO9001/TL9000 certifications. She has a BS in computer engineering from Federal University of São Carlos (UFSCar) and a post-graduation degree in quality engineering from University of Campinas (Unicamp), Brazil.

Deepak Ramadas is manager of the Process Excellence Group at TCS. He has more than eight years of cross-functional experience in operations, project management, Six Sigma, and Lean deployment. He is a certified Six Sigma Master Black Belt from Indian Statistical Institute and an ASQ-certified Black Belt. Prior to joining TCS, Deepak served as Black Belt at Sitel Corporation. Deepak has a bachelor's degree in electronics engineering from Bangalore

University, India. Deepak enjoys working on socioeconomic development and community excellence.

Anshuman Tiwari is the lead in business excellence at ANZ Support Services, India (ANZ). ANZ is part of the Australia and New Zealand Banking Group, Limited. At ANZ Anshuman is responsible for improving quality and productivity metrics for the Australia operations. Prior to ANZ, Anshuman was head of the business excellence program at Infosys Technologies, Limited and a principal consultant with Qimpro Consultants. At Qimpro, four of his clients won the Deming Application Prize. Anshuman is a certified manager of quality/OE from ASQ and is a trainer for examiners of the IMC Ramakrishna Bajaj National Quality Award. He has played a leadership role with the India Chapter for American Society for Quality (ASQ) and is currently a member leader for Bangalore. Anshuman has a bachelor's degree in industrial engineering from Nagpur University, India, and an MBA in operations management from K. J. Somaiya Institute of Management Studies and Research, India.

Prakash Viswanathan is a unit quality head for the Product Lifecycle and Engineering Services unit at Infosys. He has diverse industry experience spanning more than 19 years, including design of automotive accessories, aerospace guidance and launch vehicle applications, product design, and software process improvement. He holds a product patent with GE and several product design disclosures. He is a recipient of the QIMPRO Bronze Medal for lifetime contributions to Six Sigma in an individual capacity. He has trained thousands of professionals in Six Sigma, published a couple of papers with ASQ, and presented at Six Sigma conferences in India. Prakash is a certified Black Belt from ASQ; certified Master Black Belt; trained and certified in DMAIC, DFSS, Design for Reliability, New Product and Service Introduction processes; a senior member of ASQ; chartered engineer; and a member of Institute of Engineers (India) Chapter. He has a bachelor's degree in mechanical engineering from Osmania University, Hyderabad, and a master's degree in engineering design from Bharathiyar University, Coimbatore, India.

Company Profiles

Cisco

Cisco Systems, Inc., is the worldwide leader in networking solutions, creating hardware, software, and service offerings that are used to create Internet solutions that allow individuals, companies, and countries to increase productivity, improve customer satisfaction, and strengthen competitive advantage. The Cisco name has become synonymous with the Internet, as well as with the productivity improvements that Internet business solutions provide. Based in San Jose, California, Cisco serves a variety of market segments to include enterprise, service provider, commercial, small, and consumer customers across the globe. For more information, visit www.cisco.com.

Convergys

Convergys Corporation is a global leader in relationship management. It provides solutions that drive more value from the relationships its clients have with their customers. Convergys turns these everyday interactions into a source of profit and strategic advantage for its clients.

For more than 30 years, Convergys' unique combination of domain expertise, operational excellence, and innovative technologies has delivered process improvement and actionable business insight to marquee clients all over the world.

Convergys has approximately 70,000 employees in 82 customer contact centers and other facilities in the United States, Canada, Latin America, Europe, the Middle East, and Asia, and global headquarters in Cincinnati, Ohio. For more information, visit www.convergys.com.

EMC

EMC Corporation is the world's leading developer and provider of information infrastructure technology and solutions that enable organizations of all sizes to transform the way they compete and create value from their information. Headquartered in Hopkinton, Massachusetts, EMC employs approximately 40,000 people worldwide with revenue of $14 billion in 2009.

EMC provides the technologies and tools that can help customers release the power of their information. EMC helps customers design, build, and manage flexible, scalable, and secure information infrastructures. And with these infrastructures, customers are able to intelligently and efficiently store, protect, and manage their information so that it can be made accessible, searchable, shareable, and, ultimately, actionable. Customers can also use EMC information infrastructure as the foundation for implementing information lifecycle management strategies, securing critical information assets, leveraging content for competitive advantage, automating data center operations, reducing power and cooling costs, and more. For more information, visit www.EMC.com.

IBM

IBM is a global corporation with more than 380,000 employees and annual revenue of $98 billion, successfully delivering transformational solutions and services to clients for more than 100 years. IBM Global Business Services (GBS) is the world's largest consulting services organization, with consultants offering strategic business and information technology consulting with expertise in virtually every discipline and industry in the world. The Lean Six Sigma Practice within GBS dates back to 1996 and is focused on tackling improvement in highly complex fluid sectors in technology, healthcare, financial, and public sectors, including Department of Defense. For more information, visit www.ibm.com.

Infosys

Infosys defines, designs, and delivers IT-enabled business solutions for its clients to win by leveraging its domain and business expertise in banking and capital markets, insurance, healthcare, life sciences, automotive, aerospace, transportation, resources, energy and utilities, high-tech and discrete manufacturing, communication services providers, media and entertainment. It went public in India in 1993 and in the same year was listed on NASDAQ.

The service offerings span business and technology consulting, application services, systems integration, product engineering, custom development, maintenance, re-engineering, independent testing and services, IT infrastructure services, and business process outsourcing. Infosys services span the application lifecycle and combine rigorous processes based on industry-standard frameworks (CMMi, Six Sigma) with domain expertise, best-of-breed methodologies, structured knowledge management, tools and intellectual property developed by its Centers of Excellence and the Global Delivery Model to deliver world-class solutions. For more information, visit www.infosys.com.

Motorola

Motorola has been at the forefront of communication inventions and innovations for more than 80 years. With 53,000 employees located around the world, the company is organized into two business units. The Mobile Devices and Home business unit (Motorola Mobility) is positioned to lead the convergence of mobility, media, and the Internet. The Enterprise Mobility Solutions and Networks business unit (Motorola Solutions) offers a comprehensive end-to-end portfolio of products and solutions, including rugged two-way radios, mobile computers, secure public safety systems, barcode scanning, RFID readers, and wireless network infrastructures to enterprises and governments, as well as 4G broadband infrastructure, devices, and services to network operators globally. In 2009, Motorola had sales of US$22 billion. Headquartered in Schaumburg, Illinois, Motorola is a global company leading the next wave of innovations that enable people, enterprises, and governments to be more connected and more mobile. For more information, visit www.motorola.com.

Seagate

Seagate is the worldwide leader in the design, manufacture, and marketing of hard disk drives, providing products for a wide range of enterprise, desktop, mobile computing, and consumer electronics applications. Seagate's business model leverages technology leadership and world-class manufacturing to deliver industry-leading innovation and quality to its global customers. In addtion, the company aims be the low-cost producer in all markets in which it participates while maintaining its commitment to providing award-winning products, customer support, and reliability to meet the world's growing demand for information storage. For more information, visit www.seagate.com.

Tata Consultancy Services (TCS)

Tata Consultancy Services—BFS (Banking and Financial Services) BPO (Business Process Outsourcing) holds a premier position as the largest financial services BPO in India. It brings with it a strong financial services domain expertise and the ability to provide transaction processing, voice, and analytics services for the complete suite of financial products in the corporate, consumer, and private banking domains. It was established in 1968 and is currently Asia's largest software exporter. It has a global presence in 42 countries with 139 offices and 105 solution centers. It has a total employee strength of 123,092 (75 nationalities, 30 percent women). Its overall revenue is US$6 billion (FY09). In 2000, Forbes Global recognized TCS as one of the top 10 global IT service players. It is the first company in the world to receive an integrated enterprise-wide CMMI Level 5 and PCMM Level 5 assessment. For more information, visit www.tcs.com.

Thomson Reuters

Thomson Reuters is the world's leading source of intelligent information for businesses and professionals. A significant part of its business concerns the accurate, timely, and intelligent provision of financial and market information, as well as the software and tools to access, use, and analyze this information. This part of the business is managed by the Markets Division and it accounts for 53 percent of total company revenue. Within the Markets Division there are three content-producing organizations (also known as verticals within Thomson Reuters): Sales & Trading Content (S&T), Investment & Advisory Content (I&A), and Enterprise Content. The data provided by Thomson Reuters' content organizations span some 3 million organizations and 12.5 million instruments. It is estimated that every ten seconds, Thomson Reuters processes almost 340,000 updates. More than 5,000 content production staff works around the clock, across the world, to ensure the highest level of data accuracy, timeliness, and completeness. For more information, visit www.thomsonreuters.com.

Xerox

Xerox Corporation is a $22 billion leading global enterprise for business process and document management. Through its broad portfolio of technology and services, Xerox provides the essential back-office support  that clears the way for clients to focus on what they do best: their real business. Headquartered in Norwalk, Connecticut, Xerox provides leading-edge document technology, services, software, and genuine Xerox supplies for graphic communication and office printing environments of any size. Through ACS, which Xerox acquired in February 2010, Xerox also offers extensive business process outsourcing and IT outsourcing services, including data processing, HR benefits management, finance support, and customer relationship management services for commercial and government organizations worldwide. The 130,000 people of Xerox serve clients in more than 160 countries. For more information, visit www.xerox.com.

INDEX

Accuracy, 5–7
Affinity diagrams, 35, 246
Alternate hypothesis, 84, 105, 203, 297, 425
Analyze phase, 9, 10, 36–38
ANOVA, 83
ARMI, 74
As-is process map, 125, 179
Attribute data, 170, 262
Autocorrelation, 262
Availability, 491–508

Baselining, 247
BB. *See* Black Belt
Benchmarking, 380
Benefit realization, 66, 94, 118, 140, 157,
        172, 186, 212, 229, 268, 288, 306, 335,
        357, 406, 427, 442, 461, 486, 506
Best practice, 15, 33, 493
Big Y, 9–11, 243, 435
Black Belt, 11–15, 533, 544
Bottleneck, 164, 248
Box-Cox transform, 83
Box plots, 110, 111, 135, 209, 425, 426,
        457
Brainstorming, 39, 41, 130
Breakthrough, 9, 557
Burning platform, 32, 50, 388
Business case, 32–34
Business case for change, 386–388

C-chart, 44
Cause-effect diagram, 77
Cause-effect matrix, 35, 130
CCR. *See* Critical customer requirements
CDOC, 367, 374

CDOV, 367–369
Champion, 49, 50
Change agent, 14, 526, 554
Charter, 32–34
Classification of defects, 7, 202
Common cause, 185
Communications plan, 391
Complexity, 17, 313, 324
Control chart, 6, 7
Control limits, 7, 41, 94
Control phase, 9, 10, 31, 41–44
Control plan, 31, 44
COPIS, 396
COPQ. *See* Cost of poor quality
Core defect, 276–277
Corrective actions, 7, 10, 21–43
Correlation, 37, 38
Cost of poor quality, 4, 8
Cp and Cpk, 9, 10, 36, 80, 104, 116, 164,
        175, 181, 215, 219, 221, 223, 247, 262,
        369, 376, 378, 468, 557
Critical customer requirements, 9, 10
Critical parameter flow-down, 366
Critical parameter flow-up, 366
Critical parameter management, 366–375
Critical parameters, 366–375
Critical success factor, 40, 162
Critical to quality, 9, 102
CSF. *See* Critical success factor
CTQ. *See* Critical to quality
CUMSUM. *See* Cumulative sum control
        chart
Cumulative eRPN, 61
Cumulative sum control chart, 44
Cycle time, 11

Defects per million opportunities, 8–12
Define phase, 9, 10, 30–34
Delighter, 450
Design width, 4, 5
DFMEA, 48
DFSS, 8–11, 365–378
Distribution, 4, 5
DMADOV, 367, 374
DMADV, 10, 367–374
DMAIC, 8–16, 27–45
Document value map, 292, 301
DPMO. *See* Defects per million opportunities

Effort-impact matrix, 88, 92, 225, 244
Eight types of waste, 237
eRPN, 60–66
eRPNr, 61–66
External failure, 51, 55

Fishbone diagram, 35, 77, 78, 168, 204, 224
5 Ss, 237–238
5 Whys, 335, 352
Fraud detection, 191–212
FTE time, 316, 322, 324, 327

Gap analysis, 317, 400
Gauge R&R, 35, 246, 532
GB. *See* Green Belt
Generate project ideas, 18
Goal, 16, 17
Goal question metric paradigm, 218
Goal statement, 33
GQM. *See* Goal question metric paradigm
Green Belt, 17, 516–558

H_0. *See* Null hypothesis
H_a. *See* Alternate hypothesis
Heijunka, 311–361
House of quality, 370–372, 451
Hypothesis testing, 31, 84, 115

IDOV, 11, 367, 374, 464–485
Improve phase, 9, 10, 39–41

In/out frame, 73
Initiating enough projects, 524
Initiating the right projects, 524
Inputs, 77, 103
Inspection effort rate, 417
Intellect, 237
Internal failure, 51, 55
Inventory, 237
Ishikawa diagram. *See* Fishbone diagram

Kaizen, 233, 238, 239, 274–306, 312–362,
    510
Kaizen blitz, 233, 239, 254, 292–306, 361
Kaizen event. *See* Kaizen blitz
Kanban, 11, 249, 311–337
Kano analysis, 195, 244
Key performance indicator, 32
KJ analysis, 449, 450
KPI. *See* Key performance indicator
KPIV, 223, 493, 498–508
Kruskal-Wallis test, 83–85

LCL. *See* Lower control limit
Lean concepts, 237
Lean Six Sigma, 11, 235–251
Line balancing, 248
Location, 80
Lower control limit, 7
Lower specification limit (LSL), 4–7

Management commitment, 13–15
Mann–Whitney test, 83, 137, 171, 424–427
Master Black Belt (MBB), 14, 15, 528–542
Mean, 4–7
Measure phase, 9, 10, 31, 34–36
Measurement systems analysis, 245–247
Mistake proofing, 41, 44, 233, 239
Monitor chart, 139
Monte Carlo simulation, 41, 307, 308,
    368–377, 422, 436–439, 458–461
Motion, 237
MSA. *See* Measurement systems analysis
Muda, 237, 243, 248, 320, 321

Needs analysis, 515, 516
Nominal group technique, 39
Non-value-added, 11, 241, 291–309
Nonparametric tests, 38, 83, 170
Normal distribution, 4, 79
Null hypothesis, 84, 85, 87, 105, 132, 170, 185, 203, 226, 297, 425

Objective, 7, 11, 18–20, 33, 34, 48, 50, 72, 101, 108, 121, 147, 162, 164, 174, 176, 179, 215, 216, 218, 219, 235, 278, 315, 357, 385, 387, 410, 416, 421, 424, 445, 538–550
Operational metrics definitions, 35
Opportunity statement, 32, 51, 72, 102, 123, 148, 198, 258, 294, 344, 383, 413, 434, 447, 495, 542
Organizational change management, 385, 386, 388, 391
Out-of-control process, 7, 139
Outputs, 77, 103, 150, 242, 346, 394, 398, 401, 464, 469, 471, 474, 475, 477, 479, 481, 483, 484, 486, 488
Overproduction, 237, 243, 300, 302

P-charts, 44
P-value, 80–81, 84, 85, 87, 95, 105, 108, 133, 138, 170, 185, 186, 203, 210, 226, 297, 298, 420, 421, 424–426, 455, 458, 461
Pareto, 38, 49, 77, 81, 104, 113, 115, 121, 130, 131–133, 149, 152, 153, 156, 167, 175, 181, 183, 246, 262, 263, 277, 279–282, 298, 299, 414, 415
Payoff tables, 36
Pearson coefficient, 297, 419–421
PFMEA, 49
Pilot, 10, 21, 22, 39, 120, 136, 140, 175, 178, 184, 249, 317, 327, 330, 399, 400, 401, 410, 411, 416, 417, 424, 439, 446, 478, 517
Poka-yoke, 41, 233, 239, 249, 284, 361, 381
Precision, 5–7, 246, 464
Prioritizing projects, 244

Problem statement, 9, 13, 21, 32, 146, 162, 174, 177, 216, 258, 274, 276, 277, 392, 433, 542
Process capability index. *See* Cp and Cpk
Process cycle efficiency 282, 284, 324
Process map, 77, 116, 125, 136, 151, 162, 164–166, 176, 179, 180, 214, 216, 220, 259, 260, 315, 346, 414, 416, 423
Process owner, 74, 79, 94, 345, 346, 472
Process width, 4–7
Processing, 23, 28, 32, 35, 65, 73, 103, 112, 160, 163, 183, 193, 204–208, 210, 237, 241, 242, 247, 249, 274, 277, 301, 315, 346, 351, 374, 375, 467
Productivity, 145, 147–149, 153, 157, 189, 191, 192, 194, 196–198, 200–205, 209–217, 219–221, 223–229, 231, 248, 269, 289, 338, 358, 380, 427, 429, 464, 465, 467, 487
Project charter, 9, 21, 32–34, 50, 72, 101, 102, 104, 122, 126, 161, 162, 164, 166, 174, 176, 178, 179, 197, 214, 216–218, 220, 257, 294, 382–386, 388, 394, 412, 433, 447, 494, 534, 543
Project reviews, 18, 166, 167, 183, 185, 186, 220, 222
Project schematic, 435
Prototyping, 41, 329, 330, 368, 373, 445, 447, 448
Pugh matrix, 249, 250
Pull system, 326
Push system, 11, 249

Quality function deployment (QFD), 363, 368, 373, 449, 509
Quick wins, 17, 345, 356, 357, 430

RADI, 444–446, 449, 459
RADIOV, 367–369, 374, 378
Range chart (R chart), 94
RCA. *See* Root cause analysis
Regression analysis, 37, 38, 368, 374, 422, 501, 509

Response plan, 43, 44
Return on investment. *See* ROI
Rework, 4, 51, 54, 55, 66, 119, 140, 215,
    221–223, 226, 227, 229, 237, 243, 282,
    284, 315, 319, 320, 322, 324, 325, 328,
    341, 343, 344, 349, 350, 357, 393, 414,
    415, 423, 465–468, 471, 481, 488
ROI, 14, 33, 44, 49, 53, 54, 74, 94, 176, 244,
    249, 268, 293, 298, 306, 411, 461, 463,
    506
Rollout/deployment plan, 21, 22, 44, 368, 371
Root cause analysis, 8–9, 35, 81, 100, 131,
    146, 156, 160, 167, 172, 179, 181, 192,
    219, 248, 335, 361, 492
RPN, 56, 58–63, 66, 67, 175, 182–184, 204,
    205, 499

Satisfier, 195, 244,
Scatter diagram, 38, 112
SDFSS, 446
Seiketsu. *See* Standardize
Seiri. *See* Sort
Seiso. *See* Shine
Seiton. *See* Store
Sensitivity analysis, 38, 483
Shine, 238
Shitsuke. *See* Sustain
Sigma level, 4, 7, 8, 10, 11, 15, 36, 80, 81,
    105, 116, 172, 181
Sigma score, 247, 248
SIPOC, 34, 77, 103, 104, 150, 199, 200, 242,
    342, 345, 396
Six Sigma
    approach, 9
    countercultures, 158
    financial benefits, 25
    history, 16
    improvement methodologies, 13
    infrastructure, 19
    philosophy, 9
    program success factors, 17
    project characteristics, 22
    project selection approach, 22

resources, 19
training, 20
what is, 9
SMART, 33, 51, 163, 178, 217, 313, 416, 471
Solution matrix, 469
Sort, 238, 244
Source of variation (SOV), 35, 472
Spaghetti diagram, 242, 243
SPC. *See* Statistical process control
Special cause, 7, 202
Sponsor assessment, 386, 388, 389
Spread, 4, 6, 7, 14, 15, 80, 111
Stakeholder analysis/stakeholder
    management plan, 34, 149, 349, 358,
    386, 390, 391
Standard deviation, 5, 9, 458, 461, 547
Standardize, 238, 532
Statistical process control, 24, 42, 44, 247,
    266, 361, 505
Store, 238
Student's t-test, 83, 185, 209
Sustain, 104, 238
SWOT, 75, 76

T-test, 9, 38, 83, 105, 138, 185, 203, 209, 210,
    226, 227, 468
Takt time, 248
TAT. *See* Turn-around time
Tests of normality, 38, 79–81, 83, 94, 298,
    458, 461
Theory of constraints, 368, 371, 530
Time and motion study, 76, 235
Time series analysis, 38, 262, 525
TOC. *See* Theory of constraints
Total defect density, 417, 418
Training plan, 22, 44, 228, 386, 391, 398,
    403, 404, 539, 540
Transportation, 237, 241, 320, 321
Tree diagram, 453, 454
Trend control chart, 38, 246
TRIZ, 39, 368, 371, 373
Turn-around time, 69, 71, 73, 75, 77, 79, 81,
    83, 85, 87, 89, 91, 93, 95

Two-proportions test, 133
Two-sample t-test. *See* Student's t-test

U chart, 44
Upper control limit (UCL), 7, 95, 117, 171, 186, 202, 209, 228, 251, 279, 287, 336, 424, 505
Upper specification limit (USL), 4–7, 181, 222, 227, 377, 458, 461, 486

Value added, 76, 204, 241, 276, 282, 292, 295, 299–303, 305, 306, 309, 318–320, 325, 337, 379, 406
Value stream mapping, 11, 34, 35, 233, 240, 274, 282, 292, 299, 302, 309, 310, 317
Variation, 4–8, 16, 23, 35, 70, 75, 79, 80, 81, 83, 93, 94, 121, 134, 164, 167, 170, 181, 186, 202, 209, 219, 222, 223, 225, 226, 236, 247, 279, 286, 288, 315, 322, 325, 335, 365, 375, 376, 378, 419, 436, 438, 468, 472, 477, 479, 484, 485, 488, 509, 547–549
Vision, 22, 24, 394, 488, 538, 539
Vital few, useful many, 81, 223, 371, 373, 455
Vital X, 94, 243, 435

Voice of the customer (VOC), 9, 18, 34, 72, 76, 102, 120, 243, 269, 292, 296, 304, 346, 361, 365, 371, 372, 433, 444, 449, 450, 467, 486
Volume, 23, 72, 90, 99, 100, 102, 106, 108–113, 116–118, 147, 156, 169, 172, 184, 191, 196, 199, 201, 205, 211, 267, 277, 322, 336, 495, 506

Waiting, 11, 237, 243, 247, 265, 284, 302, 315, 319, 321, 326, 345, 500
Waste, 11, 29, 42, 51, 121, 146, 233, 235–237, 243–249, 251, 271, 274, 276, 277, 288, 292, 300, 302, 309, 312, 317, 319–322, 335, 338, 341, 345, 361, 384, 392, 393, 481, 514, 550, 557
Weighted attribute analysis, 39, 41, 249
Work in process (WIP), 237, 315, 316, 318, 335, 337, 346, 496, 500

X moving average charts, 44
Xbar-R charts, 44, 251
Xbar-S charts, 44

Zero defects, 4